# The Power of Antioxidants - Unleashing Nature's Defense Against Oxidative Stress

*Edited by Ana Novo Barros and Ana Cristina Santos Abraão*

Published in London, United Kingdom

The Power of Antioxidants - Unleashing Nature's Defense Against Oxidative Stress
http://dx.doi.org/10.5772/intechopen.1001782
Edited by Ana Novo Barros and Ana Cristina Santos Abraão

Contributors
Alfredo Saavedra-Molina, Alica Navrátilová, Ana Abraão, Ana Novo Barros, Andrea Ramalho, Angel Gil-Izquierdo, Asdrubal Aguilera-Méndez, Baiq Amarwati Tartillah, Benevides C. Pessela, Christian Cortés-Rojo, Cinthia Landa-Moreno, Claudia García-Berumen, Claudia Teresa Bento, David Villanueva-Bermejo, Deng Yiran, Ehab M. Abdalla, Farshid Parvini, Federico Ferreres, Flávio José Rodrigues Cruz, Frederick Lia, Gustavo Alberto De la Riva, Honória S. Chipaca-Domingos, Irene Gouvinhas, Jenaro Lemus-de la Cruz, Juan Carlos Hernández González, Karen Attard, Lucia Klongová, Ma Lifang, Marek Kovár, Marina Murillo-Villicaña, Miroslava Požgajová, Mohammad Ali Farazi Fard, Ngoc-Anh Le, Nuriye Arslansoy, Ozkan Fidan, Parisa Naji, Paulina Gomes Gonçalves, Poonam Tripathi, Rafael Salgado-Garciglia, Rocío Montoya-Pérez, Rolando García González, Rolando Morán Valdivia, Salvador Manzo-Avalos, Samar A. Aly, Sofia Kimi Uehara, Soraia Santos, Tu Nguyen, Wang Jiayi



First published in London, United Kingdom, 2025 by IntechOpen
IntechOpen is the global imprint of INTECHOPEN LIMITED, registered in England and Wales, registration number: 11086078, 167-169 Great Portland Street, London, W1W 5PF, United Kingdom

For EU product safety concerns: IN TECH d.o.o., Prolaz Marije Krucifikse Kozulić 3, 51000 Rijeka, Croatia, info@intechopen.com or visit our website at intechopen.com.

British Library Cataloguing-in-Publication Data
A catalogue record for this book is available from the British Library

The Power of Antioxidants - Unleashing Nature's Defense Against Oxidative Stress
Edited by Ana Novo Barros and Ana Cristina Santos Abraão
p. cm.

This title is part of the Biochemistry Book Series, Volume 60
Topic: Chemical Biology
Series Editor: Andrei Surguchov
Topic Editors: Murat Şentürk and Deniz Ekinci

Print ISBN 978-0-85466-206-7
Online ISBN 978-0-85466-205-0
eBook (PDF) ISBN 978-0-85466-207-4
ISSN 2632-0983

If disposing of this product, please recycle the paper responsibly.

IntechOpen Book Series

# Biochemistry

Volume 60

## Aims and Scope of the Series

Biochemistry, the study of chemical transformations occurring within living organisms, impacts all of the life sciences, from molecular crystallography and genetics, to ecology, medicine and population biology. Biochemistry studies macromolecules - proteins, nucleic acids, carbohydrates and lipids –their building blocks, structures, functions and interactions. Much of biochemistry is devoted to enzymes, proteins that catalyze chemical reactions, enzyme structures, mechanisms of action and their roles within cells. Biochemistry also studies small signaling molecules, coenzymes, inhibitors, vitamins and hormones, which play roles in the life process. Biochemical experimentation, besides coopting the methods of classical chemistry, e.g., chromatography, adopted new techniques, e.g., X-ray diffraction, electron microscopy, NMR, radioisotopes, and developed sophisticated microbial genetic tools, e.g., auxotroph mutants and their revertants, fermentation, etc. More recently, biochemistry embraced the 'big data' omics systems. Initial biochemical studies have been exclusively analytic: dissecting, purifying and examining individual components of a biological system; in exemplary words of Efraim Racker, (1913 –1991) "Don't waste clean thinking on dirty enzymes." Today, however, biochemistry is becoming more agglomerative and comprehensive, setting out to integrate and describe fully a particular biological system. The 'big data' metabolomics can define the complement of small molecules, e.g., in a soil or biofilm sample; proteomics can distinguish all the proteins comprising e.g., serum; metagenomics can identify all the genes in a complex environment e.g., the bovine rumen.
This Biochemistry Series will address both the current research on biomolecules, and the emerging trends with great promise.

# Meet the Series Editor

Andrei Surguchov, Ph.D., joined Baylor College of Medicine, Houston, TX, as a faculty member in 1992, where he studied the mechanisms of the genetic control of lipid metabolism. At the University of Utah, his research interests focused on cloning new genes encoding retinal proteins. He studied molecular and cellular mechanisms of neurodegenerative diseases and retinal degeneration at Washington University, St. Louis. Currently, his research focuses on the structure-function relationship of proteins involved in neurodegeneration and ocular diseases. Andrei Surguchov is an Editor-in-Chief at Biochemistry Research International and Associate Editor in several biomedical journals.

# Meet the Volume Editors

Ana Novo Barros is an Associate Professor of Habilitation at the University of Trás-os-Montes and Alto Douro in Portugal. She was the Director of the Centre for the Research and Technology of Agro-Environmental and Biological Sciences (CITAB) for 5 years (May 2017-July 2022). Her research targets are mainly in the identification, separation and recovery of functional molecules from different natural products, as well as their implementation as ingredients and bioactive compounds in food, with an ultimate goal to extract high added-value molecules and re-use them in the food chain (from agriculture to the consumer). She has been awarded several prizes and distinctions in the last few years and has published over 130 scientific documents (34 h-index, SCOPUS, November 2024). She is the coordinator of the European project Wasteless. For a year, she was responsible for a cosmetic company's research and innovation department.

Ana Cristina dos Santos Abraão is a full-time integrated researcher at CITAB and Inov4Agro, contributing to various R&D projects. Ana holds a Ph.D. in Agrarian Food Chains- From Fork to Farm, and her research focuses on the microbiological, chemical, and nutritional composition of food matrices, sustainability, and the development of innovative food products. She has participated in key projects, including Nutridouro, EUROLEGUME and Sambucus Fresh, and has published 9 JCR-indexed papers and a book chapter. Ana has supervised students and is actively involved in academic activities, including organizing jury roles and scientific events. She is currently developing her research in the frame of the VIIAFOOD project, emphasizing healthy food and sustainability. Ana also collaborates with national and international research groups.

# Contents

# Preface

Despite substantial changes at the scientific forefront, there is more significant concern about health in today's world due to oxidative stress, which is largely hidden and contributes to aging processes at the cellular level, along with several chronic diseases. Such adverse consequences are not irreversible and can be avoided by so-called antioxidants.

A particular focus of the volume *The Power of Antioxidants – Unleashing Nature's Defense Against Oxidative Stress* is to discuss how knowledge of antioxidants improves health and is beneficial for sustainable practice. At the same time, this volume provides scientists and health professionals with practical knowledge. Such approaches allow for dealing with oxidative stress and antioxidants, as well as understanding their status and use in biological systems and various applications. We believe this highlights the need for further research and advancements in the area of antioxidants and the fight against oxidative stress.

**Ana Novo Barros and Ana Cristina Abraão**
CITAB – Centre for the Research and Technology of Agro-Environmental and Biological Sciences,
UTAD -University of Trás-os-Montes e Alto Douro,
Vila Real, Portugal

## Section 1

# Natural Sources of Antioxidants

Chapter 1

# Borututu (*Cochlospermum angolense welw*) Extracts with Antioxidant Activity: Pressurized Liquid Extraction Used

*Honória S. Chipaca-Domingos, Paulina Gomes Gonçalves, Federico Ferreres, Angel Gil-Izquierdo, David Villanueva-Bermejo and Benevides C. Pessela*

## Abstract

Borututu (*Cochlospermum angolense welw*) is a plant native to Africa. The infusion obtained from borututu roots by decoction with water has been traditionally consumed by many African communities for the treatment of malaria due to its hepatoprotective properties. The search for ecological alternatives to obtain functional borututu extracts based on the application of techniques linked to new technologies, such as pressurized liquid extraction (PLE), was studied in comparison to the most commonly applied conventional solid-liquid extraction processes. Using PLE was used to evaluate different green/ecological solvents (water, ethanol, and ethyl acetate), providing lower consumption, different extraction temperatures (50–200°C), short extraction times (10 minutes), obtaining extraction yields with high concentrations of total phenolics compounds (TPC) for root (57–21%) and flower (47–16%) extracts. Generating greater antioxidant capacity at a temperature of 150°C with water in both root extracts (1413, 3645 µmol/g) and flower extracts (2029–4457 µmol/g) obtained by the DPPH (2,2-diphenyl-1-picrylhydrazyl) and ABTS (2,2′-azino-bis (3-ethylbenzthiazoline-6-sulfonic acid)) method, surpassing extracts obtained by the traditional infusion method (1179–1698 µmol/g) root and (1225–3097 µmol/g) flowers. The borututu plant has demonstrated potential for obtaining bioactive extracts for the production of functional ingredients on an industrial scale.

**Keywords:** borututu, *Cochlospermum angolense welw*, pressurized liquids extraction, antioxidant activity, ecological solvents and phenolics compounds

## 1. Introduction

In traditional medicine across many cultures, plants are considered the primary source of remedies, and for this reason, they have aroused interest in studying their chemical composition, bioactive properties, and the mechanisms by which they affect health. Many of these plants are widely consumed as infusions (hot herbal water) as dietary supplements [1].

In some rural communities, traditional therapeutic practices have been carried out based on beliefs, in an empirical way, which is transmitted between generations, providing greater demand for products of natural origin. This interest is due to the wide variety of chemical compounds distributed across various parts of plants (seeds, floral buds, fruits, leaves, stems, and roots) [2]. Among these, the borututu plant tree reaches 6 m in height and 20 cm in diameter, belongs to the Bixaceae family, genus *Cochlospermeaceae*, which has great potential as a source of bioactive extracts commonly known as Borututu or Brututu (*Cochlospermum angolense welw) C. angolense* is an African tropical tree native to Angola [3–5]. There are also authors who also refer to its presence in some places in the Democratic Republic of the Congo [6] and on a smaller scale in some areas in southern Guinea [7].

Borututu (*Cochlospermum angolense welw*) has been traditionally used by different communities in Africa as a natural remedy against various pathologies, generating benefits for health when incorporated as a food in the diet, as it has been described in the literature as a source of bioactive compounds with a potential effect on health, among them, with antioxidant capacity. The consumption of this plant has been carried out by obtaining extracts through infusion for traditional therapeutic purposes due to its medicinal and nutritional properties. **Figure 1** shows some of the parts of the plant (root, leaves, flowers, fruit) [8].

Borututu roots are currently marketed in Europe primarily by herbalists and commercial establishments in Portugal, packaged as tea or nutritional supplements due to nutritional and medicinal properties, described by several authors [6, 9–14], with claims regarding antioxidant, antimalarial, antimicrobial, and hepatoprotective or neuroprotective and antitumor activity, beneficial health properties, related to its consumption. Likewise, Presber [15] demonstrates antiplasmodic activity of extracts of the root husk of the species *C. angolense in vitro*, such as influenza virus inhibition and the treatment of diseases caused by plasmodium [15] and worms [16]. Malaria, a disease caused by plasmodium in Africa, particularly in Angola, is the cause of morbidity and mortality for thousands of people of different ages, posing a threat to public health. Some studies have reported three endemic African species, *C. angolense;*

**Figure 1.**
*Parts of the borututu plant. (a) roots, (b) leaves and flowers, (c) flowers and fruit.*

*C. planchonii; and C. tinctorium,* for their hepatoprotective properties and their use in the treatment of malaria [12]. Antiplasmodial activity has been demonstrated through both in vivo and in vitro tests, mainly in essential oils from extracts of roots and leaves of *C. planchonii* and *C. tinctorium* [7]. However, most of the studies carried out were based on obtaining extracts through conventional methods, mainly with boiling water, while the use of efficient new extraction techniques and green solvents to obtain functional ingredients rich in bioactive compounds remains underexplored [6], carried out the fractionation of a borututu extract obtained with methanol, using several solvents of different polarity, with the aim of determining its antioxidant capacity and trying to relate it to its chemical composition, using the extraction technique with pressurized liquids, reported for the first time, whose objective differs from that presented in this study. In recent decades, there has been growing interest in the use of plant matrices through the application of a systematic approach, based on scientific evidence, to obtain bioactive plant extracts, with application in the food and pharmaceutical industries. The pressurized liquid technique (PLE) has been successfully applied as a green extraction technology to extract bioactive compounds from different plant materials [17–19]. The basic principle is based on the use of solvents at temperatures above their boiling point. For this, a certain pressure is applied that can keep the solvent in a liquid state (at subcritical conditions) during the extraction process [20, 21].

The PLE technique was recently investigated for the production of bioactive extracts from borututu roots and flowers using different temperatures (50–200°C) and different solvents: water, ethanol, and ethyl acetate [8].

## 2. Materials and methods

### 2.1 Obtaining natural extracts of borututu (*Cochlospermum angolense welw*): Collection and selection of the vegetable matrix, roots, and flowers

Borututu roots and flowers were harvested in the northern region of Angola and dried at room temperature to approximate the conditions used traditionally. They were then packaged in plastic bags and transported to the laboratory for analysis. Subsequently, the samples were ground, and particles smaller than 250μm were obtained using a laboratory sieves column.

### 2.2 Obtaining natural borututu extracts (*Cochlospermum angolense welw*) by PLE (pressurized liquids extraction)

The natural extracts of borututu roots and flowers were obtained using the PLE technique (using accelerated solvent extraction equipment) (ASE35 0 Dionex Corporation, Sunnyvale, CA, USA), equipped with a solvent control unit [21, 22]. This method allowed the evaluation of three ecological solvents (water, ethanol, and ethyl acetate) at different temperatures (50, 100, 150, and 200°C) with a static extraction mode for 10 minutes, applying a fixed pressure of 10 MPa to keep the solvent in a liquid state above its boiling point. The system was automatically washed with fresh solvent and purged with nitrogen to promote the recovery of the extract, which was collected in a vial. **Figure 2** shows the simplified extraction scheme for PLE [8].

For instance, water, ethanol, and ethyl acetate were chosen as solvents due to their varying polarities, allowing the extraction of a broad range of bioactive compounds.

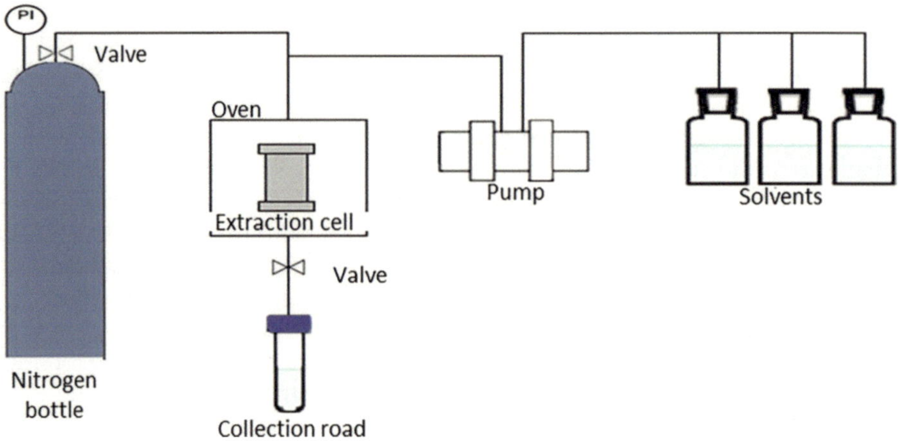


**Figure 2.**
*Simplified extraction diagram using the method PLE. Valve; nitrogen bottle; oven; extraction cell; collection road; pump and solvents.*

The temperature range of 50, 100, 150, and 200°C was selected to optimize the extraction of different compounds, as higher temperatures generally improve the solubility of bioactive molecules.

PLE is a highly efficient extraction method that uses high pressure and temperature to increase the solubility of compounds, allowing for faster and more complete extractions compared to traditional methods.

### 2.3 Obtaining extracts by infusion

For the traditional infusion extraction process, the extracts were obtained using a solvent/sample ratio of 200mL/g, using boiling water and allowed to rest for 5 minutes. All samples obtained with organic solvents were evaporated using a rotavapor (R-210 Buchi) and turbo Vap to eliminate the solvent. Aqueous samples were lyophilized in a Telstar LyoBeta freeze dryer.

### 2.4 Extraction yields

The extraction yields (extract mass/plant mass) obtained are shown in **Table 1**, indicating their dependence on the variables under analysis. As shown in **Table 1**, the extraction yield increased with the temperature for all solvents and for the two matrices (root and flowers).

## 3. Results and discussion

The highest extraction yields were obtained with the aqueous solvent, i.e., pure water. From both the roots (57.32%) and flowers (47.33%), both extracted at a maximum temperature of 200°C. It can be deduced that the extracts in general contain a large amount of water-soluble compound. Especially the root extracts, which could explain why they are used as an infusion in African countries and are currently expanding to Europe.

| Extraction yields (%) | | | Temperature (°C) | | | |
|---|---|---|---|---|---|---|
| | | Solvent | 50 | 100 | 150 | 200 |
| PLE | Root | water | 17.94 ± 2.60 | 19.25 ± 2.89 | 43.63 ± 0.10 | 57.32 ± 0.83 |
| | | Ethanol | 4.58 ± 0.27 | 7.17 ± 0.76 | 10.24 ± 0.54 | 18.89 ± 1.10 |
| | | Ethyl acetate | 3.80 ± 0.18 | 3.73 ± 0.16 | 4.42 ± 0.22 | 6.43 ± 0.25 |
| | | Root infusion | | 28.39 ± 2.1 | | |
| PLE | Flowers | water | 17.05 ± 2.4 | 24.79 ± 6.58 | 39.44 ± 2.36 | 47.33 ± 5.69 |
| | | Ethanol | 6.71 ± 0.90 | 13.21 ± 1.65 | 17.09 ± 2.30 | 18.80 ± 6.28 |
| | | Ethyl acetate | 5.62 ± 1.85 | 8.03 ± 0.60 | 12.42 ± 0.75 | 15.44 ± 1.41 |
| | | Flower infusion | | 18.64 ± 2.3 | | |

**Table 1.**
*Extraction yields by PLE of the root and flowers (g of extract/g plant).*

It was also observed that the yields obtained by infusion of root extracts (28.39%) were higher than those obtained at a temperature of 100°C (19.25%) with pure water, while for flower infusion extracts (18.64%), they presented a lower yield than that obtained by PLE at 100°C (24.79%). We observed a yield greater than 1.5 times the yield obtained with flowers, demonstrating the same behavior with extracts obtained by PLE at 100°C, with a greater presence of more water-soluble compounds in root extracts. On the other hand, as the polarity of the solvents is reduced, the extraction yields by PLE decreased for both vegetable matrices, presenting lower values obtained with ethyl acetate for root extracts varying (3.80–6.43%) while flower extracts will present higher values than the previous ones (5.62–15.44%). If we compared both parts, in flower extracts obtained with this solvent, the yield was between 1.5 and 2.8 times higher than that obtained from the root, suggesting the presence of a greater quantity of non-polar compounds in this raw material. The use of high temperatures allowed the obtaining of extracts from borututu roots and flowers with high extraction yields, as expected when producing an increase in the solubility of solutes and also, to a lesser extent, a decrease in viscosity and surface tension of the solvent with increasing temperature, which favors better penetration of the solvent into the vegetable matrix [21, 23]. The extraction yield also increases with the polarity of the solvent. Therefore, the greatest yields were obtained with water at 200°C (57.32% in the case of roots and 47.33% in the case of flowers), as we mentioned previously, which shows the greatest number of polar compounds containing this material, which allows high yields to be obtained with this technique, just like what has been described from other plants [18, 24, 25]. This is possibly the reason why infusion with water has been used as a common practice by communities for the traditional therapeutic use of borututu [2]. These results in **Table 1** were achieved within just 10 minutes of extraction time. The increase in extraction yield produced with the increase in temperature for each of the solvents, in the case of the root (**Table 1**), the largest increases are produced when high temperatures are used, which is why the PLE technique was especially efficient to obtain extracts with high extraction yields. The extraction yield at 200°C using water increased 3.2 times compared to that obtained at 50°C with the same solvent. From 50 to 100°C, the increase was only 1.1, but this increased considerably before increasing from 100 to 150°C (increase of 2.3), leading to the production of an extraction yield that was 3 times greater when raising the temperature from 100°C (maximum that, theoretically, can be reached with water in a conventional process where pressure is not applied) to 200°C. Something similar occurred with the remaining solvents, especially from 150°C. In the case of ethanol, the extraction yield increased linearly between 50 and 150°C. In this temperature range, the yield increased around 1.4 times for each 50°C increase in extraction temperature. However, when rising from 150 to 200°C, the yield almost doubled. Therefore, the PLE technique allowed for obtaining high extraction yields, which could not have been achieved by conventional extraction methods carried out at room temperature, where the maximum extraction temperature applied is limited by the boiling temperature of the solvent.

### 3.1 Concentration of phenolic compounds

In the case of the root (**Table 2**), the highest concentration of TPC was obtained at temperatures of 50 and 100°C (≈21.4%), after which it slightly decreased and remained stable at 150°C (17.30%).

However, the lowest concentrations were generally obtained with ethyl acetate, especially at 50°C (5.27%). Therefore, the presence of water facilitated the extraction of higher TPC values from the root (21–17%), which suggests a higher presence of water-soluble phenolic compounds or relatively high polarity in this material. In contrast, with ethanol and ethyl acetate, there was a progressive increase in concentration as the extraction temperature rose, with the highest results being obtained at 200°C for both the root and flower matrices. For the root extracts, the TPC was 15.28 and 16.30% for ethanol and ethyl acetate, respectively. The root was thus suggested as the best raw material to obtain extracts with a high concentration of TCP, based on the solvents and conditions applied. Regarding the flowers, a different pattern was observed (**Table 2**). The highest concentration was obtained with ethanol at 200°C (16.27%), which was considerably lower than the highest concentrations obtained from the root extracts.

The lower concentrations of phenolic compounds obtained with water from the flowers, compared to the root, suggest that flowers contain fewer water-soluble phenolic compounds. No major differences were observed between the flowers and the root in the maximum concentration of phenolic compounds obtained in the ethanol extracts, which were 16.27 and 15.28% at 200°C.

| TPC (%) | | | Temperature (°C) | | | |
|---|---|---|---|---|---|---|
| | | **Solvent** | **50** | **100** | **150** | **200** |
| PLE | Root | Water | 21.39 ± 0.87 | 21.37 ± 0.83 | 17.30 ± 0.28 | 17.16 ± 0.59 |
| | | Ethanol | 7.24 ± 0.28 | 11.07 ± 0.22 | 13.09 ± 0.58 | 15.28 ± 0.10 |
| | | Ethyl acetate | 5.27 ± 0.04 | 7.84 ± 0.06 | 8.77 ± 0.17 | 16.30 ± 0.29 |
| | | Root infusion | | 17.52 ± 0.13 | | |
| PLE | Flowers | Water | 9.27 ± 0.04 | 11.77 ± 0.07 | 9.01 ± 0.05 | 7.06 ± 0. 16 |
| | | Ethanol | 5.47 ± 0.28 | 7.25 ± 0.38 | 9.08 ± 0.16 | 16.27 ± 0.13 |
| | | Ethyl acetate | 4.77 + 0.07 | 6.96 ± 0.20 | 7.18 ± 0.06 | 11.17 ± 0.16 |
| | | Flower infusion | | 28.79 ± 0.11 | | |

**Table 2.**
*Concentration of total phenolic compounds (TPC) of borututu root and flower extracts, expressed as % by weight (mg GAE/100mg extract).*

For the flower extracts, the maximum concentration obtained with ethyl acetate at 200°C was 11.17%, which was significantly lower than the 16.30% obtained from the root with the same solvent at the same temperature.

In the extracts obtained by infusion, the concentration of phenolic compounds from the root was 17.52%, similar to that obtained in the PLE extracts with water at 100°C (21.37%) (**Table 2**).

These large differences could be due to the higher solvent/feed ratio used in the case of infusion.

### 3.2 Identification of the phenolic profile in the root and flower extracts of borututu (*Cochlospermum angolense welw*)

Chromatographic analyses were performed using an HPLC-DAD-ESI (Ion Trap) -MSn system, as described by Herrero et al. [20]. The exact mass determination was carried out using UPLC-ESI-QTOF-MS2 with an Agilent 1290 Infinity LC system coupled to a 6550 Accurate-Mass QTOF (Agilent Technologies, Waldbronn, Germany) with an electrospray interface."

The analyses were performed according to the method described by Garcia-Mendoza et al. [26]. The compounds identified in the two matrices (**Figure 3**) are as follows: roots (A) and flowers (B).

In the identification of phenolic compounds, the different root extracts showed very similar profiles at 250 nm (**Figure 3A**), regardless of the solvent used, and the same occurred with those of the flowers at 340 nm (**Figure 3B**). It was possible to confirm the presence of the same compounds in the extracts obtained from each of the matrices.

Therefore, the solvent did not seem to exert a considerable action on the type of compound extracted.

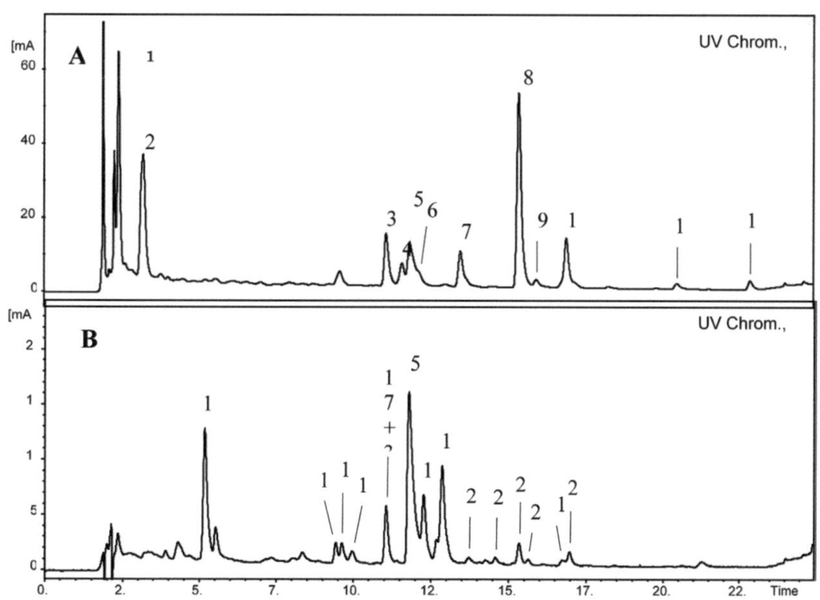


**Figure 3.**
*UV-Visible chromatograms of borututu extracts obtained by PLE with water, at 100°C, root extracts (250 nm) (*__Figure 3A__*) and flower extracts (340 nm) (*__Figure 3B__*).*

The compounds shown in **Figure 3A** correspond to the root extract from the UV-Visible chromatogram at 250 nm, with the main compounds (peaks 3–12) derived from ellagic acid.

These peaks presented UV spectra characteristic of these compounds (~248, 300 sh, 350 sh, 356 nm). In the case of compound 1 (2.3 minutes retention time; UV: 270 nm; MS: [M-H]−: 169.0148 [C7H5O5]−, MS2 [M-H]−: 125.0242 [C5H5O3, (M-H) − CO2]−), this was identified as possibly gallic acid, indicating they could be the pentoside (3) and rhamnoside (6) of ellagic acid (**Figure 3A**).

Compound 5 presented a deprotonated molecular ion m/z 300.9979, with the formula C14H6O8, corresponding to ellagic acid, whose retention time and fragmentation profile coincided with the pure pattern of this compound (**Figure 3A**). Compounds 3 and 6, with deprotonated molecular ions 132 and 146 atomic mass units (amu) higher than compound 5, respectively, presented in their mass fragments a base peak with an m/z ratio of 300.9970, corresponding to the deprotonated ion of ellagic acid, so they could be the pentoside (3) and rhamnoside (6) of ellagic acid (**Figure 3A**). The remaining compounds (4 and 7–12) showed in their spectra the presence of an ion with an m/z ratio of 315.0145, which could correspond to deprotonated methyl ellagic acid (molecular formula [C5H7O8]). Still in the same **Figure 3A**, in 4 and 7–10, it is the base peak, and it is produced by the loss of a 162 amu fragment (hexosyl radical) in compound 4, of a 146 amu fragment (rhamnosyl radical) in compound 10, and of a 132 amu fragment (pentosyl radical) in compounds 7–9. Therefore, these compounds could be the hexoside (4) and rhamnoside (10) of methyl ellagic acid, as well as isomers of the pentoside of methyl ellagic acid (7–9).

Compounds 11 and 12 (**Figure 3A**) exhibit the same deprotonated molecular ion with m/z 599.0690 (molecular formula C27H20O16).

Both present, in addition to the deprotonated methyl ellagic acid ion, an ion (base peak) with m/z 447.0567 ([C20H15O12]), as a consequence of the loss of the 152 amu fragment ([C7H4O4], galloyl radical). Fragmentation of this ion (MS2 [599 → 447]) gives rise to the deprotonated methyl ellagic acid ion. All this led to indicate that these compounds were galloyl isomers of methyl ellagic acid pentosides (**Figure 3A**).

**Table 3** lists the phenolic compounds identified in the root extracts.

In this case, ellagic acid, methyl ellagic acid, and their respective glycosides (mainly hexosides and pentosides) are the most representative compounds commonly found in borututu root, depending on the origin of the plant and the extraction technique.

In the case of flowers, the identified compounds were flavonoids, which can be divided into two main categories: flavonoid-C-glycosides (**Table 4**)—mainly luteolin glycosides and apigenin—and flavonoid-O-glycosides (**Table 5**), primarily luteolin glycosides, quercetin, and myricetin. These compounds are common constituents in flowers.

*Flavonoids-C-glucosides* (compounds 14-15-17-18-21-24), as shown in **Figure 3B** and **Table 4**, compounds 14 and 15 presented UV spectra of luteolin derivatives and their deprotonated molecular ions (m/z447.0940), with molecular formula C21H20O11, indicated to be, possibly, luteolin-hexoside isomers. Based on their fragment masses, ions with m/z of 357 and 327 ([(MH) - 90/−120]-), characteristic of C-hexosylation, were observed, which in mono-C-glucosyl flavones coincide with [aglycone +71/+41]- ions, which confirmed luteolin (286 + 71/+41) as the aglycone. Moreover, the higher abundance of the aglycone+71 ion in the 6-C-glucosyl derivatives compared to the 8-C-isomer, as well as the presence of the ([(M-H) − 18]−) ion, indicated the C-glucosylation position [10].

| Compound | Identification* | Rt | Molecular formula | [M-H]⁻ *(m/z)* | $MS^2$ *(m/z)* |
|---|---|---|---|---|---|
| 1 | Gallic acid | 2.3 | C5H5O3 | 169.0148 | 125.0242 |
| 2 | Protocatechuic acid | 3.1 | C5H5O2 | 153.0193 | 109.0293 |
| 3 | EA pentoside | 11.0 | C19H14O12 | 433.0402 | 300.9966 |
| 4 | MEA hexoside | 11.6 | C21H18O13 | 477.0682 | 315.0140 |
| 5 | EA | 11.8 | C14H6O8 | 300.9979 | |
| 6 | EA rhamnoside | 12.1 | C20H16O12 | 447.0569 | 300.9970 |
| 7 | MEA pentosid | 13.4 | C20H16O12 | 447.0570 | 315.0145 |
| 8 | MEA pentosid | 15.3 | C20H16O12 | 447.0574 | 315.0142 |
| 9 | MEA pentosid | 16.0 | C20H16O12 | 447.0569 | 315.0147 |
| 10 | MEA rhamnoside | 16.8 | C21H18O12 | 461.0732 | 315.0147 |
| 11 | MEA galloyl pentoside | 20.4 | C27H20O16 | 599.0689 | 447.0567<br>315.0146 |
| 12 | MEA galloyl pentoside | 22.8 | C27H20O16 | 599.0691 | 447.0569<br>315.0144 |

**EA: ellagic acid; MEA: methylellagic acid.*

**Table 3.**
*Retention time (Rt), molecular formula, and main ions generated in the analysis of phenolic compounds from root extracts of borututu.*

| Compound | Identification | Rt | Molecular formula | $[M-H]^-$ *(m/z)* | $MS^2$ *(m/z)* |
|---|---|---|---|---|---|
| 14 | luteolin –8-*C*- hexoside | 9.4 | C21H20O11 | 447.0939 | 357.0617 |
| 15 | luteolin –6-*C*- hexoside | 9.6 | C21H20O11 | 447.0941 | 357.0615 |
| 17 | Apigenin-8-6-*C*- hexoside | 11.1 | C21H20O10 | 431.0992 | 341.0663 |
| 18 | Apigenin-6-*C*- hexoside | 12.3 | C21H20O10 | 431.0989 | 341.0665 |
| 21 | Apigenin-8-*C*-(6- Acetyl) hexoside | 14.5 | C22H22O11 | 473.1095 | 341.0660 |
| 24 | Apigenin-6-*C*-(6- Acetyl) hexoside | 16.9 | C22H22O11 | 473.1091 | 341.0662 |

**Table 4.**
*Retention time (Rt), molecular formula, and main ions generated in the analysis of derivados Flavonides-C-glucosidos from borututu flower extracts.*

The phenolic compound profiles were markedly different between the root and flower extracts. Only three compounds were common to both matrices: ellagic acid, its pentoside, and the rhamnoside of methyl ellagic acid."

Similarly, isomers 17 and 18 differed from the previous ones in their deprotonated molecular ions, as well as in the lower fragment ions 16 amu. On the other hand, the aglycone+71/+41 ions (341/311), characteristic of mono-C-glucosyl apigenin derivatives, allowed them to be formulated as apigenin-8-C-hexoside and apigenin-6-C-hexoside, which could be vitexin (apigenin-8-C-glucoside) (17) and isovitexin (apigenin-6-C-glucoside) (18) (**Table 4**). For compounds 21 and 24, the characteristic mono-C-glucosyl-apigenin ions (341/311) were also detected, as in 17 and 18, but instead of the [(M-H)-90/−120]-ions, losses of-132(90 + 42)/−162(120 + 42)-from

| Compound | Identification | Rt | Molecular formula | $[M-H]^-$ (m/z) | $MS^2$ (m/z) |
|---|---|---|---|---|---|
| 16 | Myricetin-3-O-glucuronide | 9.9 | C21H18O14 | 493.0634 | 317.0307 |
| 19 | Quercetin-3-O-glucuronide | 12.8 | 21H18O13 | 477.0669 | 301.0355 |
| 20 | Quercetin-3-O-pentoside | 13.7 | C20H18O1 | 433.0776 | |
| 22 | Isorhamnentin-3-O-pentoside | 15.3 | C21H20O11 | 447.0945 | |
| 23 | Luteolin-7-O-glucuronide | 16.6 | C21H18O12 | 461.0729 | 285.0395 |

*Source: Ref. [8].*

**Table 5.**
*Retention time (Rt), molecular formula, and main ions generated in the analysis of phenolic compounds from borututu flowers extracts.*

the [M-H]-, which indicated the presence of an acetyl radical (42 amu) on the C-glucosylation sugar, were observed. Moreover, this acetylation was at the 6″-position of the hexose, and the [(M-H) -(120 + 42)]-ion implies an overall loss of the acetyl radical plus the remaining sugar moiety from C-glycosylation, which contains carbons 6″-3″. Therefore, these compounds could be named as apigenin-8-C-(6-acetyl) hexoside and apigenin-6-C-(6-acetyl) hexoside, so they could be 6″-acetyl-vitexin (apigenin-8-C-(6-acetyl) glycoside) (21) and 6″-acetyl-isovitexin (apigenin-6-C-(6-acetyl) glycoside) (24).

b) *Flavonoids-O-glucosides* The rest of the compounds (16, 19, 20, 22, and 23) also show the peaks identified in **Figure 3** and **Table 5**, which present deprotonated molecular ions of flavonoid monoglucosides and fragmentations corresponding to O-glucosides, in which the losses of the glucuronyl (−176, in compounds 16, 19, and 23) and pentosyl (−132, in compounds 20 and 22) radicals are observed to give rise to deprotonated ions of their respective aglycones, specifically, myricetin (16), quercetin (19 and 20), isorhamnetin (22), and luteolin (23).

Their spectra indicated that 16, 19, 20, and 22 are substituted at position 3, so they could be myricetin-3-O-glucuronide (16), quercetin-3-O-glucuronide (19), quercetin-3-O-pentoside (20), isorhamnetin-3-O-pentoside (22), and, possibly, luteolin-7-glucuronide (23). The phenolic compound profile was very different from that determined in the root extracts. Only three compounds were found in common in both plant matrices (roots and flowers): ellagic acid, its pentoside, and the rhamnoside of methyl ellagic acid.

### 3.3 Quantification of phenolic compounds in root and flower extracts

The quantification of phenolic compounds identified in the root (**Table 6**) and flower (**Table 7**) extracts, obtained by pressurized liquid extraction (PLE) with water at 100°C, is shown in **Table 6**.

Differences were observed in the abundance of these compounds in each of the extracts. For the analyses, the samples were injected at the same concentration, so that a greater area of one of the identified compounds meant a greater abundance of that compound in the extract with respect to the others. In general, the concentration of most compounds decreased as the polarity of the solvent decreased. Thus, the higher amount, in turn, shows a higher concentration of the identified compounds, which was obtained with water, both for root and flower extracts. The type of extraction (infusion and PLE with water at 100°C) did not seem to exert any effect, obtaining practically identical abundances for each of the

| Peak | Compound | (mg/g extract) | CV (%) |
|---|---|---|---|
| 3 | Ellagic acid pentoside | 0.37 ± 0.01 | 2.18 |
| 4 | Methyl ellagic acid hexoside | *<LOQ | — |
| 5 + 6 | Ellagic acid and rhamnoside of ellagic acid | 0.66 ± 0.00 | 0.68 |
| 7 | Methyl ellagic acid pentoside | 0.19 ± 0.01 | 5.26 |
| 8 | Methyl ellagic acid pentoside | 1.51 ± 0.04 | 2.50 |
| 9 | Methyl ellagic acid pentoside | *<LOQ | — |
| 10 | Methyl ellagic acid rhamnoside | 0.39 ± 0.01 | 1.34 |
| 11 | Galloyl pentoside of methyl ellagic acid | *<LOQ | — |
| 12 | Galloyl pentoside of methyl ellagic acid | *<LOQ | — |
| Total | | 3.12 | |

*Source: Ref. [8].*

**Table 6.**
*Concentration (mg/g) of phenolic compounds identified in root extracts obtained by PLE at 100°C, with water.*

| Peak | *Compound | (mg/g extract) | CV (%) |
|---|---|---|---|
| 14 | Luteolin-8-C-hexoside | 0.29 ± 0.02 | 7.21 |
| 15 | Luteolin-6-C-hexoside | 0.49 ± 0.06 | 12.98 |
| 16 | Myricetin-3-O-glucuronide | 0.13 ± 0.03 | 22.08 |
| 5 | Ellagic Acid | 5.28 ± 0.01 | 0.10 |
| 18 | Apigenin-6-C-hexoside | 4.27 ± 0.17 | 4.02 |
| 19 | Quercetin-3-O-glucuronide | 5.22 ± 0.08 | 1.47 |
| 20 | Quercetin-3-O-pentoside | *<LOQ | — |
| 21 | Apigenin-6-C-(6-acetyl) hexoside | 0.06 ± 0.03 | |
| 22 + 23 | Isohamentin-3-O-pentoside and Luteolin-7-O-glucuronide | 0.99 ± 0.07 | 7.56 |
| 10 + 24 | Rhamnoside-methyl ellagic acidand Apigenin-6-C-(6-acetyl) hexoside | 0.30 ± 0.06 | 19.78 |
| Total | | 17.03 | |

**<LOQ: Below limit of quantification. Source: Ref. [8].*

**Table 7.**
*Concentration (mg/g) of phenolic compounds identified in flower extracts obtained by PLE at 100°C, with water.*

compounds. In contrast, the lowest abundances were obtained with the solvent of lower polarity, namely ethyl acetate.

Among the phenolic compounds identified, a higher content of phenolic compounds was determined in the flower extracts, with a total of 17.03 mg/g, being more than 5 times higher than in the root extracts, which presented a total of 3.12 mg/g, as shown in **Table 6** for the root extracts and for the flower extracts (**Table 7**).

The coefficient of variation of each of the compounds identified from the root (**Table 6**) was less than 6%, showing a good repeatability of the results. In this case, the highest variability (5.26%) was shown by the methyl ellagic acid pentoside (7), and the lowest (0.68%) corresponded to the sum of the ellagic acids and their rhamnoside structure (5 + 6). As for their concentration in the extract, the methyl ellagic acid pentosides (7–9) are the compounds present in the highest amount (1.7 mg/g), followed by the sum of ellagic acid and its rhamnoside (0.66 mg/g). In contrast, the galloylated forms of methyl ellagic acid (11 and 12) were found in concentrations below the limit of quantification. Although ellagic acid and its derivatives were the main phenolic compounds identified in root extracts (3.12 mg/g), the amount of ellagic acid (5.28 mg/g) in flower extracts (**Table 7**) was 1.7 times higher (**Table 7**).

In the case of flowers, along with ellagic acid, the other two compounds (18 and 19) that occurred in high concentration were quercetin-3-O-glucuronide (5.22 mg/g), followed by apigenin-6-C-hexoside (4.27 mg/g). Based on the low coefficients of variation (less than 4.1%), the reproducibility obtained in relation to the concentration of these three major compounds in the flower extracts was high. For the rest of the compounds quantified, the concentrations were less than 1 mg/g.

Flavonoids, including the flavonols quercetin and myricetin and the flavones luteolin and apigenin, which were identified in the PLE extracts of flowers (**Figure 3B** and **Table 7**), have also been described as potent antioxidants [27–30]. These compounds could also be responsible, together with ellagic acid, for the antioxidant capacity of these extracts. Pereira et al. [12], observed a positive correlation between the content of phenolic compounds and the antioxidant capacity, determined by DPPH of an extract obtained by water infusion of borututu root. Moreover, although their antioxidant action in Cochlospermum extracts is discrete, especially compared to that provided by phenolic compounds, the presence of specific apocarotenoids of this genus, such as cocloxanthin and dihydrococloxanthin, would help to increase the antioxidant capacity of root and flower extracts [6, 8, 31]. In a general way, phenolic compounds are bringing much interest due to scientific evidence suggesting potential antioxidant effects related to cardiovascular activities, anticancer, reduction of inflammatory damage, and also as a modulator in immune response among other diseases [32]. Other authors consider that phenolic compounds, such as tannins and flavonoids, are responsible for the antioxidant defense of the plant when it suffers environmental stress or when it undergoes different morphological stages of development that cause an increase in ROS, helping the chemical adaptation of the plant to these conditions [33]. This adaptation occurs through genes that activate the biosynthesis of antioxidant mechanisms, in particular, secondary metabolism genes that help in plant development [33]. It should be noted that the levels of polyphenols and bioactive compounds present in plant extracts can be affected by many factors such as plant growth conditions (soil, climate) [34].

### 3.4 Evaluation of the biological activity of borututu root and flower extracts

The antioxidant activity of the extracts obtained by PLE and infusion by in vitro methods was determined by two methods: by neutralization of the DPPH radical (**Table 8**), SET method, mediated by one-electron transfer reactions (described by Brand-Williams et al. [35]), and by ABTS (**Table 4**), the method mediated by one-hydrogen atom transfer reactions described by Re et al. [36]. In **Table 8**, the results for the case of root extracts obtained with pressurized water at 50°C and by conventional infusion were quite similar (1208 and 1179 μmol/g, respectively). At 100°C it was slightly higher (1286 μmol/g). However, the antioxidant capacity was lower when using pressurized water at 150°C and reached the maximum antioxidant activity (1413 μmol/g).

| AO/DPPH | | Solvent | Temperature (°C) | | | |
|---|---|---|---|---|---|---|
| PLE | | | 50 | 100 | 150 | 200 |
| | Root | Water | 1208 ± 1 | 1286 ± 1 | 1413 ± 2 | 1276 ± 1 |
| | | Ethanol | 1108 ± 3 | 1156 ± 1 | 1300 ± 2 | 1386 ± 4 |
| | | Ethyl acetate | 770 ± 4 | 919 ± 1 | 948 ± 1 | 1051 ± 4 |
| | Flowers | Water | 1295 ± 4 | 1587 ± 2 | 2029 ± 2 | 1820 ± 6 |
| | | Ethanol | 205 ± 21 | 210 ± 23 | 671 ± 23 | 735 ± 9 |
| | | Ethyl acetate | 99 ± 63 | 98 ± 23 | 241 ± 23 | 171 ± 6 |
| Infusion | | Flowers | | 1225 ± 2 | | |
| | | Root | | 1179 ± 2 | | |

*Source: Ref. [8].*

**Table 8.**
*Antioxidant activity of borututu root and flower extracts determined by the ABTS method and expressed as Trolox equivalents (μmol Trolox/g extract).*

A progressive decrease in antioxidant capacity can be observed as solvent polarity decreases, with the lowest antioxidant capacity values being obtained with ethyl acetate (770 μmol/g, at 50°C). Regarding the behavior at different temperatures by PLE, in general, the antioxidant capacity was relatively similar, even increasing with increasing temperature for the different solvents. In the case of water, this is true up to a temperature of 150°C, after which the TEAC value of the extracts decreases slightly. However, in the case of ethanol and ethyl acetate, the highest antioxidant capacity using these solvents was obtained at 200°C. At that temperature, there was a 25% increase in the antioxidant capacity of the extract with respect to that obtained at 50°C using ethanol. In the case of ethyl acetate, this increase was 36%. As for the results obtained from flowers (**Table 8**), the use of ethanol and ethyl acetate generated extracts with considerably lower TEAC values than those obtained with the same solvents from the root. Considering the flower extracts with the highest antioxidant capacity obtained with each of these two solvents, the highest TEAC values using ethanol and ethyl acetate were 735 and 171 μmol/g, respectively, while the values obtained from the root were 1386 μmol/g with ethanol and 1051 μmol/g with ethyl acetate, which means antioxidant capacities are 1.9 and 6.1 times higher than in the flower extracts using the same solvents. Based on these results, the root gave rise to extracts with considerably higher TEAC values when organic solvents were used. However, the highest TEAC values were obtained with the extracts obtained with pressurized water from the flowers.

In this case, at 150°C it was possible to obtain an extract with a TEAC value of 2029 μmol/g, which was 36% higher than the extract with the highest antioxidant capacity from the root (**Table 8**). As for the effect of temperature on the TEAC values of the flower extracts, their behavior was similar to that observed with the root, with an increase in antioxidant capacity as the extraction temperature increased. In the case of water and ethyl acetate, the highest TEAC values were obtained at 150°C and decreased somewhat as the temperature was increased to 200°C. On the contrary, using ethanol, the highest value was obtained at the maximum temperature of 200°C. When comparing PLE extraction with water at 100°C with respect to traditional infusion, it was observed that the results obtained with pressurized water were 23% higher than those obtained by infusion (1587 and 1225 μmol/g, respectively). As in the case of the root, the temperature that allows PLE extraction led to obtaining extracts from flowers with higher TEAC values (2029 μmol/g with water at 150°C) than those obtained by infusion (1225 μmol/g), which meant a 66% increase in antioxidant capacity. As with the results obtained in the DPPH method, the antioxidant capacity determined by the ABTS method (**Table 9**) showed that the increase in temperature in the extractions from the root generated an increase in the antioxidant capacity when pure water was used with higher TEAC values, obtained at 150°C, producing a decrease at the maximum extraction temperature, while for ethanol and ethyl acetate, the maximum values were reached at 200°C. In any case, the increase in temperature above the boiling point of the solvent that allows extraction by PLE to be achieved generated substantial increases (**Table 9**).

The TEAC value of the extract obtained by root infusion (1698 μmol/g) was considerably lower than that obtained with pressurized water at 100°C (3475 μmol/g) and those obtained with pressurized water at the four temperatures (2812–3591 μmol/g). In the case of the extracts obtained from borututu flowers (**Table 9**), a behavior similar to that determined for the root extracts was observed. The decrease in solvent polarity led to a reduction in the TEAC value of the extracts so that the extracts with the lowest antioxidant capacity were obtained with ethyl acetate, followed by ethanol (**Table 9**).

The extract with the highest antioxidant capacity was obtained with water at 150°C (4457 μmol/g), which coincides with the behavior observed on the basis of the results obtained with the DPPH method (**Table 8**). Also, as for the results obtained in the DPPH method, the TEAC values obtained with water in the ABTS from flowers (**Table 9**) were considerably higher than those obtained with organic solvents. In the case of flowers, there were large differences in the values obtained between solvents and the antioxidant capacity increased with solvent polarity in both test methods, with the highest values obtained with water at 150°C (2029 μmol/g, determined by DPPH and 4457 μmol/g by ABTS) (**Table 9**).

Therefore, the highest values obtained, both of root and flowers, were reached at temperatures above the boiling point of the solvent (in subcritical conditions) in short periods of time, obtaining higher extraction yields, resulting in a lower use of solvent volume and minimal environmental impact. This demonstrates that PLE was an efficient technique for obtaining extracts with high antioxidant capacity. This can also be verified by comparing the results obtained with an infusion with respect to those of PLE with water. When the temperature was raised to 150°C, the increase in antioxidant

| AO/ABTS | | Solvent | Temperature (°C) | | | |
|---|---|---|---|---|---|---|
| PLE | | | 50 | 100 | 150 | 200 |
| | Root | Water | 2812 ± 49 | 3475 ± 22 | 3645 ± 52 | 3590 ± 9 |
| | | Ethanol | 639 ± 27 | 1005 ± 25 | 1366 ± 42 | 1693 ± 59 |
| | | Ethyl acetate | 301 ± 1 | 553 ± 7 | 445 ± 9 | 1322 ± 22 |
| | Flowers | Water | 2597 ± 18 | 2952 ± 21 | 4457 ± 36 | 3919 ± 56 |
| | | Ethanol | 483 ± 6 | 594 ± 7 | 728 ± 3 | 1156 ± 5 |
| | | Ethyl acetate | 275 ± 0 | 281 ± 2 | 263 ± 3 | 809 ± 9 |
| Infusion | | Flowers | | 3097 ± 13 | | |
| | | Root | | 1698 ± 12 | | |

**Table 9.**
*Antioxidant activity of borututu one and flower extracts determined by the ABTS method and expressed as Trolox equivalents.*

| Variable | | | | | | | | |
|---|---|---|---|---|---|---|---|---|
| | Y (%) Flower | Y (%) | TPC | TPC | ABTS | ABTS | DPPH | DPPH |
| | | Root | Flowers | Root | Flower | Root | Flower | Root |
| p-valor* | $0.000^{1}$ | $0.000^{1}$ | $0.000^{1}$ | $0.000^{1}$ | $0.000^{1}$ | $0.000^{1}$ | $0.000^{1}$ | $0.000^{1}$ |
| Treatments | Averages (homogeneous groups)** | | | | | | | |
| Water 50°C | $17.047^{bcd}$ | $17.917^{d}$ | $9.287^{d}$ | $21.393^{i}$ | $2596.670^{g}$ | $2812.000^{e}$ | $1295.057^{g}$ | $1208.360^{g}$ |
| Water 100°C | $24.795^{e}$ | $19.251^{d}$ | $11.773^{f}$ | $21.270^{i}$ | $2951.330^{h}$ | $3474.330^{f}$ | $1587.103^{h}$ | $1285.727^{h}$ |
| Water 150°C | $39.438^{f}$ | $43.120^{f}$ | $9.010^{d}$ | $17.297^{h}$ | $4456.330^{j}$ | $3645.000^{fg}$ | $2028.870^{j}$ | $1412.920^{k}$ |
| Water 200°C | $47.315^{g}$ | $57.325^{h}$ | $7.060^{c}$ | $17.157^{h}$ | $3918.670^{i}$ | $3589.330^{f}$ | $1823.117^{i}$ | $1276.040^{h}$ |
| Ethanol 50°C | $6.807^{a}$ | $4.583^{ab}$ | $5.470^{b}$ | $7.540^{b}$ | $481.670^{b}$ | $638.670^{ab}$ | $205.150^{c}$ | $1107.827^{e}$ |
| Et. 100°C | $13.237^{bcd}$ | $7.173^{b}$ | $7.253^{c}$ | $11.067^{d}$ | $598.000^{c}$ | $1005.000^{bc}$ | $209.777^{c}$ | $1155.653^{f}$ |
| Et. 150°C | $17.090^{bcd}$ | $10.237^{c}$ | $9.083^{d}$ | $13.087^{e}$ | $728.000^{d}$ | $1366.000^{cd}$ | $671.437^{e}$ | $1299.860^{h}$ |
| Et. 200°C | $18.803^{de}$ | $18.888^{d}$ | $16.267^{g}$ | $15.280^{f}$ | $1153.330^{f}$ | $1692.670^{d}$ | $735.283^{f}$ | $1386.360^{j}$ |
| EA 50°C | $5.613^{a}$ | $3.870^{a}$ | $4.727^{a}$ | $5.270^{a}$ | $275.330^{a}$ | $301.330^{a}$ | $99.093^{a}$ | $770.380^{a}$ |
| EA 100°C | $8.023^{a}$ | $3.730^{a}$ | $6.932^{c}$ | $7.840^{b}$ | $281.330^{a}$ | $553.670^{ab}$ | $97.737^{a}$ | $919.123^{b}$ |
| EA 150°C | $12.417^{b}$ | $4.420^{a}$ | $7.182^{c}$ | $8.770^{c}$ | $263.670^{a}$ | $444.670^{ab}$ | $241.247^{d}$ | $947.973^{c}$ |
| EA 200°C | $15.352^{bc}$ | $6.433^{ab}$ | $11.173^{e}$ | $16.300^{g}$ | $807.330^{e}$ | $1321.330^{cd}$ | $171.353^{b}$ | $1051.087^{d}$ |

*Et. -Ethanol; EA. -Ethyl acetate.*

**Table 10.**
*Analysis of variance of the results obtained for each of the extraction conditions (different letters indicate significant differences between means at level)* ($P < 0.05$).

capacity of the extract obtained by PLE was 20% higher (DPPH assay) (**Table 8**) and even 2 times higher (ABTS assay) (**Table 9**) than that of the extract obtained by infusion from the root. In the case of flowers, this increase with respect to the extract obtained by infusion was also considerable (66%, determined by DPPH, and 44% by ABTS). In the case of the antioxidant capacity determined by ABTS from the flowers, water at 150°C allowed the highest values to be obtained; likewise, for the root, the most appropriate treatment to achieve the TEAC value with water was at 150°C.

Likewise, for the results determined in the DPPH assay from the flower extracts, as for the ABTS, the best treatment corresponded to water at 150°C, which generated the extract with the highest antioxidant capacity. **Table 10** shows a summary of the main results obtained and their statistical treatment with Duncan's post hoc and Games-Howell tests as appropriate, which made it possible to determine r ($p < 0.05$). Both for the yield obtained from flowers and from the root, the highest significant values were obtained with water at 200°C obtained when water was used as a solvent and water at 100°C. (**Table 10**).

## 4. Conclusions

- Extraction yields increased with solvent polarity and extraction temperature, reaching a maximum value at 200°C of 57.32% in water extracts from the root and 47.33% in those from flowers.

- The main phenolic compounds identified in the root extracts were gallic, protocatechuic, ellagic, methyl ellagic acids, and the glycosides of the latter two. The glycosides were mainly hexosides and pentosides; one also found rhamnoside forms and a recently described galloyl pentoside derivative.

- The main phenolic compounds identified in the flower extracts were ellagic acid, ellagic acid and methyl ellagic acid glycosides, and flavonoids, further divided into C-glycosides of luteolin and apigenin and O-glycosides of luteolin, quercetin, and myricetin.

- The extracts obtained from each part of the plant presented the same profile of phenolic compounds, independent of the solvent used, varying only the relative abundance of each one of them. This abundance decreased as the solvent polarity decreased.

- The highest antioxidant capacity presented the extract obtained by PLE (1413 μmol/g in DPPH and 3645 μmol/g in ABTS) for roots and (2029 μmol/g in DPPH and 4457 μmol/g in ABTS) for flowers, both obtained at a temperature of 150°C, using water as a solvent. Both matrices present a similar profile and a great potential for obtaining antioxidant ingredients. These values were higher than those obtained by infusion, which for root extracts were (1179 μmol/g for DPPH and ABTS 1698 μmol/g) and (1225 μmol/g for DPPH and 3097 μmol/g for ABTS) for flowers, respectively.

- The PLE extraction technique allowed for obtaining extracts with high antioxidant capacity, both from the root and the flowers, as well as a considerable concentration and yield of phenolic compounds, using water in only 10 minutes of extraction. These extracts were obtained at a temperature of 150°C, which is above the boiling point of the solvent. The use of water is postulated as an efficient and clean alternative for obtaining extracts with antioxidant capacity from the borututu plant (roots and flowers).

## Acknowledgements

First of all, to God, who has given me a life full of joy and learning, allowing me to live a very pleasant experience. To my father, in memoriam, for always feeling your light and your protection guiding my steps; to my great warrior mother, Esmeralda, to you, a unique tribute, protective companion, giving strength to continue and fight for my dreams. To my great family for always being by my side, making me the person I am, lifting me up when I fall, my loves, even when distant, always present. To the Directorate of the Ministry of Higher Education, Science, Technology and Innovation and the Directorate of the National Center for Scientific Research, and to all my colleagues for the opportunity given to me for my first steps dedicated to training and research, always present. To the Instituto de Investigación en Ciências de la Alimentación (CIAL)/CSIC in Madrid, Kingdom of Spain, and to all my professors of the Institute, in particular to Dr. Tiziana Fornari, Dr. Monica Rodrigues, Professors José Manuel Guisán, Federico Ferreres, to my doctoral thesis directors, Benevides Pessela and David Vilanueva-Bermejo, who taught me to walk through the world of research, for always being available to contribute with their teachings and sharing experiences.

## Acronyms and abbreviations

| | |
|---|---|
| ABTS | 2,2′-azino-bis (3-ethybenzothiazoline-6-sulfonic acid) |
| ASE | accelerated solvent extraction |
| AMU | atomic mass unit |
| AO | antioxidant activity |
| C. angolense | cochlospermum angolense |
| DPPH | 2,2-diphenyl-1-picryl hydrazyl |
| EA | ellagic acid |
| MEA | methyl ellagic acid |
| GAE | gallic acid equivalent |
| PLE | pressurized liquid extraction |
| TEAC | trolox antioxidant capacity |
| TPC | total phenolic compounds |
| Trolox | 6-hydroxy-2,5,2,8-tetramethylchroman-2-carboxylic acid with 97% of purity |
| Rt | retention time |

## Author details

Honória S. Chipaca-Domingos[1], Paulina Gomes Gonçalves[1], Federico Ferreres[2], Angel Gil-Izquierdo[3], David Villanueva-Bermejo[4] and Benevides C. Pessela[5,6]*

1 National Center for Scientific Research (CNIC/MESCTI), Luanda, Angola

2 Department of Food Technology and Nutrition, Molecular Recognition and Encapsulation (REM) Group, Catholic University of Murcia, UCAM, Múrcia, Spain

3 Research Group on Quality, Safety and Bioactivity of Plant Foods, Department of Food Science and Technology, CEBAS-CSIC, Múrcia, Spain

4 Department of Production and Characterization of Novel Foods, Institute of Food Science Research, CIAL (CSIC-UAM), Section of Food Sciences, Faculty of Science, Autonomous University of Madrid (UAM), CEI UAM+CSIC, Madrid, Spain

5 Department of Food Microbiology and Biotechnology, Institute of Food Science Research, CIAL (CSIC-UAM), Madrid, Spain

6 Higher Polytechnic Institute of Technologies and Sciences, ISPTEC, Talatona-Belas-Luanda, Angola

*Address all correspondence to: b.pessela@csic.es

## References

[1] Pereira C, Calhelha RC, Barros L, Queiroz MJRP, Ferreira ICFR, Al E. Synergisms in antioxidant and antihepatocellular carcinoma activities of artichoke, milk thistle and borututu syrups. Industrial Crops and Products. 2013

[2] Chandrasekara A, Shahidi F, Al E. Herbal beverages: Bioactive compounds and their role in disease risk reduction-A review. Journal of Traditional and Complementary Medicine. 2018;**8**(4):451-458. DOI: 10.1016/j.jtcme.2017.08.006

[3] Sanfilippo M. Trinta árvores e Arbustos Do Miombo Angolano. 86. Santos, D. T., Veggi, P. C., and Meireles, M. A. A. Optimization and Economic Evaluation of Pressurized Liquid Extraction of Phenolic Compounds–4522014. DOI: 10.1016/J.JFOODENG.2011.08.022

[4] Videira C, Pedro JM, Nery SV. Rastreio etnobotânico nas comunas de Caxito, Mabubas e Úcua (Província do Bengo): resultados preliminares. 2011. Available from: https://repositorio.ipl.pt/handle/10400.21/261

[5] Vinha AF, Costa A, Soares M, Pires P. (Borututu). 11o Encontro de Química Dos Alimentos Atividade, 2-5. Brangaça/Portugal; 2012

[6] Abourashed E, Fu H, Al E. Hydroxybenzoic acids are significant contributors to the antioxidant effect of Borututu bark, *Cochlospermum angolensis welw*. Ex Oliv. Antioxidants. 2017;**6**(1):9. DOI: 10.3390/antiox6010009

[7] Leonardi M, Giovanelli S, Cioni PL, Flamini G, Pistelli L. Evaluation of volatile constituents of Cochlospermum angolense. Natural Product Communications. 2012;**7**(5):1934578X1200700. DOI: 10.1177/1934578x1200700522

[8] Domingos HSC. Tesis Doctoral. Obtención y Caracterización de la actividad Bioquimica de Borututu (*Cochlospermum angolense welw*) Para el desenho de nuevos alimentos Bioactivos. Spain: Universidad Autónoma de Madrid; 2021

[9] Correia CS, Directores P, Isabel D, Fernandes C, Ferreira R, Santos-Buelga C, et al. Tesis Doctoral. Aportaciones científicas para el uso de suplementos dietéticos. Un estudio de caso de plantas protectoras. Facultad de Farmacia, Universidad de Salamanca; 2015. Available from: https://gredos.usal.es/jspui/bitstream/10366/129635/3/DQANB_PereiraCorreia_Suplementosdietéticos.pdf

[10] Ferreres F, Grosso C, Gil-Izquierdo A, Valentão P, Andrade PB, Al E. Ellagic Acid and Derivatives from Cochlospermum Angolensis Welw. Extracts: HPLC-DAD-ESI/MSn Profiling, Quantification and in Vitro Anti. Murcia, Spain: Research Group on Quality, Safety and Bioactivity of Plant Foods, Department of Food Science and Technology, CEBAS (CSIC), Campus University Espinardo; 2013. DOI: 10.1002/pca.2429

[11] Pereira C, Calhelha RC, Barros L, Ferreira ICFR, et al. Antioxidant properties, anti-hepatocellular carcinoma activity and hepatotoxicity of artichoke, milk thistle and borututu. Industrial Crops and Products. 2013;**49**:61-65. DOI: 10.1016/j.indcrop.2013.04.032

[12] Pereira C, Calhelha RC, Barros L, Queiroz MJRP, Ferreira ICFR, Al E.

Synergisms in antioxidant and anti-hepatocellular carcinoma activities of artichoke, milk thistle and borututu syrups. Food and Funtion. 2014

[13] Pereira C, Barreira JCM, Calhelha RC, Queiroz MJRP, Barros L, Ferreira ICFR, et al. New insights into the effects of formulation type and compositional mixtures on the antioxidant and cytotoxic activities of dietary supplements based-on hepatoprotective plants. Food and Function. 2014b;**5**(9):2052-2060. DOI: 10.1039/c4fo00387j

[14] Presber H-W, Hegenscheid B, Hernandez-Alvarez H, Herrmann D, Brendel C, Al E. Inhibition of the growth of plasmodium falciparum and plasmodium berghei in vitro by an extract of *Cochlospermum angolense (Welw.)*. Acta Tropica. Institute of Medical Microbioloy, University Berlin and Institue of Tropical Medicine, Havana, Cuba. 1992;**50**(4):331-338. DOI: 10.1016/0001-706X(92)90068-9

[15] Presber H-W; Rer Nat Beyer G, Karl, Med Hegenrlscheied B, Rer Nat Hillet K, Med Zahner H. Patents Anthelmintic agent to treat diseases due to worms-contg. carotenoid drug extracted from roots of cochlospermum angolense. German: University Berlin; 1991a. (Patent No. DE4021428A1)

[16] Presber H.-W. Patentes Antiprotozoal Compsns. -Contain Root Extract of Cochlospermum Angolense (Patent No. DE4021427A1). 1991b

[17] Hossain MB, Brunton NP, Martin-Diana AB, Barry-Ryan C. Application of response surface methodology to optimize pressurized liquid extraction of antioxidant compounds from sage (*Salvia officinalis L.*), basil (*Ocimum basilicum L.*) and thyme (*Thymus vulgaris L.*). Food and Function. School of Food Science and Environmental Health, Dublin Institute of Technology, Dublin, Ireland. 2010;**1**(3):269-277. DOI: 10.1039/c0fo00021c

[18] Otero P, López-Martínez MI, García-Risco MR. Application of pressurized liquid extraction (PLE) to obtain bioactive fatty acids and phenols from Laminaria ochroleuca collected in Galicia (NW Spain). Journal of Pharmaceutical. 2019

[19] del PA S-C, Pleite N, Herrero M, Cifuentes A, Ibáñez E, Gilbert-López B, et al. New approaches for the selective extraction of bioactive compounds employing bio-based solvents and pressurized green processes. The Journal of Supercritical Fluids. 2017;**128**:112-120. DOI: 10.1016/j.supflu.2017.05.016

[20] Herrero M, Castro-Puyana M, Mendiola JA, Ibañez E, Al E. Compressed fluids for the extraction of bioactive compounds. TrAC—Trends in Analytical Chemistry. 2013;**43**(2):67-83. DOI: 10.1016/j.trac.2012.12.008

[21] Mustafa A, Turner C, Al E. Pressurized liquid extraction as a green approach in food and herbal plants extraction: A review. Analytica Chimica Acta. Lund, Sweden: Center for Analysis and Synthesis. Department of Chemistry, Lund University; 2011;**703**(1):8-18. DOI: 10.1016/j.aca.2011.07.018

[22] Scientific TF. Dionex ASE 150 Accelerated Solvent Extractor Operator's Manual (Issue 065207). 2011

[23] Santos DT, Veggi PC, Meireles MAA. Optimization and Economic Evaluation of Pressurized Liquid Extraction of Phenolic Compounds–4522012. DOI: 10.1016/J.JFOODENG.2011.08.022

[24] Miron TL, Plaza M, Bahrim G, Ibáñez E, Herrero M. Chemical composition of bioactive pressurized extracts of Romanian aromatic plants. Journal of Chromatography A. 2011;**1218**(30):4918-4927. DOI: 10.1016/j.chroma.2010.11.055

[25] Bermejo DV, Mendiola JA, Ibáñez E, Regl G, Fornari T. Extraction of high-value compounds from Arthrospira platensis (Spirulina) with CO2 + pressurized ethanol as cosolvent. The Journal of Supercritical Fluids. 2015;**96**:211-216. DOI: 10.1016/j.supflu.2014.10.021

[26] M Del P G-M, Espinosa-Pardo FA, Baseggio AM, Barbero GF, Maróstica Junior MR, Rostagno MA, et al. Extraction of phenolic compounds and anthocyanins from juçara (Euterpe edulis Mart.) residues using pressurized liquids and supercritical fluids. The Journal of Supercritical Fluids. 2017;**119**:9-16. DOI: 10.1016/J.SUPFLU.2016.08.014

[27] Ferreira ICFR, Martins N, Barros L. Phenolic compounds and its bioavailability: In vitro bioactive compounds or health promoters? Advances in Food and Nutrition Research. 2017;**82**:1-44. DOI: 10.1016/bs.afnr.2016.12.004

[28] Ganesan K, Xu B, Al E. A critical review on phytochemical profile and health promoting effects of mung bean (Vigna radiata). Food Science and Human Wellness. 2018;7(1):11-33. DOI: 10.1016/j.fshw.2017.11.002

[29] Kamali H, Ahmadzadeh Sani T, Mohammadi A, Alesheikh P, Khodaverdi E, Hadizadeh F. A comparison between pressurized hot water and pressurized liquid extraction for optimizing phenolic and antioxidants capacity of the wooden layer between of walnut seed. The Journal of Supercritical Fluids. 2018;**133**:535-541. DOI: 10.1016/J.SUPFLU.2017.10.01

[30] TF De M P, Bonamigo TR, Da Silva J, Vasconcelos P, Ramon M, Félix JM, et al. Chemical constituents of Cochlospermum regium (Schrank) Pilg. Root and its antioxidant, antidiabetic, antiglycation, and anticholinesterase effects in Wistar rats. Biomedicine and Pharmacotherapy. 2019;**111**:1383-1392. DOI: 10.1016/j.biopha.2019.01.005

[31] Lamien-Meda A, Kiendrebeogo M, Compaoré M, Meda RNT, Bacher M, Koenig K, et al. Quality assessment and antiplasmodial activity of west African Cochlospermum species. Phytochemistry. 2015;**119**:51-61. DOI: 10.1016/j.phytochem.2015.09.006

[32] Clifford MN. Chlorogenic acids and other cinnamates – Nature, occurrence, dietary burden, absorption, and metabolism. Journal of the Science of Food and Agriculture. 2000;**80**(7):1033-1043. DOI: 10.1002/(sici)1097-0010(20000515)80:7<1033:aid-jsfa595>3.3.co;2-k

[33] Araujo TA. de S, de Almeida e Castro VTN, da Silva Solon LG, da Silva GA, das Graças Almeida M, da Costa JGM, de Amorim ELC, Albuquerque UP et al. Does rainfall affect the antioxidant capacity and production of phenolic compounds of an important medicinal species? Industrial Crops and Products. 2015;**76**:550-556. DOI: 10.1016/j.indcrop.2015.07.008

[34] Moore J, Yousef M, Tsiani E. Anticancer effects of rosemary (*Rosmarinus officinalis L.*) extract and rosemary extract polyphenols. Nutrients. 2016;**8**(11). DOI: 10.3390/nu8110731

[35] Brand-Williams W, Cuvelier ME, Berset C. Respostas Perceptivas E. LWT - Food Science and Technology. 1995;**28**(1):25-30. DOI: 10.1016/S0023-6438(95)80008-5

[36] Re R, Pellegrini N, Proteggente A, Pannala A, Yang M, Rice-Evans C. Antioxidant activity applying an improved ABTS radical cation decolorization assay. Free Radical Biology and Medicine. 1999;**26**(9-10):1231-1237. DOI: 10.1016/S0891-5849(98)00315-3

Chapter 2

# Antioxidant Activity of Natural Products from Medicinal Plants

*Alfredo Saavedra-Molina, Jenaro Lemus-de la Cruz, Cinthia Landa-Moreno, Marina Murillo-Villicaña, Claudia García-Berumen, Rocío Montoya-Pérez, Salvador Manzo-Avalos, Asdrubal Aguilera-Méndez, Rafael Salgado-Garciglia and Christian Cortés-Rojo*

## Abstract

Ethnobotanical study is an important activity related to the research and development of drugs. The growing need to find alternatives for the treatment of chronic degenerative diseases, such as diabetes, hypertension, and metabolic syndrome, among others, justifies the study of medicinal plants used in traditional medicine. The therapeutic effects of plants are due to the content of different secondary metabolites such as essential oils, tannins, phenolic acids, sesquiterpenes, and flavonoids—for example, several reports about the beneficial effects of a wide range of plants to treat diabetes. In Mexico, most of the traditional knowledge about medicinal plants comes from pre-Hispanic times, and different ethnic groups still retain it.

**Keywords:** medicinal plants, oxidative stress, antioxidants, Eryngium, Justicia, Potentilla

## 1. Introduction

The ethnobotanical study is an important activity in the research and development of drugs. The growing need to find alternatives for the treatment of chronic degenerative diseases, such as diabetes, hypertension, and metabolic syndrome, among others, justifies the study of medicinal plants used in traditional medicine. The therapeutic effects of plants are due to the content of different secondary metabolites such as essential oils, tannins, phenolic acids, sesquiterpenes, flavonoids, among others [1]. In the literature are several reports about the beneficial effects of various plants in treating diabetes. In Mexico, most of the traditional knowledge about medicinal plants comes from pre-Hispanic times, and different ethnic groups still retain it. Also, the country's biodiversity regions range from semi-desert regions in Northern Mexico to tropical and rainy Southern regions with many plants that grow under different extreme climate conditions, producing many chemical compounds. Some plants, such as *Eryngium carlinae* F. Delaroche, commonly known as "Frog herb," a perennial herb plant that is distributed in forests of fir, pine and pine-oak hillsides, and canyons, deep soils rich in an organic matter [1]. The genus Eryngium (Picos, Eryngo, and Sea holly), spread throughout the world, representing one of the largest genera of the Apiaceae family [1], develops mainly in the pine-oak forest of the high parts of the country. It is frequently found in the open places on the edges of plots, around de houses, on degraded surfaces, and as weeds. There are eight species of Eryngium reported as medicinal in the central-western region of Mexico: *E. cymosum, E. longifolium, E. fluitans, E. beecheyanum, E. carlinae, E. comosum, E. heterophyllum*, and *E. nasturtiifolium*. The extracts contain various metabolites (e.g., tannins, phenolic acids, saponins, and terpenoids) that have shown biological activities such as hypoglycemic, hypocholesterolemic, renoprotective, anti-inflammatory, antibacterial, antioxidant, and others [2]. The hexanic extract of *Eryngium carlinae* possesses activities as an antioxidant, antilipidemic, hypoglycemic at the serum level, and in mitochondria from left ventricle increased nitric oxide levels and glutathione reduced, which may eventually reduce the risk of developing diabetic cardiomyopathy [3]. Also, this hexanic extract of *E. carlinae* reduced lipid peroxidation, protein carbonylation, and reactive oxygen species (ROS) in the liver, the kidneys, and the brain of streptozotocin-induced diabetes rats [4], and protection has been reported when cells of the yeast *Saccharomyces cerevisiae* reported were exposed to hydrogen peroxide-induced stress by reducing lipid peroxidation and protein carbonylation [4]. Under the organic solvent utilized, the chemical composition of the compounds identified, that is, in the hexanic extract of *E. carlinae*, was mainly identified (Z)β-farnesene and β-pinene [4].

*Potentilla indica* is a scientific synonym of *Fragaria indica* Andr. and *Duchesnea indica* Andr., is a plant belonging to the Rosaceae family, commonly known as "false strawberry," "Indian strawberry," or "wild strawberry." It is an herbaceous perennial plant; it flowers from May to October, and the seeds ripen from July to October. Traditionally, *Potentilla indica* leaves are used as a decoction or applied externally as a poultice after the fresh leaves are crushed. The flowers are used in infusions, and the fruit is edible [2].

*Justicia spicigera* (Acanthaceae), Muicle or Mexican honeysuckle, has antioxidant properties due to significant concentrations of flavonoids [5]. It contains over 250 flowering plant species, commonly used as edible and medicinal plants in different countries. *Potentilla indica* is a perennial herb belonging to the Rosaceae family and native to Asia. It is widely used in traditional Asian medicine to treat leprosy, tissue inflammation, congenital fever, cancer, and diabetes mellitus [2]. *J. spicigera* stands out due to its high antioxidant capacity. Recent studies have demonstrated the therapeutic effects of this plant attributed to the phenolic secondary metabolites obtained with solvents of different polarities that can be used as a potential treatment in different pathologies such as cancer, diabetes, and hypertension. It should be noted that these effects may be due to the combination of compounds in the extracts, generating interactions that can be classified as antagonistic or synergistic [3], in which different mechanisms are involved that serve as points of opportunity for new research.

## 2. Diabetes mellitus and oxidative stress

Diabetes mellitus (DM) is a metabolic disorder of multiple etiologies characterized by chronic hyperglycemia and alterations in the metabolism of carbohydrates, lipids, and proteins because of defects in the secretion or action of insulin or a combination of both [6]. The prevalence of DM is growing exponentially worldwide being one of the main challenges for health systems [7], in fact, according to IDF data [6], it is estimated that there are 537 million people with DM worldwide, and by 2045 the figure will increase to 783 million. There are different types of diabetes: type 1 DM (T1DM), type 2 DM (T2DM), and gestational diabetes. T1DM represents approximately 5–10% of all cases, the pathophysiology of this type of diabetes results from the destruction of pancreatic beta cells, resulting in decreased insulin levels [8]. T2DM is the most prevalent type of diabetes, accounting for 90% of all cases and is associated with both deficiency and tissue resistance to insulin [9]. Gestational diabetes is caused by hormonal variation during pregnancy, which results in the appearance of chronic hyperglycemia [10].

Chronic hyperglycemia in non-insulin-dependent tissues causes a higher glycolytic rate, which leads to a greater conversion of pyruvate to acetyl-CoA, which feeds the Krebs cycle, and this, in turn, produces a greater amount of NADH y FADH2 than they donate their electrons to mitochondrial electron transport chain (ETC), and this causes an increase in the leak of electrons resulting in greater generation of reactive oxygen species (ROS) [11]. In this sense, the main source of ROS in the cell is the mitochondria, specifically complex I and III of ETC [12]. Chronic hyperglycemia, in addition, increases the activity of other metabolic pathways that together lead to the increase in generation of ROS and consequently the development of oxidative stress [13]. Among these, pathways related to hyperglycemia is overactivation of protein kinase C (PKC) due to the increase in *de novo* synthesis of diacylglycerol (DAG), and due to a higher availability of glycolytic intermediate glyceraldehyde-3-phosphate, this activation results in the increase of several pro-oxidant enzymes, such as nicotinamide adenine dinucleotide phosphate (NADPH) oxidase (NOX) [14]. The polyol pathway metabolizes approximately 30% of total glucose during hyperglycemia, the rate-limiting step is the reduction of glucose into sorbitol catalyzed by aldose reductase, using NADPH as a cofactor, this reaction decreases the cytosolic NADPH/NADP+ ratio, NADPH depletion leads to a decrease in the reduced/oxidized glutathione (GSH/GSSG) ratio, and GSH reduction contributes to oxidative stress [15]. Hexosamine pathway is activated by the increase in concentration of glycolytic intermediate fructose-6-phosphate and results in the synthesis of N-acetylglucosamine, which is a potent protein glycosylation agent [16]. Excess ROS leads to a decrease/alteration in the activity of the enzymes of antioxidant system and in turn damage to cellular macromolecules, such as respiratory chain complexes or membrane lipids, thus culminating in oxidative damage [17–19]. This imbalance between overproduction of ROS and the inability of the antioxidant system leads to a state of oxidative stress that exacerbates the complications and damage caused by DM per se [20, 21].

### 2.1 *Eryngium carlinae* F. Delaroche

Currently, different drugs are available for the treatment of diabetes [22]. However, these drugs only act by reducing glucose levels and not oxidative stress, and furthermore, there are different drawbacks to their use such as severe hypoglycemia, weight gain, lower therapeutic efficacy, ineffective dosage, solubility and permeability problems, low potency, and altered side effects due to drug metabolism [23]; due to these effects, the use of medicinal plant extracts that can exert hypoglycemic and antioxidant effects is of importance in the research of effective therapies against DM and oxidative stress.

Among the compounds that have been isolated from *E. carlinae* and have demonstrated antioxidant activity *in vivo* are gallic acid, protocatechuic acid, rutin, epicatechin, and quercetin, for this reason, *E. carlinae* is a plant that has great potential to be used as a treatment against DM and oxidative stress [24].

## 2.2 Antioxidant, hypoglycemic, and hypolipidemic effects of *E. Carlinae* extracts

Pérez-Ramírez et al. [25] obtaining an aqueous decoction from the flowers, performed the phytochemical identification using the HPLC-ESI/MSD technique, where the presence of phenolic acids, flavonoids, phytosterols, and saponins was found. The main components of this extract were ellagic acid (38.3 ± 1.8 mA), campesteryl β-D-glucopyranoside (28.9 ± 1.4 mA), and caffeic acid (20.3 ± 1.5 mA); administration of this decoction in obese male Wistar rats induced with a high fat and fructose diet (13% protein, 18% lipids [6% saturated fat] and 43% carbohydrates [14% fructose]) at a dose of 0.6 g/day; this treatment reduces oxidative stress in kidneys because lipid peroxidation and protein carbonylation decreased by 29 and 18%, respectively, in the treated obese group compared to the control obese group. Trejo-Hurtado et al. [26] from an ethyl acetate extract from the inflorescences, carried out the identification and quantification of secondary metabolites where the main compounds were rosmarinic acid (3473.79 ± 146.18 μg/g of dried extract), chlorogenic acid (64.92 ± 1.24 μg/g of dried extract), and kaempferol-3-O-glucoside (50.42 ± 1.72 μg/g of dried extract); oral administration of this extract in streptozotocin-induced (45 mg/kg of body weight) diabetic male Wistar rats at a dose of 30 mg/kg of body weight for 60 days; this treatment did not show hypoglycemic effect; however, antioxidant effects were observed in the liver since the treated group showed a two-fold decrease in ROS generation and lipid peroxidation compared to diabetic group. These results were related to the restoration of the activity of antioxidant enzyme catalase in the treated group. Noriega-Cisneros et al. [27] identified the compounds present in the ethanolic extract of aerial part by gas chromatography/mass spectrometry (GC/MS), which were β-selinene (26.04% of abundance), α-selinene (17.54% of abundance), and stearic acid (14.54% of abundance); oral administration of this extract in streptozotocin-induced (45 mg/kg of body weight) diabetic male Wistar rats at a dose of 30 mg/kg of body weight for 40 days; this treatment showed hypolipidemic activity since it significantly decreased the levels of total cholesterol from 74 ± 7 to 55 ± 4 mg/dl, triglycerides from 224 ± 40 mg/dl, and non-HDL cholesterol from 61 ± 7 to 35 ± 4 mg/dl in the serum of the diabetic group with respect to the group administered respectively. Peña-Montes et al. [4] obtained a hexanic extract of inflorescences where identified the main compounds by GC/MS that were (Z)β-Farnesene (38.79% of abundance), β-pinene (17.53% of abundance), and calamenene (13.3% of abundance); oral administration of this extract in streptozotocin-induced (45 mg/kg of body weight) diabetic male Wistar rats at two doses of 3 and 30 mg/kg of body weight for 7 weeks; only the group treated with the dose of 30 mg/kg body weight showed antioxidant and protective activity against oxidative damage in brain, liver, and kidney because it significantly decreased lipid peroxidation, protein carbonylation, and ROS generation with respect to diabetic group, and blood glucose levels also decreased from 503.3 mg/dl in the diabetic group to 410 mg/dl in the group treated with the 30 mg dose. Furthermore, García-Cerrillo et al. [3] obtained a hexanic extract of inflorescences oral administration of this extract in streptozotocin-induced (45 mg/kg of body weight) diabetic male Wistar rats at a dose of 30 mg/kg of body weight for 7 weeks; this treatment showed hypoglycemic and hypolipidemic effects in serum because it significantly decreased glucose levels from 355.2 to 120 mg/dl and triacylglycerides from 274.8 to 111.8 mg/dl in the diabetic group compared with respect to administered group, respectively.

On the other hand, in recent years, nanoformulations have presented a new therapeutic alternative in DM treatment. Nanoparticles increase several key pharmacological characteristics of drugs such as solubility, rapid onset of action, controlled release, increased half-life, and optimized bioavailability [28]. Among all nanoparticle synthesis techniques is the green synthesis technique. This technique is based on the reduction of mono- or divalent metal ions, nucleation, and stabilization using biological species, such as fungi and bacteria, especially plant extracts [29]. The advantage of this method is that the compounds of the extract used first carry out the synthesis of nanoparticles, and subsequently these metabolites remain attached to the surface of these nanoparticles; this characteristic allows nanoparticles to increase the pharmacokinetic characteristics of secondary metabolites when they are administered in a biological system [30]. Lemus de la Cruz et al. [31], using an aqueous extract of aerial part of *E. carlinae*, synthesized silver nanoparticles (AgNPs) through green synthesis technique from which they obtained structures with a spherical morphology and an approximate size of 10–100 nm. They also carried out the qualitative characterization of secondary metabolites present in the extract where it was found that the most abundant metabolites were flavonoids. After oral administration for 45 days of both the extract and extract-AgNPs combination at a dose of 30 mg/kg body weight in streptozotocin-induced diabetic rats, both treatments showed antioxidant activity in mitochondria isolated from the brain since they significantly decreased ROS generation and lipid peroxidation, also restored the activity of antioxidant enzymes SOD2 and glutathione peroxidase (GPx), and also restored the activity of complexes I, II, and III of respiratory chain; these results were related to presence of flavonoids in both treatments. On the other hand, only the extract-AgNPs combination had hypoglycemic and hypolipidemic activity in serum, which was associated with the biological activity of AgNPs per se.

## 3. *Potentilla indica* (Andrews) Th.Wolf

*Potentilla indica*, which has scientific synonymy with *Fragaria indica* Andr. and *Duchesnea indica* Andr., is a plant belonging to the Rosaceae family, commonly known as "false strawberry," "Indian strawberry," or "wild strawberry." It is an herbaceous perennial plant; it flowers from May to October, and the seeds ripen from July to October. This species grows in temperate biomes, humid, and semi-shady areas, and presents a height of 10 cm, the foliage is semi-deciduous with trifoliate leaves, leaflets with serrated edge, and a petiole up to 5 cm long. Its flowers are yellow with five nonoverlapping petals, and its fruits are red, which are ovoid or globose, 1 to 1.5 cm in diameter with a great similarity to strawberries (**Figure 1**) [32].

This species is native to East Asia; however, it is widely distributed in the Asian, European, and American continents, which converts it into a plant with a favorable environmental adaptability.

The use of plants as a complementary therapy is an ancestral practice that nowadays continues to spread all over the world. Traditionally, *Potentilla indica* leaves are used as a decoction or applied externally as a poultice after the fresh leaves are crushed. The flowers are used in infusions, and the fruit is edible. The plant has been widely used in traditional Asian medicine, mainly for the treatment of leprosy, burns, congenital fever, tissue inflammation, hematemesis, diarrhea, cancer, diabetes, among other uses [33].

Medicinal plants have been of great interest to different research groups as they may represent an important natural source of bioactive compounds for the

**Figure 1.**
*Potentilla indica plant.*

development of new and promising drugs in the search for an effective treatment with a lower degree of side effects for various diseases. These plant-derived bioactive compounds can exert their therapeutic effects through different mechanisms that may include, in general terms, their interaction with cytoplasmic membrane receptors, resulting in the modulation of distinct signaling cascades with the activation/inhibition of different regulatory kinases. Also, through their ability to regulate gene expression in target cells or by interacting directly with the aggressor agent.

It has been documented that the different extracts of *Potentilla indica* present a great variety of pharmacological effects associated with antioxidant, anti-inflammatory, immunomodulatory, anticancer, and antimicrobial activity [32, 34–36] whose effects are attributed precisely to the content of secondary metabolites (SM) that it presents.

### 3.1 Antioxidant effect of *Potentilla indica*

It is well documented that oxidative stress represents a determinant factor in the development and evolution of multiple chronic degenerative diseases such as diabetes, metabolic syndrome, cardiovascular disease, cancer, Alzheimer's, and Parkinson's disease, and its incidence has been increasing exponentially in recent years [37]. Therefore, it is necessary to search for and implement drugs with antioxidant properties as a therapeutic strategy. In this sense, numerous studies have reported a potent antioxidant activity exhibited by different extracts of certain parts of *Potentilla indica* due to their phytochemical composition. In 2009, compounds of phenolic nature were identified in the fruit by HPLC-MS/MS, comprising three anthocyanins: cyanidin 3-O-ruthinoside, peonidin 3-O-ruthinoside, and petunidin 3-O-ruthinoside [38]. Anthocyanins are phenolic compounds pertaining to a subgroup of flavonoids, which are characterized by being water-soluble pigments responsible for the range of colors in flowers, fruits, and leaves, and in addition, they have shown promising antioxidant properties directly related to their structure due to the presence of hydroxyl groups in the B ring, as well as the oxonium ion in the C ring; thus, they are phytochemicals with a great capacity to neutralize free radicals [39]. Moreover, in 2019 Shan et al.

[40] demonstrated that *Potentilla indica* fruit possesses potent antioxidant activity by neutralizing the DPPH radical and $H_2O_2$, which may be correlated to its high content of phenolic compounds. In addition, a neutral polysaccharide obtained from *Potentilla indica* has also been reported to exhibit *in vitro* antioxidant and anti-inflammatory activity [32]. Additionally, it was demonstrated that the methanolic extract of *Potentilla indica*, in which ellagic acid was identified as the major SM by HPLC, exhibits antioxidant activity at a concentration of 200 μg/mL in human skin fibroblast cells (CCD-986Sk) and at 100 μg by topical application to mouse skin. Another *in vitro* study identified two compounds (2,4-dichloro-6-hydroxy-3,5-dimethoxytoluene and 2 2-methyl-6-(4-methylphenyl)-2-hepten-4-one) isolated from an ethyl acetate fraction of *Potentilla indica* by their mass and 1H-NMR, which showed potent antioxidant activity by neutralizing the DPPH radical, in addition to significant antidiabetic activity by inhibiting α-amylase and α-glucosidase enzymes [41].

In addition to the fruit, other parts of the plant have been subjected to research and have been shown to possess antioxidant activity, both *in vitro* and in different experimental models. In 2015 Zhu et al. identified by HPLC-ESI-MS/MS and ESI-IT-MS (ion trap MS) 27 phenolic compounds in an ethanolic extract of the aerial part (stems and leaves) of *P. indica*, which comprised ellagitannins, ellagic acid and ellagic acid glycosides, hydroxybenzoic acid, hydroxycinnamic acid derivatives, and flavanols, such as quercetin 3-O-glucuronide and kaempferol, which have been shown in various studies to possess potent antioxidant activity [33]. Quercetin is a considerably prominent flavonoid due to the extensive pharmacological effects it has been demonstrated to possess as an antioxidant, antimicrobial, antidiabetic, anti-inflammatory, and anticarcinogenic. However, among the various biological properties of quercetin and its conjugated metabolites, the most remarkable property is undoubtedly the antioxidant aspect, which is directly associated with its chemical structure, where the hydroxyl groups at the 3, 5, 7, 3′, 4′ position of the A and B rings, the double bond between the second, and third carbon and the carbonyl group of the fourth carbon play a fundamental role in its antioxidant properties [42]. In the same way, Liaqat et al. [43] determined the *in vitro* antioxidant activity of extracts of different polarity of different parts of *Potentilla indica* (whole plant, leaves, stems, and root) where they showed that the aqueous and methanolic extract of all plant parts evaluated in the study and exhibit potent antioxidant activity and in which they determined a high content of phenolic compounds through spectrophotometric methods. In another study, Landa-Moreno et al. [2] demonstrated that ethyl acetate extract from the leaves and stems of the plant exhibited antioxidant activity *in vitro* and in renal mitochondria in an animal model of streptozotocin-induced type 1 diabetes after oral administration for 60 days at a dose of 25 mg/kg, and in which phenolic acids, such as ferulic acid, vanillic acid, and t-cinnamic acid, were identified by UPLC-MS/MS as the major phytochemicals. Ferulic acid is a phytochemical derived from cinnamic acid that has been shown to possess a pharmacological effect on different diseases related to oxidative stress such as neurodegenerative diseases, diabetes and its complications, cardiovascular diseases, cancer, inflammation, and even bacterial and viral infections [44]. Ferulic acid and its derivatives have demonstrated significant antioxidant activity due to their structural features attributed to their phenolic ring and unsaturated side chain. The free radical reacting with ferulic acid easily abstracts a hydrogen atom leading to the formation of the phenoxyl radical, which is rapidly stabilized by resonance, as the unpaired electron can either remain on oxygen or delocalize throughout the molecule [45]. Therefore, the generated phenoxyl radical is unable to initiate or propagate a radical chain reaction, and thus condenses with another ferulate radical to form dimerized compounds, such as curcumin [45, 46]. The curcumin dimer, whose structure presents a second phenolic hydroxyl group, can enhance antioxidant activity due to increased resonance stabilization and o-quinone formation. Additionally, ferulic acid can increase the activity of antioxidant enzymes and inhibit oxidative enzymes, which converts it into a potent natural antioxidant [45].

In 2019, ferulic acid was demonstrated to activate Nrf2 protein, a factor involved in the positive transcriptional regulation of cellular antioxidant response, *via* the PI3K signaling pathway in human umbilical vein endothelial cells (HUVEC) [47]. Furthermore, ferulic acid exerts a renoprotective effect on methotrexate-induced nephrotoxicity in rats by activation of the Nrf2/ARE/HO-1 and PPARγ pathway and suppression of NF-κB/NLRP3 inflammasome axis [47]. Chowdhury et al. [22] demonstrated that an oral treatment with ferulic acid (50 mg/kg) for 8 weeks exerts a renoprotective effect in streptozotocin-induced diabetic rats *via* modulation of the MAPK pathway (p38, JNK and ERK 1/2), NF-κB, mitochondria-dependent and -independent apoptosis, as well as induction of autophagy. In addition, it exerts a protective effect on methotrexate-induced hepatoxicity in mice by decreasing oxidative stress and inflammation [48]. Thus, plants containing ferulic acid as a chemical constituent, including Potentilla indica, show promising potential in the treatment of different diseases associated with oxidative stress.

### 3.2 Anti-inflammatory effect of *Potentilla indica*

There is accumulating evidence that dysregulation of inflammatory pathways is closely linked to the development and progression of various chronic diseases [49]. Several investigations have been reported an anti-inflammatory effect of *Potentilla indica* in different study models. In 2021, Ullah et al. [50] demonstrated that the 18-hour exposure to ethanolic extract of *Potentilla indica* leaves and stems at different concentrations (12.5, 25, 50, y 100 μg/mL) potentially attenuates coal fly ash-induced inflammation in murine alveolar macrophages by inhibiting the proinflammatory NF-κB signaling pathway, where the major SM identified in the extract by UPLC-QTOF-MS was ellagic acid. Ellagic acid is a phenolic acid well recognized for its high therapeutic potential since it has been shown in several studies to possess pronounced antioxidant, anti-inflammatory, antiproliferative, antimicrobial, antifibrotic, and antidiabetic properties [39]. This compound has been the secondary metabolite mostly identified in alcoholic extracts of *Potentilla indica*. In 2022, Lee et al. [51] demonstrated that the oral administration of ethanolic extract of the *Potentilla indica* (100 and 300 mg/kg) in which ellagic acid was identified by UPLC as the major component, ameliorates LPS-induced septic shock in mice and decreases the levels of inflammatory markers in the serum of mice in a chronic sepsis model by suppressing NF-κB translocation to the nucleus of lung cells in a dose-dependent manner. Similarly, in 2008, Zhao and colleagues [34] demonstrated that ethanolic extract of *Potentilla indica* (whole plant) displayed anti-inflammatory activity in an inflammatory cell model by addition of lipopolysaccharide (LPS, 10 ng/mL) in the RAW264.7 cell line in the presence of 10, 5, and 1 μg/mL concentrations of the extract for 24 h, blocking NF-κB activation.

### 3.3 Anticancer effect of *Potentilla indica*

Cancer is one of the leading causes of mortality worldwide [52]. Cancer cells frequently experience resistance to chemotherapeutic drugs; however, it has been reported in different cellular models that several phytochemicals are able to increase the sensitivity and efficacy to chemotherapy through different mechanisms; therefore, they may contribute considerably to decrease the mortality of this disease [53]. In this context, it has been reported in different studies that *Duchesnea indica* has anticancer activity. This effect can be attributed to epicatechin gallate (ECG) and derivatives, which are phenolic compounds that have been shown in several studies to possess potent antiproliferative and antioxidant activity in different tumor models [54, 55] and have been identified in extracts of *Potentilla indica*. In 2017, Chen et al. [56] demonstrated that *Potentilla indica* leaves extract, whose main secondary metabolite identified by HPLC was ECG and exhibits antitumor activity by significantly reducing tumor growth in a BALB/c nude mouse xenograft model orally administered with the extract (500 mg/kg). Likewise, it inhibits highly metastatic cells by reducing extracellular matrix metalloproteinase-2 (MMP-2) and urokinase-type plasminogen activator secretions. This study also demonstrated that *Potentilla indica* negatively regulates N-cadherin, fibronectin, and vimentin and increases E-cadherin expression, decreasing cell adhesion capacity. Phenolic extracts of this plant have also been shown to significantly inhibit SKOV-3 ovarian cancer cell proliferation and *in vitro* and *in vivo* cervical cancer growth through induction of apoptosis *via* mitochondrial pathway and cell cycle arrest [57, 58]. It has also been reported that an extract of *Potentilla indica* possesses antimetastatic effects on oral squamous cell carcinoma through the inhibition of the MMP-2 activity by down-regulating the MEK/ERK pathway [36].

### 3.4 Antimicrobial effect of *Potentilla indica*

Currently, infectious diseases caused by different pathogenic microorganisms represent one of the main global threats to public health. Additionally, in recent decades, drug resistance to commercially available synthetic compounds has increased considerably [59], and natural products could represent a promising source to satisfy this demand. In this context, *Potentilla indica* has been demonstrated to possess effective activity against pathogenic microorganisms. In 2021, Liaqat et al. [60] demonstrated a significant antimicrobial activity of extracts of different polarity (aqueous, chloroform, n-hexane, and methanolic) of *Potentilla indica* (whole plant, leaves, stems, and root) against *Staphylococcus aureus*, *Salmonella typhi*, *Escherichia coli*, *Klebsiella pneumoniae*, *Bacillus subtilis*, *Vibrio cholera*, and *Saccharomyces cerevisiae* and in which they determined a high content of phenolic compounds by quantitative spectrophotometric methods. The potent antimicrobial capacity of plant-derived phenolic compounds toward different microorganisms has been reported in multiple studies [61]. This antimicrobial effect has been associated to different mechanisms promoted by these phytochemicals, which include the inhibition of virulence factors; membrane disruption; inhibition of efflux pump; biofilm inhibition; inhibition of virulence enzymes: neuraminidase, sortase, coagulase, and urease y protease; cell envelop synthesis inhibition, nucleic acid synthesis inhibition, and bacterial motility inhibition [61].

On the other hand, 25 compounds have been identified by GC–MS in the essential oil of *Potentilla indica*, whose main constituents were carvacryl acetate, valencene, nona hexacontanoic acid, aistalone, dehydroaromadendrene, eicosane, 2-hexadecan-1ol, and aromadendrene, which have been demonstrated to possess antimicrobial activity [62].

### 3.5 Perspectives and limitations of using of *Potentilla indica*

The administration of the plant for specific therapeutic purposes may present some limitations in medical practice since years of clinical research are required for the establishment of a therapeutic dose with controlled studies to validate its efficacy and safety. However, no toxic side effects of *Potentilla indica* have been documented. Additionally, it is worth mentioning that thanks to the different lines of research directed to the study of medicinal plants and their effects in different experimental models, as well as the possible molecular mechanisms involved in such effect, and we can infer that these plants, among them *Potentilla indica*, contain bioactive compounds with promising therapeutic properties that can contribute significantly as adjuvant therapy to the treatment and prophylaxis of multiple chronic degenerative diseases that currently represent a public health problem worldwide, such as diabetes, metabolic syndrome, infectious diseases, cancer, among others, and may also represent an important natural source for the pharmaceutical industry for the discovery and development of new drugs with a high therapeutic potential, a wide margin of safety and fewer side effects. Further. Research in this direction is required to consolidate this pharmacological application.

## 4. *Justicia spicigera* Schltdl

*Justicia spicigera* corresponds to the kingdom Plantae, division Tracheophyta, class Magnoliopsida, order Lamiales, Family Acanthaceae, Genus Justicia L., and spicigera species Schltdl [63]. It is known by the names of "Muicle," "Mouij," "Stone indigo," "purple grass," "limanin," "micle," "mohuite," "muele," "muecle," "muille," "muite," "muitle," among others [64] and is widely distributed in Latin American countries such as Mexico, Belize, Costa Rica, and Honduras where there are optimal conditions for its development, such as dry or semi-dry, temperate climates warm, and semi warm, from 0 to 3000 meters above sea level [65, 66]. Likewise, this species of Justicia is specified as a perennial subshrub that can reach between 0.5 and 1.5 meters in height, its leaves are petiolate with an entire leaf margin approximately 0.02 meters long, and it produces orange inflorescences arranged in spikes with a bilobed rear lip. And anterior three-lobed, two stamens, a capsule with four seeds, and a sterile basal portion that are grouped at the end of the union of the stem and the leaf [65, 67].

It is a plant that has been used in Mexico since pre-Hispanic times as a source of natural pigment to dye fabrics and crafts [68]. Furthermore, in traditional Mexican medicine, *Justicia spicigera* leaves are commonly used as an infusion, mainly for the treatment of headaches, hypertension, epilepsy, stomach pain, diarrhea, dysentery, constipation, and in the treatment of chronic degenerative diseases, such as diabetes and hypertension. On the other hand, the branches of the leaves and flowers are used as therapy for dermatological conditions, tumors, or pimples that are difficult to cure [5, 67, 69]. These effects are related to the presence of different compounds in the plant, among which are phenolic secondary metabolites, such as kaempferitrin and kaempferol triamnoside that have been extracted from the leaves or tannins that have been extracted from the branches and flowers [65].

Currently, various studies have been carried out aimed at the use of *J. spicigera* extracts obtained with solvents of different polarities to identify the nature and effect of the secondary metabolites present in the plant for the treatment of various pathologies.

### 4.1 Anticancer activity

Recent research has shown that extracts of dried plant material of *Justicia spicigera* obtained with polar solvents present cytotoxic activity in certain cell lines [70]. In a study conducted in 2020 [9], it was reported that methanolic extract of *J. spicigera* 1:10 (w/v) decreased the viability of HEP-G2 hepatocellular carcinoma cells with a mean inhibitory concentration (IC 50) of 2.92 μg/mL. In addition to the above, in 2023, [70] it was shown that the 1:20 (w/v) methanolic extract inhibits the growth of tumor cells from murine lymphoma L5178Y-R at an IC 50 of 29.10 μg/mL. In the same study, the presence of coumarins, quinones, and tannins in the extract was determined, to which these effects are attributed. In both experiments, the Probit test was used to calculate the IC 50 values.

On the other hand, it was found that the ethanolic extract of *J. spicigera* also exerts cytotoxic effects on MCF-7 cells (IC50 = 28 μg/mL) and HeLa (IC50 = 17 μg/mL), which was calculated by regression analysis (% of survival versus logarithmic concentration), where it is highlighted that an effective dose ≤30 μg/mL for plant extracts is considered cytotoxic [71]. Likewise, it was found that the hydroalcoholic extract (water: ethanol; 50:50) of *J. spicigera* obtained by evaporation under reduced pressure and lyophilization, reduces the proliferation of androgen-dependent LNCaP prostate cancer cells by 62% ± 1. 7% (IC50 = 3026 ± 421 μg/ml) without affecting cell viability, that is, it was found that there is a cytostatic mechanism on the part of the extract, which prevents progression to mitosis in the cell cycle. These effects could be related to the presence of different compounds present in the plant determined by qualitative methods, where the presence of flavonoids, sterols, steroids, and steroidal sapogenins has been reported [72].

### 4.2 Hypoglycemic activity

In previous reports, the potential hypoglycemic effect present in the ethanolic extract of dried plant material of *J. spicigera* has been documented. In a study carried out by Ramírez and collaborators [73], reported that the ethanolic extract (60%) of *J. spicigera* 1:10 (w/v) inhibits alpha glucosidase in tests carried out in vitro by 0.4 ± 1.5% at a concentration of 1 mg/mL. Similarly, Ortiz-Andrade and collaborators [67] carried out in vivo tests, where they demonstrated that the ethanolic extract of *J. spicigera* at a dose of 100 mg/kg significantly reduces the glucose levels obtained in an oral tolerance test after its administration in male rats of the Wistar strain, induced to type 2 diabetes with streptozotocin (65 mg/kg) and nicotinamide (100 mg/kg). These effects were attributed to the presence of kaempferitrin, which was determined by the reverse phase HPLC method in water 2795; In this study, one-way analysis of variance (ANOVA) with Tukey's post hoc test (sample versus control) was performed. Finally, in a recent study, the effect of the ethanolic extract of fresh leaves of *J. spicigera* 1:10 (w/v) was evaluated in a treatment for 30 days at a dose of 100 mg/kg in male Wistar rats, induced to diabetes with streptozotocin (50 mg/kg), showing a significant decrease in blood glucose levels in the group treated with the extract (219 mg/dL) when compared to the control diabetic group (427.3 mg/dL) [74], which could be related to the presence of kaempferitrin.

### 4.3 Antioxidant activity

Various *in vitro* and *in vivo* studies have been carried out to evaluate the antioxidant capacity of the extracts obtained from the plants. Antioxidant activity can be determined by electron transfer methods for the reduction of a compound such as the 2,2-diphenyl-1-picrylhydrazyl (DPPH•) radical scavenging assay or by scavenging free radicals by donating from a hydrogen atom to another compound, as is the case of ABTS• + (2,2′-azino-bis-3-ethylbenzothiazoline-6-sulfonic acid) [75]. In a previous study using in vitro tests carried out with powdered leaf and stem extracts of *J. spicigera*, it was shown that the content of phenolic compounds depends on the extraction conditions influenced by the solvent used (water, ethanol, glycerol, and propylene glycol), pH (3.5, 5.5 and 7.4), temperature (25 and 60°C) and storage time (20 days). The results obtained indicated that the polar extracts have greater inhibitory activity of DPPH• and ABTS•+ [76]. According to above, a study demonstrated that the ethanolic extract of *J. spicigera,* where the majority presence of kaempferitrin was determined, presents a mean effective concentration (EC50) of 180 μg/mL in the ABTS test and 100 μg/mL in the DPPH• assay [67]. In contrast, a study showed that the CH2Cl2: MeOH 1:1 extract in the DPPH• and ABTS• + assays present an EC50 of 35.5 and 15.7 μg/mL [77]. On the other hand, a decrease in oxidative stress markers has been observed in experiments carried out *in vivo*. Awad and collaborators in 2015 [78] found that the ethyl acetate fraction of *J. spicigera*, administered two times a week at a dose of 500 mg/kg for 6 weeks in rats induced liver fibrosis with CCl4 (0.5 mL/Kg; via I.P.) improves liver functions and levels of oxidant stress markers, attributed to the phenolic compounds identified in this extract, mainly Peonidin 3,5-diglucoside.

### 4.4 Lipid-lowering and anti-inflammatory activity

Obesity is a proinflammatory disease that is related to certain chronic degenerative diseases such as type 2 diabetes, nonalcoholic fatty liver disease (NAFLD), and cancer [79]. Real-Sandoval [80], administered the ethanolic extract of *J. spicigera* powder 1:100 (w/v) at a dose equivalent to 250 μg/kg of weight daily for 8 weeks and observed a decrease in the percentage of accumulated fat and triglycerides, as well as their participation in the negative modulation of the expression of proinflammatory genes such as Tlr4, Tnf-α, Nlrp3, Caspase-1, Il-18, Il-1β, Srebp-1c, Ppar-γ, and Ucp2, in addition to the overexpression of Ppar-α in a model of obesity-induced with a high-fat diet in male rats of the Wistar strain. These effects were attributed to the presence of kaempferitrin, which was determined by the reverse phase HPLC method.

### 4.5 Antihypertensive activity

The effect of extracts of dry plant material of *J. spicigera* obtained with nonpolar solvents such as chloroform has been evaluated, where the majority presence of hesperidin, naringenin, and kaempferol was identified by reverse phase HPLC, compounds to which an effect is attributed antihypertensive at a dose of 150 mg/kg in male Wistar rats induced with L-NAME at 75 mg/kg in drinking water [81].

### 4.6 Antimicrobial activity

Recent studies have shown that *J. spicigera* compounds extracted with CH2Cl2: MeOH 1:1 inhibit the growth of *E. faecalis* with a minimum inhibitory concentration (MIC) of 0.06 mg/mL, which is considered a sensitive strain and *P. aeruginosa* with a MIC of 0.5 mg/mL, that is, it presents medium sensitivity to the extract, which showed the presence of alkaloids, flavonoids, phenols, carbohydrates, and triterpenes that were determined by qualitative methods [77].

In another sense, *J. spicigera* biomolecules have been used in foods, bioproducts, dyes, and materials synthesis [65].

In summary, Mexico is a country with a great diversity of medicinal plants, among which *J. spicigera* stands out due to its high antioxidant capacity. Recent studies have demonstrated the therapeutic effects of this plant attributed to the phenolic secondary metabolites obtained with solvents of different polarities that can be used as a potential treatment in different pathologies such as cancer, diabetes, and hypertension. It should be noted that these effects may be due to the combination of compounds in the extracts, generating interactions that can be classified as antagonistic or synergistic [82], in which different mechanisms are involved that serve as points of opportunity for new research.

## 5. Conclusions

According to the research on these medicinal plants, Eryngium native plant of Mexico, Potentilla widely in Asian traditional medicine, and Justicia distributed in Latin American, all represent an important natural source of bioactive compounds. Flavonoids it is the major compound in all plants and help improve lipid profiles, blood pressure, insulin resistance, and systemic inflammation and have important effects on various chronic degenerative diseases. However, seasonality, water, light supply, temperature, herbivory and microbes, and soil factors can be influenced by climate and can result in up to 50% increased or decreased content of secondary plant metabolites. It is for that reason that further studies need to be done to characterize new active compounds with the objective of making science and evidence-based hard claims for the functionality and efficacy of the phytonutrient blend. Synergistic activity in the compounds is what makes them have a beneficial effect because they contain each of the compounds in specific quantities demonstrating the claimed biological functionality and health benefit as specifically to the mixture of active compounds. However, such molecular blends are difficult to reproducibly generate and formulate it is because regulatory authorities are more in favor of approving single or few active principles rather than complex blends.

## Acknowledgements

The authors appreciate the partial support from Scientific Research Coordination (18070, to ASM), Michoacana University of San Nicholas of Hidalgo.

## Conflict of interest

The authors declare no conflict of interest.

## Author details

Alfredo Saavedra-Molina*, Jenaro Lemus-de la Cruz, Cinthia Landa-Moreno, Marina Murillo-Villicaña, Claudia García-Berumen, Rocío Montoya-Pérez, Salvador Manzo-Avalos, Asdrubal Aguilera-Méndez, Rafael Salgado-Garciglia and Christian Cortés-Rojo
Chemical Biological Research Institute, Michoacana University of San Nicholas of Hidalgo, Morelia, Mexico

*Address all correspondence to: francisco.saavedra@umich.mx

## References

[1] Noriega-Cisneros R, Ortiz-Avila O, Esquivel-Gutiérrez E, Clemente-Guerrero M, Manzo-Avalos S, Salgado-Garciglia R, et al. Hypolipidemic activity of *Eryngium carlinae* on streptozotocin-induced diabetic rats. Biochemistry Research International. 2012;**2012**:603501. DOI: 10.1155/2012/603501

[2] Landa-Moreno CI, Trejo-Hurtado CM, Lemus-de la Cruz J, Peña-Montes D, Murillo-Villicaña M, Huerta-Cervantes M, et al. Antioxidant effect of the ethyl acetate extract of *Potentilla indica* on kidney mitochondria of streptozotocin-induced diabetic rats. Plants. 2023;**12**:3196. DOI: 10.3390/plants12183196

[3] García-Cerrillo D, Noriega-Cisneros R, Peña-Montes D, Huerta-Cervantes M, Ríos-Silva M, Salgado-Garciglia R, et al. Asian Journal of Applied Sciences. 2018;**6**:308-314

[4] Peña-Montes DJ, Huerta- Cervantes M, Ríos-Silva M, Trujillo X, Huerta M, Noriega-Cisneros R, et al. Protective effect of the Hexanic extract of *Eryngium carlinae* inflorescences *in vitro*, in yeast, and in Streptozotocin-induced diabetic male rats. Antioxidants (Basel). 2019;**8**(3):73. DOI: 10.3390/antiox8030073

[5] Awad NE, Abdelkawy MA, Rahman EHA, Hamed MA, Ramadan NS. Phytochemical and *in vitro* screening of *Justicia spicigera* ethanol extract for antioxidant activity and *in vivo* assessment against *Schistosoma mansoni* infection in mice. Anti-Infective Agents. 2018;**16**:49-56

[6] International Diabetes Federation, editor. IDF Diabetes Atlas. 10th ed. International Diabetes Federation; 2021

[7] Mayer-Davis EJ, Lawrence JM, Dabelea D, Divers J, Isom S, Dolan L, et al. Incidence trends of type 1 and type 2 diabetes among youths, 2002-2012. The New England Journal of Medicine. 2017;**376**(15):1419-1429. DOI: 10.1056/NEJMoa1610187

[8] American Diabetes Association. Diagnosis and classification of diabetes mellitus. Diabetes Care. 2014;**37**(Supplement 1):S81-S90

[9] Galicia-Garcia U, Benito-Vicente A, Jebari S, Larrea-Sebal A, Siddiqi H, Uribe KB, et al. Pathophysiology of type 2 diabetes mellitus. International Journal of Molecular Sciences. 2020;**21**(17):6275. DOI: 10.3390/ijms21176275

[10] Maraschin JF. "Classification of diabetes," in diabetes. Advances in Experimental Medicine and Biology. 2013;**771**:12-19

[11] Misrani A, Tabassum S, Yang L. Mitochondrial dysfunction and oxidative stress in Alzheimer's disease. Frontiers in Aging Neuroscience. 2021;**18**(13):617588. DOI: 10.3389/fnagi.2021.617588

[12] Sivitz WI, Yorek MA. Mitochondrial dysfunction in diabetes: From molecular mechanisms to functional significance and therapeutic opportunities. Antioxidants & Redox Signaling. 2010;**12**(4):537-577. DOI: 10.1089/ars.2009.2531

[13] González P, Lozano P, Ros G, Solano F. Hyperglycemia and oxidative stress: An integral, updated and critical overview of their metabolic interconnections. International Journal of Molecular Sciences. 2023;**24**(11):9352. DOI: 10.3390/ijms24119352

[14] Kay AM, Simpson CL, Stewart JA Jr. The role of AGE/RAGE signaling

in diabetes-mediated vascular calcification. Journal Diabetes Research. 2016;**2016**:6809703. DOI: 10.1155/2016/6809703

[15] Dong K, Ni H, Wu M, Tang Z, Halim M, Shi D. ROS-mediated glucose metabolic reprogram induces insulin resistance in type 2 diabetes. Biochemical and Biophysical Research Communications. 2016;**476**(4):204-211. DOI: 10.1016/j.bbrc.2016.05.087

[16] Mizukami H, Osonoi S. Pathogenesis and molecular treatment strategies of diabetic neuropathy collateral glucose-utilizing pathways in diabetic polyneuropathy. International Journal of Molecular Sciences. 2020;**22**(1):94. DOI: 10.3390/ijms22010094

[17] Roy Chowdhury S, Djordjevic J, Thomson E, Smith DR, Albensi BC, Fernyhough P. Depressed mitochondrial function and electron transport complex II-mediated $H_2O_2$ production in the cortex of type 1 diabetic rodents. Molecular and Cellular Neurosciences. 2018;**23**(90):49-59. DOI: 10.1016/j.mcn.2018.05.006

[18] Ortiz-Avila O, Sámano-García CA, Calderón-Cortés E, Pérez-Hernández IH, Mejía-Zepeda R, Rodríguez-Orozco AR, et al. Dietary avocado oil supplementation attenuates the alterations induced by type I diabetes and oxidative stress in electron transfer at the complex II-complex III segment of the electron transport chain in rat kidney mitochondria. Journal of Bioenergetics and Biomembranes. 2013;**45**(3):271-287. DOI: 10.1007/s10863-013-9502-3

[19] Raza H, John A, Howarth FC. Increased oxidative stress and mitochondrial dysfunction in zucker diabetic rat liver and brain. Cellular Physiology and Biochemistry. 2015;**35**(3):1241-1251. DOI: 10.1159/000373947

[20] Maiese K. New insights for oxidative stress and diabetes mellitus. Oxidative Medicine and Cellular Longevity. 2015;**2015**:875961. DOI: 10.1155/2015/875961

[21] Niki E. Oxidative stress and antioxidants: Distress or eustress? Archives of Biochemistry and Biophysics. 2016;**595**:19-24. DOI: 10.1016/j.abb.2015.11.017

[22] Chaudhury A, Duvoor C, Reddy Dendi VS, Kraleti S, Chada A, Ravilla R, et al. Clinical review of antidiabetic drugs: Implications for type 2 diabetes mellitus management. Frontiers in Endocrinology (Lausanne). 2017;**4**(8):6. DOI: 10.3389/fendo.2017.00006

[23] Padhi S, Nayak AK, Behera A. Type II diabetes mellitus: A review on recent drug-based therapeutics. Biomedicine & Pharmacotherapy. 2020;**131**:110708. DOI: 10.1016/j.biopha.2020.110708

[24] Cárdenas-Valdovinos JG, García-Ruiz I, Angoa-Pérez MV, Mena-Violante HG. Ethnobotany, biological activities and phytochemical compounds of some species of the genus Eryngium (Apiaceae), from the Central-Western region of Mexico. Molecules. 2023;**28**(10):4094. DOI: 10.3390/molecules28104094

[25] Pérez-Ramírez IF, Enciso-Moreno JA, Guevara-González RG, Gallegos-Corona MA, Loarca-Piña G, Reynoso-Camacho R. Modulation of renal dysfunction by *Smilax cordifolia* and *Eryngium carlinae*, and their effect on kidney proteome in obese rats. Journal of Functional Foods. 2016;**20**:545-555. DOI: 10.1016/j.jff.2015.11.024

[26] Trejo-Hurtado CM, Landa-Moreno CI, la Cruz JL, Peña-Montes DJ, Montoya-Pérez R, Salgado-Garciglia R, et al. An ethyl acetate extract of *Eryngium*

*carlinae* inflorescences attenuates oxidative stress and inflammation in the liver of Streptozotocin-induced diabetic rats. Antioxidants (Basel). 2023;**12**(6):1235. DOI: 10.3390/antiox12061235

[27] Noriega-Cisneros R, Peña-Montes DJ, Huerta-Cervantes M, Torres-Martínez R, Huerta M, Manzo-Avalos S, et al. *Eryngium carlinae* ethanol extract corrects lipid abnormalities in Wistar rats with experimental diabetes. Journal of Medicinal Food. 2020;**23**(8):827-833. DOI: 10.1089/jmf.2019.0189

[28] Uppal S, Italiya KS, Chitkara D, Mittal A. Nanoparticulate-based drug delivery systems for small molecule anti-diabetic drugs: An emerging paradigm for effective therapy. Acta Biomaterialia. 2018;**81**:20-42. DOI: 10.1016/j.actbio.2018.09.049

[29] Shah M, Fawcett D, Sharma S, Tripathy SK, Poinern GEJ. Green synthesis of metallic nanoparticles via biological entities. Materials (Basel). 2015;**8**(11):7278-7308. DOI: 10.3390/ma8115377

[30] Sahu T, Ratre YK, Chauhan S, Bhaskar L, Nair MP, Verma HK. Nanotechnology based drug delivery system: Current strategies and emerging therapeutic potential for medical science. The Journal of Drug Delivery Science and Technology. 2021;**63**:102487. DOI: 10.1016/j.jddst.2021.102487

[31] Lemus-de la Cruz J, Trejo-Hurtado M, Landa-Moreno C, Peña-Montes D, Landeros-Páramo JL, Cortés-Rojo C, et al. Antioxidant effects of silver nanoparticles obtained by green synthesis from the aqueous extract of *Eryngium carlinae* on the brain mitochondria of streptozotocin-induced diabetic rats. Journal of Bioenergetics and Biomembranes. 2023;**55**(2):123-135. DOI: 10.1007/s10863-023-09963-w

[32] Xiang B, Yu X, Li B, Xiong Y, Long M, He Q. Caracterización, actividades antioxidantes y anticancerígenas de un polisacárido neutro de Duchesnea indica (Andr.) Focke. Revista de bioquímica de alimentos. 2019;**43**(7):e12899

[33] Zhu M, Dong X, Guo M. Phenolic profiling of Duchesnea indica combining macroporous resin chromatography (MRC) with HPLC-ESI-MS/MS and ESI-IT-MS. Molecules. 2015;**20**(12):22463-22475

[34] Zhao L, Zhang SL, Tao JY, Jin F, Pang R, Guo YJ, et al. Anti-inflammatory mechanism of a folk herbal medicine, Duchesnea indica (Andr) Focke at RAW264. 7 cell line. Immunological Investigations. 2008;**37**(4):339-357

[35] Hu W, Han W, Huang C, Wang MH. Protective effect of the methanolic extract from Duchesnea indica against oxidative stress *in vitro* and *in vivo*. Environmental Toxicology and Pharmacology. 2011;**31**(1):42-50

[36] Yang WE, Ho YC, Tang CM, Hsieh YS, Chen PN, Lai CT, et al. Duchesnea indica extract attenuates oral cancer cells metastatic potential through the inhibition of the matrix metalloproteinase-2 activity by down-regulating the MEK/ERK pathway. Phytomedicine. 2019;**63**:152960

[37] Sharifi-Rad J, Quispe C, Castillo CMS, Caroca R, Lazo-Vélez MA, Antonyak H, et al. Ellagic acid: A review on its natural sources, chemical stability, and therapeutic potential. Oxidative Medicine and Cellular Longevity. 2022;**2022**:3848084

[38] Qin C, Li Y, Zhang R, Niu W, Ding Y. Separation and elucidation of anthocyanins in the fruit of mock strawberry (Duchesnea indica Focke). Natural Product Research. 2009;**23**(17):1589-1598

[39] Alappat B, Alappat J. Anthocyanin pigments: Beyond aesthetics. Molecules. 2020;**25**(23):5500

[40] Shan S, Huang S, Shah MH, Abbasi AM. Evaluation of polyphenolics content and antioxidant activity in edible wild fruits. BioMed Research International. 2019;**2019**:1381989. DOI: 10.1155/2019/1381989

[41] Huneif MA, Alqahtani SM, Abdulwahab A, Almedhesh SA, Mahnashi MH, Riaz M, et al. α-Glucosidase, α-amylase and antioxidant evaluations of isolated bioactives from wild strawberry. Molecules. 2022;**27**(11):3444

[42] Azeem M, Hanif M, Mahmood K, Ameer N, Chughtai FRS, Abid U. An insight into anticancer, antioxidant, antimicrobial, antidiabetic and anti-inflammatory effects of quercetin: A review. Polymer Bulletin. 2023;**80**(1):241-262

[43] Liaqat MUHA, Kakar MMAD, Akram IU, Muhammad SHAHZAD, Kakar MU, Ahmad NADEEM, et al. Antimicrobial and phytochemical exploration of Duchesnea indica plant. Plant Cell Biotechnology and Molecular Biology. 2021;**22**:74-85

[44] Li D, Rui YX, Guo SD, Luan F, Liu R, Zeng N. Ferulic acid: A review of its pharmacology, pharmacokinetics and derivatives. Life Sciences. 2021;**284**:119921

[45] Mancuso A, Cristiano MC, Pandolfo R, Greco M, Fresta M, Paolino D. Improvement of ferulic acid antioxidant activity by multiple emulsions: *In vitro* and *in vivo* evaluation. Nanomaterials. 2021;**11**(2):425

[46] Choe E. Roles and action mechanisms of herbs added to the emulsion on its lipid oxidation. Food Science and Biotechnology. 2020;**29**(9):1165-1179

[47] Mahmoud AM, Hussein OE, Abd El-Twab SM, Hozayen WG. Ferulic acid protects against methotrexate nephrotoxicity via activation of Nrf2/ARE/HO-1 signaling and PPARγ, and suppression of NF-κB/NLRP3 inflammasome axis. Food & Function. 2019;**10**(8):4593-4607

[48] Roghani M, Kalantari H, Khodayar MJ, Khorsandi L, Kalantar M, Goudarzi M, et al. Alleviation of liver dysfunction, oxidative stress and inflammation underlies the protective effect of ferulic acid in methotrexate-induced hepatotoxicity. Drug Design, Development and Therapy. 2020:1933-1941

[49] Kunnumakkara AB, Shabnam B, Girisa S, Harsha C, Banik K, Devi B, et al. Inflammation, NF-κB, and chronic diseases: How are they linked? Critical Reviews™ in Immunology. 2020;**40**(1):1-39

[50] Ullah HA, Lee YY, Kim SD, Rhee MH. Duchesnea indica extract attenuates coal fly ash-induced inflammation in murine alveolar macrophages through the NF-kappa B pathway. Evidence-Based Complementary and Alternative Medicine. 2021;**2021**:1-11

[51] Lee YY, Yuk HJ, Saba E, Kim SD, Kim DS, Kopalli SR, et al. Duchesnea indica extract ameliorates LPS-induced septic shock in mice. Evidence-Based Complementary and Alternative Medicine. 2022;**2022**(14):5783867

[52] Deo SVS, Sharma J, Kumar S. GLOBOCAN 2020 report on global cancer burden: Challenges and opportunities for surgical oncologists. Annals of Surgical Oncology. 2022;**29**(11):6497-6500

[53] Khatoon E, Banik K, Harsha C, Sailo B, Thakur KK, Khwairakpam AD, et al. Phytochemicals in cancer cell chemosensitization: Current knowledge and future perspectives. In: Seminars in Cancer Biology. Vol. 80. Academic Press; 2022. pp. 306-339

[54] Kürbitz Heise D, Redmer T, Goumas F, Arlt A, Lemke J, et al. Epicatechin gallate and catechin gallate are superior to epigallocatechin gallate in growth suppression and anti-inflammatory activities in pancreatic tumor cells. Cancer Science. 2011;**102**(4):728-734

[55] Xu YQ, Gao Y, Granato D. Effects of epigallocatechin gallate, epigallocatechin and epicatechin gallate on the chemical and cell-based antioxidant activity, sensory properties, and cytotoxicity of a catechin-free model beverage. Food Chemistry. 2021;**339**:128060

[56] Chen PN, Yang SF, Yu CC, Lin CY, Huang SH, Chu SC, et al. Duchesnea indica extract suppresses the migration of human lung adenocarcinoma cells by inhibiting epithelial–mesenchymal transition. Environmental Toxicology. 2017;**32**(8):2053-2063

[57] Peng B, Chang Q, Wang HQ, Wang Y, Tang J, Liu X. Suppression of human ovarian SKOV-3 cancer cell growth by Duchesnea phenolic fraction is associated with cell cycle arrest and apoptosis. Gynecologic Oncology. 2008;**108**(1):173-181

[58] Peng B, Hu Q, Liu X. Duchesnea phenolic fraction inhibits in vitro and in vivo growth of cervical cancer through induction of apoptosis and cell cycle arrest. Experimental Biology and Medicine. 2009;**234**(1):74-83

[59] Dadgostar P. Antimicrobial resistance: Implications and costs. Infection and Drug Resistance. 2019;**12**:3903-3910

[60] Liaqat M, Kakar IU, Akram M, Hussain SHAHZAD, Kakar MU, Ahmad Nade EM, et al. Antimicrobial and phytochemical exploration of *Duchesnea indica* plant. Plant Cell Biotechnology and Molecular Biology. 2021;**2019**(22):74-85

[61] Biharee A, Sharma A, Kumar A, Jaitak V. Antimicrobial flavonoids as a potential substitute for overcoming antimicrobial resistance. Fitoterapia. 2020;**146**:104720

[62] Umesh BT, Thoppil JE. Comparison of chemical constituents of tissue cultured and field grown plants of Duchesnea indica, (Andr.) Focke. Journal of Pharmacognosy and Phytochemistry. 2014;**3**(3):68-70

[63] *Justicia spicigera* Schltdl. in GBIF Secretariat. GBIF Backbone Taxonomy, GBIF Secretariat. Available from: GBIF.org [Accessed: October 2, 2023]. doi: 10.15468/39omei

[64] Fonseca-Chávez RE, Rivera-Levario LA, Vázquez-García L. Guía ilustrada de plantas medicinales en el Valle de México. Meixico: Instituto Nacional de los Pueblos Indígenas; 2020

[65] Castro-Muñoz R, León-Becerril E, García-Depraect O. Beyond the exploration of Muicle (*Justicia spicigera*): Reviewing its biological properties, bioactive molecules and materials chemistry. Processes. 2022;**10**(5):1035

[66] Baqueiro-Peña I, Guerrero-Beltrán JA. Uses of *Justicia spicigera* in medicine and as a source of pigments. Functional Foods in Health and Disease. 2014;**4**(9):401-414

[67] Ortiz-Andrade R, Cabañas-Wuan A, Arana-Argáez VE, Alonso-Castro AJ, Zapata-Bustos R, Salazar-Olivo LA,

et al. Antidiabetic effects of *Justicia spicigera* schltdl (acanthaceae). Journal of Ethnopharmacology. 2012;**143**(2):455-462

[68] Baqueiro-Peña I, Guerrero-Beltrán JÁ. Physicochemical and antioxidant characterization of *Justicia spicigera*. Food Chemistry. 2017;**218**:305-312

[69] Anaya-Esparza LM, Ramos-Aguirre D, Zamora-Gasga VM, Yahia E, Montalvo-González E. Optimization of ultrasonic-assisted extraction of phenolic compounds from *Justicia spicigera* leaves. Food Science and Biotechnology. 2018;**27**:1093-1102

[70] Rodríguez-Garza NE, Quintanilla-Licea R, Romo-Sáenz CI, Elizondo-Luevano JH, Tamez-Guerra P, Rodríguez-Padilla C, et al. *In vitro* biological activity and lymphoma cell growth inhibition by selected Mexican medicinal plants. Life. 2023;**13**(4):958

[71] Jacobo-Salcedo MDR, Alonso-Castro AJ, Salazar-Olivo LA, Carranza-Alvarez C, González-Espíndola LÁ, Domínguez F, et al. Antimicrobial and cytotoxic effects of Mexican medicinal plants. Natural Product Communications. 2011;**6**(12):1925-1928

[72] Fernández-Pomares C, Juárez-Aguilar E, Domínguez-Ortiz MÁ, Gallegos-Estudillo J, Herrera-Covarrubias D, Sánchez-Medina A, et al. Hydroalcoholic extract of the widely used Mexican plant *Justicia spicigera* Schltdl. Exert- s a cytostatic effect on LNCaP prostate cancer cells. Journal of Herbal Medicine. 2018;**12**:66-72

[73] Ramírez G, Zavala M, Pérez J, Zamilpa A. *In vitro* screening of medicinal plants used in Mexico as antidiabetics with glucosidase and lipase inhibitory activities. Evidence-Based Complementary and Alternative Medicine. 2012;**2012**:701261

[74] Murillo-Villicaña M, Noriega-Cisneros R, Peña-Montes DJ, Huerta-Cervantes M, Aguilera-Méndez A, Cortés-Rojo C, et al. Antilipidemic and hepatoprotective effects of ethanol extract of *Justicia spicigera* in streptozotocin diabetic rats. Nutrients. 2022;**14**(9):1946

[75] Gulcin İ. Antioxidants and antioxidant methods: An updated overview. Archives of Toxicology. 2020;**94**(3):651-715

[76] De La Cruz-Jiménez L, Hernández-Torres MA, Monroy-García IN, Rivas-Morales C, Verde-Star MJ, Gonzalez-Villasana V, et al. Biological activities of seven medicinal plants used in Chiapas, Mexico. Plants. 2022;**11**(14):1790

[77] García-Márquez E, Román-Guerrero A, Pérez-Alonso C, Cruz-Sosa F, Jiménez-Alvarado R, Vernon-Carter EJ. Effect of solvent-temperature extraction conditions on the initial antioxidant activity and total phenolic content of muitle extracts and their decay upon storage at different pH. Revista mexicana de ingeniería química. 2012;**11**(1):1-10

[78] Awad NE, Abdelkawy MA, Hamed MA, Souleman AM, Abdelrahman EH, Ramadan NS. Antioxidant and hepatoprotective effects of *Justicia spicigera* ethyl acetate fraction and characterization of its anthocyanin content. International Journal of Pharmacy and Pharmaceutical Sciences. 2015;7(8):91-96

[79] Cordero P, Li J, Oben JA. Obesity and NAFLD. In: Obesity: Pathogenesis, Diagnosis, and Treatment. Cham, Switzerland: Springer International Publishing; 2019. pp. 179-194

[80] Real-Sandoval SA, Gutiérrez-López GF, Domínguez-López A,

Paniagua-CastroN,Michicotl-MenesesMM, Jaramillo-Flores ME. Downregulation of proinflammatory liver gene expression by *Justicia spicigera* and kaempferitrin in a murine model of obesity-induced by a high-fat diet. Journal of Functional Foods. 2020;**65**:103781

[81] Esquivel-Gutiérrez E, Noriega-Cisneros R, Arellano-Plaza M, Ibarra-Barajas M, Salgado-Garciglia R, Saavedra-Molina A. Antihypertensive effect of *Justicia spicigera* in L-NAME-induced hypertensive rats. Pharmacology Online. 2013;**2**:120-127

[82] Caesar LK, Cech NB. Synergy and antagonism in natural product extracts: When 1+ 1 does not equal 2. Natural Product Reports. 2019;**36**(6):869-888

Chapter 3

# The Antioxidant and Bioactive Potential of Olive Mill Waste

*Karen Attard and Frederick Lia*

## Abstract

Olive mill waste (OMW) is a by-product of the olive oil production process that has attracted increasing attention due to its rich composition of bioactive compounds. This chapter explores the extensive and diverse antioxidant and bioactive potential of OMW. OMW is a complex mixture comprising organic compounds, including phenolic compounds, flavonoids, polysaccharides, and various other valuable molecules. These compounds have demonstrated a wide range of applications, including their use as fertilizers, antioxidants, antifungal and antibacterial agents, cytoprotective agents, and stabilizing agents in food preservation. The chapter delves into the types of phenolic compounds found in OMW, providing detailed insights into their structures and functions. Additionally, it discusses the factors affecting the composition of OMW, such as the extraction process and processing conditions. Additionally, the chapter explores the growing interest in the health benefits associated with the consumption of bioactive compounds derived from OMW. These compounds have been linked to potential therapeutic properties, including antioxidant, anti-inflammatory, and anticancer effects. The exploration of OMW's bioactive potential opens avenues for research and innovation, offering sustainable solutions for both waste management and the development of health-promoting products.

**Keywords:** olive mill waste, antioxidants, bioactivity, olive pomace, polyphenols, olive mill wastewater

## 1. Introduction

Olive oil derived from the olive also known as *Olea europaea L.,* is one of the oldest species of domestic trees and the second most important fruit crop cultivated worldwide [1]. The prominence of olive oil in the Mediterranean diet can be known for its nutritional abundance and beneficial health effect, establishing it as a fundamental element of this dietary pattern. In fact, approximately 3 million tons of olive oil are manufactured globally, with 2 million tons of this output originating from the European Union (EU), where the EU holds the distinction of being the leading producer, exporter, and consumer of olive oil worldwide [2]. Notably, the Mediterranean region, encompassing countries like Spain, Italy, Greece, and Portugal, accounts for nearly 99% of the EU's olive oil production [3–6].

The milling industry generates substantial waste, known as OMW. In fact, during the production of olive oil it is estimated that an average volume of olive mill wastewater (OMWW), ranging from 0.3 to 1.2 $m^3$/tons of processed olives, and an average quantity of solid residue ranging from 500 to 735 kilograms per ton of processed olives has been reported always depending on the extraction methods used [7–10]. In 3-phase mills, average freshwater consumptions range from 700 to 1000 L/per ton of olives processed and the generated OMWW is 1200 L/per ton, the highest of all processes. In comparison, the 2 phase mills utilize 100–120 L/per ton of freshwater consumed and 200 L/per ton of OMWW generated [7–10]. OMWs are considered highly polluting and phytotoxic due to their acidity, high levels of biological oxygen demand (BOD), and chemical oxygen demand (COD) [11]. However, OMWs are also a source of valuable molecules including plant nutrients, flavonoids, polysaccharides, anthocyanins and phenolic compounds [12–14], with potential industrial applications such as antioxidants, antifungal and antibacterial drugs, cytoprotective agents, fertilizers, and as gelling and stabilizing agents in food preservation [12, 15].

OMWW is characterized as a dark-brown liquid with a pH ranging from 3 to 6 [16, 17]. It is essentially a stable mixture consisting of vegetative water, processed water, residues from olive oil, and fragments of olive pulp. The specific composition of OMWW varies depending on factors such as the extraction method, the olives being processed, and the conditions of processing. Generally, OMWW is primarily composed of water, constituting about 83 to 94% of its weight, along with organic compounds accounting for 4 to 18% of its weight. These organic compounds encompass a range of substances, including sugars, polysaccharides, tannins, organic acids, phenolic compounds, and lipids [17, 18]. However, due to its intricate and variable chemical composition, OMWW poses challenges for direct utilization as a raw material in industrial applications [17, 19].

Olive pomace stands as a prominent by-product generated during the olive oil production process. The chemical composition of olive pomace exhibits considerable variability, influenced by several factors. These factors include the inherent characteristics of the fruit, such as its variety and degree of ripeness, as well as the methods used for oil extraction and the subsequent solvent depletion [20]. This solid residue consists of olive skin, water (~ 25%), pulp, fruit fragments, and traces of oil (4.5–9%) [21]. The primary constituents of olive pomace encompass sugars, mainly polysaccharides, along with proteins, fatty acids such as oleic acid, and various C2-C7 fatty acids, polyalcohols, polyphenols, as well as additional pigments [22]. Additionally, it includes small amounts of organic nitrogen in the form of crude protein and a substantial proportion of dietary fiber [23]. On average, olive pomace consists of approximately 10% hemicellulose, 15% cellulose, and 27% lignin. These parameters categorize this material as a highly lignified and densely walled substrate with very limited digestibility [21].

## 2. Classification of bioactive molecules

### 2.1 Phenolic compounds

Phenolic compounds are a class of secondary metabolites which are naturally occurring in plants, and they are typically distributed non-uniformly at both subcellular and cellular levels. These compounds exist in both soluble and insoluble forms, with the majority of insoluble phenolics being components of the cell walls, while soluble phenolics are typically contained within plant cell vacuoles [24]. Phenolic compounds contain at least one aromatic ring with one hydroxyl group in their structure. There are around 8000 individual plant phenolic compounds, with great structural variability and they can be classified into two main groups namely flavonoids and non-flavonoids [25]. Flavonoids are the most abundant phenolic compounds found in fruit and vegetables as they account for nearly two-thirds of dietary phenolic compounds [26–28].

With the increase of analytical techniques, the list of olive phenolic compounds is continuously updating. Moreover, in recent years, the potential health benefits associated with the consumption of innovative olive by-product-based preparations have encouraged research efforts to pinpoint the primary compounds known for their favorable physiological effects. In fact, investigations into the phenolic composition of by-products from olive oil production have been conducted, and numerous authors have employed chromatographic methods for their isolation and analysis. These methods include fractionation techniques like solid-phase extraction and LC retention-time identification [29], preparative high-performance liquid chromatography, capillary zone electrophoresis with UV diode array detection [30] and reversed-phase liquid chromatography coupled with mass spectrometry [31].

The major individual biophenols identified in the olive pomace and OMWW were particularly characterized in five classes, namely, simple phenols, phenolic acids, derivatives secoiridoids, flavonoids, and lignans [32]. OMW is known to contain a diverse array of phenolic compounds, each with its own unique set of properties and potential health benefits. Among these compounds, hydroxytyrosol and tyrosol are particularly noteworthy, representing some of the most abundant phenolics in OMW and renowned for their potent antioxidant, antimicrobial and antitumoral properties [33].

Over the last few years, the interest in natural antioxidants, particularly phenolic compounds, in relation to their therapeutic and health beneficial properties has significantly increased. The positive impact of various plant-derived polyphenols on human and mammalian health, including their anticancer properties, has been the subject of numerous studies. Numerous in vitro and in vivo studies have shown that the combination of natural polyphenols with chemotherapeutics can reduce the side effects of chemotherapy, increase the anticancer efficacy, and overcome the chemo or radio resistance of cancer cells [34]. Phenolic compounds extracted from olive oil waste have been shown to exert a possible chemoprotective and anticancer activities in different types of cancer cells [35]: prostate [36], breast cancer [37], colon [38], promyelocytic leukemia [39], melanoma [40], and other cancer cells.

### 2.2 Phenolic acids

Phenolic acids which were discovered by Friedlieb Ferdinand Runge in 1834 [41], are compounds identified by the presence of a benzene ring, a carboxylic group, and one or more hydroxyl and/or methoxyl groups within their molecular structure [42]. They are one of the main classes of plant phenolic compound and are widely distributed in various plant-based foods, with the highest concentrations often found in seeds, fruit skins, and vegetable leaves [43]. These compounds are typically present in bound forms like amides, esters, or glycosides, and are rarely found in their free form [44]. Notably, phenolic acids possess much higher *in vitro* antioxidant activity than widely recognized antioxidant vitamins [45, 46]. Additionally, phenolic acids are divided into two sub-groups namely hydroxybenzoic acids and hydroxycinnamic acids [47].

| Class | Structure | Phenolic Acids |
|---|---|---|
| Hydroxybenzoic Acid | 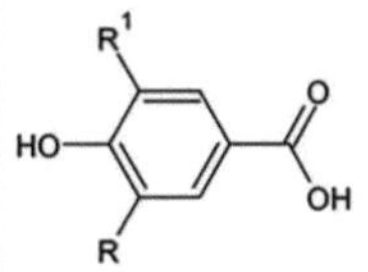 | $R_1=R=H \rightarrow$ *p*-Hydroxybenzoic Acid<br>$R_1=H, R=OH \rightarrow$ Protocatechuic Acid<br>$R_1=R=OH \rightarrow$ Gallic Acid<br>$R_1=H, R=OCH_3 \rightarrow$ Vanillic Acid<br>$R_1=R=OCH_3 \rightarrow$ Syringic Acid |
| Hydroxycinnamic Acid | 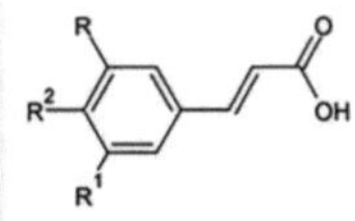 | $R^1=R=H, R^2=OH \rightarrow$ *p*-Coumaric Acid<br>$R^1=R^2=H, R=OH \rightarrow$ *o*-Coumaric<br>$R_1=H, R^2=R=OH \rightarrow$ Caffeic Acid<br>$R_1=H, R^2=OH, R=OCH_3 \rightarrow$ Ferulic Acid |

**Figure 1.**
*Major classes of phenolic acids.*

Hydroxycinnamic acids, originating from cinnamic acid, are frequently present in various foods in the form of simple esters with glucose and quinic acid [48]. Among the soluble bound hydroxycinnamic acids, chlorogenic acid is one of the most abundant, resulting from the combination of quinic and caffeic acids. The most common hydroxycinnamic acids found in OMW are typically p-coumaric, ferulic, caffeic, sinapic, vanillic acids and vanillin. [49, 50] Conversely, hydroxybenzoic acids share a common C6-C1 structure and derive from benzoic acid. They are typically found in soluble form, conjugated with organic acids or sugars [48]. Among the frequently found hydroxybenzoic acids in OMW are protocatechuic, syringic, vanillic, gallic and p-hydroxybenzoic acids (**Figure 1**) [49–51].

Phenolic acids are a key class of dietary polyphenols, renowned for their role as natural antioxidants. Their versatility is evident in their various functions, contributing to plant growth, development, and defense mechanisms. Beyond their primary functions, they also serve as an essential precursor for the production of other significant bioactive compounds, which find extensive applications in therapeutic, cosmetic, and food industries. Most notably, these dietary antioxidants play a crucial role in protecting against the progression and development of pathological conditions triggered by oxidative stress, offering a multifaceted range of benefits [43].

### 2.3 Phenylethanoids and secoiridoids

Montedoro and collaborators were the first to study the phenylethanoids and secoiridoid class in 1993. They are responsible for the structural characterization of ligstroside and oleuropein aglycones from virgin olive oils [52]. Phenylethanoids are a category of phenolic compounds synthesized by plants, characterized by a C6-C2 carbon skeleton [52]. Secoiridoids on the other hand, are a group of compounds which are usually glycosidically bounded and produce the secondary metabolism of terpenes [53]. These secoiridoids, prevalent in the Oleaceae family, which includes *Olea europaea L.*, are characterized by the presence of elenolic acid in either glycosidic or aglyconic form. Ligstroside and oleuropein stand out as the primary secoiridoids in olive fruits [54]. Both phenylethanoids and secoiridoids are esters formed from 3-hydroxytyrosol or tyrosol (p-hydroxy-phenylethyl alcohol, HPEA) and a secoiridoid corresponding to the glycosidic derivative of a carboxylic acid known as elenolic acid (**Figure 2**) [55].

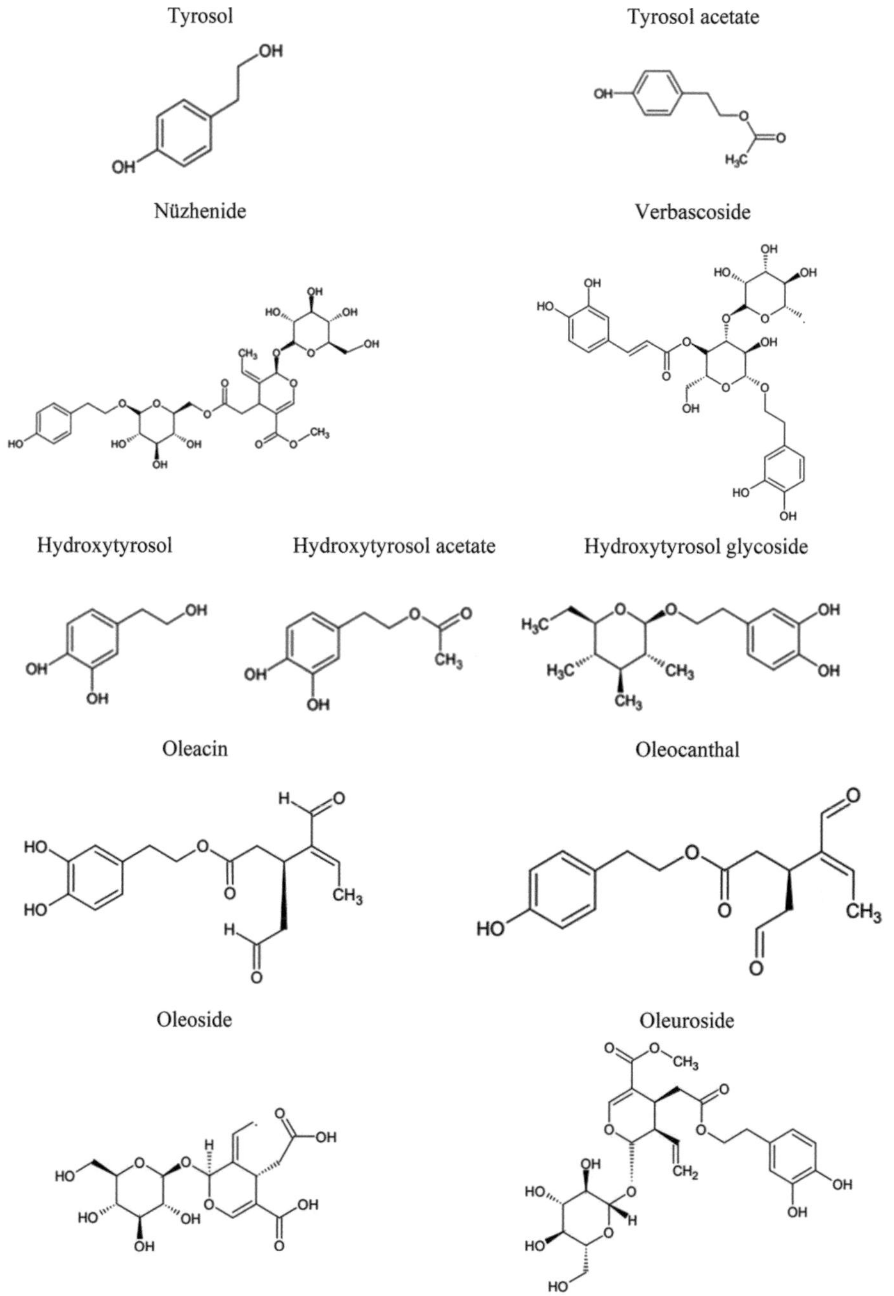


**Figure 2.**
*Phenylethanoids and Secoiridoids found in OMW.*

### *2.3.1 Hydroxytyrosol and tyrosol*

Hydroxytyrosol and tyrosol are the main compounds found in the phenolic fraction of steam-exploded olive stones and were discovered firstly by Fernandez-Bolanos and collaborators in 1998 [56]. The concentration of tyrosol and hydroxytyrosol, like other phenolic compounds, depends upon various factors inherent to the fruit [57].

These factors encompass genetic elements such as the specific cultivar responsible for production, as well as external factors like agro-pedoclimatic conditions and the degree of fruit ripeness. [58]. Tyrosol, which is also known as 2-(4-hydroxyphenyl)-ethanol, belongs to the phenylethanoids category and exhibits notable cellular antioxidant properties [59]. In fact, it has demonstrated the ability to protect Caco-2 intestinal mucosa from the harmful cytotoxic and apoptotic effects induced by oxidized LDL [60]. Moreover, hydroxytyrosol, also known as 4-(2-hydroxyethyl)-1,2-benzendiol, stands out as an exceptionally potent antioxidant compound within the phenolic compounds derived from the olive tree. Due to its molecular composition, consistent intake of hydroxytyrosol offers numerous benefits, such as serving as an antioxidant, an anti-inflammatory, and a potential anticancer agent [61–66].

### *2.3.2 Hydroxytyrosol acetate and hydroxytyrosol glycoside*

In 1999, Brenes *et al.*, found hydroxytyrosol derivative, hydroxytyrosol acetate in virgin olive oil [67]. Hydroxytyrosol acetate has been shown to effectively reduce oxidative stress, address mitochondrial dysfunction, and reduce neuronal toxicity [68]. One of the most prevalent forms of dementia among the elderly population is Alzheimer's diseases, and yet there is currently no established effective treatment. However, a study conducted by Qin *et al.* reveals that hydroxytyrosol acetate shows a promise in improving cognitive dysfunction associated with Alzheimer's diseases through a mechanism that depends on estrogen receptor β. In fact, it is evident that HT-ac exhibit a better bioactivity that hydroxytyrosol [69].

Additionally, hydroxytyrosol glucoside belongs to the class of o-glycosyl compounds, a class of organic compounds, where a sugar group is connected through one carbon to another group via O-glycosidic bond [70]. In recent studies, the glucosides of hydroxytyrosol have shown that it can improve the biological activity of the original aglycons on breast cancer cell lines [71].

### *2.3.3 Oleuropein and ligstroside*

OMW consists of a wide range of phenolic compounds, with secoiridoids being the most abundant fraction [72]. De Nino *et al.* initially detected oleuropein and ligstroside in olive leaves back in 1997, by using an MS [73]. Both compounds are secoiridoid glycosides, showcasing the methyl ester of 3,4-dihydro-2H-pyran5-carboxylic acid, substituting at positions 2, 3, and 4 with hydroxy, carboxymethyl groups, and ethylidene, respectively, as seen in **Figure 3** [75]. The anomeric hydroxy group at position 2 is converted into beta-D-glucoside, and the carboxylic acid moiety of the carboxymethyl substituent is converted into the corresponding 3,4-dihydroxyphenethyl ester [75]. Oleuropein and ligstroside, are a prominent constituent in OMW and has shown considerable potential in various health aspects, including cardio protection, anti-inflammatory properties, antioxidants, anti-cancer effects, anti-angiogenic activities, and neuroprotective functions. Therefore, they present a promising therapeutic avenue for addressing a range of human disorders [76, 77].

The aglycone form of ligstroside and oleuropein are the most abundant in secoiridoids [78]. During the maturation process of olives, oleuropein and ligstroside accumulates [79]. As shown in **Figure 3**, any damages done to the fruit during the ripening process can result in the release of esters [80] which can hydrolyse ligstroside and oleuropein into a range of compounds [81]. Oleuropein aglycone and ligstroside aglycone are formed by hydrolysis. Furthermore, the aglycones can undergo a further

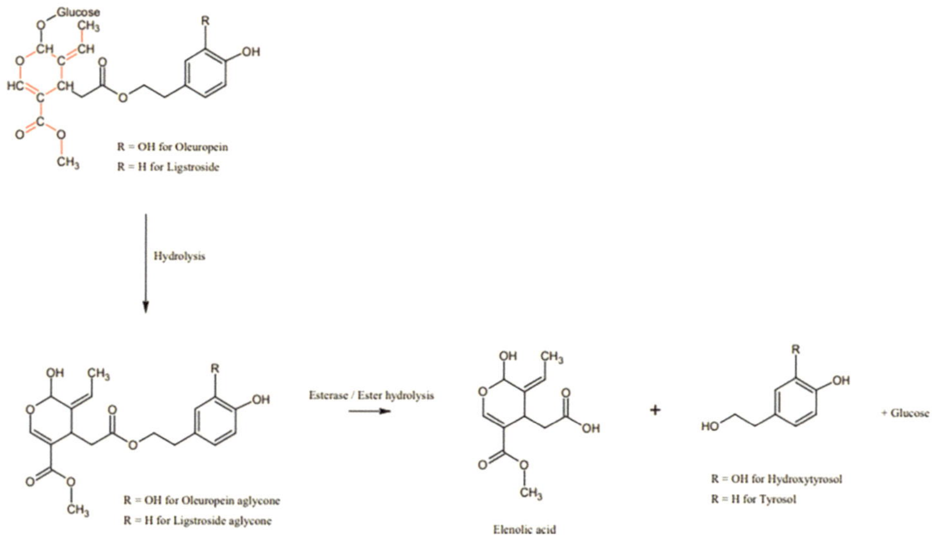


**Figure 3.**
*A figure shown by Johnson et al. [74], shows the structures of oleuropein, ligstroside and other related hydrolysis products. Oleuropein and ligstroside red bonds are the methyl ester of 3,4-dihydro-2H-pyran5-carboxylic acid, substituting positions 2, 3, and 4 with hydroxy, carboxymethyl groups, and ethylidene.*

ester hydrolysis in order to produce hydroxytyrosol or tyrosol, elenolic acid, and glucose, each of which these compounds can potentially reduce the Folin reagent [82].

### *2.3.4 Oleocanthal, oleoside and oleacin*

Oleuroside, oleocanthal, oleacin and oleoside are all found in olive pomace [51]. Oleoside secoiridoids in the Oleaceae family are typically derived from glucosides of the oleoside variety [83]. These oleosides are characterized by an exocyclic 8,9-olefinic functionality, a combination of elenolic acid and a glucosidic residue [83]. It has been discovered that secoiridoid glycosides of the oleoside type consist of a wide range of pharmacological properties, encompassing antioxidative, antitumor, anti-inflammatory, anti-diabetic, anti-obesity, neuroprotective, and cardiovascular effects [84].

Moreover, oleacin and oleocanthal are among the major phenolic compounds identified in olive oil [85]. These substances represent dialdehydic forms of elenolic acid originating from oleuropein and ligstroside, respectively. Oleacin and oleocanthal have gained significant attention due to their established benefits for health and their impact on sensory properties [86]. In terms of health benefits, they have shown positive effects in the treatment of cardiovascular conditions, specific cancer types, chronic inflammatory ailments, Alzheimer's disease, and *Helicobacter pylori* infection [87–90]. From a sensory perspective, oleacin and oleocanthal are linked to the sharpness and bitterness found in extra virgin olive oil [91].

### *2.3.5 Nüzhenide and verbascoside*

Nüzhenide, a compound discovered by Servili *et al.* [92], contains at least two glucose moieties in its chemical structure and has been exclusively detected in olive seed [92] and has been found in both olive pomace and OMWW [49]. This secoiridoid compound is known for its potent antioxidant activity [93]. In fact, it has been documented that nüzhenide exhibit anti-osteoporosis effect and numerous other pharmacological benefits, including antioxidant and neuroprotective properties [94, 95].

As for verbascoside, it is a naturally occurring phenylethanoid glycoside that can be found in various medicinal plants. This compound is one of the main hydroxycinnamic found in olives [96]. Notably, the concentration of verbascoside in olive fruit increases as the fruit undergoes maturation. Furthermore, verbascoside exhibits comprehensive pharmacological effects, such as antioxidant and antineoplastic action, and it offers a diverse array of therapeutic benefits for addressing depression [97]. Additionally, it contributes to neuroprotection in the context of Alzheimer's disease [96].

## 2.4 Flavonoids

Flavonoids, which were discovered in 1988 by Dr. Albert Szent-Gyorgyi [98], are the largest group of plant phenols which are present in glycosides. This group is represented by different several subclasses such as flavonols (quercitrin, rutin), flavones (apigenin, luteolin and their glucosides) and anthocyanins (cyanidin-3-O-glucosides) as seen in **Figure 4** [99]. These compounds are characterized by their large, planar molecular structure, with structural variations arising from modifications involving methoxylation, hydroxylation, glycosylation, and phenylation [99]. Although flavonoids share a common low molecular weight and possess a C6-C3-C6 structure, this designation encompasses a wide range of compounds with various structural variations [100].

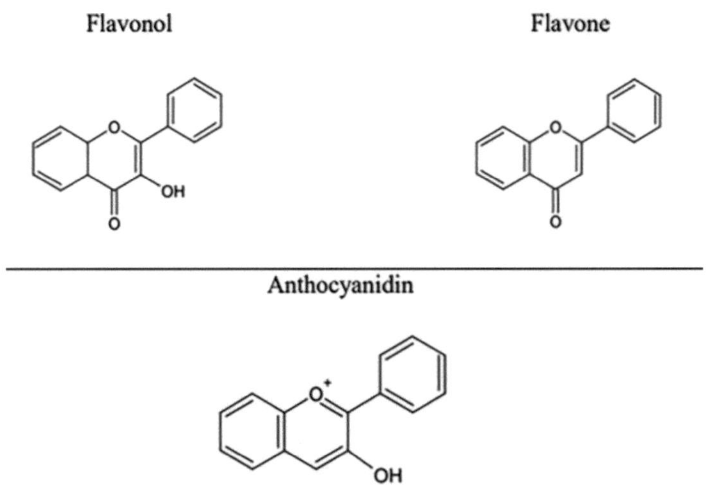


**Figure 4.**
*Simple flavonoids present in OMW.*

### *2.4.1 Flavonol*

Flavonols, as a subclass, are defined by the presence of a double bond between position 2 and 3 and an oxygen atom at position 4 of the C ring [101]. While they share similarities with flavones, they differ in the presence of hydroxyl group at position 3, making the flavanol skeleton, a 3-hydroxyflavone [101]. Flavonols are mainly represented by glycosides of quercetin, kaempferol, myricetin and isorhamnetin [102]. In fact, quercetin and rutin are the two main compounds identified in OMW [49]. The flavonols found in our dietary intake have the potential to inhibit the initiation, promotion, and progression of cancer by regulating key enzymes and receptors within signal transduction pathways associated with processes like proliferation, inflammation, angiogenesis, apoptosis, metastasis, differentiation and the reversal of multidrug resistance [103].

#### *2.4.2 Flavone*

Flavones are characterized by a structure composed of a fused A and C ring, along with a phenyl B ring attached at position 2 of the C ring [104]. A prominent compound in the flavone group is chrysin, which shares the typical flavone structure with hydroxyl groups at positions 5 and 7 of the A ring [105]. Notably, the main flavones identified in OMW are apigenin, apigenin-7-glucoside, luteolin, and luteolin-7-glucoside [49]. Flavones are increasingly recognized as effective anticancer agents due to their inherent antioxidant properties and their capacity to regulate epigenetic targets, particularly histone deacetylases [106].

#### *2.4.3 Anthocyanins*

Anthocyanins, a subgroup of flavonoids naturally found in plants, are the pigments responsible for the red, blue, and purple colors [107]. They consist of a basic chemical structure of flavylium cation which links with a hydroxyl and methoxy group, and one or more sugars [102]. The pathway through which anthocyanins enter the bloodstream differs from that of other phenolic compounds. When anthocyanins are consumed, they undergo digestion within the gastrointestinal tract. Subsequently, they are absorbed into the bloodstream and undergo metabolic processes [108]. Most anthocyanins reach the large intestine, where they undergo various transformations influenced by changing physiological conditions [109]. Additionally, anthocyanins may undergo further modifications through interactions with various enzymes in the small intestine before entering the bloodstream [110]. In contrast, the absorption of other phenolic compounds occurs in the small intestine through passive diffusion or transporters [74]. Only 5–10% of these compounds are absorbed in the colon [111]. From the colon, the resulting metabolites are subsequently transported to the liver where phenolic compounds undergo biotransformation in two phases with different enzymatic steps [112]. Once both phases are complete, the metabolites are distributed in the bloodstream with the assistance of plasma proteins such as albumin [113].

The main anthocyanins found in the olive fruit and OMWW are cyanidin, cyanidin-3-glycoside, cyanidin-3-rutinoside, cyanidin-3-caffeyglycoside, cyanidin-3-caffeylrutinoside, delphinidin 3-rhamosylglycoside-7-xyloside [4, 114, 115]. These compounds serve diverse industrial purposes, functioning as components in fertilizers, antioxidants, antifungal and antibacterial medications, cytoprotective agents, and as gelling and stabilizing agents in food preservation [15]. Anthocyanins offer demonstrated health benefits, such as anti-inflammatory and antioxidant properties [116], which help reduce oxidative stress and, consequently, contribute to the prevention of cancer and cardiovascular diseases [117].

## 2.5 Lignans

Lignans are another group of phenolic compounds which are derived from phenylalanine and consist of two phenylpropane units linked together with β, β-bonds as shown in **Figure 5**. The most abundant lignans found in OMW are (+) – pinoresinol and (+)-1- acetoxypinoresinol. These were found by Brenes *et al.* and Owen *et al.* [118, 119].

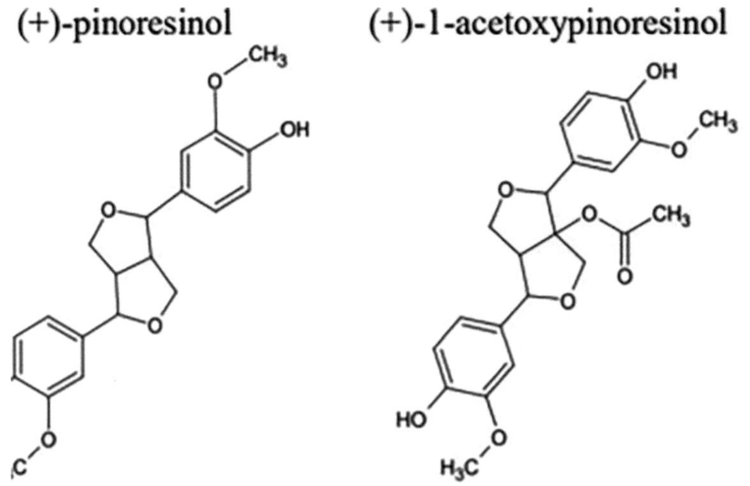


**Figure 5.**
*Lignans present in OMW.*

### *2.5.1 (+) – Pinoresinol and (+)-1- acetoxypinoresinol*

Pinoresinol exhibits a chemical structure similar to estrogen, categorizing it as a phytoestrogen [120]. Estrogen plays a fundamental role in mammary gland growth and development and has been associated with breast cancer initiation and progression, mainly due to its enhanced binding and activation of estrogen receptor α (Erα) [121]. A study conducted by López-Biedma *et al.* in 2016, suggests that pinoresinol possesses chemical antioxidant properties, implying its potential therapeutic utility in mitigating breast cancer development. This effect is achieved through the reduction of intracellular oxidative stress and DNA damage in human mammary epithelial cells. Additionally, pinoresinol elevates ROS levels in breast cancer cells upon exposure to $H_2O_2$ treatment. Consequently, pinoresinol exerts anti-tumor effects at low concentrations, facilitating cytotoxic, antiproliferative, and pro-oxidant actions in breast cancer cells, irrespective of their estrogen receptor status [120]. Regarding acetoxyinoresinol, it exhibits biological activities, such as its ability to inhibit FASN in human SKBR3 and MCF7 breast cancer cells. Moreover, it reduces HER2 expression and activation in HER2-dependent on the breast cancer cells [122].

## 2.6 Tocopherols

Tocopherols, a major form of vitamin E, are a group of fat-soluble phenolic compounds. Each tocopherol consists of a chromanol ring and a 16-carbon phytyl chain [123].

Depending on the number and position of methyl groups on the chromanol ring, tocopherols are designated as α, β, δ and γ [123]. α-Tocopherols are trimethylated at the fifth, seventh, and eighth positions of the chromanol ring, β-Tocopherol is dimethylated at the fifth and eighth positions, γ-Tocopherol is dimethylated at the seventh and eighth positions, and δ-Tocopherol is methylated at the 8th position [124]. In a study conducted by Russo *et al.* in 2020, it was discovered that all four tocopherols were present in both OMW and OMWW [51] Typically, the levels of tocopherols in oil can vary due to various factors, including the year of harvest, climatic conditions, storage duration, extraction methods, soil characteristics, and the spacing between olive trees [125].

Due to their potent antioxidant properties, there is a suggested potential for tocopherols to reduce the risk of cancer [126]. Multiple studies have indicated an association between lower vitamin E nutritional status and an elevated risk of cancer [127, 128]. Among tocopherols, α-Tocopherol is recognized as the most biological active form of tocopherols that makes up the family of vitamin E molecule [129]. Consequently, α-tocopherol has been the most commonly utilized form in studies focusing on cancer prevention. Nevertheless, findings from *in vitro* and *in vivo* research conducted by Das Gupta and Suh in 2016 propose that γ-tocopherol and δ-tocopherol might exhibit stronger anticancer properties compared to α-tocopherol (**Figure 6**) [130].

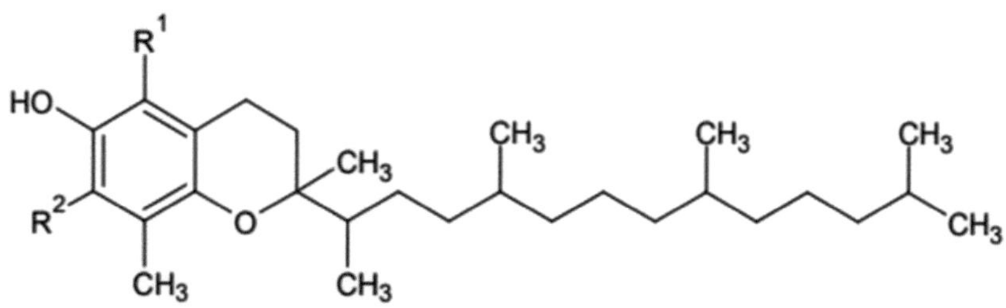


| Tocopherols | $R_1$ | $R_2$ |
|---|---|---|
| α- | $CH_3$ | $CH_3$ |
| β- | $CH_3$ | H |
| γ- | H | $CH_3$ |
| δ- | H | H |

**Figure 6.**
*Tocopherol structures with the number and position of the methyl groups on the aromatic ring.*

## 2.7 Xanthophylls and carotenoids

Olive pomace is a useful source of carotenoids [131]. Carotenoids are tetraterpene pigments that exhibit yellow, orange, red and purple colors. They are the most widely distributed pigments in the natural world and can be found in photosynthetic bacteria, certain archaea and fungi, algae, plants, and animals [132]. Typically, carotenoids are composed of eight isoprene units, forming a 40-carbon structure [132]. Their fundamental structures usually consist of a polyene chain featuring nine conjugated double bonds, with end groups at both ends of the polyene chain, as illustrated in **Figure 7** [133, 134].

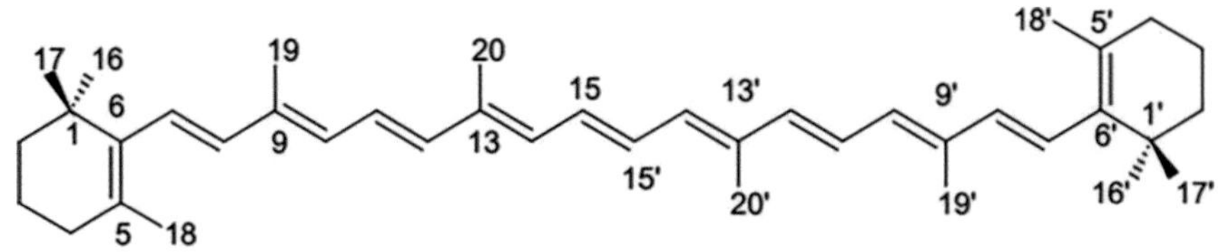


**Figure 7.**
*Basic structure of carotenoids.*

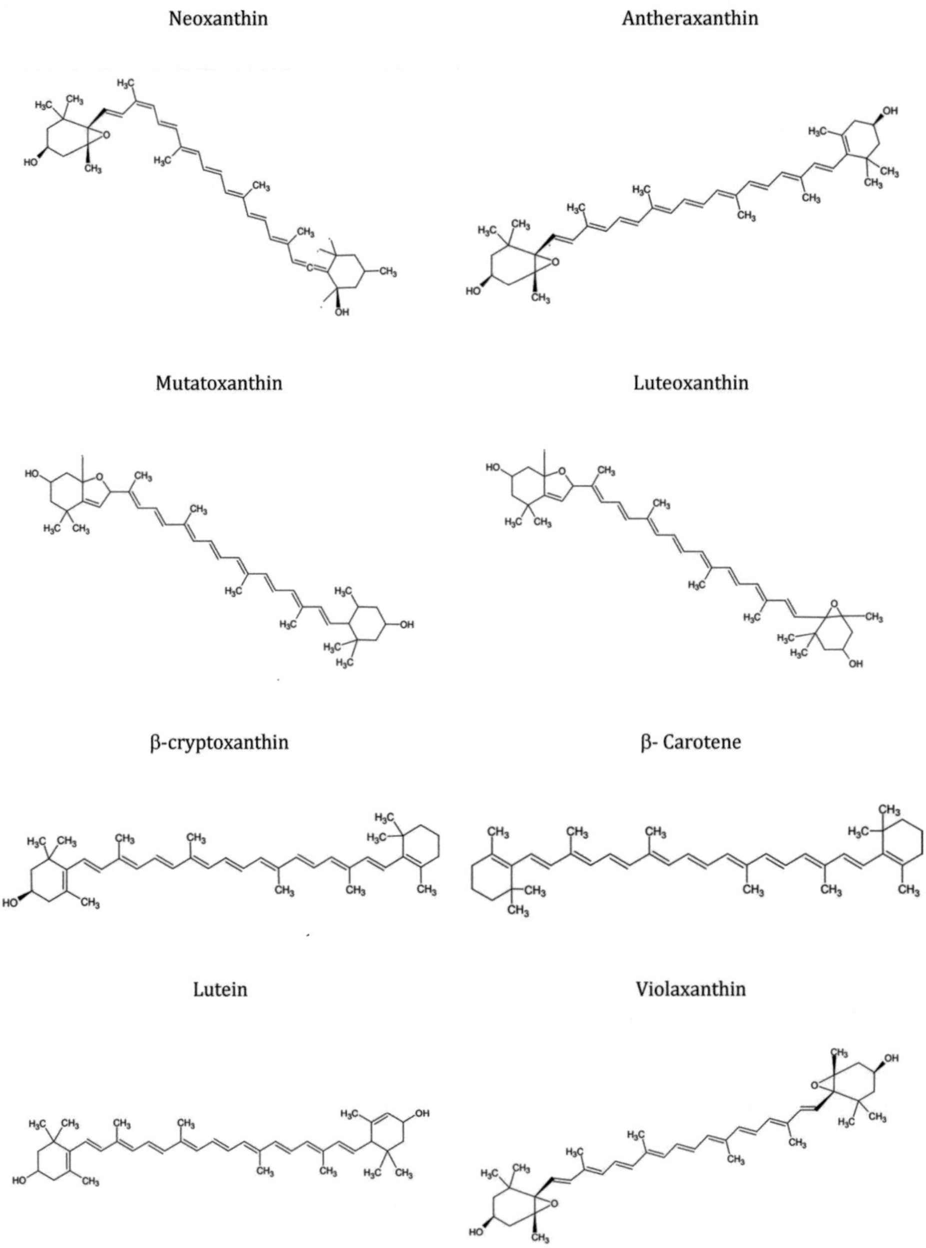


**Figure 8.**
*Basic carotenoids and xanthophylls found in olive oil and OMW.*

Carotenoids can be classified into two main categories: carotenes and xanthophylls. In nature, there are about 50 different carotenes, including α-carotene, β-carotene, β, ψ-carotene (γ-carotene), neoxanthin, mutatoxanthin, and lycopene. These carotenes are primarily hydrocarbons [133], and some of them can be found in olive oil [135]. Notably, various compounds have been identified, with β-carotene playing a significant role in protecting plants against photooxidative processes [136]. Carotenes are biologically important to prevent oxidative damage to lipid membranes, and some of them are also precursors of vitamin A. Their intake through nutrition, promotes the reduction of diseases of the skin and eyes as well as the reduction of cardiovascular problems such as oxidation of low-density cholesterol [134].

On the other hand, xanthophylls, such as β-cryptoxanthin, lutein, luteoxanthin, zeaxanthin, antheroxanthin, fucoxanthin, violaxanthin, and peridinin, incorporate oxygen atoms in the form of hydroxy, carbonyl, aldehyde, carboxylic, epoxide, and furanoxide groups within their molecular structures. Some of these xanthophylls can also be found in olive oil [135]. Additionally, certain xanthophylls exist as fatty acid esters, glycosides, sulfates, and as constituents of protein complexes, resulting in a wide range of structural diversity. As of 2018, approximately 800 distinct xanthophylls had been documented in nature (**Figure 8**) [133].

## 3. Traditional extraction methods

The initial and crucial step in OMW is the extraction process, as it is essential to isolate the specific chemical components from the plant materials to enable further separation and characterization. This process encompasses several key steps, including filtering, freeze-drying and grinding to achieve a homogenous sample. Proper actions must be implemented to ensure that the potential active constituents are not lost, distorted, or degraded during the extraction process from the plant samples [137]. In fact, the addition of sulfites, such as sulfur dioxide, can serve as a potent inhibitor of phenoloxidases, an enzyme responsible for the degradation of phenolic compounds. This can be an effective way to prevent the oxidation of bioactive compounds [138]. Furthermore, the incorporation of natural antioxidants like ascorbic acid (vitamin C) or tocopherol (vitamin E) can protect bioactive compounds from oxidative degradation [139]. Maintaining the pH at an optimal level is also important, since it will prevent the breakdown of bioactive compounds, given that many enzymes are sensitive to pH variations [140]. Additionally, storing OMW at lower temperatures is advisable, as it can retard enzymatic and chemical reactions, thereby contributing to the preservation of bioactive compounds [141].

The selection of a solvent system primarily relies on the specific characteristics of the targeted bioactive compound. Different solvent systems are available to extract the bioactive compound from natural product [142]. Obied *et al.* found that a water/methanol mixture was the most effective [143], while Allouche *et al.* identified ethyl acetate as the best solvent for recovering phenolic monomers [144]. Chanioti *et al.* further expanded on this by introducing natural deep eutectic solvents, with choline chloride and citric acid showing the highest extraction efficiency [145]. Lafka *et al.* also highlighted the potential of ethanol for extracting phenols, particularly hydroxytyrosol. These studies collectively suggest that a combination of water/methanol, ethyl acetate, and natural deep eutectic solvents may be the most effective solvents for extracting phenolic compounds from olive mill waste [146]. On the other hand, more lipophilic compounds are extracted using dichloromethane or a mixture of dichloromethane and methanol in a 1:1 ratio [142]. This was seen in a study conducted by Cequie-Sanchez *et al.* where dichloromethane was used as a solvent for lipid extraction [147]. In certain situations, hexane may be utilized for the removal of chlorophyll [142]. Giving that the target compounds may be non-polar to polar and thermally labile, the suitability of the methods of extraction must be considered. Various techniques, including sonification, soxhlet extraction, heating under reflux and others, are commonly employed for the plant sample extraction. Additionally, plant extracts can be prepared by macerating or percolating fresh green plants or dried powdered plant material in water and/or organic solvent systems. [148].

### 3.1 Advanced extraction processes

Traditional extraction techniques like, digestion [149], percolation [150], maceration [151] and the preparation of decoctions and infusions [152] are being replaced by advanced extraction methods to enhance the efficiency and specificity of extracting bioactive compounds [153]. These advanced approaches employ diverse techniques including microwaves, ultrasound, supercritical fluids, and more [153]. These innovative extraction methods yield final extracts that are notably enriched in desired compounds while minimizing the formation of unwanted byproducts. Furthermore, they are often characterized by their simplicity, speed, eco-friendliness, and full automation, making them an alternative to traditional extraction procedures to meet the rising market demand [154].

#### *3.1.1 Microwave assisted extraction*

Microwave-assisted extraction (MAE) has been successfully applied to olive mill waste, particularly in the extraction of valuable compounds such as triterpenoic acids [155], hydroxytyrosol, mannitol, and proteinaceous material [156]. This method has also been used to accelerate the extraction of biophenols from olive leaves, achieving complete extraction in just 8 minutes [157]. Furthermore, the performance of a microwave assisted Soxhlet extractor for olive seeds has been evaluated, with significant reduction in extraction time and labor intensity [158]. These studies collectively demonstrate the potential of MAE in the valorization of olive mill waste, offering a more efficient and sustainable approach to the extraction of valuable compounds.

#### *3.1.2 Pulsed electric field extraction*

Pulsed electric field (PEF) processing is an efficient non-thermal food processing technique using short, high voltage pulses. These electrical pulses create pores in the cells of plants, animals, and microorganisms, leading to cell disintegration and the deactivation of microbes [154]. PEF offers several advantages, such as enhancing the extraction of juices, sugars, colorants, and other active compounds, as well as significantly prolonging the shelf life of food products. It accelerates diffusion processes, like the removal of water from plant or animal tissues and the absorption of marinades, spices, and other additives, thus saving valuable time in production [154].

When applied to olive oil production, PEF brings various benefits. It increases oil yield by breaking down the cell walls of olives, making oil extraction easier. This also enhances the quality of the olive oil by increasing nutrient levels, such as polyphenols. Moreover, it improves stability by reducing the amount of free fatty acids that can lead to oil spoilage over time. [159, 160]. The application of pulse electric field (PEF) extraction on olive mill waste has been explored in several studies. Pappas *et al.* optimized the use of PEF for the recovery of high-value compounds from fresh olive leaves, achieving a significant increase in polyphenol extractability [161]. Yassine *et al.* also found success in treating olive mill wastewater using electrocoagulation, a process that involves the application of an electric current [162]. In the context of olive oil extraction, Tamborrino *et al.* and Romaniello*a et al.* both demonstrated the potential of PEF technology to improve extractability and enhance oil quality. These studies collectively suggest that PEF extraction can be a promising method for valorising olive mill waste [163, 164].

### *3.1.3 Ultrasonic assisted extraction*

The use of ultrasonic assisted extraction (UAE) technology is recommended for the recovery of valuable organic compounds. This method produces sonic waves to create intense shear forces that enhance the extraction process by facilitating mass transfer [165]. This enhancement is achieved through cavitation and streaming, along with an internal convection motion of solutes within the solvent, occurring within the pores of the materials [166]. UAE stands out as an environmentally friendly non-thermal technology. It offers various benefits, including shortened extraction times, reduced energy consumption, and lower solvent usage [167].

Conventional extraction techniques are frequently used when extracting compounds from OMW, with the choice of technique depending on the specific characteristics of the material and the targeted compounds [168]. One commonly used approach is solvent extraction, which involves the use of Generally Recognized as Safe (GRAS) solvents. However, this method has the drawback of extended extraction durations (typically around 48 hours) and the application of high temperatures exceeding 80°C, which can potentially result in the denaturation or degradation of the compounds [169]. To overcome this limitation, ultrasound-assisted extraction (UAE) is an alternative [167]. UAE relies on the cavitation effect, which facilitates the damage of the plant cell walls, making it easier to release bioactive compounds into the solvent [170, 171].

Research has shown that ultrasonic treatment can significantly improve the biogas and methane production from OMW [172, 173]. This is achieved by increasing the solubilization of organic matter in the waste, leading to a higher production of biogas and methane. Additionally, the combination of high-power ultrasound and electro-Fenton processes has been found to be effective in reducing contaminants in OMW, making it more suitable for byproduct utilization [174]. Furthermore, ultrafiltration has been identified as an efficient pretreatment method for OMW, particularly in enhancing the extraction of phenolic compounds and improving the effectiveness of secondary treatment [175]. These findings collectively suggest that ultrasonic treatment, in combination with other methods, can significantly enhance the treatment and utilization of OMW.

### *3.1.4 Super critical fluid extraction*

Supercritical fluid extraction (SFE) is an eco-friendly extraction method that offers a twofold solution: it can purify OMWW and simultaneously isolate high-quality oil from this wastewater [176]. SFE ranks among the most commonly utilized techniques for extracting active compounds. Typically, carbon dioxide ($CO_2$) is commonly used as a supercritical fluid, with extraction conditions maintained above the critical temperature of 31°C and critical pressure of 74 bars. Small adjustments in temperature and pressure above these critical points have a substantial impact on the extraction properties, enhancing the supercritical fluid's capacity to penetrate and dissolve targeted molecules, particularly lipophilic substances [177]. Dali *et al.*, states that freeze-drying process can also be optimized in order to improve the economic process [175].

Supercritical fluid extraction has proven successful in treating olive mill waste, yielding antioxidant-rich extracts [178]. This method has also been used in the gasification of olive oil mill waste, producing hydrogen-rich gas and removing organic carbon and chemical oxygen demand [179]. Furthermore, supercritical water oxidation has been employed for the treatment of olive oil mill wastewater, achieving high efficiency and total mineralization [180].

This green extraction process allows to preserve the quality of the extracted product. Carbon dioxide is an exceptionally eco-friendly option, as it is non-toxic, non-explosive, and readily available in high purity at a low expense. It can be easily recycled within the process and does not leave any solvent residues in the end-product [181]. Numerous studies and reports have indicated that SFE represents a viable alternative to traditional methods, such as solvent extraction, for extracting high-value natural compounds [182]. In fact, Dali *et al.*, states that supercritical extraction is more advantages compared to the conventional solvent extraction because it is ecofriendly and safer to use [175].

## 4. Olive mill waste treatment

Several approaches have been reported in the literature for addressing the treatment and disposal of OMWW, including anaerobic digestion, aerobic fermentation, and composting. However, these methods generally result in the loss or degradation of various functional compounds [183–185]. Therefore, methods such as the biological and physicochemical treatment of OMW, can also be utilized.

Anaerobic digestion is a promising method for treating olive mill waste, as it can generate energy and reduce the environmental impact of these wastes [186, 187]. However, the process can be inhibited by certain compounds, such as phenols and organic acids, which are present in high concentrations in olive mill waste [188]. To enhance the efficiency of anaerobic digestion, pretreatment methods to remove or modify these inhibitors are necessary. Additionally, the resulting digestate from anaerobic digestion can be further processed through composting to produce a high-quality soil amendment or fertilizer [189].

Aerobic fermentation of OMW has been explored in several studies. Boari *et al.* and Erguder *et al.*, both found that anaerobic treatment of olive mill waste is feasible, with high removal efficiencies and methane gas production [190, 191]. Tziotzios *et al.* further demonstrated the potential of aerobic biological treatment, achieving significant removal of phenolic compounds and COD [192]. Benitez *et al.* proposed an improvement to anaerobic biodegradation through ozonation pretreatment, which led to a significant increase in methane yield. These studies collectively suggest that both anaerobic and aerobic fermentation processes can be effective in treating olive mill waste [193].

### 4.1 Biological and physiochemical treatment of OMW

The biological treatment of OMW aims to reduce organic matter and inorganic nutrients, necessitating the selection and application of microorganisms. It's worth noting that the presence of phenolic substances in OMW can inhibit microorganisms [9]. Biological methods can encompass aerobic or anaerobic digestion treatments. However, this process is time-consuming and not commonly used as a valorisation technique, primarily involving the conversion of complex organic matter into simpler molecules like biogas [194, 195]. Moreover, physicochemical treatments for OMW include processes such as dilution, evaporation, sedimentation, filtration, centrifugation, and adsorption [196]. These processes are employed to reduce the organic load and toxicity of OMW to acceptable levels. Dilution is a common pretreatment to reduce the toxicity of OMW. Evaporation and sedimentation, in particular, can concentrate OMW by 70–75%. Further treatment is necessary for both the concentrated residue and the supernatant, in the case of sedimentation [196].

Numerous biotechnological applications have been developed for the utilization of OMW. One such method is lagooning, which employs a natural purification process to reduce the load rejection in organic matter and polyphenol content existing in the OMW, thus producing treated water meeting the required physicochemical quality standards [197]. Composting is another technique utilized for OMW, primarily aimed at enhancing the physical, chemical, and biological properties of soil. This process centers on the decomposition of organic matter into stable compounds rich in humic substances, making it suitable for use as a fertilizer. Despite environmental restrictions on the lagooning process, OMW can still find application as a fertilizer due to its high organic load and the concentration of soluble nutrients, making it valuable for agriculture [198]. Another way for OMW utilization is as animal feed. However, the high sodium and phenolic compound content in OMW can lead to digestive issues in ruminants [199]. To address this concern, research has focused on reducing phenols through specific processes, such as the Dalmolive process, which decreases phenol content to acceptable levels. Additionally, OMW can be harnessed for biogas production. The anaerobic digestion process relies on the biochemical conversion of organic matter, resulting in the generation of carbon dioxide and methane [200].

## 5. Potential nutraceuticals of OMW

Polyphenolic compounds, well-known for their beneficial effects on human health, due to their antioxidant, cardioprotective, anticancer, anti-inflammatory, and antimicrobial properties [201, 202] are nowadays widely recognized as valuable molecules in pharmaceutical and nutraceutical fields [203]. Pharmaceuticals offer immediate relief and treatment for specific medical conditions, whereas nutraceuticals focus on preventive care and overall well-being [204].

Olive polyphenols, including hydroxytyrosol, tyrosol, oleuropein, and verbascoside, have been extensively studied for their pharmacological properties. These compounds have been found to possess antioxidant, anti-inflammatory, cardiovascular, immunomodulatory, gastrointestinal, respiratory, autonomic, central nervous system, antimicrobial, anticancer, and chemopreventive properties [205]. The health benefits of these polyphenols, particularly in the context of the Mediterranean diet, have been well-documented, with a focus on their molecular mechanisms of action [206]. Their potential in the prevention and treatment of chronic diseases, including cardiovascular diseases, diabetes, neurodegenerative disorders, and cancer, has been highlighted [207]. Furthermore, the specific anti-cancer effects of hydroxytyrosol, including its chemopreventive, cytotoxic, and apoptotic effects, have been explored [208].

In general, whilst polyphenols show a potential as therapeutic agents, they also have limitations that hinder their clinical applications. These include poor stability, bioavailability, and membrane permeability [209]. Despite their potential, their use should be carefully considered due to potential toxic effects [210]. Furthermore, the protective effects of polyphenols in chronic diseases are still not fully understood, and better in vivo and in vitro studies are needed [211]. The low oral bioavailability of polyphenols is a major challenge, and their bioactivity should be re-evaluated based on their metabolites [212]. Various studies have explored the use of nanodelivery systems and encapsulation technologies to enhance the bioavailability of polyphenols [213–215]. These technologies aim to improve the solubility, stability, and controlled release of polyphenols, thereby increasing their absorption and potential health benefits. Additionally, the design of specific synergistic interactions between polyphenols has been proposed to improve their oral bioavailability [216].

## 6. Conclusion

In conclusion, this chapter emphasizes the remarkable antioxidant and bioactive potential concealed within OMW. The rich polyphenolic content, notably hydroxytyrosol and oleocanthal, highlights the valuable compounds present in these residues. As discussed, their diverse health benefits, ranging from antioxidant and anti-inflammatory properties to their potential applications in combating various health conditions, highlighting the importance of realizing the undiscovered capabilities within OMW. This includes avenues such as clinical studies, technological advancements, and sustainable applications of these bioactive compounds, fostering a broader understanding of their capabilities. Additionally, proposing practical real-world applications for these compounds could engage industries seeking novel solutions. Moreover, by exploring innovative and eco-friendly methods for extraction and utilization, we pave the way for a more sustainable and health-conscious approach to managing this waste stream. The chapter's findings offer a compelling case for the continued investigation and harnessing of OMW as a valuable resource, both in terms of health and sustainability.

## Funding

This research was funded by the Malta Council for Science & Technology and the Scientific Technological Research Council of Turkey (Tubitak) through the MCST-Tubitak 2021 Joint Call for R&I Projects under grant agreement No MCST-TUBITAK-2021-2104 for OPobicell project.

## Author details

Karen Attard* and Frederick Lia
Institute of Applied Science, Malta College of Arts, Science and Technology, Paola, Malta

*Address all correspondence to: karen.attard.b42517@mcast.edu.mt

## References


[1] Tsagaraki E, Lazarides HN, Petrotos K. Chapter 8 - Olive mill wastewater treatment. In: Utilization of by-Products and Treatment of Waste in the Food Industry. Iceland: Springer; 2006. DOI: 10.1007/978-0-387-35766-9_8

[2] Europea, Cleaner Controlled and Cost-Efficient Olive Oil Production [Internet]. 2005. Available from: https://cordis.europa.eu/article/id/80625-cleaning-olive-oil-mill-waste-products

[3] European Commission. Olive Oil [Internet]. 2020. Available from: https://agriculture.ec.europa.eu/farming/crop-productions-and-plant-based-products/olive-oil_en

[4] Alkhalidi A, Halaweh G, Khawaja MK. Recommendations for olive mills waste treatment in hot and dry climate. Journal of the Saudi Society of Agricultural Sciences. 2023;**22**(6):361-373. DOI: 10.1016/j.jssas.2023.03.002

[5] Azzam MOJ, Al-Malah KI, Abu-Lail NI. Dynamic post-treatment response of olive mill effluent wastewater using activated carbon. Journal of Environmental Science and Health. Part A, Toxic/Hazardous Substances & Environmental Engineering. 2004;**39**:269-280. DOI: 10.1081/ESE-120027383

[6] Galanakis CM. Chapter 2 - Olive Mill Waste Recent Advances for Sustainable Management. In: Olive mill waste, Recent Advances for Suitable Management 1st Edition. Greece: Elsevier; 2017

[7] Khdair A, Abu-Rumman G. Sustainable environmental management and valorization options for olive mill byproducts in the Middle East and North Africa (MENA) region. PRO. 2020;**8**:671. DOI: 10.3390/pr8060671

[8] Khdair IA, Abu-Rumman G. Pollution estimation from olive wastewater in Jordan. Heliyon. 2019;**5**:e02386. DOI: 10.1016/j.heliyon.2019.e02386

[9] Caputo AC, Scacchia F, Pelagagge PM. Disposal of by-products in olive oil industry: Waste-to-energy solutions. Applied Thermal Engineering. 2003;**23**:197-214. DOI: 10.1016/S1359-4311(02)00173-4

[10] Obied HK, Allen MS, Bedgood DR, Prenzler PD, Robards K, Stockmann R. Bioactivity and analysis of biophenols recovered from olive mill waste. Journal of Agricultural and Food Chemistry. 2005;**53**:823-837. DOI: 10.1021/jf048569x

[11] Souilem S, El-Abbassi A, Kiai H, Hafidi A, Sayadi S, Galanakis CM. Olive oil production sector: Environmental effects and sustainability challenges. In: OMW. The Netherlands: Elsevier: Amsterdam; 2017. pp. 1-28. DOI: 10.1016/B978-0-12-805314-0.00001-7

[12] Dermeche S, Nadour M, Larroche C, Moulti-Mati F, Michaud P. OMWs: Biochemical characterizations and valorization strategies. Process Biochemistry. 2013;**48**:1532-1552. DOI: 10.1016/j.procbio.2013.07.010

[13] Nadour M, Laroche C, Pierre G, Delattre C, Moulti-Mati F, Michaud P. Structural characterization and biological activities of polysaccharides from OMW water. Applied Biochemistry and Biotechnology. 2015;**177**:431-445. DOI: 10.1007/s12010-015-1753-5

[14] Aggoun M, Arhab R, Cornu A, Portelli J, Barkat M, Graulet B. OMW water microconstituents composition

according to olive variety and extraction process. Food Chemistry. 2016;**209**:72-80. DOI: 10.1016/j.foodchem.2016.04.034

[15] Ochando-Pulido JM, Pimentel-Moral S, Verardo V, Martinez-Ferez A. A focus on advanced physico-chemical processes for OMW water treatment. Separation and Purification Technology. 2017;**179**:161-174. DOI: 10.1016/j.seppur.2017.02.004

[16] Angelidaki I, Ahring BK, Deng H, Schmidt JE. Anaerobic digestion of olive mill effluents together with swine manure in UASB reactors. Water Science and Technology. 2002;**45**:213-218. DOI: 10.2166/wst.2002.0334

[17] Tundis R, Conidi C, Loizzo MR, Sicari V, Cassano A. OMW water polyphenol-enriched fractions by integrated membrane process: A promising source of antioxidant. Hypolipidemic and Hypoglycaemic Compounds. Antioxidants. 2020;**9**:602. DOI: 10.3390/antiox9070602

[18] Ammary BY. Treatment of olive mill wastewater using an anaerobic sequencing batch reactor. Desalination. 2005;**177**:157-165. DOI: 10.1016/j.desal.2004.12.006

[19] Lafi WK, Shannak B, Al-Shannag M, Al-Anber Z, Al-Hasan M. Treatment of olive mill wastewater by combined advanced oxidation and biodegradation. Separation and Purification Technology. 2009;**70**:141-146. DOI: 10.1016/j.seppur.2009.09.008

[20] Ferhat R, Laroui S, Zitouni B, Lekbir A, Abdeddaim M, Smaili N, et al. Experimental study of solid waste olive's mill: Extraction modes optimization and physicochemical characterization. Journal of Natural Product and Plant Resources. 2014;**4**:16-23

[21] Tada ZMS, Faid IAKIM, Hassani S, Salmaoui S. Caractérisation physico-chimique et microbiologique des grignons d' olive de 26 huileries traditionnelles de la région de Beni Mellal (Maroc) physicochemical and microbiological characterization of the olive residue of 26 traditional oil mills in Beni. Desalination. 2010;**5**:4-9

[22] Mateos R, Sarria B, Bravo L. Nutritional and other health properties of olive pomace oil. Critical Reviews in Food Science and Nutrition. 2020;**60**(20):3506-3521. DOI: 10.1080/10408398.2019.1698005

[23] Varelis P. Food chemistry and analysis. In: Reference Module in Food Science. Amsterdam: Elsevier; 2016. DOI: 10.1016/B978-0-08-100596-5.03341-2

[24] Pridham JB. Oligosaccharides and associated glycosidases in aspen tissues. Biochemistry Journal. 1960;**76**:13-20. DOI: 10.1042/bj0760013

[25] Dai J, Mumper RJ. Plant phenolics: Extraction, analysis and their antioxidant and anticancer properties. Molecules. 2010;**15**(10):7313-7352. DOI: 10.3390/molecules15107313

[26] Burak M, Imen Y. Flavonoids and their antioxidant properties. Turkiye Klin Tip Bil Derg. 1999;**19**:296-304

[27] Castaneda A, Pacheco-Hernandez L, Paez E, Rodriguez JA, Galan-Vidal GA. Chemical studies of anthocyanins: A review. Food Chemistry. 2009;**113**:859-871. DOI: 10.1016/j.foodchem.2008.09.001. DOI: 10.1016/j.foodchem.2008.09.001

[28] Lee YK, Yuk DY, Lee JW, Lee SY, Ha TY, Oh KW, et al. Epigallocatechin-3-gallate prevents lipopolysaccharide-induced elevation of β-amyloid

generation and memory deficiency. Brain Research. 2009;**1250**:164-174. DOI: 10.1016/j.brainres.2008.10.012

[29] Bouaziz M, Sayadi S. Isolation and evaluation of antioxidants from leaves of a Tunisian cultivar olive tree. The European Journal of Lipid Science and Technology. 2005;**107**:497-504. DOI: 10.1002/ejlt.200501166

[30] Pérez-Bonilla M, Salido S, Van Beek TA, Linares-Palomino PJ, Altarejos J, Nogueras M, et al. Isolation and identification of radical scavengers in olive tree (Olea europaea) wood. Journal of Chromatography A. 2006;**1112**:311-318. DOI: 10.1002/ejlt.200501166

[31] Ammar S, Contreras MDM, Gargouri B, Segura-Carretero A, Bouaziz M. Biosurfactant production by the crude oil degrading Stenotrophomonas sp. B-2: Chemical characterization, biological activities and environmental applications. Environmental Science and Pollution Research. 2007;**24**:3769-3779. DOI: 10.1007/s11356-016-8064-4

[32] Leouifoundi I, Harnafi H, Zyad A. OMW extracts: Polyphenols content, antioxidant, and antimicrobial activities. Advances in Pharmacological and Pharmaceutical Sciences; 2015;**2015**:714138. DOI: 10.1155/2015/714138

[33] Posadino AM, Cossu A, Giordo R, Piscopo A, Abdel-Rahman A, Piga A, et al. Antioxidant properties of OMW water polyphenolic extracts on human endothelial and vascular smooth muscle cells. Foods. 2021;**10**(4):800. DOI: 10.3390/foods10040800

[34] Nurgali RT, Jagoe R, Abalo R. Editorial: Adverse effects of cancer chemotherapy: Anything new to improve tolerance and reduce sequelae? Frontiers in Pharmacology. 2018;**9**:245. DOI: 10.3389/fphar.2018.00245

[35] Menendez A, Vazquez-Martin R, Colomer J, Brunet A, Carrasco-Pancorbo R, GarciaVillalba A, et al. Olive oil's bitter principle reverses acquired autoresistance to trastuzumab (HerceptinTM) in HER2-overexpressing breast cancer cells. BMC Cancer. 2007;**7**. DOI: 10.1186/1471-2407-7-80

[36] Zubair A, Bhardwaj A, Ahmad SK, Srivastava MA, Khan GK, Patel S, et al. Hydroxytyrosol induces apoptosis and cell cycle arrest and suppresses multiple oncogenic signaling pathways in prostate cancer cells. Nutrition in Cancer Care. 2017;**69**:932-942. DOI: 10.1080/01635581.2017.1339818

[37] Calahorra E, Martínez-Lara C, De Dios C, Siles E. Hypoxia modulates the antioxidant effect of hydroxytyrosol in MCF-7 breast cancer cells. PLoS One. 2018;**13**(9):e0203892. DOI: 10.1371/journal.pone.0203892

[38] Hashim Y, Worthington J, Allsopp P, Ternan NG, Brown EM, McCann MJ, et al. Virgin olive oil phenolics extract inhibit invasion of HT115 human and mammalian colon cancer cells in vitro and in vivo. Food & Function Journal. 2014;**5**:1513-1519. DOI: 10.1039/C4FO00090K

[39] Fabiani R, Rosignoli P, Bartolomeo AD, Fuccelli R, Servili M, Montedoro GF, et al. Oxidative DNA damage is prevented by extracts of olive oil, hydroxytyrosol, and other olive phenolic compounds in human and mammalian blood mononuclear cells and HL60 cells. The Journal of Nutrition. 2008;**138**:1411-1416. DOI: 10.1093/jn/138.8.1411

[40] Song DY, Lim JI, Jung HJ, Cho SY, Park GT, Kwon YH, et al. Dietary

oleuropein inhibits tumor angiogenesis and lymphangiogenesis in the B16F10 melanoma allograft model: A mechanism for the suppression of high-fat diet-induced solid tumor growth and lymph node metastasis. Oncotarget. 2017;**8**:32027-32042. DOI: 10.18632/oncotarget.16757

[41] Runge FF. Ueber einige Produkte der Steinkohlendestillation. Annalen der Physik und Chemie. 1834;**31**:65-78. DOI: 10.1002/andp.18341070502

[42] Yang CS, Landau JM, Huang MT, Newmark HL. Inhibition of carcinogenesis by dietary polyphenolic compounds. Annual Review of Nutrition. 2001;**21**:381-406. DOI: 10.1146/annurev.nutr.21.1.381

[43] Kumar N, Goel N. Phenolic acids: Natural versatile molecules with promising therapeutic applications. Biotechnol Rep (Amst). 2019;**20**:24:e00370. DOI: 10.1016/j.btre.2019.e00370

[44] Pereira DM, Valentão P, Pereira JA, Andrade PB. Phenolics: From chemistry to biology. Molecules. 2009;**14**(6):2202-2211. DOI: 10.3390/molecules14062202

[45] Tsao R, Deng Z. Separation procedures for naturally occurring antioxidant hytochemicals. Journal of Chromatography. B, Analytical Technologies in the Biomedical and Life Sciences. 2004;**812**:85-99. DOI: 10.1016/j.jchromb.2004.09.028

[46] Shetty K, Wahlqvist ML. A model for the role of the proline-linked pentose-phosphate pathway in phenolic phytochemical bio-synthesis and mechanism of action for human health and environmental applications. Asia Pacific Journal of Clinical Nutrition. 2004;**13**(1):1-24

[47] Clifford MN. Chlorogenic acids and other cinnamates - nature, occurrence, and dietary burden. Journal of the Science of Food and Agriculture. 1999;**79**:362-372

[48] Strack D. Phenolic metabolism. In: Dey PM, Harborne JB, editors. Plant Biochemistry. London: Academic Press; 1997. p. 387

[49] Leouifoudi I, Harnafi H, Zyad A. OMW extracts: Polyphenols content, antioxidant, and antimicrobial activities. Advanced Pharmacology Science. 2015;**2015**:714138. DOI: 10.1155/2015/714138

[50] Cuffaro D, Bertolini A, Berini S, Ricci C, Cascone MG, Danti S, et al. OMW water as source of polyphenols with nutraceutical properties. Nutrients. 2023;**15**(17):3746. DOI: 10.3390/nu15173746

[51] Russo M, Bonaccorsi IL, Cacciola F, Dugo L, De Gara L, Dugo P, et al. Distribution of bioactives in entire mill chain from the drupe to the oil and wastes. Natural Products. 2020;**35**(21):4182-4187. DOI: 10.1080/14786419.2020.1752208

[52] Montedoro GF, Servili M, Baldioli M, Miniati E. Simple and hydrolizable phenolic compounds in virgin olive oil. 1. Their extraction, and quantitative and semiquantitative evaluation by HPLC. Journal of Agricultural and Food Chemistry. 1992;**40**:1571-1576. DOI: 10.1021/jf00021a019

[53] Siler B, Misic D. Biologically active compounds from the genus Centaurium S.L. (Gentianaceae): Current knowledge and future prospects in medicine. Studies in Natural Products Chemistry. 2016:363-397. DOI: 10.1016/B978-0-444-63601-0.00011-9

[54] Marković AK, Torić J, Barbarić M, Brala CJ. Hydroxytyrosol, tyrosol and derivatives and their potential effects on human health. Molecules. 2019;**24**:2001-2039. DOI: 10.3390/molecules24102001

[55] Preedy VR, Watson RR. Olive Oil in Health and Disease Prevention. 2nd edition. Burlington: Academic Press; 2010

[56] Fernandez-Bolanos J, Flizon B, Brenes M, Guillen R, Heredia A. Hydroxytyrosol and tyrosol as the main compounds found in the phenolic fraction of steam-exploded olive stones. Journal of American Oil Chemists Society. 1998;**75**:1643-1649. DOI: 10.1007/s11746-998-0106-8

[57] Ghanbari R, Anwar F, Alkharfy KM, Gilani A-H, Saari N. Valuable nutrients and functional bioactives in different parts of olive (Olea europaea L.)—A review. International Journal of Molecular Sciences. 2012;**13**(3):3291-3340. DOI: 10.3390/ijms13033291

[58] González-Rodríguez J, Pérez-Juan P, Luque de Castro MD. Superheated liquids for extraction of solid residues from winemaking processes. Analytical and Bioanalytical Chemistry. 2003;**7–8**:1190-1195. DOI: 10.1007/s00216-003-2194-5

[59] Karković Marković A, Torić J, Barbarić M, Jakobušić BC. Hydroxytyrosol, tyrosol and derivatives and their potential effects on human health. Molecules. 2019;**24**(10). DOI: 10.3390/molecules24102001

[60] Giovannini C, Starface E, Modesti D, Coni D, Cantafora A, De Vincenzi M, et al. Tyrosol, the major olive oil biophenol, protects against oxidized-LDL-induced injury in Caco-2 cells. The Journal of Nutrition. 1999;**129**(7):1269-1277. DOI: 10.1093/jn/129.7.1269

[61] Martinez L, Ros G, Nieto G. Hydroxytyrosol: Health benefits and use as functional ingredient in meat. Medicine. 2018;**5**(1):13. DOI: 10.3390/medicines5010013

[62] De la Puerta R, Rutierrez VR, Haolt RS. Inhibition of leukocyte 5-lipoxygenase by phenolics from virgin olive oil. Biochemical Pharmacology. 1999;**57**:445-449. DOI: 10.1016/s0006-2952(98)00320-7

[63] Di Benedetto R, Varì R, Scazzocchio B, Filesi C, Santangelo C, Giovannini C, et al. Tyrosol, the major extra virgin olive oil compound, restored intracellular antioxidant defences in spite of its weak antioxidative effectiveness. Nutrition, Metabolism and Cardiovascular Diseases. 2006;**17**:535-545. DOI: 10.1016/j.numecd.2006.03.005

[64] Esterbauer H, Gebiki J, Puhl H, Jurgens G. The role of lipid peroxidation and antioxidants in oxidative modification of LDL. Free Radical Biology and Medicine. 1992;**13**:341-390. DOI: 10.1016/0891-5849(92)90181-f

[65] Berliner JA, Schwartz DS, Territo MC, Andalibi A, Almada L, Lusis AJ, et al. Induction of chemotactic cytokines by minimally oxidized LDL. In: The Chemokines Lindley. Advances in Experimental Medicine and Biology. US: Springer; 1993. pp. 13-18. DOI: 10.1007/978-1-4615-2952-1_2

[66] Holvoet P, Collen D. Oxidized lipoproteins in atherosclerosis and thrombosis. The FASEB Journal. 1994;**8**:1279-1284. DOI: 10.1096/fasebj.8.15.8001740

[67] Brenes M, Garcıa A, Garcıa P, Rios JJ, Garrido A. Phenolic compounds in Spanish olive oils. Journal of Agricultural and Food Chemistry.

1999;**47**:3535-3540. DOI: 10.1021/jf990009o

[68] Granados-Principal S, El-Azem N, Pamplona R, Ramirez-Tortosa C, Pulido-Moran M, Vera-Ramirez L, et al. Hydroxytyrosol ameliorates oxidative stress and mitochondrial dysfunction in doxorubicin-induced cardiotoxicity in rats with breast cancer. Biochemical Pharmacology. 2014;**90**(1):25-33. DOI: 10.1016/j.bcp.2014.04.001

[69] Qin C, Hu S, Zhang S, Zhao D, Wang Y, Li H, et al. Hydroxytyrosol acetate improves the cognitive function of APP/PS1 transgenic mice in ERβ-dependent manner. Molecular Nutrition and Food Research. 2021;**65**(3). DOI: 10.1002/mnfr.202000797

[70] Foods. Showing Compound Hydroxytyrosol 1-O-Glucoside (FDB020893). Canadian Institutes of Health Research, Canada Foundation for Innovation, and The Metabolomics Innovation Centre (TMIC), Canada. Available from: https://foodb.ca/compounds/FDB020893

[71] Méndez-Líter JA, Tundidor I, Nieto- Domínguez M, Fernández de Toro B, González Santana A, Isabel de Eugenio L, et al. Transglycosylation products generated by talaromyces amestolkiae GH3 β-glucosidases: Effect of hydroxytyrosol, vanillin and its glucosides on breast cancer cells. Microbial Cell Factories. 2019;**18**:97. DOI: 10.1186/s12934-019-1147-4

[72] Baccouri O, Guerfel M, Baccouri B, Cerretani L, Bendini A, Lercker G. Chemical composition and oxidative stability of Tunisian monovarietal virgin olive oils with regard to fruit ripening. Food Chemistry. 2008;**109**:743-754. DOI: 10.1016/j.foodchem.2008.01.034

[73] De Nino A, Lombardo N, Perri E, Procopio A, Raffaelli A, Sindona G. Direct identification of phenolic glucosides from olive leaf extracts by atmospheric pressure ionization tandem mass spectrometry. Journal of Mass Spectrometry. 1997;**32**:533-541. DOI: 10.1002/(SICI)1096-9888(199705)32:5<533::AID-JMS506>3.0.CO;2-9

[74] Johnson RL, Mitchell AE. Reducing phenolics related to bitterness in table olives. In: Quality of Phenolic Compounds: Occurrence, Health Benefits, and Applications in Food Industry. Hindawi; 2018. DOI: 10.1155/2018/3193185

[75] EMBL-EBI. ChEBI-oleuropein [Internet]. 2017. Available from: https://www.ebi.ac.uk/chebi/searchId.do?chebiId=CHEBI:7747

[76] Sun W, Frost B, Liu J. Oleuropein, unexpected benefits! Oncotarget. 2017;**8**(11):17409. DOI: 10.18632/oncotarget.15538

[77] De Andres MC, Meiss MS, Sanchez-HidalgoM,Gonzalez-BenjumeaA, Fernandez-Bolanos JG, Alarcon-de-la-Lastra C, et al. Osteoarthritis treatment with a novel nutraceutical acetylated ligstroside aglycone, a chemically modified extra-virgin olive oil polyphenol. Journal of Tissue Engineering. 2020;**11**:2041731420922701. DOI: 10.1177/2041731420922701

[78] Oliveras-López MJ, Innocenti M, Giaccherini C, Ieri F, Romani A, Mulinacci N. Study of the phenolic composition of Spanish and italian monocultivar extra virgin olive oils: Distribution of lignans, secoiridoidic, simple phenols and flavonoids. Talanta. 2007;**73**:726-732. DOI: 10.1016/j.talanta.2007.04.045

[79] Mumm R, Hilker M. Direct and indirect chemical defence of pine against folivorous insects. Trends Plant Science. 2006;**11**(7):351-358. DOI: 10.1016/j.tplants.2006.05.007

[80] Marsilio V, Seghetti L, Iannucci E, Russi F, Lanza B, Felicioni M. Use of a lactic acid bacteria starter culture during green olive (Olea europaea L cv 230 Ascolana tenera) processing. Journal of the Science of Food and Agriculture. 2005;**85**(7):1084-1090. DOI: 10.1002/jsfa.2066

[81] Jemai H, Bouaziz M, Sayadi S. Phenolic composition, sugar contents and antioxidant activity of Tunisian sweet olive cultivar with regard to fruit ripening. Journal of Agricultural and Food Chemistry. 2009;**57**:2961-2968. DOI: 10.1021/jf8034176

[82] Gutierrez-Rosales F, Romero MP, Casanovas M, Motilva MJ, Mínguez Mosquera MI. Metabolites involved in oleuropein accumulation and degradation in fruits of Olea europaea L.: Hojiblanca and Arbequina varieties. Journal of Agricultural and Food Chemistry. 2010;**58**(24):12924-12933. DOI: 10.1021/jf103083u

[83] Omar SH. Oleuropein in olive and its pharmacological effects. Scientia Pharmaceutica. 2010;**78**(2):133-154. DOI: 10.3797/scipharm.0912-18

[84] Bulotta S, Oliverio M, Russo D, Procopio A. Biological activity of Oleuropein and its derivatives. Natural Products. 2013:3605-3638. DOI: 10.1007/978-3-642-22144-6_156

[85] Obied HK, Prenzler PD, Ryan D, Servili M, Taticchi A, Esposto S, et al. Biosynthesis and biotransformations of phenol-conjugated oleosidic secoiridoids from Olea europaea L. Natural Product Reports. 2008;**25**(6):1167-1179. DOI: 10.1039/b719736e

[86] Andrews P, Busch JLHC, De Joode T, Groenewegen A, Alexandre H. Sensory properties of virgin olive oil polyphenols: Identification of deacetoxy-ligstroside aglycon as a key contributor to pungency. Journal of Agricultural and Food Chemistry. 2003;**51**(5):1415-1420. DOI: 10.1021/jf026042j

[87] Romero C, Medina E, Vargas J, Brenes M, De Castro A. In vitro activity of olive oil polyphenols against helicobacter pylori. Journal of Agricultural and Food Chemistry. 2007;**55**(3):680-686. DOI: 10.1021/jf0630217

[88] Menendez JA, Vazquez-Martin A, Garcia-Villalba R, Carrasco-PancorboA, Oliveras-FerrarosC, Fernandez-Gutierrez A, et al. tabAnti-HER2 (erbB-2) oncogene effects of phenolic compounds directly isolated from commercial extra-virgin olive oil (EVOO). BMC Cancer. 2008;**8**:377. DOI: 10.1186/1471-2407-8-377

[89] Lacono A, Gomez R, Sperry J, Conde J, Bianco G, Meli R, et al. Effect of oleocanthal and its derivatives on inflammatory response induced by lipopolysaccharide in a murine chondrocyte cell line. Arthritis and Rheumatism. 2010;**62**(6):1675-1682. DOI: 10.1002/art.27437

[90] Michel T, Termentzi A, Gikas E, Halabalaki M, Smith AB, Skaltsounis AL. Oleacin and oleocanthal: Two olive oil bioactives in multiple chemical forms. Planta Medica. 2012;**78**(11):PJ43. DOI: 10.1055/s-0032-1321203

[91] Mateos R, Cert A, Perez Camino MC, Garcia JM. Evaluation of virgin olive oil bitterness by quantification of Secoiridoid derivatives. Journal

of the American Oil Chemists' Society;**81**(1):71-75. DOI: 10.1007/s11746-004-0859-x

[92] Servili M, Baldioli M, Selvaggini R, Macchioni A, Montedoro G. Phenolic compounds of olive fruit: One- and two-dimensional nuclear magnetic resonance characterization of Nu1 zhenide and its distribution in the constitutive parts of fruit. Journal of Agricultural and Food Chemistry. 1999;**47**:12-18

[93] Tóth G, Barabás C, Tóth A, Kéry Á, Béni S, Boldizsár I, et al. Characterization of antioxidant phenolics in Syringa vulgaris L. flowers and fruits by HPLC-DAD-ESI-MS. Biomedical Chromatography. 2016;**30**:923-932. DOI: 10.1002/bmc.3630

[94] Chen Q, Yang L, Zhang G, Wang F. Bioactivity-guided isolation of antiosteoporotic compounds from Ligustrum lucidum. Phytotherapy Research. 2013;**27**:973-979. DOI: 10.1002/ptr.4820

[95] Silva S, Gomes L, Leitao F, Coelho A, Boas LV. Phenolic compounds and antioxidant activity of Olea europaea L. fruits and leaves. Food Science and Technology International. 2006;**12**:385-395. DOI: 10.1177/1082013206070166

[96] Chen S, Liu H, Wang S, Jiang H, Gao L, Wang L, et al. The neuroprotection of verbascoside in Alzheimer's disease mediated through mitigation of neuroinflammation via blocking NF-κB-p65 signaling. Nutrients. 2022;**14**(7):1417. DOI: 10.3390/nu14071417

[97] Zhao Y, Wang S, Pan J, Ma K. Verbascoside: A neuroprotective phenylethanoid glycosides with anti-depressive properties. Phytomedicine. 2023;**120**:155027. DOI: 10.1016/j.phymed.2023.155027

[98] Sheabar FZ, Neeman I. Separation and concentration of natural antioxidants from the rape of olives. Journal of the American Oil Chemists' Society. 1988;**65**(6):990-993. DOI: 10.1007/BF02544526

[99] Gorica N. Olive Fruit Phenols in Olive Oil Processing: The Fate and Antioxidant Potential. Nova Gorica: University of Nova Gorica; 2014

[100] Benito M, Manuel Lasa J, Gracia P, Oria R, Abenoza M, Cristina Sánchez-Gimeno A. Evolution of phenols and pigments in extra virgin olive oil from irrigated super-intensive orchard. European Journal of Lipid Science and Technology. 2012;**114**(5):558-567. DOI: 10.1002/ejlt.201100078

[101] Panche AN, Diwan AD, Chandra SR. Flavonoids: An overview. Journal of Nutritional Science. 2016;**5**:e47. DOI: 10.1017/jns.2016.41

[102] Tazzini N. Flavonols: Definition, structure, food sources', Tuscany diet. Phytochemicals. 2014;**30**:301-313

[103] Kubina R, Krzykawski K, Kabala-Dzik A, Wojtczka RD, Chodurek E, Dziedzic A. Fisetin, a potent anticancer flavonol exhibiting cytotoxic activity against neoplastic malignant cells and cancerous conditions: A scoping, comprehensive review. Nutrients;**14**(13):2604. DOI: 10.3390/nu14132604

[104] Yin KB, Chua SL, Balaram P. Apoptotic effects of chrysin in human cancer cell lines. International Journal of Molecular Science. 2010;**11**(5):2188-2199. DOI: 10.3390/ijms11052188

[105] Boon Yin K, Ling Chua S, Balaram P. Apoptotic effects of chrysin in human cancer cell lines. International Journal of Molecular Sciences. 2010;**11**(5):2188-2199

[106] Ganai SA, Sheikh FA, Baba ZA. Plant flavone chrysin as an emerging histone deacetylase inhibitor for prosperous epigenetic-based anticancer therapy. Phytotherapy Research. 2020;**35**(2):823-834. DOI: 10.1002/ptr.6869

[107] Kong JM, Chia LS, Goh NK, Chia TF, Brouillard R. Analysis and biological activities of anthocyanins. Phytochemistry. 2003;**64**:923-933. DOI: 10.1016/S0031-9422(03)00438-2

[108] Han F, Yang P, Wang H, Fernandes I, Mateus N, Liu Y. Digestion and absorption of red grape and wine anthocyanins through the gastrointestinal tract. Trends in Food Science and Technology. 2019;**83**:211-224. DOI: 10.1016/j.tifs.2018.11.05

[109] Kay CD, Pereira-Caro G, Ludwig IA, Clifford MN, Crozier A. Anthocyanins and flavanones are more bioavailable than previously perceived: A review of recent evidence. Annual Review of Food Science and Technology. 2017;**8**:155-180. DOI: 10.1146/annurev-food-030216-025636

[110] Sandhu AK, Miller MG, Thangthaeng N, Scott TM, Shukitt-Hale B, Edirisinghe I, et al. Metabolic fate of strawberry polyphenols after chronic intake in healthy older adults. Food & Function. 2018;**9**:96-106. DOI: 10.1039/c7fo01843f

[111] Cardona F, Andrés-Lacueva C, TulipaniS,TinahonesFJ,Queipo-OrtuñoMI. Benefits of polyphenols on gut microbiota and implications in human health. The Journal of Nutritional Biochemistry. 2013;**24**:1415-1422. DOI: 10.1016/j.jnutbio.2013.05.001

[112] Bilal-Hussain M, Hassan S, Waheed M, Javed A, Adil-Farooq M, Tahir A. Bioavailability and metabolic pathway of phenolic compounds. In: Plant Physiological Aspects of Phenolic Compounds. London, UK: IntechOpen; 2019. pp. 1-18. DOI: 10.5772/intechopen.84745

[113] Quiñones M, Miguel M, Aleixandre A. Los polifenoles, compuestos de origen natural con efectos saludables sobre el sistema cardiovascular. Nutrición Hospitalaria. 2012;**27**:76-89

[114] Foti P, Romeo FV, Russo N, Pino A, Vaccalluzzo A, Caggia C, et al. OMW water as renewable raw materials to generate high added-value ingredients for agro-food industries. Applied. Science. 2021;**11**(6):7511. DOI: 10.3390/app11167511

[115] Kirirsakis AK, Kiritsakis KA, Tsitsipas CK. A review of the evolution in the research of antioxidants in olives and olive oil during the last four decades. Journal of Food Bioactives. 2020. DOI: 10.31665/JFB.20xx.000xx

[116] Wolniak M. Mechanisms of antioxidative action of anthocyanins. Farmacja Polska;**58**:931-933

[117] Pool-Zobel B, Bub A, Schröder N, Rechkemmer G. Anthocyanins are potent antioxidants in model systems but do not reduce endogenous oxidative DNA damage in human colon cells. European Journal of Nutrition. 1999;**38**:227-234. DOI: 10.1007/s003940050065

[118] Brenes M, Garcia A, Garcia P, Garrido A. Rapid and complete extraction of phenols from olive oil and determination by means of a coulometric electrode array system. Journal of Agricultural and Food Chemistry. 2000;**48**:5178-5183. DOI: 10.1021/jf000686e

[119] Owen RW, Giacosa A, Hull WE, Haubner R, Spiegelhader B, Bartsh H.

The antioxidant/anticancer potential of phenolic compounds isolated from olive oil. European Journal of Cancer. 2000;**36**:1235-1247. DOI: 10.1016/s0959-8049(00)00103-9

[120] Lopez-BiedmaA,Sanchez-QuesadaC, Beltran G, Delgado-Rodriguez M, Gaforio JJ. Phytoestrogen (+)-pinoresinol exerts antitumor activity in breast cancer cells with different oestrogen receptor statuses. BMC Complementary and Alternative Medicine. 2016;**61**(1):350. DOI: 10.1186/s12906-016-1233-7

[121] Zhou Z, Qjao JX, Shetty A, Wu G, Huang Y, Davidson NE, et al. Regulation of estrogen receptor signaling in breast carcinogenesis and breast cancer therapy. Cellular and Molecular Life Sciences. 2015;**71**(8):1549. DOI: 10.1007/s00018-013-1376-3

[122] Qusa MH, Abdelwahed KS, Meyer SA, El Sayed KA. Olive oil lignan (+)-acetoxypinoresinol peripheral motor and neuronal protection against the tremorgenic mycotoxin penitrem a toxicity via STAT1 pathway. ACS Chemical Neuroscience. 2020;**11**:3575-3589. DOI: 10.1021/acschemneuro.0c00458

[123] Constantinou C, Papas A, Constantinou AI. Vitamin E and cancer: An insight into the anticancer activities of vitamin E isomers and analogs. International Journal of Cancer. 2008;**123**:739-752

[124] Lu GX, Lee M-J, Liu AB, Yang Z, Lin Y, Shih WJ, et al. δ-Tocopherol is more active than α- or γ-tocopherol in inhibiting lung tumorigenesis in vivo. Cancer Prevention Research (Philadelphia, PA.). 2012;**4**(3):404-413. DOI: 10.1158/1940-6207.CAPR-10-0130

[125] Rodrigues N, Casal S, Peres AM, et al. Effect of olive trees density on the quality and composition of olive oil from cv. Arbequina. Scientia Horticulturae. 2018;**238**:222-233. DOI: 10.1016/j.scienta.2018.04.059

[126] Ju J, Picinich SC, Yang Z, Zhao Y, et al. Cancer preventive activities of tocopherols and tocotrienols. Carcinogenesis. 2010;**31**:533-542

[127] Taylor PR, Qiao Y-L, Abnet CC, Dawsey SM, et al. Prospective study of serum vitamin E levels and esophageal and gastric cancers. Journal of the National Cancer Institute. 2003;**95**:1414-1416

[128] Mahabir S, Schendel K, Dong YQ, Barrera SL, et al. Dietary -, –, – and -tocopherols in lung cancer risk. International Journal of Cancer. 2008;**123**:1173-1180

[129] Brigelius-Flohe R, Kelly FJ, Salonen JT, Neuzil J, et al. The European perspective on vitamin E: Current knowledge and future research. The American Journal of Clinical Nutrition. 2002;**76**:703-716

[130] Das Gupta S, Suh N. Tocopherols in cancer: An update. Molecular Nutrition & Food Research. 2016;**60**:1354-1363. DOI: 10.1002/mnfr.201500847

[131] Roselló-Soto E, Koubaa M, Moubarik A, Lopes RP, Saraiva JA, Boussetta N, et al. Emerging opportunities for the effective valorization of wastes and by-products generated during olive oil production process: Non-conventional methods for the recovery of high-added value compounds. Trends in Food Science and Technology. 2015;**45**:296-310. DOI: 10.1016/j.tifs.2015.07.003

[132] Maoka T. Carotenoids as natural functional pigments. Journal of

Natural Medicines. 2020;**74**(1):1-16. DOI: 10.1007/s11418-019-01364-x

[133] Britton G, Liaaen-Jensen S, Pfander H. Carotenoids Hand Book. Basel: Birkhäuser; 2004. DOI: 10.1007/978-3-0348-7836-4

[134] Jaswir I, Noviendri D, Rewo FH, Octavianti F. Carotenoids: Source, medical properties and their application in food and nutraceutical industry. Journal of Medicinal Plants Research. 2011;**5**:7119-7131. DOI: 10.5897/JMPRx11.011

[135] Mateos R, Garcia-Mesa JA. Rapid and quantitative extraction method for the determination of chlorophylls and carotenoids in olive oil by high-performance liquid chromatography. Analytical and Bioanalytical Chemistry. 2006;**385**:1247-1254. DOI: 10.1007/s00216-006-0472-8

[136] Stahl W, Sies H. Antioxidant activity of carotenoids. Molecular Aspects of Medicine. 2003;**24**:345-351. DOI: 10.1016/s0098-2997(03)00030-x

[137] Fabricant DS, Farnsworth NR. The value of plants used in traditional medicine for drug discovery. Environmental Health Perspectives. 2001;**109**:69-75. DOI: 10.2307/3434847

[138] Sayavedra-soto LA, Montgomery MW. Inhibition of polyphenoloxidase by sulfite. Journal of Food Science. 2006;**51**(65):1531-1536. DOI: 10.1111/j.1365-2621.1986.tb13852.x

[139] Miazek K, Beton K, Sliwinska A, Brozek-Puska B. The effect of β-carotene, tocopherols and ascorbic acid as anti-oxidant molecules on human and animal In vitro/In vivo studies: A review of research design and analytical techniques used. Biomolecules. 2022;**12**(8):1087. DOI: 10.3390/biom12081087

[140] Robinson PK. Enzymes: Principles and biotechnological applications. Essays in Biochemistry. 2015;**59**(1):1-41. DOI: 10.1042/bse0590001

[141] El Zbakh H, Abbassi A. Potential use of olive mill wastewater in the preparation of functional beverages: A review. Journal of Functional Foods. 2012;**4**(1):53-65. DOI: 10.1016/j.jff.2012.01.002

[142] Cos P, Vlietinck AJ, Berghe DV, Maes L. Anti-infective potential of natural products: How to develop a stronger in vitro 'proof-of-concept'. Journal of Ethnopharmacology. 2006;**106**:290-302. DOI: 10.1016/j.jep.2006.04.003

[143] Obied H, Allen M, Bedgood D, Prenzler P, Robards K. Investigation of Australian olive mill waste for recovery of biophenols. Journal of Agricultural and Food Chemistry. 2005;**53**(26):9911-9920. DOI: 10.1021/JF0518352

[144] Allouche N, Fki I, Sayadi S. Toward a high yield recovery of antioxidants and purified hydroxytyrosol from olive mill wastewaters. Journal of Agricultural and Food Chemistry. 2004;**52**(2):267-273. DOI: 10.1021/JF034944U

[145] Chanioti S, Tzia C. Extraction of phenolic compounds from olive pomace by using natural deep eutectic solvents and innovative extraction techniques. Innovative Food Science & Emerging Technologies. 2018;**8**:228-239. DOI: 10.1016/J.IFSET.2018.07.001

[146] Lafka T-I, Lazou A, Sinanoglou VJ, Lazos ES. Phenolic and antioxidant potential of olive oil mill wastes. Food Chemistry;**125**(**1**):92-98. DOI: 10.1016/j.foodchem.2010.08.041

[147] Cequier-Sanchez E, Rodriguez C, Ravelo AG, Zarate R. Dichloromethane

as a solvent for lipid extraction and assessment of lipid classes and fatty acids from samples of different natures. Journal of Agricultural and Food Chemistry. 2008;**56**(12):4297-4303. DOI: 10.1021/jf073471e

[148] Sasidharan S, Chen Y, Saravanan D, Sundram KM, Yoga LL. Extraction, isolation and characterization of bioactive compounds from Plants' extracts. African Journal of Traditional, Complementary, and Alternative Medicines. 2011;**8**(1):1-10

[149] Doellinger J, Schneider A, Hoeller M, Lasch P. Sample preparation by easy extraction and digestion (SPEED) - a universal, rapid, and detergent-free protocol for proteomics based on acid extraction. Molecular & Cellular Proteomics. 2020;**19**(1):209-2022. DOI: 10.1074/mcp.TIR119.001616

[150] Wang W-Y, Qu H-B, Gong X-C. Research progress on percolation extraction process of traditional Chinese medicines. Zhongguo Zhong Yao Za Zhi. 2020;**45**(5):1039-1046. DOI: 10.19540/j.cnki.cjcmm.20191221.305

[151] Gueboudji Z, Kadi K, Mahmoudi M, Hannachi H, Nagaz K, Addad S, et al. Maceration and liquid-liquid extractions of phenolic compounds and antioxidants from Algerian olive oil mill wastewater. Environmental Science and Pollution Research. 2023;**30**(2):3432-3439. DOI: 10.1007/s11356-022-22482-2

[152] Daswani PG, Ghadge AA, Brijesh S, Birdi TJ. Preparation of decoction of medicinal plants: A self-help measure? Journal of Alternative and Complementary Medicine. 2011;**14**(12):1099-1100. DOI: 10.1089/acm.2011.0217

[153] Belwal T, Ezzat SM, Rastrelli L, Bhatt ID, Daglia M, Baldi A, et al. A critical analysis of extraction techniques used for botanicals: Trends, priorities, industrial uses and optimization strategies. Trends in Analytical Chemistry. 2018;**100**:82-102. DOI: 10.1016/j.trac.2017.12.018

[154] PulseMaster. What is pulsed electric field processing? [Internet]. 2009. Available from: https://www.pulsemaster.us/about-pulsemaster

[155] Özkan G, Karacabey E, Arslan N, Odabaşı N. Optimisation of microwave-assisted extraction of triterpenoic acids from olive mill waste using response surface methodology. Quality Assurance and Safety of Crops & Foods. 2016;**9**(2):179-185. DOI: 10.3920/QAS2015.0783

[156] Gómez-Cruz I, Contreras M, Romero I, Castro E. Optimization of microwave-assisted water extraction to obtain high value-added compounds from exhausted olive pomace in a biorefinery context. Food. 2022;**11**(14). DOI: 10.3390/foods11142002

[157] Japón-Luján R, Luque-Rodríguez JM, Luque de Castro MD. Multivariate optimisation of the microwave-assisted extraction of oleuropein and related biophenols from olive leaves. Analytical and Bioanalytical Chemistry. 2006;**385**:753-759. DOI: 10.1007/S00216-006-0419-0

[158] García-Ayuso LE, Luque de Castro MD. A multivariate study of the performance of a microwave-assisted Soxhlet extractor for olive seeds. Analytica Chimica Acta. 1999;**382**(3):309-316. DOI: 10.1016/S0003-2670(98)00795-8

[159] Tamborrino A, Urbani S, Servili M, Romaniello R, Perone C, Leone A. Pulsed electric fields for the treatment of olive pastes in the oil extraction process.

Applied Sciences. 2020;**10**(1):114. DOI: 10.3390/app10010114

[160] Leone A, Tamborrino A, Esposto S, Berardi A, Servili M. Investigation on the effects of a pulsed electric field (PEF) continuous system implemented in an industrial olive oil plant. Food. 2022;**11**(18):2758. DOI: 10.3390/foods11182758

[161] Pappas VM, Lakka A, Palaiogiannis D, Athanasiadis V, Bozinou E, Ntourtoglou G, et al. Optimization of pulsed electric field as standalone "green" extraction procedure for the recovery of high value-added compounds from fresh olive leaves. Antioxidants (Basel). 2021;**10**(10):1554. DOI: 10.3390/antiox10101554

[162] Yassine W, Akazdam S, Zyade S, Gourich B. Treatment of olive mill wastewater using electrocoagulation process. Journal of Applied Surface and Interfaces. 2018;**4**(1-3):24-30

[163] Tamborrino A, Urbani S, Servili M, Romaniello R, Perone C, Leone A. Pulsed electric fields for the treatment of olive pastes in the oil extraction process. Applied Sciences. 2019;**10**(1):114. DOI: 10.3390/app10010114

[164] Romanielloa R, Tamborrinob A, Leonea A. Use of ultrasound and pulsed electric fields technologies applied to the olive oil extraction process. Chemical Engineering Transactions;**75**:13-18. DOI: 10.3303/CET1975003

[165] Chemat F, Rombaut N, Sicaire AG, Meullemiestre A, Fabiano-Tixier AS, AbertVian M. Ultrasound assisted extraction of food and natural products. Mechanisms, techniques, combinations, protocols and applications. A review. Ultrasonics Sonochemistry. 2017;**34**:540-560. DOI: 10.1016/j.ultsonch.2016.06.035

[166] Allaf T, Tomao V, Ruiz K, Chemat F. Instant controlled pressure drop technology and ultrasound assisted extraction for sequential extraction of essential oil and antioxidants. Ultrasonics Sonochemistry. 2013;**20**:239-246. DOI: 10.1016/j. ultsonch.2012.05.013

[167] Chemat F, Rombaut N, Meullemiestre A, Turk M, Perino S, Fabiano-Tixier AS, et al. Review of green food processing techniques. Preservation, transformation, and extraction. Innovative Food Science & Emerging Technologies. 2017;**41**:357-377. DOI: 10.1016/j.ifset.2017.04.016

[168] Skaltsounis AL, Argyropoulou A, Aligiannis N, Xynos N. Recovery of high added value compounds from olive tree products and olive processing byproducts. Olive and Olive Oil Bioactive Constituents. 2015:333-356. DOI: 10.1016/B978-1-63067-041-2.50017-3

[169] Karen A, Halil Oztop M, Lia F. The effect of hydrolysis on the antioxidant activity of OMW. Applied Science. 2022;**12**(23):12187. DOI: 10.3390/app122312187

[170] Medina-Torres N, Ayora-Talavera T, Espinosa-Andrews H, Sánchez-Contreras A, Pacheco N. Ultrasound assisted extraction for the recovery of phenolic compounds from vegetable sources. Agronomy. 2017;7:47. DOI: 10.3390/agronomy7030047

[171] Patist A, Bates D. Ultrasonic innovations in the food industry: From the laboratory to commercial production. Innovative Food Science and Emerging Technologies. 2008;**9**:147-154. DOI: 10.1016/j.ifset.2007.07.004

[172] Oz NA, Uzun AC. Ultrasound pretreatment for enhanced biogas production from olive mill wastewater.

Ultrasonics Sonochemistry. 2015;**22**:565-572. DOI: 10.1016/j.ultsonch.2014.04.018

[173] Kheiredine B, Derbal K, Charbit K, Amrani M. Biogas production by an anaerobic co-digestion process from olive mill waste: Effect of ultrasonic pre-treatment. Desalination and Water Treatment. 2022;**246**:139-145. DOI: 10.5004/dwt.2022.28011

[174] Mostafa H, Iqdiam BM, Abuagela M, Marshall MR, Pullammanappallil P, Goodrich-Schneider R. Treatment of olive mill wastewater using high power ultrasound (HPU) and electro-Fenton (EF) method. Chemical Engineering and Processing - Process Intensification. 2018;**131**:131-136. DOI: 10.1016/J.CEP.2018.07.015

[175] El-Abbassi A, Kiai H, Raiti J, Hafidi A. Application of ultrafiltration for olive processing wastewaters treatment. Journal of Cleaner Production. 2014;**65**(15):432-438. DOI: 10.1016/j.jclepro.2013.08.016

[176] Dali I, Aydi A, Stamenic M, Kolsi L, Ghachem K, Zizovic I, et al. Extraction of lyophilized olive mill wastewater using supercritical $CO_2$ processes. Alexandria Engineering Journal. 2022;**61**:237-246

[177] Sapkale GN, Patil SM, Surwase US, Bhatbhage PK. A review -supercritical fluid extraction. International Journal of Chemical Sciences. 2010;**8**(2):729-743

[178] Caballero AS, Romero-Garcia JM, Castro E, Cardona CA. Supercritical fluid extraction for enhancing polyphenolic compounds production from olive waste extracts. Journal of Chemical Technology and Biotechnology. 2018;**95**(2):356-362. DOI: 10.1002/jctb.5907

[179] Casademont P, Cardozo-Filho L, Meurer E, Sánchez-Oneto J, Portela JR. Gasification of olive oil mill waste by supercritical water in a continuous reactor. The Journal of Supercritical Fluids. 2018;**142**:10-21. DOI: 10.1016/J.SUPFLU.2018.06.001

[180] Rivas R, Gimeno O, Portela JR, De La Ossa EM, Beltrán F. Supercritical water oxidation of olive oil mill wastewater. Industrial and Engineering Chemistry Research. 2001;**40**(16):3670-3674. DOI: 10.1021/ie000995u

[181] De Lucas A, De La Martinez Ossa E, Rincón J, Blanco M, Gracia I. Supercritical fluid extraction of tocopherol concentrates from olive tree leaves. The Journal of Supercritical Fluids. 2002;**22**(3):221-228. DOI: 10.1016/S0896-8446(01)00132-2

[182] Rui PFF, Rocha-Santos TAP, Duarte AC. Supercritical fluid extraction of bioactive compounds. Trends in Analytical Chemistry. 2016;**76**:40-51. DOI: 10.1016/j.trac.2015.11.013

[183] Fiorentino A, Gentili A, Isidori M, Lavorgna M, Parrella A, Temussi F. Olive oil mill wastewater treatment using a chemical and biological approach. Journal of Agricultural and Food Chemistry. 2004;**52**:5151-5154. DOI: 10.1021/jf049799e

[184] Günay A, Çetin M. Determination of aerobic biodegradation kinetics of olive oil mill wastewater. International Biodeterioration & Biodegradation. 2013;**85**:237-242. DOI: 10.1016/j.ibiod.2013.07.012

[185] Papadimitriou EK, Chatjipavlidis I, Balis C. Application of composting to OMW water treatment. Environmental Technology. 1997;**18**:101-107. DOI: 10.1080/09593331808616517

[186] Roig A, Cayuela ML, Sánchez-Monedero MA. An overview on olive mill wastes and their valorisation

methods. Waste Management. 2006;**26**(9):960-969. DOI: 10.1016/J.WASMAN.2005.07.024

[187] Serrano A, Villa-Gómez D, Fermoso FG, Alonso-Fariñas B. Is anaerobic digestion a feasible alternative to the combustion of olive mill solid waste in terms of energy production? A critical review. Biofuels, Bioproducts and Biorefining. 2020;**15**(1):150-162. DOI: 10.1002/bbb.2159

[188] Hamdi M. Anaerobic digestion of olive mill wastewaters. Process Biochemistry. 1996;**31**(2):105-110. DOI: 10.1016/0032-9592(95)00035-6

[189] Abu Khayer MD, Chowdhury MB, Akratos CS, Vayenas DV, Pavlou S. Olive mill waste composting: A review. International Biodeterioration & Biodegradation. 2013;**85**:108-119. DOI: 10.1016/j.ibiod.2013.06.019

[190] Boari G, Brunetti A, Passino R, Rozzi A. Anaerobic digestion of olive oil mill wastewaters. Agricultural Waste. 1984;**10**(3):161-175. DOI: 10.1016/0141-4607(84)90057-X

[191] Erguder TH, Güven E, Demirer GN. Anaerobic treatment of olive mill wastes in batch reactors. Process Biochemistry. 2000;**36**(3):243-248. DOI: 10.1016/S0032-9592(00)00205-3

[192] Tziotzios G, Michailakis S, Vayenas DV. Aerobic biological treatment of olive mill wastewater by olive pulp bacteria. International Biodeterioration & Biodegradation. 2007;**60**(4):209-214. DOI: 10.1016/j.ibiod.2007.03.003

[193] Benitez FJ, Beltran-Heredia J, Torregrosa J, Acero JL. Improvement of the anaerobic biodegradation of olive mill wastewaters by prior ozonation pretreatment. Bioprocess Engineering. 1997;**17**:169-175. DOI: 10.1007/s004490050371

[194] Rozzi A, Malpei F. Treatment and disposal of olive mill effluents. International Biodeterioration & Biodegradation. 1996;**38**(3-4):135-144. DOI: 10.1016/S0964-8305(96)00042-X

[195] Beccari M, Bonemazzi F, Majone M, Riccardi C. Interaction between acidogenesis and methanogenesis in the anaerobic treatment of olive oil mill effluents. Water Research. 1996;**30**(1):183-189. DOI: 10.1016/0043-1354(95)00086-Z

[196] Azzam MOJ. Olive mills wastewater treatment using mixed adsorbents of volcanic tuff, natural clay and charcoal. Journal of Environmental Chemical Engineering. 2018;**6**(2):2126-2136. DOI: 10.1016/j.jece.2018.03.009

[197] Borja R, Alba J, Mancha A, Martin A, Alonso V, Sánchez E. Comparative effect of different aerobic pretreatments on the kinetics and macroenergetic parameters of anaerobic digestion of olive mill wastewater in continuous mode. Bioprocess Engineering. 1998;**18**(2):127-134. DOI: 10.1007/s004490050422

[198] Paredes C, Bernal MP, Cegarra J, Roig A. Bio-degradation of olive mill wastewater sludge by its co-composting with agricultural wastes. Bioresource Technology. 2002;**85**(1):1-8. DOI: 10.1016/S0960-8524(02)00078-0

[199] Hamdi M. Future prospects and constraints of olive mill wastewaters use and treatment: A review. Bioprocess Engineering. 1993;**8**(5-6):209-214. DOI: 10.1007/bf00369831

[200] Martilotti E. Use of Olive by-Products in Animal Feeding in Italy. Rome, Italy: Division de la Production et de la Santé Animale. FAO; 1983

[201] Fraga CG, Croft KD, Kennedy DO, Tomás-Barberán FA. The effects of polyphenols and other bioactives on human health. Food & Function. 2019;**10**:514-528. DOI: 10.1039/C8FO01997E

[202] Li AN, Li S, Zhang YJ, Xu XR, Chen YM, Li HB. Resources and biological activities of natural polyphenols. Nutrients. 2014;**6**:6020-6047. DOI: 10.3390/nu6126020

[203] Smiljković M, Dias MI, Stojković D, Barros L, Bukvički D, Ferreira ICFR, et al. Characterization of phenolic compounds in tincture of edible Nepeta nuda: Development of antimicrobial mouthwash. Food & Function. 2018;**9**:5417-5425. DOI: 10.1039/C8FO01466C

[204] Larian B. Exploring the differences between pharmaceutical vs nutraceutical supplements. 2023. Available from: https://www.mdbiowellness.com/blogs/doctors-desk/exploring-the-differences-between-pharmaceutical-vs-nutraceutical-supplements#:~:text=Pharmaceuticals% 20provide%20immediate%20relief% 20andcare%20and%20overall%20well-being

[205] Obied HK, Prenzler PD, Omar SH, Ismael R, Servili M, Esposto S, et al. Pharmacology of olive biophenols. Advances in Molecular Toxicology. 2012;**6**:195-242. DOI: 10.1016/B978-0-444-59389-4.00006-9

[206] Crespo MC, Tomé-Carneiro J, Dávalos A, Visioli F. Pharma-nutritional properties of olive oil phenols. Transfer of new findings to human nutrition. Food. 2018;**7**(6):90. DOI: 10.3390/foods7060090

[207] Bucciantini M, Leri M, Nardiello P, Casamenti F, Stefani M. Olive polyphenols: Antioxidant and anti-inflammatory properties. Antioxidants. 2021;**10**(7):1044. DOI: 10.3390/antiox10071044

[208] Nan J, Ververis K, Bollu S, Rodd AL, Swarup O, Karagiannis T. Biological effects of the olive polyphenol, hydroxytyrosol: An extra view from genome-wide transcriptome analysis. Hellenic Journal of Nuclear Medicine. 2014;**1**:62-9

[209] Oliver S, Vittorio O, Cirillo G, Boyer C. Enhancing the therapeutic effects of polyphenols with macromolecules. Polymer Chemistry. 2016;**7**:1529-1544. DOI: 10.1039/C5PY01912E

[210] Miranda AR, Albrecht C, Cortez MV, Soria EA. Pharmacology and toxicology of polyphenols with potential as neurotropic agents in non-communicable diseases. Current Drug Targets. 2018;**19**(2):97-110. DOI: 10.2174/1389450117666161220152336

[211] Del Rio D, Rodriguez-Mateos A, Spencer J, Tognolini M, Borges G, Crozier A. Dietary (poly)phenolics in human health: Structures, bioavailability, and evidence of protective effects against chronic diseases. Antioxidants and Redox Signaling. 2013;**18**(14):1818-1892. DOI: 10.1089/ars.2012.4581

[212] Teng H, Chen L. Polyphenols and bioavailability: An update. Critical Reviews in Food Science and Nutrition. 2019;**59**(13):2040-2051. DOI: 10.1080/10408398.2018.1437023

[213] Aguirre A, Borneo R. Improving bioavailability of polyphenols using nanodelivery systems based on food polymers. Polyphenols in Plants. 2019:59-65. DOI: 10.1016/B978-0-12-813768-0.00004-9

[214] Vivekanandhan DK, Verma P, Singh S. Emerging technologies for improving bioavailability of polyphenols. Current Nutrition & Food Science. 2016;**12**(1):12-22. DOI: 10.2174/1573401311666151015213704

[215] Hu B, Liu X, Zhang C, Zeng X. Food macromolecule based nanodelivery systems for enhancing the bioavailability of polyphenols. Journal of Food and Drug Analysis. 2017;**25**(1):3-15. DOI: 10.1016/j.jfda. 2016.11.004

[216] Scheepens A, Tan K, Paxton JW. Improving the oral bioavailability of beneficial polyphenols through designed synergies. Genes & Nutrition. 2009;**5**:75-87. DOI: 10.1007/s12263-009-0148-z

Chapter 4

# Medicinal Herbs and Probiotics: Antioxidant Activities for Damaged DNA Protection and Cytotoxicity toward Cancer Treatment

*Tu Nguyen*


## Abstract

Oxidative stress is an imbalance between oxidation and antioxidant responses caused by the accumulation of free radicals in the body. Cells produce free radicals that cause oxidative damage such as aging, atherosclerosis, arthritis, cancer, and neurological diseases. Antioxidants are substances that can neutralize free radicals, thereby reducing the damage to the body. Antioxidants are classified as synthetic and natural antioxidants. Natural products, mainly extracted from some medicinal plants, have strong antioxidant activity and low toxicity and side effects. Moreover, antioxidant activities are also detected in probiotics. Therefore, this chapter summarized the roles of antioxidants and then suggested antioxidants from natural sources have good prospects in the prevention and treatment of various diseases related to oxidative stress. In this study, antioxidant activities were mentioned in plants such as *Basella alba*, *Cistanche* sp., and representative probiotics such as *Bifidobacterium breve* ATCC 15700, *Lactobacillus rhamnosus* PN04, and *Lactococcus lactis* PN05 which could serve in DNA damage protection and cytotoxicity oriented for cancer treatment. With the results obtained in these medicinal herbs and probiotics, the antioxidant activities of more and more medicinal herbs and probiotics should be exploited in health care and pharmaceutical product development.


**Keywords:** antioxidant activities, medicinal herbs, probiotics, DNA damage protection, cytotoxicity, cancer

## 1. Introduction

Cells maintain reactive oxygen species (ROS) homeostasis through various systems, such as the superoxide dismutase (SOD) system, catalase (CAT) system, and glutathione peroxidase (GSH-Px) system [1]. Antioxidants are a group of substances that help trap and neutralize free radicals, thereby reducing the damage to the body caused by free radicals. Additionally, antioxidants play a protective role against several diseases including inflammation and cancer caused by oxidative stress [2]. Antioxidants are classified as synthetic antioxidants and general natural antioxidants.

However, synthetic antioxidants sometimes have toxicity and side effects, especially with the long-term use [3]. The exploration of natural antioxidants that are safe and free of side effects has become a hot spot in the past few years. Some natural products, mainly extracted from some medicinal plants and microorganisms, have strong antioxidant activity and low toxicity and side effects [3]. For this reason, antioxidants from natural sources have good application prospects in the prevention and treatment of various diseases related to oxidative stress. Due to reducing the chemical drugs' negative effects, herbal medicines have gained their attractiveness in recent years. The natural sources contain vast components having antioxidant activities which are essential in cancer prevention, damaged DNA protection, and preventing lipid peroxidation [4]. For example, phenylethanoid glycosides (PhGs) in medicinal herbs act as a hydrogen donor to reductive radicals, allowing them to scavenge free radicals and regulate antioxidant enzymes involved in in vivo free radical metabolism [5].

Oxidation affects humans causing many diseases. Importantly, DNA is easy to spoil, leading to health problems in humans. DNA, the genetic material of all living organisms on the planet, is often influenced by a variety of variables, including exogenous and endogenous agents, which may alter its activity inside cells and ultimately lead to dramatic changes in living animals. The main exogenous factors can include UV light, ionizing radiation, toxic chemicals, and so on, whereas the primary endogenous factor is commonly caused by ROS [6]. The endogenous agents include superoxide anion, hydroxyl (OH) radicals, and hydrogen peroxide radicals. Furthermore, endogenous agent-caused damage occurs more often than exogenous agent-caused harm. If the damage is not addressed or corrected, it may lead to mutations and other health issues including aging, cancer, diabetes, reperfusion, cataracts, and lung failure.

Because of these, the therapeutic approach to alternative conventional medicine has expanded, such that increased research focused on the use of natural metabolites produced by medicinal herbs and lactic acid bacteria.

## 2. Roles of antioxidants

Life needs oxygen element for many biological activities. When oxygen damages the cells, oxidation occurs. Cells use oxygen to produce energy and free radicals. Reactive oxygen species (ROS) and reactive nitrogen species (RNS) are the by-products produced by cellular redox processes. These reactive species play dual roles in humans toxically and beneficially. At low or moderate levels, reactive species have beneficial effects on cellular and immune redox signaling function; however, high concentrations cause oxidative stress which can produce harmful substances, leading to spoiling cell function and structure. The toxic properties may be free radicals [6]. Free radicals play a role in the aging process. Free radicals cause oxidative stress, involving cardiovascular, neurological, and other human disorders. Antioxidants are our important defense against free radicals. ROS species such as superoxide, hydroxyl, and hydrogen peroxide are equivalent to the reduced oxygen levels. Oxidation stress is due to unhealthy lifestyles, exposure to chemicals, pollution, smoking, drugs, illness and sol. Antioxidants from plants, animals, and mineral sources have been shown to be beneficial to human health and effectively reduce the incidence of free radicals. Antioxidants are also involved in reducing free radicals, improving the normal function, and treating those diseases. Low levels of antioxidants or low levels of antioxidants in the blood increase the risk of diseases such as cancer. So, antioxidants in a healthy diet and natural antioxidants are now being recognized for protecting health from oxidative stress. Oxidative stress is through four important steps including membrane lipid peroxidation, protein oxidation, DNA damage, and disorders in reduction cellular equivalent, leading to cell destruction and changes in signaling pathways. Consequently, these mechanisms are implicated in the pathogenesis of diseases such as mitochondrial diseases in which cells cannot produce energy, leading to death [7]. Every biological molecule presenting in our body is at risk of free radicals. Such damaged cellular molecules may impair cellular function or may result in cell death. Lipids in the membranes of subcellular organelles are very susceptible to free radical damage. Lipid peroxidation leads to the generation of large amounts of toxic substances. Lipid peroxidation in cell membranes can damage the cell membranes by disrupting fluidity and permeability. Lipid peroxidation begins by removing hydrogen atoms from the methylene group, leading to the formation of a single electron on the carbon atom or carbon radical. Proteins are also susceptible to free radicals. Free radicals cause the destruction of many proteins and loss of enzyme activities. Protein hydroperoxide can be generated by ROS/RNS-induced protein oxidation, forming radicals. Although most oxidized proteins are essentially nonfunctional and eliminated quickly, some may accumulate gradually over time and thus contribute to aging-related damage as well as various diseases [5].

## 3. Antioxidant activities in DNA damage protection and cytotoxicity

### 3.1 Medicinal herbs

#### *3.1.1 Phytochemicals served as antioxidants*

Antioxidants are found in many vegetables, fruits, grains, tea, legumes, nuts, herbs and so on. Antioxidants may provide protection against diseases caused by free radicals. Phytochemicals or plant components are the main sources of antioxidants. The majority of these phytochemicals are redox-active molecules; therefore, they work to maintain a redox balance that is defined as an antioxidant. Phytochemicals are often classified as primary or secondary metabolites. The primary metabolites are sugars, amino acids, proteins, purines, and pyrimidines of nucleic acids, chlorophyll, etc., while secondary components include alkaloids (derived from amino acids), terpenes (a group of lipids), and phenolics (derived from carbohydrates). Plants produce an extremely impressive range of antioxidant compounds such as carotenoids, flavonoids, cinnamic acid, benzoic acid, folic acid, ascorbic acid, tocopherols, and tocotrienols to prevent oxidation of sensitive substrates. The plant-based antioxidants are believed to have better biological effects than synthetic ones because the plant components are part of the physiological processes of living flora, and thus, they are believed to have better compatibility with the human body. In this chapter, the plant extracts were studied for antioxidant activities. These plants have been used as daily vegetables in Vietnam.

#### *3.1.2 Representative medicinal herbs*

##### *3.1.2.1* Basella alba *(*B. alba*)*

To clarify the roles of antioxidants, the study focused on the daily vegetable that can be used as a medicinal herb named *Basella alba* (*B. alba*). *B. alba* is one of the typical vegetables in Vietnam (**Figure 1**). *B. alba* is a good source of fiber, vitamins A, B, and C, thiamine, riboflavin, niacin, minerals such as calcium, magnesium, and iron, and several antioxidants. It also contains essential amino acids such as arginine, isoleucine, leucine, lysine, threonine, and tryptophan and a low percentage of soluble oxalates. *B. alba* has saponins, peptides, and phenolic compounds. *B. alba* leaves have anthocyanins, flavonoids, and phenolic acids which are good sources of phytochemicals and are also a potential source of antioxidants [8]. Because this vegetable is prepared in Vietnamese soup, the aqueous extract was performed. The extract was not soluble in methanol and ethanol; therefore, the antioxidant activities of the vegetable were determined by applying $Fe^{3+}$ reducing power assay, reducing the power of the cucurbits which was determined by the previous method [9]. The reducing capacity of the polyphenolic extract of the *B. alba* leaves was measured by its ability to transform $Fe^{3+}$ to $Fe^{2+}$ at various concentrations (10, 25, 50, 75, 100, and 125 μg/mL). The results revealed that the reducing activity significantly increased ($R^2$ = 0.9916), with the increased concentration at 250 μg/mL. The reductive capacity of a compound depends on the presence of reductones. The reducing power of the reference compound (gallic acid) was higher than the *B. alba* aqueous extract.

*3.1.2.2* Cistanches *sp.*

Cistanches belong to the family Orobanchaceae, and based on different chemical positions and genetic variations, the accepted genus *Cistanche* includes *C. deserticola*, *C. tubulosa*, *C. sinensis*, *C. salsa*, and *C. salsa* var. Albiflora [10].

*C. derserticola* can produce PhGs which belong to a group of natural compounds with many pharmacological properties, including antibacterial, antiviral, anticancer, anti-inflammatory, neuroprotective, antioxidant, and hepatoprotective activities. Individual PhGs have been shown to reduce apoptosis of neurons induced by various agents in in vitro and in vivo tests. Therefore, individual PhGs may have promising options for treating common neurodegenerative diseases including dementia and Parkinson's disease [11]. Polysaccharides include 10 simple sugar bioactive components in *Cistanche* species including cinnamic acid, beta-glucopyranose, and hydroxyphenyl ethyl moieties, connected by ester and glycosidic bonds. Polysaccharides are

**Figure 1.**
Basella alba *and antioxidant activity test.*

responsible for enhancing immune function and fighting cancer. Furthermore, the free radical scavenging ability of polysaccharides that function as singlet and triplet oxygen scavengers, peroxide decomposers, enzyme inhibitors, and activators [12]. PhGs, and polysaccharides are also found in other plants which are considered as antioxidant sources. Moreover, *Cistanche* and others also contain ginsenosides including triterpene saponins and steroid glycosides called ginseng compounds. The common ginsenosides are Rb1, Rb2, Rg3, Rh2, Rh3, Rg1, Rg2, and Rh1. Typically, *C. derserticola* has ginsenosides (Rb1) and apigenin which are antioxidants and were detected by thin layer chromatography (TLC) (**Figure 2**).

They play a pivotal role in strong antioxidant effects in animal models and cell cultures. For instance, ginsenoside Rb1 exerts its antioxidant activity through two major mechanisms: chemical scavenging of certain types of ROS and a receptor-mediated genomic effect on antioxidant protein expression, such as SOD [13]. Secondarily, ginseng saponins have been shown to stimulate the formation of blood vessels and improve blood circulation in the brain, thereby enhancing memory and cognitive ability [14]. Additionally, ginsenoside Rg1 inhibits platelet activation via the inhibition of the ERK pathway and attenuates arterial thrombus formation *in vivo* [15].

Of all the flavonoids, apigenin (4′,5,7-trihydroxyflavone) is one of the most widely distributed flavonoids in the plant kingdom, and one of the most studied phenolics [16]. Apigenin can be found in significant amounts in vegetables, herbs, plants, and plant-based beverages. Like other flavonoids, apigenin reported numerous functional activities such as antioxidant properties [17], anti-hyperglycemic [18], anti-inflammatory, and (in myocardial ischemia) anti-apoptotic effects [19]. Therefore, the various biological activities of apigenin and several methods of extraction from natural sources, or also with the use of modern extractive approaches, are further analyzed.

ROS/RNS interferes with DNA, leading to oxidative damage. DNA is vulnerable to free radicals such as the hydroxyl group (−OH). The C4-C5 double bond of pyrimidine is very sensitive to hydroxyl group attack, which generates a series of oxidative pyrimidine destruction products, such as thymine glycol, uracil glycol, urea residue, 5-hydroxydeoxyuridine, 5-hydroxydeoxycytidine, hydantoin, and others. Likewise, purines are also susceptible to the hydroxyl group, leading to the formation of 8-hydroxydeoxyguanosine (8-OHdG), 8-hydroxydeoxyadenosine, formamidopyrimidines, and purine products of oxidation. Free radical attack also caused the activation of the enzyme poly(ADP-ribose) synthetase; thus, DNA fragmentation

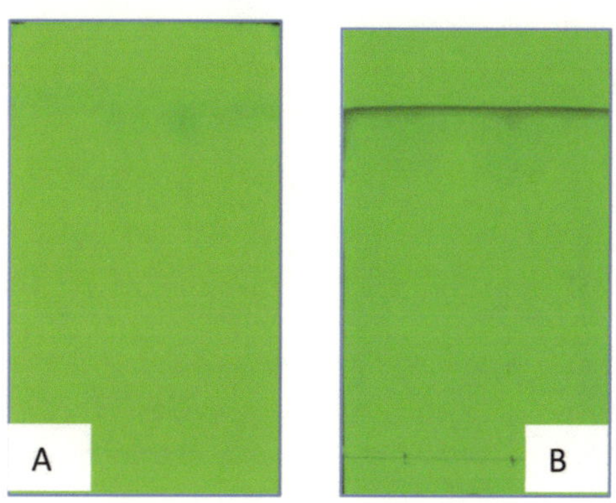


**Figure 2.**
C. derserticola *extracts checked for ginsenoside Rb1 (A) and apigenin (B).*

and programmed cell death occur. This process is due to exhausting $NAD^+$ levels in the cells, thereby disrupting the electron transport chain function. Free radicals such as the hydroxyl group react with carbohydrates. Plant extracts gave antioxidant activities. Based on this activity, the study evaluated the ability of the extract for the protection of DNA from UV/$H_2O_2$ effects.

### *3.1.3 DNA damage prevention and cytotoxicity*

Different amounts of dry extracts showed similar protection levels, while there was no band of DNA exposed in UV/$H_2O_2$ (**Figure 3**). **Figure 3** shows the loss of DNA under UV effect, but still remained in plant treatments.

DNA damaged by UV was recovered because ginsenoside accelerated nucleotide excision repair (NER). NER is the important pathway that mammals can use to eliminate DNA lesions [20]. Rg2 contributed to p53 upregulation, but the mode of activation had not been reported [21]. Upregulation of p53 caused posttranslational modification that includes targeting murine double minute 2 (MDM2). MDM2 is an E3 ubiquitin ligase. MDM2 has oncogenic activities that might be dependent and in-dependent on p53. MDM2 was increased in many kinds of human cancer. Another protein also involves p53 that is wild-type p53-induced phosphatase 1 (WIP1) [22]. Moreover, the antioxidant property of ginseng was also reported to be involved in cell cycle arrest. Ginsenosides arrest the cell cycle via the c-Myc/p53 pathway [23]. The Myc gene is one of the oncogenes relating a nuclear protein involved in p53 expression and cell cycle. It is also reported that ginsenoside Rg3 reduces DNA damage by alkylating agents in normal cells [24]. In this study, tubulin was selected as the target protein to investigate different mechanisms in cell cycle arrest. Beside these tumor proteins, $O^6$-methylguanine-DNA methyl transferase (MGMT) was also used to study on the interaction with ligands. Ligand-protein complexes were investigated on potential inhibition on MDM2, a primarily negative regulator of p53 and MGMT.

Additionally, tubulin beta class I (TUBB) protein belonging to globular protein is the major component of the cytoskeleton. It takes part in many important processes including cell mobility, cellular transport, and DNA segregation (i.e., mitosis). Residues involved in the inhibitor-binding site of TUBB validated by taxol molecule are Val23, Asp26, Leu217, Leu219, His229, Leu230, Ala233, Ser236, Phe272, Pro274, Leu275, Thr276, Ser277, Arg278, Pro260, Arg369, Gly370, and Leu371. Among them, the

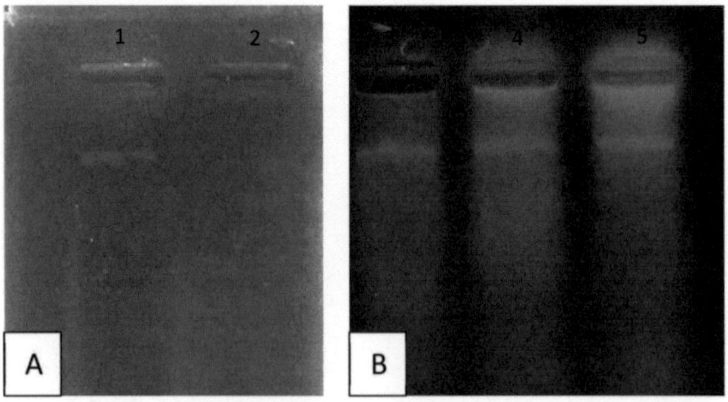


**Figure 3.**
*DNA damaged protection of* C. derserticola *extracts. (A) DNA before and after UV treatment; (B) DNA protection in extracts.*

| Rank | Ligands | Protein target | Binding energy value (kcal/mol) | Residues involved in interaction |
|---|---|---|---|---|
| 1 | Ginsenoside Rh4 | TUBB | −8.12 | Val23, Asp26, His229, Ala233, Phe272, Pro274,Arg284, Leu286, and Leu371 |
| 2 | Ginsenoside Rh2 | TUBB | −6.34 | Asp26, Leu217, Lys218, Leu219, Phe272, Arg278,Gly370, and Leu371 |
| 3 | Ginsenoside Rg5 | TUBB | −6.17 | Glu22, Asp26, His229, Pro274, Arg284, Leu286, and Leu371 |
| 4 | Panaxydol | TUBB | −6.0 | Phe272, Leu275, Thr276, Pro360, Arg369, Gly370, and Leu371 |
| 5 | Falcarindiol | TUBB | −5.99 | Leu217, His229, Leu230, Pro274, Leu275, Thr276,Arg284, Leu286, and Leu371 |
| 6 | Ginsenoside Rg3 | TUBB | −5.67 | Glu22, Asp26, His229, Thr276, Leu286, Leu371, andLys372 |
| 7 | Quercetin | TUBB | −4.61 | Phe272, Pro274, Thr276, Pro360, Leu371, and Gly370 |
| 8 | Panaxynol | TUBB | −4.36 | Leu217, His229, Leu230, Ala233, Phe272, Leu275,Ser277, and Arg278 |

**Table 1.**
*Binding energy of compounds docked into tubulin beta protein.*

residues Pro274, Leu275, Thr276, Ser277, and Arg278 are claimed to be important for the taxol molecule. The selected constituents showed interaction to maximum two out of five aforementioned residues. Docking results showed that all the ligands mainly displayed hydrophobic interaction with the active residues. Only four ginsenosides could display hydrogen bonds with Asp26, mimicking the interaction between Taxotere and TUBB (**Table 1**, **Figure 4**). Ginsenoside Rh4 had the lowest binding energy as it showed three H-bonds with both Asp26 and Pro274. Despite occupying four H-bonds including Asp26 and Thr276, ginsenoside Rg3 was ranked 6 as it lacked hydrophobic interaction to stabilize. Due to the limit of hydrogen bonds, PN constituents showed weak interaction with TUBB featuring protein-ligand complexes that may be reversible. This confirmed PN compounds only temporarily halt the cell cycle allowing DNA to repair.

MGMT, commonly known as the suicide repair enzyme, plays a pivotal role in maintaining genome stability. MGMT repairs DNA-alkylated lesions at the $O^6$ position of guanine. Such lesions may lead to DNA double-strand breaks or DNA mismatch resulting in mutations. The binding pocket of this enzyme is located at Tyr114, Pro138, Pro140, Cys145, Val148, Asn157, Tyr158, Ser159, and Lys165. The residue Cys145 is considered an important site for the occurrence of alkyl transfer. Only quercetin and the ligand group of polyacetylene could form hydrogen bonds with Cys145, whereas the ginsenoside showed H-interaction mainly with Ser159. Plus, the ginsenosides Rh2 and Rh4 were ranked second and third, respectively, but they only occupied one or two orientations with key residues, leaving most of the pocket empty. Although quercetin possessed the best binding energy, it created an unfavorable connection with Val128.

MDM2 is a regulator suppressing the expression of p53 protein [25]. At homeostasis, p53 is maintained at a low level via the MDM2-p53 complex (**Table 2**, **Figure 5**). This complex is mediated by hydrophobic interaction via three key residues of p53,

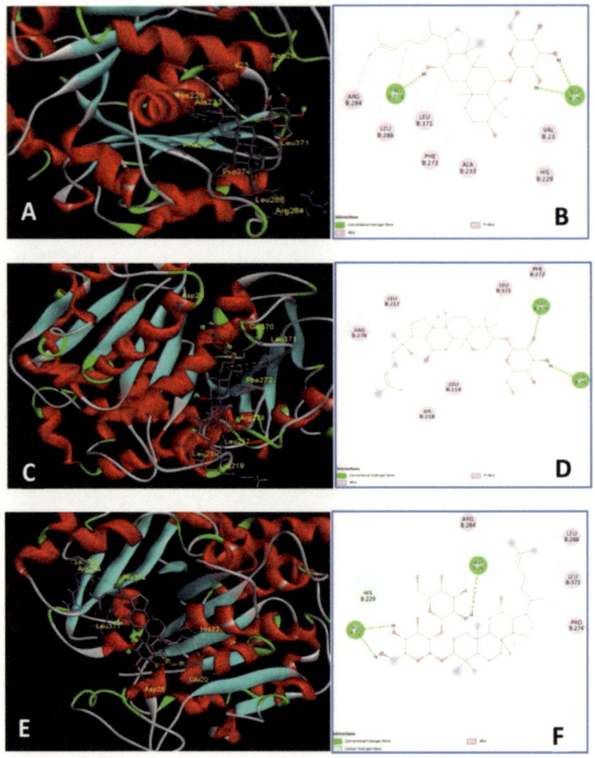

**Figure 4.**
*Best docking conformations and interactions between TUBB (shown in colored cartoon mode) and ginsenoside Rh4, Rh2, and Rg5 (gray stick), respectively. The 3D illustration demonstrated the interaction of TUBB with ginsenoside Rh4 (A), Rh2 (C), and Rg5 (E). 2D diagrams depicted interaction types of ginsenoside Rh4 (B), Rh2 (D), and Rg5 (F) and the active sites of TUBB.*

| Rank | Ligands | Protein target | Binding energy value (kcal/mol) | Residues involved in interaction |
|---|---|---|---|---|
| 1 | Ginsenoside Rh2 | MDM2 | −8.16 | Gln24, Lys51, Leu54, Leu57, Ile61, Met62,Tyr67, Val93, and Ile99 |
| 2 | Ginsenoside Rh4 | MDM2 | −7.64 | Lys51, Leu54, Leu57, Ile61, Met62, Tyr67, Ala69,and Val93 |
| 3 | Panaxydol | MDM2 | −6.3 | Lys51, Leu54, Leu57, His96, and Ile99 |
| 4 | Ginsenoside Rg5 | MDM2 | −6.23 | Gln24, Lys51, Met62, Tyr67, Val93, and Tyr100 |
| 5 | Falcarindiol | MDM2 | −5.57 | Ile19, Lys51, Leu54, Leu57, His96, Ile99, and Tyr100 |
| 6 | Panaxynol | MDM2 | −5.3 | Ile19, Lys51, Leu54, Leu57, Phe91, Ile99, His96and Tyr100 |

**Table 2.**
*Binding energy of compounds docked into murine double minute 2 protein.*

namely, Phe19, Trp23, and Leu26. Inhibition of this negative regulator leads to the stability of p53 by mimicking p53 hydrophobic interaction with MDM2. Residues of MDM2 involved in the p53-MDM2 complex are Leu54, Leu57, Ile61, Met62, Gln72,

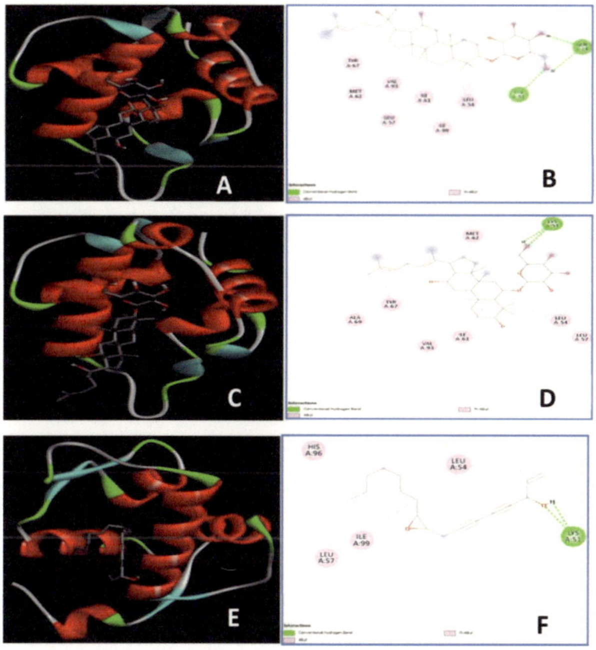


**Figure 5.**
*Best docking conformations and interactions between MDM2 (shown in colored cartoon mode) and ginsenoside Rh2, Rh4, and panaxydol (gray stick), respectively. The 3D illustration demonstrated the interaction of MGMT with ginsenoside Rh2 (A), Rh4 (C), and panaxydol (E). 2D diagrams depicted interaction types of ginsenosideRh2 (B), Rh4 (D), panaxynol (F) and the active sites of MDM2.*

His73, Val75, Val93, His96, Ile99, and Tyr100. MDM2 binding pocket can be divided into one large and main pocket Phe19/Trp23 and the subpocket Leu26. One ring of Rh2 formed hydrophobic bonds with Leu54, Leu57, Ile61, Met62, Val93, and Ile99 occupying pocket Phe19/Trp23. It also showed an H-interaction with Gln24 and a hydrophobic bond with Ile99 which was similar to Leu26 subpocket. Ginsenoside Rh4 ranked the second displayed linkage and interactions to the main Phe19/Trp23 pocket but left the subpocket Leu26 mostly empty (no interactions were found with Ile99). All the ginsenosides, flavonoids, and polyene ligands displayed strong affinity to MDM2; however, as this is an in silico test, further in vitro and in vivo experiments need to be performed to confirm this mode of interaction.

From the results in **Tables 1** and **2** and **Figures 4** and **5**, ginsenosides were predicted to interact with the oncogenes (MDM2 and TUBB). Then, via antioxidant activities of ginsenosides, these genes will suppress p53 and cell cycle [23, 24]. The prediction was also performed with MGMT to help with the effect of ginsenosides in DNA damage prevention and cancer treatment. Together with the roles of medicinal herbs in antioxidation relating to cancer treatment toward, probiotics are also discussed.

## 3.2 Probiotics

### *3.2.1 Probiotic concepts*

Lactic acid bacteria are important organisms honored for their fermentative capability as well as their health and nutritive benefits. They produce colorful composites similar to organic acids, diacetyl, hydrogen peroxide, and bacteriocin

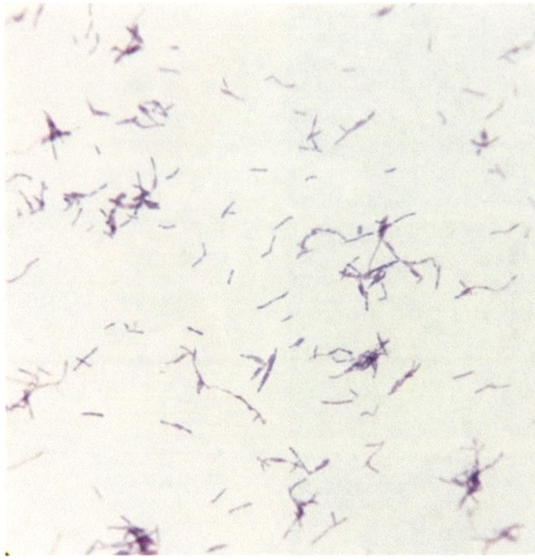

**Figure 6.**
*The morphology of Bifidobacterium breve ATCC 15700 after gram-staining.*

or bactericidal proteins during lactic restlessness. The lactic acid bacteria used commonly are *Bifidobacterium* strains. The *Bifidobacterium* genus belongs to the phylum Actinobacteria, class Actinobacteria, order Bifidobacteriales, and family Bifidobacteriaceae [26]. A more detailed description of *Bifidobacterium* genus is identified as a genus of Gram-positive bacteria, nonmotile [27]. Cells often stain irregularly with methylene blue and often branched anaerobic bacteria species (**Figure 6**), except *Bifidobacterium psychrophilum*, *Bifidobacterium scardovii*, and *Bifidobacterium tsurumi* tense which can tolerate $O_2$ [28]. To date, there are 80 species (sub-species) in the genus *Bifidobacterium*, and this number is growing every year [29].

To date, bifidobacteria are regulated as dietary supplements and specific products. These species are also called as "generally recognized as safe" (GRAS) microorganisms [30]. A number of clinical studies claim that *Bifidobacterium breve* and other species (sub-species) in the genus *Bifidobacterium* can induce multiple health-promoting effects on its host through a variety of metabolites, including vitamins, polyphenols, conjugated linoleic acid, and short-chain fatty acids (SCFAs), proteins, or peptides [31].

Besides, the genera *Bifidobacterium*, *Lactobacillus*, *Leuconostoc*, *Pediococcus*, *Enterococcus*, and so on were used in food fermentation [32]. Only a few strains of *Lactobacillus* are used mainly in pseudomembrous colitis prevention such as *Lactobacillus acidophilus*. They produce acid, aroma, exopolysaccharides, proteins, and a variety of metabolites [32]. Therefore, these bacteria can involve antioxidant activity, DNA damage, and cancer prevention in a certain.

To find out the point of view, the study showed the antioxidant activity, DNA damage, and cancer prevention of *Bifidobacterium breve* ATCC 15700, *Lactobacillus rhamnosus* PN04, and *Lactococcus lactis* PN05 as representatives besides medicinal plants.

### *3.2.2 Representative probiotics*

This chapter gave the research as an illustration of lactic acid bacteria benefits, especially the antioxidant activities involving DNA damage protection

| Components | | % Survival | |
|---|---|---|---|
| | | CE (20 mg/mL) | |
| | | $H_2O_2$ (1 mM) + HU (40 mM) | $H_2O_2$ (1 mM) + HU (100 mM) |
| Case 1 | (hH)1 | 78.1003 ± 1.4394 | 54.2366 ± 1.4394 |
| | (hH)2 | 61.5917 ± 1.9819 | 47.5754 ± 0.79 |
| | (hH)3 | 67.8872 ± 2.8716 | 41.8788 ± 2.9496 |
| Case 2 | (hH)1E | 67.214 ± 0.6169 | 53.8313 ± 0.0353 |
| | (hH)2E | 69.2042 ± 0.2447 | 45.266 ± 0.3657 |
| | (hH)3E | 68.4472 ± 2.5178 | 38.3967 ± 1.35927 |
| Case 3 | (hHE)1 | 68.1475 ± 1.2307 | 65.5729 ± 1.4192 |
| | (hHE)2 | 67.9151 ± 0.273 | 59.2573 ± 0.6829 |
| | (hHE)3 | 69.9547 ± 2.7488 | 69.5909 ± 1.1288 |
| Case 4 | E(hH)1 | 79.5539 ± 2.7549 | 68.9468 ± 0.9499 |
| | E(hH)2 | 74.7305 ± 0.7937 | 68.8953 ± 0.7768 |
| | E(hH)3 | 67.5318 ± 1.5285 | 40.75504 ± 0.217604735 |

*Case 1:* S. cerevisae *was treated in hydroxyurea (HU) and $H_2O_2$ in 1, 2, and 3 hours. Case 2: extracts were added into yeast after adding $H_2O_2$ – HU (100mM) and incubated at 1 h named (hH)1; 2 h named (hH)2; and 3 h named (hH)3. Case 3: extract, HU, and $H_2O_2$ were added in yeast culture and incubated in 1 h, 2 h, and 3 h. Case 4: extract was incubated with yeast for 1, 2, and 3 h and then added with HU and $H_2O_2$ followed more incubation.*

**Table 3.**
*Survival percentage of* Saccharomyces cerevisiae *in oxidative stress and antioxidant effects.*

and cytotoxicity toward anticancer activity. **Table 3** shows the survival rate of *Saccharomyces cerevisiae* (*S. cerevisiae*) compared to negative control sample (untreated sample). Based on results, hydroxyurea (HU) has a significant effect at a concentration of 100 mM. Besides, incubation time had influence on the protection ability of *B. breve* protein extracts. All extracts provided better DNA protection when better DNA protection when mixed with *S. cerevisiae* for 1–2 hours before treatment with $H_2O_2$ – HU complex. However, longer incubation time did not change the percentage of survival of *S. cerevisiae*. *Bifidobacterium* could prevent DNA damaged by hydroxyurea via antioxidant activity. The study performed in *S. cerevisae* which could play *in vivo* test which can be replaced for working in mice. The study also did *in vitro* test on DNA as **Table 3**. As a result, plant and LAB could prevent DNA damage via antioxidant agents.

When comparing with case 1, the survival rate was significant higher. Furthermore, gradually increasing incubation time resulted in increased survival (53.8314 ± 0.0353% to 57.9562 ± 1.5168% when extracts were added and incubated at 1–2 hours after adding $H_2O_2$ – HU (100 mM) (case 2) and 65.5729 ± 1.4192% to 69.5909 ± 1.1288% when adding extracts and $H_2O_2$ – HU (100 mM) at the same time (case 3). For case 4, extract was added in yeast culture to protect the yeast survival capacity, and then the mixture was contacted with $H_2O_2$ – HU (40 mM and 100 mM). It seemed that the yeast survival rate was improved than case 1, case2, and case 3 during the 2-hour incubation. Especially, $H_2O_2$ – HU (100 mM) caused the lower survival rates, but case 4 could improve survival rate when comparing to cases 1 and 2. However, the survival rate was low in case 4 during the 3-hour incubation. Probably, any active component of bacteria was degraded by yeast during longer incubation.

Actually, the incubation time of the extract in case 4 was longer than 3 h because the extract was incubated with yeast before and then prolonged 3 h incubation in HU and $H_2O_2$. Anyway, the bacterial extract could prevent the toxicity of HU and $H_2O_2$ by means of the antioxidant activity.

The protection of incentive cells by the expert under the effect of $H_2O_2$ – HU is harmonious with the compliances above, suggesting the possible part of the excerpt in the junking of hydrogen peroxide ($H_2O_2$) revolutionaries. This supports the excerpt's antioxidant exertion by demonstrating its part in guarding cells from DNA dangerous chemicals that the release of OH- revolutionaries from $H_2O_2$ – HU complex is the main reason for converting DNA damage in cells [33, 34]. The protection of Bifidobacteria for prokaryotic and eukaryotic cells involved in numerous DNA mismatch form (DMR) protein. DMR's top function is to ameliorate replication dedication by repairing replication-associated base-base mismatches. Modulating DNA recombination and easing DNA damage signaling are significant primary functions [35]. DMR blights are connected with cancer vulnerability, including heritablenon-polyposis colorectal cancer, resistance to certain chemotherapeutic agents, and anomalies in meiosis and sterility in mammalian systems [36]. Inactivation of DMR in cells causes sporadic malice in humans. In response to some forms of DNA damage, the DMR system is necessary for cell cycle arrest and/or programmed cell death. This capability was inactivated when exposed to high attention of $H_2O_2$ (1 mM). As a result, when adding HU– a medicine applied to treat neoplastic conditions but has a side effect of inhibits DNA conflation, *S. cerevisiae* vitality was being dropped. Former study showed that *S. cerevisiae* mutants lacking CuZnSOD have been shown to be hypersensitive to hypertonic medium and to parade elevated oxidative damage [37]. This study aimed to estimate the capability of the excerpt to recover damage to *S. cerevisiae* under stressors and to estimate the antioxidant capacity. *S. cerevisiae* came weak and unable of resistance to DNA mutagens called hydroxyurea (HU). Numerous exploration studies have concentrated on anticancer and antioxidative parcels as defensive adjuncts against a host of conditions [38]. Although colorful synthetic agents parade anticancer and antioxidative conditioning, pointing the agents which are not only precious, but stable and safe is important. Multitudinous reports have indicated that both LAB and investigated milk as anticancer goods [39]. Data from epidemiological and experimental studies have also indicated that the ingestion of certain LAB strains, or of fermented dairy products, might palliate the threat of certain type of cancers and inhibit the growth of excrescences. In an *in vitro* environment, the anticancer and antioxidative parcels of heat-killed (HK) cells and fractionations of HK cells of several *Lactobacillus* strains.

Lactic acid bacteria including *Lactobacillus acidophilus* and *Lactobacillus casei* to test the apoptosis of human cancer colon cell line [40, 41]. In recent years, there have been many studies using animal models, which have clearly demonstrated a protective effect of dietary supplements of LAB against colon tumor development [42]. The antioxidant activity of the extracts of the culture of *Lactobacillus rhamnosus* PN04 and *Lactococcus lactis* PN05 showed the DPPH (2,2′-diphenyl-1-picrylhydrazyl radical) scavenging assay [43]. The effect of two LAB strains to cancer cell lines is fairly similar. Indeed, the stationary phase produces the highest amount of antioxidant activity (**Table 4**). So, only culture supernatant and polysaccharide at this phase were performed test antitumor activity on HeLa and Hep G2 cells. Depending on the different effect mechanism to cancer cell lines, the cytotoxicity percentage of each sample is

| Growth phase | % DPPH scavenging activity of culture supernatant |
|---|---|
| Early exponential phase | 93.17 ± 0.075 |
| Late exponential phase | 94.33 ± 0.117 |
| Stationary phase | 94.69 ± 0.075 |
| Death phase | 94.31 ± 0.075 |

**Table 4.**
*Antioxidant activities of Lactobacillus rhamnosus PN04 in culture phases.*

also different. For example, culture supernatant of *Lactobacillus rhamnosus* PN04 (64.40) was more effective than polysaccharide (58.78) on Hep G2 cells. However, there is a converse on HeLa cells with values of 46.99 and 86.16 respectively. Similarly, the culture supernatant of *Lactococcus lactis* PN05 (66.09) was more effective than polysaccharide (50.36) on Hep G2 cells. However, there is a converse on HeLa cells with values of 34.60 and 86.86, respectively. These results indicated that both culture supernatant and polysaccharide of *Lactobacillus rhamnosus* PN04 and *Lactococcus lactis* PN05 potently inhibited the viability of each cell line. Consequently, lactic acid bacteria can show cytotoxicity due to antioxidant activities. Lactic acid bacteria can protect DNA based on antioxidant action, leading to anticancer activity prospectively. The curious question is whether lactic acid bacteria can involve DNA repair/protection that makes lactic acid bacteria kill cancer cells, but not normal cells. The study brings out the important effects of antioxidant activities for health prevention that should be clarified more.

Consequently, antioxidant activities of probiotics/lactic acid bacteria could involve in DNA damage protection. Many metabolites were produced in the bacteria; however, exopolysaccharides were studied most in cytotoxicity. This activity of lactic acid bacteria was like other species such as seaweeds [44]. It is undoubtable. Therefore, the lactic acid bacteria could be exploited for cancer therapy development by finding their antioxidants in the future.

## 4. Conclusion

Antioxidant conditioning can be detected in both medicinal plants and lactic acid bacteria. Antioxidant agents could prevent cytotoxicity and DNA damage. This chapter showed the antioxidant activities of daily vegetables such as *Basella alba* or *Cistanche* sp., a source of functional food together with a few representative probiotics such as *Bifidobacterium breve* ATCC 15700, *Lactobacillus rhamnosus* PN04, and *Lactococcus lactis* PN05 which could be served in DNA damage protection and cytotoxicity oriented for cancer treatment. The hydroxyl-free revolutionaries are known to beget damage to cellular DNA. UV photolysis of $H_2O_2$ is responsible for utmost oxidative damage to proteins and DNA. The study mentioned the representatives as the daily food or functional food. Thus, medicinal plants and probiotics for damaged DNA protection and cytotoxicity toward cancer treatment should be focused more. This chapter recommends that medicinal plants and lactic acid bacteria in health forestallment should be exploited more for DNA damage prevention and cancer treatment based on antioxidant activities.

## Acknowledgements

The study was developed by private fund.

## Conflict of interest

The authors declare no conflict of interest.

## Author details

Tu Nguyen[1,2]

1 School of Biotechnology, International University, Hochiminh City, Vietnam

2 Vietnam National University, Hochiminh City, Vietnam

*Address all correspondence to: nhktu@hcmiu.edu.vn

## References

[1] He L, He T, Farrar S, Ji L, Liu T, Ma X. Antioxidants maintain cellular redox homeostasis by elimination of reactive oxygen species. Cell Physiology and Biochemistry. 2017;**44**(2):532-553

[2] Cao L, Xu X, Cao LL, Wang RH, Coumoul X, Kim SS, et al. Absence of full-length Brca1 sensitizes mice to oxidative stress and carcinogen-induced tumorigenesis in the esophagus and forestomach. Carcinogenesis. 2007;**28**:1401-1407

[3] Mahmoud AM, Wilkinson FL, Sandhu MA, Dos Santos JM, Alexander MY. Modulating oxidative stress in drug-induced injury and metabolic disorders: The role of natural and synthetic antioxidants. Oxidative Medicine and Cellular Longevity. 2019;**2019**:3206401

[4] Conrad M, Kagan VE, Bayir H, et al. Regulation of lipid peroxidation and ferroptosis in diverse species. Genes and Development. 2018;**32**(9-10):602-619

[5] Knecht KT, Bradford BU, Mason RP, Thurman RG. In vivo formation of a free radical metabolite of ethanol. Molecular Pharmacology. 1990;**38**(1):26-30

[6] Collins AR. Oxidative DNA damage, antioxidants, and cancer. BioEssays. 1999;**21**(3):238-246

[7] Rahaman MM, Hossain R, Herrera-Bravo J, et al. Natural antioxidants from some fruits, seeds, foods, natural products, and associated health benefits: An update. Food Science & Nutrition. 2023;**11**(4):1657-1670

[8] Deshmukh SA, Gaikwad DK. A review of the taxonomy, ethnobotany, phytochemistry and pharmacology of *Basella alba* (Basellaceae). Journal of Applied Pharmaceutical Sciences. 2014;**4**(01):153-165

[9] Güder A, Korkmaz H. Evaluation of in-vitro antioxidant properties of hydroalcoholic solution extracts *Urtica dioica* L., Malva neglecta Wallr. and their mixture. Iranian Journal of Pharmaceutical Research. 2012;**11**(3):913-923

[10] Song Y, Zeng K, Jiang Y, Tu P. Cistanches Herba, from an endangered species to a big brand of Chinese medicine. Medicinal Research Reviews. 2021;**41**(3):1539-1577

[11] Huang J, Zhao D, Cui C, Hao J, Zhang Z, Guo L. Research progress and trends of phenylethanoid glycoside delivery systems. Food. 2022;**11**(5):769

[12] Sui ZF, TingMin G, Liu B, Peng SW, Zhao ZL, Li L, et al. Water-soluble carbohydrate compound from the bodies of Herba Cistanches: Isolation and its scavenging effect on free radical in skin. Carbohydrate Polymers. 2011;**85**(1):75-79

[13] Salekdeh GH, Siopongco J, Wade LJ, Ghareyazie B, Bennett J. Proteomic analysis of rice leaves during drought stress and recovery. PROTEOMICS: International Edition. 2022;**2**(9):1131-1145

[14] Lü JM, Jiang J, Jamaluddin M, Liang Z, Yao Q, Chen C. Ginsenoside Rb1 blocks ritonavir-induced oxidative stress and eNOS downregulation through activation of estrogen receptor-beta and upregulation of SOD in human endothelial cells. International Journal of Molecular Sciences. 2019;**20**(2):294

[15] Oliynyk S, Oh S. Actoprotective effect of ginseng: Improving mental and

physical performance. Journal of Ginseng Research. 2023;**37**(2):144-166

[16] Zhou Q, Jiang L, Xu C, et al. Ginsenoside Rg1 inhibits platelet activation and arterial thrombosis. Thrombosis Research. 2014;**133**(1):57-65

[17] Salehi B, Venditti A, Sharifi-Rad M, Kręgiel D, Sharifi-Rad J, Durazzo A, et al. The therapeutic potential of apigenin. International Journal of Molecular Sciences. 2019;**20**(6):1305

[18] Villa-Rodriguez JA, Kerimi A, Abranko L, et al. Acute metabolic actions of the major polyphenols in chamomile: An in vitro mechanistic study on their potential to attenuate postprandial hyperglycaemia. Scientific Reports. 2018;**8**(1):5471

[19] Lim R, Barker G, Wall CA, Lappas M. Dietary phytophenols curcumin, naringenin and apigenin reduce infection-induced inflammatory and contractile pathways in human placenta, foetal membranes and myometrium. Molecular Human Reproduction. 2013;**19**(7):451-462

[20] Zhou Z, Zhang Y, Lin L, Zhou J. Apigenin suppresses the apoptosis of H9C2 rat cardiomyocytes subjected to myocardial ischemia-reperfusion injury via upregulation of the PI3K/Akt pathway. Molecular Medicine Reports. 2018;**18**(2):1560-1570

[21] Cai BX, Jin SL, Luo D, Lin XF, Gao J. Ginsenoside Rb1 suppresses ultraviolet radiation-induced apoptosis by inducing DNA repair. Biological and Pharmaceutical Bulletin. 2009;**32**(5):837-841

[22] Chung YH, Jeong SA, Choi HS, Ro S, Lee JS, Park JK. Protective effects of ginsenoside Rg2 and astaxanthin mixture against UVB-induced DNA damage. Animal Cells and Systems. 2018;**22**(6):400-406

[23] Ladds M, Laín S. Small molecule activators of the p53 response. Journal of Molecular Cell Biology. 2019;**11**(3):245-254

[24] He NW, Zhao Y, Guo L, Shang J, Yang XB. Antioxidant, antiproliferative, and pro-apoptotic activities of a saponin extract derived from the roots of *panax notoginseng* (Burk.) F.H. Journal of Medicinal Food. 2012;**15**(4):350-359

[25] Li T, Sun W, Dong X, et al. Total ginsenosides of Chinese ginseng induces cell cycle arrest and apoptosis in colorectal carcinoma HT-29 cells. Oncology Letter. 2018;**16**(4):4640-4648

[26] Zhang YH, Li HD, Li B, Jiang SD, Jiang LS. Ginsenoside Rg3 induces DNA damage in human osteosarcoma cells and reduces MNNG-induced DNA damage and apoptosis in normal human cells. Oncology Reports. 2014;**31**(2):919-925

[27] Hidalgo-Cantabrana C, Delgado S, Ruiz L, Ruas-Madiedo P, Sánchez B, Margolles A. Bifidobacteria and their health-promoting effects. Microbiology Spectrum. 2017;**5**(3):10

[28] Alessandri G, van Sinderen D, Ventura M. The genus bifidobacterium: From genomics to functionality of an important component of the mammalian gut microbiota running title: Bifidobacterial adaptation to and interaction with the host. Computational and Structural Biotechnology Journal. 2021;**19**:1472-1487

[29] Whitman WB, Rainey F, Kämpfer P, Trujillo M, Chun J, DeVos P, et al., editors. Bergey's Manual of Systematics of Archaea and Bacteria. 1st ed. Wiley; 2015

[30] Turroni F, Duranti S, Milani C, Lugli GA, van Sinderen D, Ventura M. *Bifidobacterium bifidum*: A key member of the early human gut microbiota. Microorganisms. 2019;**7**(11):544

[31] Petersen AB, Gniadecki R, Vicanova J, Thorn T, Wulf HC. Hydrogen peroxide is responsible for UVA-induced DNA damage measured by alkaline comet assay in HaCaT keratinocytes. Journal of Photochemistry and Photobiology. B, Biology. 2000;**59**(1-3):123-131

[32] Wang Y, Wu J, Lv M, Shao Z, Hungwe M, Wang J, et al. Metabolism characteristics of lactic acid bacteria and the expanding applications in food industry. Frontiers in Bioengineering and Biotechnology. 2021;**9**:612285

[33] Koziol S, Zagulski M, Bilinski T, Bartosz G. Antioxidants protect the yeast *Saccharomyces cerevisiae* against hypertonic stress. Free Radical Research. 2005;**39**(4):365-371

[34] Fakruddin M, Hossain MN, Ahmed MM. Antimicrobial and antioxidant activities of *Saccharomyces cerevisiae* IFST062013, a potential probiotic. BMC Complementary and Alternative Medicine. 2017;**17**(1):64

[35] Golla U, Bhimathati SS. Evaluation of antioxidant and DNA damage protection activity of the hydroalcoholic extract of *Desmostachya bipinnata* L. Stapf. Scientific World Journal. 2014;**2014**:215084

[36] Garbacz K. Anticancer activity of lactic acid bacteria. Seminars in Cancer Biology. 2022;**86**(Pt 3):356-366

[37] Davies BW, Kohanski MA, Simmons LA, Winkler JA, Collins JJ, Walker GC. Hydroxyurea induces hydroxyl radical-mediated cell death in *Escherichia coli*. Molecular Cell. 2009;**36**(5):845-860

[38] Li GM. Mechanisms and functions of DNA mismatch repair. Cell Research. 2008;**18**(1):85-98

[39] Stojic L, Brun R, Jiricny J. Mismatch repair and DNA damage signalling. DNA Repair (Amst). 2004;**3**(8-9):1091-1101

[40] Zhong L, Zhang X, Covasa M. Emerging roles of lactic acid bacteria in protection against colorectal cancer. World Journal of Gastroenterology. 2014;**20**(24):7878-7886

[41] Baldwin C, Millette M, Oth D, Ruiz MT, Luquet FM, Lacroix M. Probiotic *Lactobacillus acidophilus* and *Lactobacillus casei* mix sensitize colorectal tumoral cells to 5-fluorouracil-induced apoptosis. Nutrition and Cancer. 2010;**62**(3):371-378

[42] Kim SH, Moon JY, Lim YJ. Dietary intervention for preventing colorectal cancer: A practical guide for physicians. Journal of Cancer Prevention. 2022;**27**(3):139-146

[43] Nguyen DT, Nguyen TH. Detection on antioxidant and cytotoxicity activities of exopolysaccharides isolated in plant-originated *Lactococcus lactis*. Biomedical and Pharmacology Journal. 2014;**7**(1):33-38

[44] Pradhan B, Bhuyan PP, Ki JS. Immunomodulatory, antioxidant, anticancer, and pharmacokinetic activity of Ulvan, a seaweed-derived sulfated polysaccharide: An updated comprehensive review. Marine Drugs. 2023;**21**(5):300

Chapter 5

# From Bees to Wellness: Exploring Nutraceutical Properties and Health-Boosting Potentials of Portuguese Honeys

*Soraia Santos, Ana Abrão, Irene Gouvinhas and Ana Novo Barros*


## Abstract

Honey is a natural sweet substance produced from secretions of living parts of plants or excretions of nectar by *Apis mellifera* bees. There is a diversity of possible botanical sources responsible for the different types of honey. Portugal, in particular, is renowned for its abundant and varied melliferous flora, influenced by geographical regions and climatic conditions, resulting in the production of distinct monofloral and multifloral varieties. These honeys are intricately linked with their organoleptic characteristics and chemical composition, including phenolic compounds. These phytochemicals are secondary metabolites that are commonly responsible for diverse biological properties, such as antioxidant, antimicrobial, and anti-aging activities, and its multifaceted applicability extends across several industries, thus underscoring its pivotal role in the economic landscape. As a result of these benefits, the scientific community focuses its attention on this remarkable food matrix. Regrettably, honey production has witnessed a decline in numerous countries, including Portugal. This concerning trend has attracted the attention of researchers seeking to understand the underlying factors and explore potential solutions. This chapter will present a comprehensive review of Portuguese honey and its production process, focusing on chemical composition and biological properties. It will also highlight the influence of geographical origin and botanical sources.


**Keywords:** honey, chemical composition, phenolic compounds, nutraceutical properties, biological activities, botanical source, geographical origin

## 1. Introduction

Honeybees play a crucial role in the environment, serving as vital contributors to biodiversity and essential agents of pollination for a wide variety of crops, affecting the quality and quantity of several fruits and seeds [1, 2]. In this sense, they contribute to the genetic diversity, ensuring the natural plant resources required for the sustenance of ecosystems on a global scale [3]. It is also important to highlight that the absence of honeybee pollination would result in a staggering 90% reduction in the production of numerous fruits [4]. Consequently, this pollinating agent, of enormous relevance for agriculture and ecology, assumes an unparalleled position within human culture [5].

Beekeeping is the art and science of breeding and managing swarms of bees in artificial hives, yielding substantial economic and cultural advantages. This ancient tradition enables human contribution in the increase of biodiversity and food production, thereby embodying a noteworthy exemplar of an environmentally sustainable production model. Its significance extends to the realms of biodiversity preservation, agriculture, and the food industry, underscoring its indispensable role in fostering ecological equilibrium and ensuring the availability of nourishment for future generations [1, 6, 7].

This activity is the establishment and diligent upkeep of colonies of social bees belonging to the genus *Apis*, for which the beekeeper obtains profit. While honey stands as the primary product derived from these carefully tended beehives, this endeavor also yields a diverse range of valuable co-products. These include pollen, propolis extract, royal jelly, beeswax, and even apitoxin (venom). Thus, beekeeping serves as a multifaceted venture, yielding not only the honey but also an array of valuable substances that find application in various industries, symbolizing the bounty that emanates from the world of bees [1, 7–10].

Beekeeping holds a rich historical legacy, making it part of many cultural heritages, where some of them have religious significance [11]. The use of honey by human species is recorded since the Stone Age [12]. Remarkably, the oldest written reference to honey can be found in ancient Egyptian records, originating as far back as 5500 BC [13]. However, the true emergence of beekeeping, as we recognize it today, occurred when humans acquired the knowledge and skills to construct specialized structures that provided a protected environment for bee colonies, allowing for meticulous care and control of the beehives. This practice, defined as the maintenance of honeybees in purpose-built hives and its comprehensive management, signified a pivotal advancement in the preservation and future sustainability of these remarkable creatures [10].

Until the industrial production of sugar began after the eighteenth century, honey was an important carbohydrate source and the only available natural sweetener [12], being also highly valued for its diverse medical potential [11]. In the twentieth century, its significance was further highlighted as a promising sugar substitute with nutraceutical properties [13–15].

According to several authors, there exists a variety of potential functional properties in honey, in particular the ones associated with health benefits and pharmaceutical applications, such as antimicrobial, anti-inflammatory, anticarcinogenic, antithrombotic, and antioxidant, analgesic, as well as the treatment of burns, gastrointestinal disorders, asthma, infected wounds, and skin ulcers [11, 16–20].

In summary, honey is a carbohydrate-saturated solution containing components that are essential for metabolism, with several nutraceutical properties. Its popularity has been increasing due to its health benefits, sweetness, and versatile applications in various industries. In fact, Óses and collaborators [21] referred to the fact that honey popularity is increasing as a sugar substitute, for example, in candies, confectionery, and bakery products, but it is also utilized in different types of foods to improve their bioactive properties or extend their shelf life, such as sausages, fruits, and vegetables, among others.

The European Union produces about 250,000 tonnes of honey per year, about half China's production (around 500,000 tonnes), making China the world's leading honey producer [22]. Particularly, Portugal produces several types of honey between 10,000 and 15,000 tonnes annually [22], being the production, a traditional practice well implemented in several regions [16]. In fact, according to the *Instituto Nacional de Estatística* (INE), the annual honey production in Portugal has been decreasing in the last few years. Consequently, the limited availability of this product leads to an increase in the commercial value of honey. Anklam [23] affirmed that this increase has provided a heightened interest in the adulteration of honey, making the authentication of honey an important concern [24].

In this sense, beekeeping holds remarkable relevance for the production of honey and for the secondary products originated from the hive, which provide economical returns. Beyond these aspects, beekeeping exerts a far-reaching influence on various sectors, encompassing social dimensions, rural development, and environmental preservation as important pollination agents [6, 12]. Moreover, the implications extend to global food security, as the decline or disappearance of pollinators, particularly honeybees, poses a profound threat. Pollination is an essential process for the sustenance and proliferation of plants and biodiversity, emphasizing the critical role played by honeybees in maintaining the delicate balance of our ecosystems and ensuring food security [6, 9].

## 2. Honey

Honey is established as a pure natural product by food regulations worldwide, and its definition is the major key to understanding this food product [25]. According to the Portuguese Decree-Law n° 214/2003 from 18th September, honey is a natural sweet substance produced by *Apis mellifera* bees from the nectar of plants or from secretions of living parts of plants, which bees collect, transform by combining with endogenous substances, deposit, dehydrate, store, and leave in honeycombs to ripen and mature. Also, according to APPENDIX I of this Decree-Law, the main types of honey can depend on the origin, being nectar honey or flower honey, when it is obtained by plants nectar, or honeydew, when it is obtained essentially by plant-sucking insects excretions that remain on the plants living parts or directly from the living parts plants secretions, or can depend on the mode of production and/or presentation, such as honey in combs, honey with pieces of combs, centrifuged honey, grained honey, pressed honey, or filtered honey. Generally, and also in accordance with the Portuguese law, it is established that honey is categorized into distinct types based on its origin. Blossom honey, derived from the nectar of flowering plants, is the most prevalent and widely consumed variety. On the other hand, honey can be categorized as honeydew honey, characterized by its higher disaccharide content and lower levels of monosaccharides [26]. Honeys sourced from both nectar and honeydew are collectively referred to as blend honeys, encompassing a range of intriguing flavor profiles and compositional attributes [27].

Furthermore, this product can be consumed unprocessed and is easily digested by humans [28]. However, it is a very complex product due to its organoleptic characteristics (color, aroma, flavor) and chemical composition, depending not only on the nectar but also on other factors such as geographical regions, climate, and honeybee species involved in the production. Processing, manipulation, packaging, and storage time are also factors that contribute to its characteristics [24, 29].

During honey commercialization, it is commonly classified into two types based on its botanical source: multifloral honey and monofloral honey [7, 29, 30].

Multifloral honey refers to honey produced from the nectar of various types of flowers in a way where neither one of the types predominates, after storage, resulting in a blend of flavors and aromas [31, 32], the most commercially available type [30]. Monofloral honey, on the other hand, is derived predominantly from the nectar of a single species of flower to an extent of at least 45%, offering a distinct and characteristic taste profile specific to that particular floral source. These classifications serve to distinguish the unique characteristics and flavor profiles of different types of honey available in the market [16, 28, 31]. This variability is harnessed for honey typification, serving as a means to enhance the value and distinctiveness of the product [33]. However, it is important to note that pure monofloral honey is rare, as honeybees typically gather nectar from multiple plant species within their foraging vicinity [34].

Within Europe, an extensive range of over one hundred plant species has been identified as capable of yielding monofloral honeys [35]. This particular type of honey is often analyzed, directing researchers attentions within the realm of scientific research, prompting numerous studies aimed at analyzing its distinct composition and properties [24]. The diverse array of compounds found within monofloral honey contributes to its unique and discernible characteristics, offering a fascinating exploration of the botanical origins and chemical profiles associated with each distinct monofloral variety.

### 2.1 Honey extraction: From beehive to packaging

Honeybees, as well as several insects, have a sophisticated colonial organization based on labor division among nestmates in numerous aspects to be cooperative in food acquisition and storage for the maintenance of the colony. Bees live in colonies that consist of three main types of bees: the drones, the queen bee, and worker bees (**Figure 1**) [36, 37].

The nutrition of a bee colony hinges upon the collection of two vital sources: nectar and pollen. Nectar, rich in carbohydrates, serves as an essential energy source for the bees. In contrast, pollen, a valuable source of protein, is diligently gathered and deposited within the honeycomb. When both substances are mixed in the honeycomb, it originates bee's bread, the main source of food for honeybee workers and larvae. Together, the collection and utilization of nectar and pollen form the cornerstone of the colony's nutrition, ensuring the energy and nutritional requirements essential for their survival and thriving [39–41].

To generate honey, foreign worker bees collect the nectar from the flowers through proboscises and storage in the crop (specific stomachs of honeybees). Endogenous

**Figure 1.**
*The three main types of bees in a colony (adapted from [38]).*

enzymes from the hypopharyngeal glands catalyze physicochemical transformations of nectar, where sucrose is inversed into two simple sugars, dextrose and levulose. Once the forager bees arrive at the beehive, it is delivered to storer bees, who then distribute to nestmates, a process known as trophallaxis. To obtain the honey, worker bees regurgitate the nectar into honeycomb cells with the enzymes, and in order to increase the sugar concentration, the water is eliminated. Then, the honeybees seal the honeycomb cells with beeswax to protect the honey, guaranteeing, simultaneously a reduced moisture content, which helps prevent spoilage [34, 36, 42, 43].

The unit operations in this product industry associated with honey extraction involve the removal of honey from combs to be isolated as a pure liquid, using modern practices such as harvesting and centrifugation [44]. The process starts with the remotion of the frames with combs from the beehive. Then, they are transported to an extraction room, where the surface layer of wax created by the bees to close the combs must be removed, with a knife, fork, or mechanically, depending on the size and production of the honey company [8, 9, 45]. Afterwards, the honey must be extracted from the combs, usually by centrifugation, where the frames filled with honey are fixed on radial supports attached to a central axis. The centripetal force makes the honey detach itself from the combs, being thrown against the wall of the extractor, from where it flows by gravity to the bottom of the equipment, where it is channeled for filtration [9, 45]. This process should be performed at a temperature between 25°C and 35°C, which is similar to the beehive [46]. The extracted honey should be strained to remove impurities, such as wax, bees, and other debris, and the process may be repeated more than once through finer mesh [8, 9, 45]. The honey, after extraction, is bottled in a closed container that can either be composed of iron or food-grade plastic, where the waste rises by density and the honey is cleaned on the bottom. Finally, in a specific cleaned room, usually, the honey is bottled into a glass jar before sale [45].

## 3. Botanical source and geographical origin of honey

Bees are continuously engaged in interactions with their environment, as they can cover an area of about 7 $km^2$. This expansive territory serves as their foraging grounds, allowing them to collect essential resources from different plants, which highlights their role as pollinators, facilitating the reproduction of plants, and, importantly, contributing to the balance of nature, sustaining biodiversity and ecosystem health [47]. As a consequence, the produced honey can have different and unique organoleptic and chemical characteristics. Therefore, the identification of honey's origin, both in terms of botanical and geographical attributes, is imperative. By discerning the specific botanical sources and geographical regions associated with a particular honey sample, a comprehensive understanding of its distinct qualities can be attained. Several approaches have been used to characterize honeys of particular botanical and geographical origins; however, nowadays, it has not yet been established one single and universal methodology, so multiple approaches can be used to understand the honey's sources [35, 48]. These methods aim to ensure transparency, traceability, and the preservation of the honey's unique characteristics, allowing consumers to make informed choices and fostering trust in the honey market.

The botanical origin of honey holds significant importance as it profoundly influences the properties and characteristics of this natural product [49]. The botanical sources from which monofloral honey originates contribute to the wide variability observed in terms of organoleptic and chemical composition [35, 50].

Concerning the organoleptic characteristics of the honey, the color varies from transparent to dark due to non-volatile compounds that include glycosides and phenolic compounds. Consequently, the colors depend on several factors, such as suspended particles, pigment composition, source, storage, and age, being the most common amber or yellow colored. However, some specific botanical sources provide other types of colors, for example, eucalyptus, which usually, confers different colors such as grayish, and chestnut, which offers reddish color, among others [14, 51].

Furthermore, various studies indicate that the medicinal properties found in honey are transmitted by the plants from which they are produced, making honey a vehicle for transporting these properties. In this sense, monofloral honey usually is in higher demand from the consumer and, in some cases, can be considered a premium product, which means that it has potentially higher commercial value for the producers than multifloral honeys [16, 28, 31]. Moreover, as the market for monofloral honeys expands, reliable methods for determining geographical origin and botanical authenticity are required [35].

Identifying the botanical origin of honey is challenging due to its complex matrix and content, which are impacted by soil conditions and climate [52]. However, it is important to improve the knowledge of the botanical origins to avoid commercial frauds [53]. The discrimination of monofloral honeys has traditionally been approached through physicochemical, melissopalynological, and sensory analysis [15, 30]. Nowadays, the most widely utilized methodology to determinate the botanical origin of honeys is pollen analysis, specifically melissopalynology [47, 54]. However, these analyses are protracted and require specialized experts with appropriate knowledge and experience in pollen morphology [55]. Within this methodology, the primary botanical material under investigation consists of pollen grains and fungal spores. To associate a particular species or group of plants, various characteristics are analyzed, such as size, type, color and grain number of pollen [56]. However, methodologies such as Fourier-Transform Infrared Spectroscopy (FTIR) also allow the identification of botanical origin, being a faster and more practical approach [57].

Also, different components of honey depend on the botanical source. This dependence is based on the nutrients present in the plants and their variability in the soil, influencing the quantity and variety of sugars, sugar ratio, proteins and amino acids, bioactive compounds, organic acids, minerals, among others [29, 50, 52, 58–60]. It is important to emphasize that within the various compounds present in honey, the bioactive compounds also depend on botanical sources and geographical location due to multiple factors, such as climatic conditions, soil composition, flora diversity, environmental conditions, and the contribution of beekeepers. Nevertheless, these elements can vary even if the honey comes from the same floral source and can change due to seasonal climatic variations [23]. In fact, it is reported that the phenolic profile of honeys originated from different botanical sources has differences, which suggests that this is a consequence of the natural variability of compounds in the respective plants [61].

In addition to the geographical origin, which is not widely regarded as a prominent consideration when discussing honeys derived from the same botanical source, multiple researchers have observed that the geographical area exerts an influence and potentially affects various physicochemical parameters. These parameters encompass water content, pH, conductivity, minerals, organic acids, and, also, sugar composition [18, 58, 60]. Whereas, in the 70s, according to Davies [62, 63], it was founded that the ratios between certain honey amino acids can differ, depending on geographical origin, as well as the antioxidant capacity that has been commonly associated to the honey botanical origin, as previously mentioned, but, in some cases, to its geographical origin

[30]. In contrast, according to Seraglio et al. [30], the acidity in honey has high variability, independently of the botanical source and geographic origin.

As previously referred, pollen analysis is also effective to distinguish honeys from different areas but laborious and time-consuming, which limits its application. In addition to this method, there are others such as stable isotope fingerprinting of bio-elements analysis and FTIR-ATR spectroscopies (Fourier-Transform Infrared Spectroscopy—Attenuated Total Refraction) accompanies chemometric analysis that have the potential to discriminate different honeys according to provenance and analyze the geographical location through statistical methods, relating pH, electrical conductivity, moisture, ash content, total acidity, dynamic viscosity, color, and sugar composition [34].

It is common for autochthonous honey to have a higher price than imported honey, which can lead to false origin labeling. Therefore, investigating the geographical origin of honey becomes crucial to ensure authenticity and prevent misleading practices [34].

Portugal possesses a very diverse natural heritage due to its geographical location and geophysical conditions [6]. Indeed, this country has a richness and diversity of melliferous flora, wild species, and cultivated plants, such as chestnut and eucalyptus, which means that there is a great variety of monofloral honey throughout the country [33].

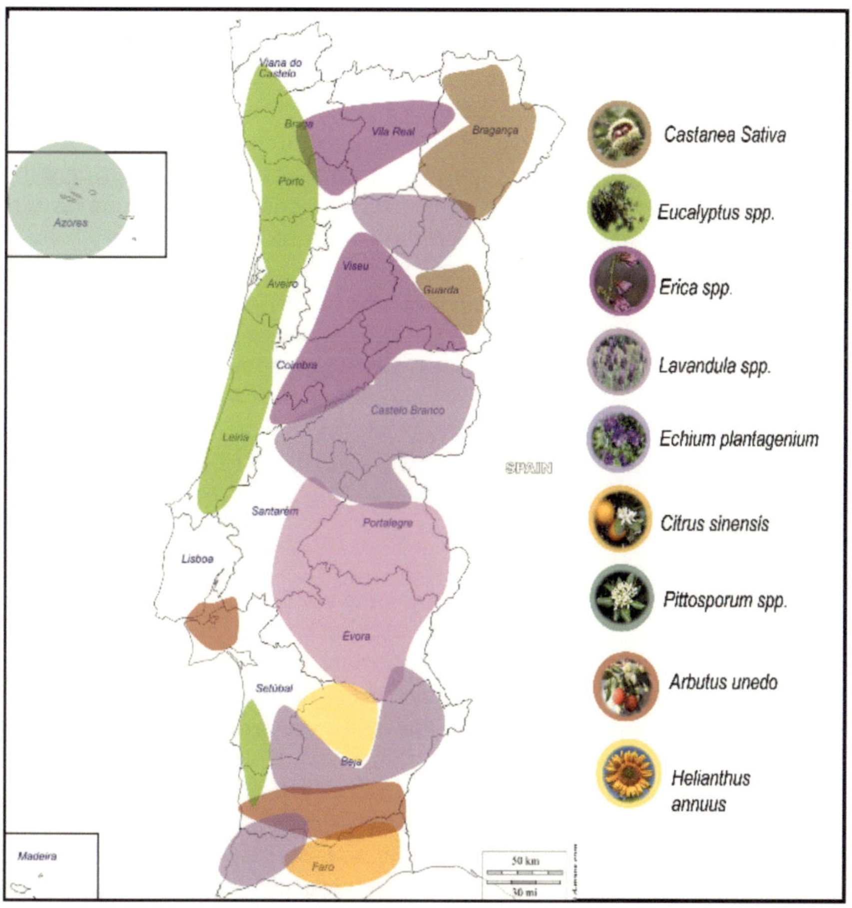


**Figure 2.**
*Map of the main melliferous flora species in Portugal (adapted from [33]).*

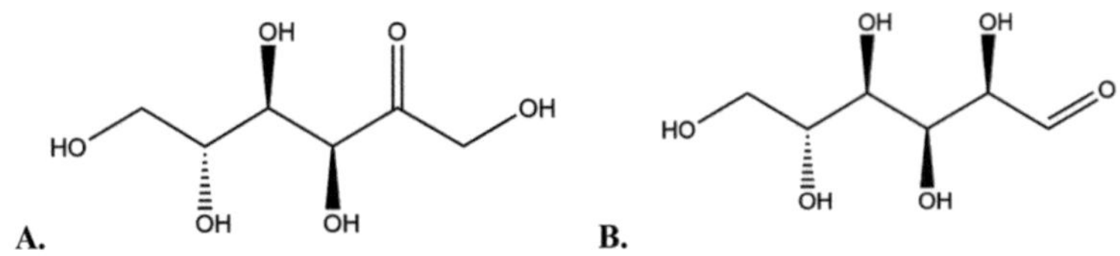


**Figure 3.**
*Chemical structure of D-fructose (A) and D-glucose (B).*

The different types of honey vary from the north to the south of the country and may include different species of plants, namely, *Rosmarinus officinalis*, *Arbutus unedo*, *Echium plantagineum*, *Mentha pulegium*, *Citrus sinensis*, *Carlina racemose*, *Eucalyptus* spp., and *Helianthus annuus*. However, according to the *Federação Nacional de Apicultores do Norte* [33], Portugal has three types of monofloral honey that can be considered the most emblematic, namely lavender honey (*Lavandula stoechas*), heather honey (*Erica umbellate*), and chestnut honey (*Castanea sativa*) [28, 33].

In Azores Islands, two types of honey, incense honey and multifloral honey, have a Protected Designation of Origin (PDO), called "MEL DOS AÇORES," with its specifications, including geographical origin and organoleptic characteristics, among others, available in the Official Journal of the European Union (OJEU) Reg. no. 1107/96 L 148 on 21/6/199 and modified on 12/11/2019 (2019/C 384/09) [64, 65].

The location of the different plants where it could be easier to obtain monofloral honey is demonstrated in **Figure 2**. The three emblematic types of monofloral honey previously mentioned, and as it is possible to observe in **Figure 3**, are predominant in certain regions. Namely, rosemary honey is originated from the north, center, and south, in areas below 400 m of altitude; heather honey is from the north and center, in regions up to the elevation of 900 m; and chestnut honey is also from the north and center, in mountain areas whose altitude ranges between 700 and 1200 m. Furthermore, it is important to highlight that arbutus (*Arbutus unedo*) and citrus (*Citrus sinensis*) honey present a high level of production in the Algarve region (south of Portugal), while eucalyptus (*Eucalyptus* spp.) honey is predominant in the northern region of Minho, and the honey from *Pittosporum undulatum* Vent. (incense honey) came exclusively from the Azores [33, 64, 66].

## 4. Chemical composition of honey

The honey components are influenced by some biotic and abiotic factors [67] and can have different origins: from plants, others added by honeybees, and others as a result of biochemical reactions in honey maturations [27].

Raw honey contains different components, such as pollen, traces of wax, and a variable number of sugar-tolerant yeasts. These microorganisms can cause fermentation if the moisture content of the honey is higher than 17% [23].

The sweet substance has a high nutritional value (330 kcal/100 g), and its composition is essentially based on simple sugars or monosaccharides, whose main compounds are glucose and fructose (65%) and water (18%) (**Table 1**). On the other hand, proteins (enzymes), free amino acids, organic acids, vitamins (including, niacin, pantothenic acid, pyridoxine, riboflavin, and thiamine), essential minerals (such as calcium, copper, iron, magnesium, manganese, phosphorus, potassium,

| Components | | Amount in 100 g of honey |
|---|---|---|
| Water | | 16.9–18 g |
| Carbohydrates | Total | 64.9–73.1 g |
| | Fructose | 35.6–41.8 g |
| | Glucose | 25.4–28.1 g |
| | Maltose | 1.8–1.21 g |
| | Sucrose | 0.23–1.21 g |
| Proteins, amino acids, vitamins, and minerals, secondary metabolites (phenolic compounds) | | 0.50–1.0 g |

**Table 1.**
*General chemical composition of the most commonly consumed types of honey (adapted from [50]).*

sodium, and zinc), and secondary metabolites (phenolic compounds) constitute minor components of this food [29, 47, 58].

As previously referred to, honey is a supersaturated sugar solution, and the remaining honey components also depend on the geographical and botanical origins [12, 29]. The carbohydrates comprise about 95% of the honey dry weight but have fast absorption on consumption [12, 29, 47]. Typically, natural sugars consist of molecules containing either 6 or 12 carbon atoms [68]. Owing to their low molecular weight, significant hydrophilicity, and solubility, they are capable of generating highly concentrated solutions with elevated osmolality. This characteristic holds significant importance for preservation purposes while also contributing to the sweet taste associated with these sugars [68, 69]. Essentially, honey contains a concentrated water solution of fructose (**Figure 3A**), which is, usually, the dominant compound, and glucose (**Figure 3B**). In other words, the mostly represented molecules of sugar in honey are monosaccharides (75%), followed by disaccharides (10–15%), and about 25 different oligosaccharides have been detected [12, 29, 50, 68].

The aforementioned compounds are accountable for endowing honey with specific characteristics, encompassing energy value, viscosity, hygroscopicity, and granulation [29]. Subsequent to the ingestion of honey, the primary monosaccharides, namely fructose and glucose, are expeditiously transported into the bloodstream, where they become readily available for meeting the energy needs of the human body [12].

Sucrose (**Figure 4**) is found in relatively low concentrations, yet it holds significant importance as a disaccharide within this food. Comprising fructose and glucose linked together via an α-1-4 linkage, sucrose contributes to the overall sweetness of the food. In terms of sweetness intensity, sucrose ranks lower than fructose but slightly higher than glucose [68].

The concentration and relative proportion of fructose and glucose serve as pivotal indicators utilized to categorize monofloral honey. This ratio arises from the nectar processing of the specific flower and the time taken for honey to undergo crystallization. Honey that remains in a liquid state for an extended period typically exhibits a fructose/glucose ratio greater than 1.33, whereas rapidly crystallizing honey tends to possess a fructose/glucose ratio lower than 1.11. However, the rate at which glucose crystallizes in honey is also influenced by the balance between glucose and water content. In essence, the crystallization process hinges on the interplay of moisture, temperature, and the concentration and ratio of sugars (fructose and glucose) [29, 50, 68, 70].

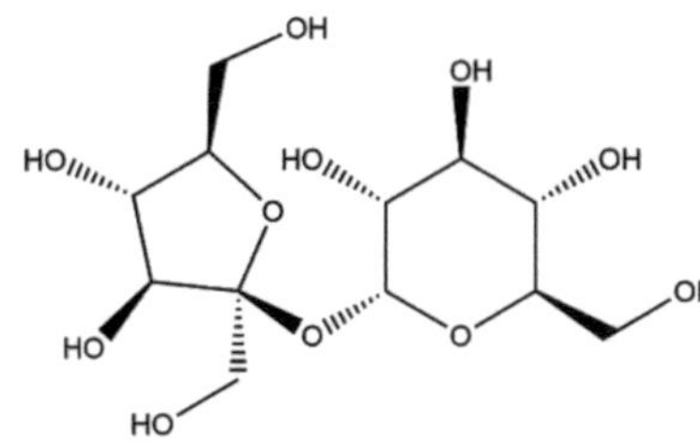


**Figure 4.**
*Chemical structure of sucrose.*

The amino acids comprise less than 1% (w/w) of honey's composition, constituting the major contributor (50–85%) of the total amino groups [52].

The main source of proteins and amino acids in honey emanates from pollen, derived from botanical origins. Additionally, fluids and nectar secretions of the salivary glands and pharynx of honeybees (animal source) can also contribute to its content [29, 71]. Notably, specific quantities of proteins may exhibit variations based on the honeybee species [72]. The relative proportion of 26 amino acids in honey is contingent upon its origin, while storage time and thermal treatments significantly impact the composition and efficacy of these components. With the passage of time, the effectiveness of amino acids diminishes due to their distinct reactivity, leading to changes in their relative proportions [52]. There are a few examples of amino acids found in honey, including alanine, glutamic acid, isoleucine, leucine, phenylalanine, and tyrosine [60]. However, the main amino acid present in honey is proline, which accounts for 50–85% of the total amino acid content and is derived from the salivary secretions of honeybees. Proline commonly serves as a parameter for measuring the maturation of honey. Literature reports indicate that these amino acid components not only contribute to the honey's composition but also act as reliable indicators to discern the blossom or honeydew origin of honeys, providing valuable information for characterizing and classifying different varieties of honey [27].

The proteins found in honey predominantly exist in the form of enzymes and free amino acids [60]. Despite their relatively low concentration, these proteins have garnered significant attention due to their crucial role. Among the noteworthy enzymes present are diastase, which includes α-amylase and β-amylase, as well as α-glucosidase, glucose-oxidase, catalase, and acid phosphatase. These enzymes stem from various sources, such as nectar, salivary fluids, and pharyngeal gland secretions of honeybees [73, 74]. Their presence and diverse origins underscore their importance and potential impact on honey's properties and potential benefits, such as α-amylase, which is responsible for hydrolyzing the starch chain at random locations, forming a variety of dextrins, β-amylase hydrolyzes the starch chain at the end, forming maltose [73, 74], and glucose-oxidase produces hydrogen peroxide, which has antibacterial capacity and converts glucose into δ-gluconolactone, which is hydrolyzed to gluconic acid (acidity regulator). Furthermore, enzymes in honey have been studied over the years as a quality and purity factor, such as diastase, which denatures at rising temperatures [29].

Honey composition can have many changes that occur as a response of different chemical reactions, including fermentation, oxidation, and thermal processing during storage [29]. The reaction of the carbonyl group of a reducing sugar with a free amino group from amino acids, peptides, or proteins produces numerous compounds [27]. For example, chemical reactions, such as caramelization and *Maillard* reaction, are responsible for 5-hydroxymethylfurfural (HMF) formation, which is formed through two reactions: the acidic decomposition of monosaccharides or the condensation of carbohydrates that have free amine groups (*Maillard* reaction). In fresh honey, this compound exists in low concentrations and increases during storage and/or elevated temperatures. Therefore, HMF level is a good indicator for honey freshness, and, in some concentrations, it may be associated with toxicity, so its control is important [75–77].

Organic acids are also present in low concentrations and are responsible for or have their effects associated with honey acidity, flavor, and antimicrobial activity. These components result from the transformation of nectar in honey by honeybees through the enzymatic reaction with sugar or nectar. Aspartic acid, acetoglutaric, butyric, citric, acetic, formic, fumaric, galacturonic, formic, gluconic, glutamic, and glutaric are some examples of organic acids present in this matrix [60, 78]. Similarly, to the previous components, the type and content of organic acids depend on their botanical and/or geographical origin [29].

Generally, honey contains small amounts of vitamins, being the most common derived mainly from pollen, namely, vitamins of the vitamin B complex, such as, thiamine (B1), riboflavin (B2), nicotinic acid (B3), pantothenic acid (B5), pyridoxine (B6), biotin (B8), and folic acid (B9) [79].

Ascorbic acid (vitamin C) is present as well and has been studied for its importance in antioxidant effects. However, this component is vulnerable to chemical and enzymatic oxidation, light, oxygen, or heat, making it difficult to be determined [80].

The mineral or ash content of honey ranges from 0.02 to 1.03% [67] and depends on its botanical and geographical origin [81], environmental conditions, and processing [60]. It is reported that honeys can be categorized botanically based on the estimation of their mineral content and can be an indicative of geographical origin [67], being also an important environmental indicator, tracing pollutants, emissions, and others [14]. In the matrix, some authors believe that minerals influence the color and the taste [58, 67]. Mineral function is associated with biomedical activities, being involved in biological functions such as maintenance of intracellular oxidative balance, metabolic reactions, and others [71]. In contrast to vitamins and amino acids, the minerals cannot be degraded by exposure to light, heat, oxidizing agents, extreme pH, or other factors [69], which is an advantage for the cells and body function [82]. These minerals include macro and microelements, such as potassium, magnesium, calcium, iron, phosphorus, sodium, manganese, iodine, zinc, and lithium, among others. According to several authors, the mineral in major abundance is potassium (45–85%), accounting for almost one-third of the total mineral composition [67, 83].

Besides that, it is important to highlight that metals associated with toxicity, in particular heavy metals such as arsenic, cadmium, and mercury, may exist if the established maximum value is exceeded [29]. However, it was proven that honeybees can act as a biofilter against toxic metal contamination, particularly in cadmium [84].

## 5. Phenolic composition

Phenolic compounds are secondary metabolites with numerous chemical structures and functions, synthetized by plants through two pathways, namely, the shikimate pathway and/or phenylpropanoid metabolism [85, 86], as a response against biotic and abiotic stresses [87], such as aggression by biological (pathogens, parasites, and predators) and physical hazards (ultraviolet radiation), among others [88]. Indeed, these phytochemicals are produced in response to an overwhelmed natural antioxidant defense system—both enzymatic and non-enzymatic—due to excessive activity of reactive oxygen species or prooxidants, leading to damage of cellular and extracellular macromolecules [89].

Structurally, phenolic compounds are composed by one or more aromatic rings, which contain one or more hydroxyl groups, and may differ from a simple to a complex molecule phenolic polymer of high molecular weight [90]. The referred compounds can be found linked to one or more sugars (glycosides) or in their basic structure (aglycones), originating functional compounds with different biological effects [90, 91], being partially responsible for the several organoleptic characteristics [28, 66, 88].

The classification of these phytochemicals can be divided into two main groups, one composed by non-flavonoids, which are divided into phenolic acids (hydroxycinnamic acids and hydroxybenzoic acids) and stilbenes, and the other, represented by flavonoids composed by flavonols, flavones, isoflavones, anthocyanidins, chalcones, and flavanols [90].

The phenolic acids (phenol carboxylic acid) compounds comprise molecules with one phenolic ring and at least one organic carboxylic acid function. These molecules can be categorized according to their structure in C6-C3, C6-C2, and C6-C1 structures. Generally, most of these compounds are bound to organic acids such as glucose or flavonoids and lignin and cellulose from plants' structural components [17, 60]. The beneficial effects in human health of this set of compounds are vast, namely, antidiabetic, anti-inflammatory, anticarcinogenic, radioprotective potential, and neuroprotective [92].

Regarding flavonoids, these compounds refer to a group of phenolic compounds constituted by low molecular weight and mainly water-soluble [60, 87]. The basic flavonoid structure (**Figure 5**) contains two benzene rings, alternated by a linear chain of three atoms of carbon (C6-C3-C6) that rearranges itself, forming 15 carbon atoms (C15) arranged on three rings labeled A, B, and C and generally having at least two phenolic groups (OH), being biogenetically derived from phenylpropane (C6-C3) and three acetate units (C6) [60, 88, 90]. These chemical compounds are aromatic substances frequently associated with glycosides, mainly glucose with galactose, xylose, rhamnose, arabinose, and rutinoside, among others, but can also be present in an aglycone form [60].

In humans, flavonoids play an important role in health, and it is considered that the daily ingestion through the diet may help in the prevention of some pathologies, a fact that is related to the biological properties of this set of compounds, such as antioxidant, anti-inflammatory, anti-allergic, anti-hepatotoxic, anti-ulcerogenic, antiplatelet, antimicrobial, antiviral, and anticarcinogenic, among others [90, 93, 94].

## 5.1 Phenolic composition of honey

Phenolic compounds are widely distributed throughout various plant organs, playing a crucial role in shaping the chemical composition of nectar [90]. Honeybees, during their collection of nectar or honeydew, can transfer these phytochemicals into honey. The presence of aglycones in honey might be justified by the glucosidases

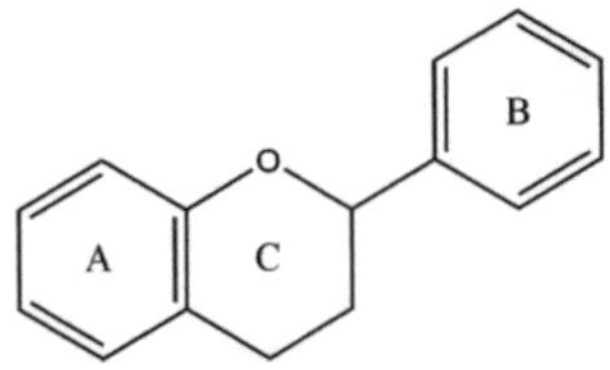


**Figure 5.**
*Flavonoid basic structure.*

present in the salivary glands of bees. These compounds offer an advantage as they can be readily absorbed through intestinal barriers, enhancing their bioavailability when compared to the same glycosylated flavonoid [95]. Aside from their beneficial effects on the immune system response, these phytochemicals also serve as signaling molecules, attracting pollinators and other animals [90, 96].

In literature, numerous studies from different authors analyzed the phenolic profile of honey and identified several phenolic compounds. Generally, apigenin, vanillic acid, syringic acid, ferulic acid, quercetin, kaempferol, myricetin, 3-hydroxybenzoic acid, chlorogenic acid, 4-hydroxybenzoic acid, hesperetin, benzoic acid, apigenin, and chrysin, among others, are the main phenolic compounds founded [51, 59, 61, 66, 97, 98].

In **Table 2**, are represented some phenolic compounds identified in different types of honey according to their botanical source.

| Monofloral honey type | Phenolic compound | Reference |
|---|---|---|
| Eucalyptus honey (*Eucalyptus camaldunensis*) | Gallic acid, chlorogenic acid, caffeic acid, *ρ*-coumaric acid, benzoic acid, ferulic acid, quercetin, epicatechin, catechin, gallocatechin, rutin, myricetin, apigenin, kaempferol, chlorogenic acid | [99–101] |
| Rubus honey (*Rubus* spp.) | Caffeic acid, *ρ*-coumaric acid, ellagic acid, pinocembrin, kaempferol | [102] |
| Rapeseed honey (*Brassica* spp.) | Quercetin, 8-methoxykaempferol, protocatechuic acid, gallic acid, kaempferol, chlorogenic acid, caffeic acid, vanillic acid, syringic acid, *ρ*-coumaric acid, benzoic acid, salicylic acid, apigenin | [97, 99] |
| Heather honey (*Erica* spp.) | Gallic acid, ellagic acid, protocatechuic acid, catechin, caffeic acid, *ρ*-coumaric acid, quercetin, apigenin, myricetin, kaempferol, chlorogenic acid | [59, 99, 101] |
| Chestnut honey (*Castanea sativa*) | Gallic acid, chlorogenic, acid, protocatechuic acid, vanillic acid, caffeic acid, *ρ*-coumaric acid, ferulic acid, benzoic acid, quercetin, apigenin, epicatechin, catechin, gallocatechin, rutin, myricetin, hesperetin | [59, 100, 101] |
| Pine honey (*Pinus* spp.) | Protocatechuic acid, catechin, vanillic acid, caffeic acid, syringic acid, *ρ*-coumaric acid, ferulic acid, rutin, quercetin, kaempferol | [59] |
| Rosemary honey (*Rosmarinus officinalis*) | 8-Methoxykaempferol, kaempferol | [99, 103] |
| Citrus honey (Citrus sinensis) | Hesperetin, caffeic acid, *ρ*-coumaric acid, ferulic acid, apigenin, quercetin, chlorogenic acid | [99, 101, 103, 104] |
| Lime honey (*Tilia europaea*) | Protocatechuic, caffeic acid, *ρ*-coumaric acid, gallic acid, rutin, chlorogenic acid, vanillic acid, benzoic acid, salicylic acid, apigenin | [59, 97] |
| Sunflower honey (*Helianthus annuus*) | Gallic acid, quercetin, protocatechuic acid, 4-hydroxybenzoic acid, caffeic acid, vanillic acid, syringic acid, apigenin, chrysin | [97, 99, 103] |
| Lavender honey (*Lavandula* spp.) | Catechin, caffeic acid, epicatechin, *ρ*-coumaric acid, rutin | [59] |
| Acacia honey (*Robinia pseudoacacia*) | Ellagic acid, caffeic acid, syringic acid, *ρ*-coumaric acid, apigenin, myricetin, kaempferol, quercetin, isohamnetin | [59, 99, 101] |
| Clover honey (*Trifollium* spp.) | *ρ*-Coumaric acid, caffeic acid, protocatechuic acid, phloroglucinol, gallic acid | [59, 97] |

**Table 2.**
*Summary of phenolic compounds identified in different types of monofloral honeys.*

Indeed, in the United States Department of Agriculture (USDA) database is registered that different varieties of honey, from different botanical sources and geographical origins (different countries), commonly contain apigenin, isorhamnetin, kaempferol, luteolin, quercetin, and myricetin [105].

Regarding the non-flavonoids, it has been firmly established that phenolic acids play a central role in determining the honey type, with ellagic acid and hydroxycinnamates being particularly significant in this regard. These compounds are commonly employed as floral markers for identifying heather honey and chestnut honey, respectively. Furthermore, their utility extends to geographical identification, enabling the discernment and classification of honey based on its specific floral sources and geographical origin [103, 106, 107].

In this food matrix, the main functional components are flavonoids, being the amount of these phenolic compounds dependent on the honey type [17, 66]. In honey, these phytochemicals are commonly present in three sub-classes, namely, flavonols, flavones, and flavanones that have a significant influence in this food matrix, affecting the color, taste, flavor, as well as beneficial effects on health [108, 109].

In the case of colors, a substantial part of these phytochemicals has conjugated chromophores present in the molecules, which can range from yellow to a red color. However, specific flavonoids such as anthocyanidins, in particular, cyanidin, are commonly associated with colors from red to magenta, whereas perlargonidin is associated with orange to red. Flavones and flavonols usually demonstrate a red color, while flavanones tend to be colorless, white, or brownish [110].

## 6. Biological activities

Nowadays, synthetic drugs and chemicals can have numerous negative effects on human health, which has increased the interest from the scientific community to find natural products with these important compounds [59], including new sources of functional ingredients from food [111], in order to minimize several health problems, such as multidrug resistance, including antibacterial resistance, among others [107].

Apitherapy is a therapeutic practice with various health purposes, including the improvement of the human system through the biological properties of bee products [59]. Honey contains bioactive compounds that have positive effects for human health [29], such as anti-inflammatory, antimicrobial, anticancer, anti-aging, and antidiabetic (**Figure 6**) [110] and protective effects such as protection of cardiovascular, nervous, respiratory, and gastrointestinal systems (**Figure 6**) [60]. The bioavailability of the phytochemical components, as well as how they are absorbed and metabolized, determine how honey affects human health [60].

### 6.1 Antioxidant capacity

Uncontrolled free radicals can affect different biomolecules, including proteins, DNA, and lipids, among others, and cause cell damage [112]. These oxidative stress effects have attracted the interest of scientists and the general public, whereby antioxidant capacity has been considered as a solution for this problem, being one of the primary requirements for disease prevention; however, it also has other purposes, playing, for example, an important role in food shelf life [113].

Initially, these effects were associated with the deleterious actions caused by the ability of specific polyphenols to bind and precipitate macromolecules, such as,

**Figure 6.**
*Honey biological activities (adapted from [60]).*

digestive enzymes and carbohydrates, but recently, they are also associated with the antioxidant capacity and free radical scavenging abilities [90].

Since ancient times, the antioxidant potential of honey is well-known, due to its associated medical properties, being related to different substances present in this food [17, 61, 114]. Indeed, it has been reported that in this food matrix this effect is caused by enzymatic and non-enzymatic sources (phenolic compounds), such as, glucose-oxidase and catalase, and L-ascorbic acid, flavonoids, and phenolic acids, respectively [51, 98].

To achieve these effects, antioxidants have two basic pathways: free radical inactivation and electron transfer [110, 115]. In the first mechanism, a hydrogen atom from the antioxidant can be removed by a free radical, becoming a radical. In the second one, the free radical receives an electron from the antioxidant, becoming a cation radical [110]. In this sense, they can act in different mechanisms, such as free radical scavenging, which can be responsible for the inhibition of hydrolytic and oxidative enzymes and, for example, can modulate lipid peroxidation involved in some diseases such as thrombosis and carcinogenesis [116].

It is important to highlight that the antioxidant action of flavonoids is established by the structure-activity relation (SAR) (**Figure 7**), number and positions of hydroxyl groups, presence of double bonds, glycosylation, and the presence of substituents in the rings [29, 110, 117]. Indeed, if the molecule is composed of three hydroxyl groups, it is suggested that the activity is suppressed, while when it is composed of two hydroxyl groups, the antioxidant effect is improved. Furthermore, other structural groups besides hydroxyl groups are also associated with the capacity to scavenge radicals; the 2,3-double bond conjugated with the 4-oxo function in the C-ring, the *ortho*-di-hydroxyl group, in the B-ring, and the existence of a double bond between $C_2$ and $C_3$ in conjunction with a 4-carbonyl group in the C-ring also affect this property (**Figure 7**), which reinforces the importance of the structure of the molecule for antioxidant activity [118–120].

In the glycosylation case, the activity is weaker if flavonoids are linked with glycosides than the respective aglycones [110].

According to several authors, and as mentioned above, there are numerous different flavonoids present in honey, such as apigenin, chrysin, and kaempferol, which present antioxidant activity according to their functional groups and consequently are responsible for their SAR (**Figure 8**) [60, 101, 110].

The *ortho*-diphenol structure displays high antioxidant activity through the formation of intramolecular hydrogen bonds formed during the reaction with free radicals,

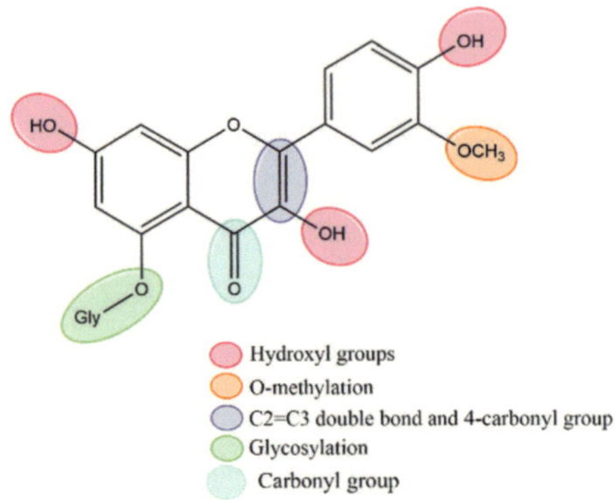


**Figure 7.**
*Summary of SAR for flavonoids-induced antioxidants (adapted from [110]).*

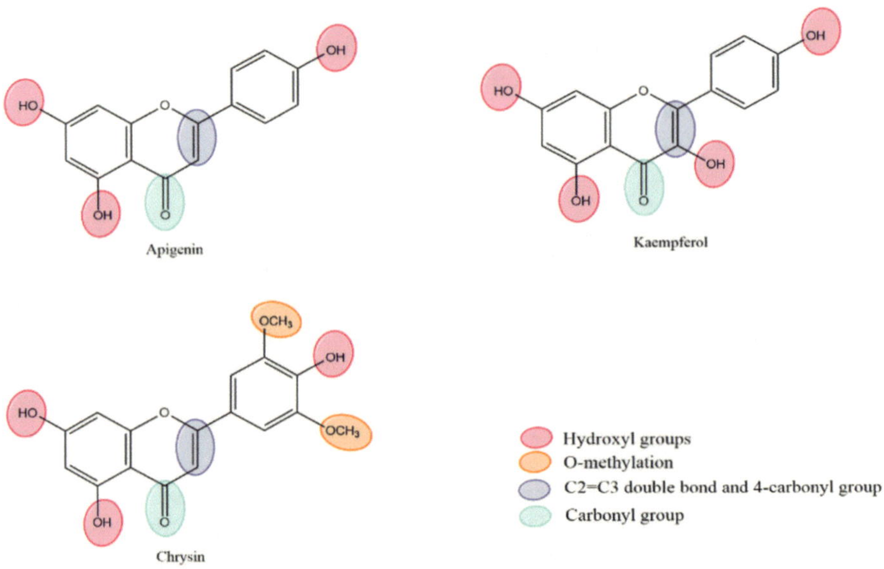


**Figure 8.**
*Examples of flavonoids founded in honey and their respective functional groups (adapted from [60, 110]).*

improving their stability. For example, due to its pyrogallol structure (3,4,5-OH), gallic acid is one of the most effective scavengers and has a noticeable H-donating ability [118].

Phenolic acids are also known for their potent antioxidant properties, which can be attributed to diverse mechanisms, notably encompassing radical scavenging activity, thus earning them recognition as chain-breaking antioxidants. These compounds exhibit the capacity to donate hydrogen or electrons, as well as the ability to stabilize phenoxyl radicals within their molecular structure. Moreover, the antioxidant activity of phenolic acids tends to escalate with a higher degree of hydroxylation, while it is reduced with the substitution of hydroxyl groups at the 3- and 5-positions with methoxy groups [92, 118, 121]. However, besides the antioxidant capacity, several studies state that some phenolic acids are also responsible for antimicrobial activity, such as chlorogenic acid and ferulic acid, among others [122, 123].

*DOI: http://dx.doi.org/10.5772/intechopen.112872*

Additionally, the phenolic compounds influence the activity of several enzymes and cell detectors, affecting free radical species [98]. Phenols also have the ability to chelate metals, especially iron and copper ($Fe^{2+}$, $Cu^{2+}$), preventing their involvement in reactions and suppressing metal-catalyzed free radical formations [118].

The antioxidant capacity of honey has garnered considerable attention from researchers who have extensively analyzed various types of honey originating from different botanical and geographical sources using spectrophotometric methodologies. Significant differences in antioxidant capacity have been observed between these honeys. These disparities can be attributed to the distinct phenolic profiles associated with each honey, as evidenced in **Table 2**, which are influenced primarily by the phenolic compounds and vitamins derived from the nectar. The variations in antioxidant capacity among different honeys can be explained by a multitude of factors. The climatic conditions and botanical sources play a crucial role, as the nectar and excretions of plants encompass diverse polyphenolic compounds that contribute to the honey's antioxidant activity. Additionally, bees are responsible for transferring the bioactive compounds from plants to honey during the honey-making process. Apart from natural factors, certain human-induced elements can also impact the antioxidant capacity of honey. The methodology employed for honey extraction from the wax, the type of packing material used, and the storage time can all influence the final antioxidant properties of the honey [17, 30, 124].

Overall, the variations in antioxidant capacity observed in different types of honey stem from the complex interplay of natural and anthropogenic factors that govern the composition and properties of honey from various sources. Furthermore, it was reported that dark honeys present more phenolic compounds and, consequently, higher antioxidant activity than lighter ones [17].

### 6.2 Anti-aging activity

Skin aging is one of the most common skin issues nowadays; in this sense, its care and treatment are vital to reverse these effects [125]. Skin aging is a biological process that can occur by intrinsic factors such as hormonal disorders, genetic changes, vitamin deficiencies, and extrinsic factor such as environmental toxins, inadequate care, and UV radiation [126]. Another significant component that forms free radicals and influences both factors is oxidative stress [112, 125].

Enzyme inhibitory activity may be an answer as a clinical nutraceutical tool when found as a component in food [112]. These types of products find common use in traditional medicine, particularly within traditional Chinese and Arabic medicine. In these ancient practices, honey has long been recognized for its beneficial effects on enhancing the general appearance of the skin and combating fungal infections. Moreover, in Northern Pakistan, honey is applied to the eyelids and used for facial protection, as well as for treating freckles and white spots on the skin [127].

In current days, honey's moisturizing, nourishing, emollient, and antibacterial properties have made it a popular ingredient in cosmetics. It has also been reported to have the ability to inhibit certain enzymes like tyrosinase and elastase, which may be related to the presence of polyphenols and flavonoids [64, 112, 127, 128].

In fact, enzyme inhibitors are in their majority bioactive secondary metabolites, linking enzymes and reducing their biological and catalytic activities. Moreover, an enzyme's deactivation can reduce or eliminate pathogens as well as treat metabolic disorders [128].

In pharmacology, the inhibition of human degradative skin enzymes has been used as a technique to fight pathological disorders and has replaced synthetic and chemical drugs. The enzyme inhibitors are usually synthesized as part of pharmacology and biochemistry and also occur naturally [112, 128, 129].

The main enzymes that have been studied as potential targets include tyrosinase and α-amylase, among others [112].

Tyrosinase is an important enzyme involved in melanogenesis, specifically by the production of melanin, which is responsible for skin pigmentation and protection of the damage caused by UV radiation. This amino acid absorbs UV rays and removes reactive oxygen species, acting as the primary defense mechanism of human skin. However, when there is an overproduction of tyrosinase, it causes an excessive accumulation of epidermal pigmentation, which can lead to various disorders, such as flecks, ephelides, and age spots, among others. In this sense, inhibitors of this enzyme should be clinically useful for the treatment of dermatological disorders, including aging [127, 130, 131].

Tyrosinase acts as a catalyzer of two distinct reactions. Initially, L-tyrosine is hydroxylated to 3,4-dihydroxyphenylalanine (L-DOPA, *ortho*-diphenol product) and thereafter oxidized to the corresponding *ortho*-quinone [127]. To catalyze this reaction, this enzyme contains copper in its active site, which can be blocked by bioactive compounds [130].

Also associated with the skin matrix, elastin is an extracellular protein that provides elasticity and resilience, being a constituent and the main responsible for connective tissue [130]. One of the most popular aging characteristics is associated with the hydrolyze of this amino acid by elastase, causing loss of skin firmness [125, 132, 133]. In 2023, a significant report highlighted the remarkable properties of incense honey from Portugal, demonstrating its ability to inhibit a specific enzyme associated with skin aging, the elastase. This finding underscores the critical importance of this particular honey variety as a potential matrix for effectively preventing skin aging [64].

## 7. Conclusion

Nowadays, consumers interest in natural products with beneficial properties is increasing. In this sense, honey has emerged as a natural food source of phenolic compounds with biological properties that are commonly associated with its chemical composition. Portugal, in particular, is celebrated for its abundant melliferous flora, resulting in distinct monofloral and multifloral honey varieties, each influenced by unique geographical regions and climatic conditions.

The organoleptic characteristics and chemical composition of Portuguese honey, including phenolic compounds, play a vital role in defining its properties. These phytochemicals are secondary metabolites, being the main responsibles for diverse bioactive activities such as antioxidant, antimicrobial, and anti-aging activities. Their ability to combat oxidative stress, a significant factor in various diseases, enhances the nutritional and health value of honey.

As consumer demand for honey continues to grow, its commercial value has risen significantly, not only as a source of energy in the human diet but also as a nutraceutical product offering potential health benefits. The multifaceted applicability of honey extends across several industries, including food, pharmaceuticals, and cosmetics, enhancing their impact on the national economy.

## Conflict of interest

The authors declare no conflict of interest.

## Funding

This work is supported by National Funds by FCT—Portuguese Foundation for Science and Technology, under the project UIDB/04033/2020 (https://doi.org/10.54499/UIDB/04033/2020, https://doi.org/10.54499/LA/P/0126/2020 (accessed on 4 March 2024)).

## Author details

Soraia Santos, Ana Abraão, Irene Gouvinhas and Ana Novo Barros*
Centre for Research and Technology of Agro-Environmental and Biological Sciences, CITAB, Inov4Agro, University of Trás-os-Montes and Alto Douro, Vila Real, Portugal

*Address all correspondence to: abarros@utad.pt

## References

[1] Gaggìa F, Baffoni L, Alberoni D. Probiotics for Honeybee's health. In: Probiotics and Prebiotics in Animal Health and Food Safety. Bologna: Springer, University of Bologna; 2018. pp. 219-245. DOI: 10.1007/978-3-319-71950-4_9

[2] Hung KLJ, Kingston JM, Albrecht M, Holway DA, Kohn JR. The worldwide importance of honey bees as pollinators in natural habitats. Proceedings of the Royal Society B: Biological Sciences. 1870;**2018**(285):20172140. DOI: 10.1098/rspb.2017.2140

[3] Abrol DP. Wild bees and crop pollination. In: Pollination Biology. India: Springer, University of Agricultural Sciences and Technology; 2012. pp. 111-184. DOI: 10.1007/978-94-007-1942-2_6

[4] Royan M. Mechanisms of probiotic action in the honeybee. Critical Reviews in Eukaryotic Gene Expression. 2019;**29**(2):95-103. DOI: 10.1615/CritRevEukaryotGeneExpr.2019025358

[5] Webster MT. Apis mellifera. Trends in Genetics. 2019;**35**(11):880-881. DOI: 10.1016/j.tig.2019.08.003

[6] Estevinho LM, Feás X, Seijas JA, Pilar Vázquez-Tato M. Organic honey from Trás-Os-Montes region (Portugal): Chemical, palynological, microbiological and bioactive compounds characterization. Food and Chemical Toxicology. 2012;**50**(2):258-264. DOI: 10.1016/J.FCT.2011.10.034

[7] Martinotti S, Bucekova M, Majtan J, Ranzato E. Honey: An effective regenerative medicine product in wound management. Current Medicinal Chemistry. 2019;**26**(27):5230-5240. DOI: 10.2174/0929867325666180510141824

[8] Crane E. Beekeeping. In: Encyclopedia of Insects. 2007. pp. 66-71. DOI: 10.1016/B978-0-12-374144-8.00019-9

[9] Sperandio G, Simonetto A, Carnesecchi E, Costa C, Hatjina F, Tosi S, et al. Beekeeping and honey bee colony health: A review and conceptualization of beekeeping management practices implemented in Europe. Science of the Total Environment. 2019;**696**:133795. DOI: 10.1016/J.SCITOTENV.2019.133795

[10] Aylanc V, Falcão SI, Ertosun S, Vilas-Boas M. From the hive to the table: Nutrition value, digestibility and bioavailability of the dietary phytochemicals present in the bee pollen and bee bread. Trends in Food Science and Technology. 2021;**109**:464-481. DOI: 10.1016/j.tifs.2021.01.042

[11] Erban T, Shcherbachenko E, Talacko P, Harant K. The unique protein composition of honey revealed by comprehensive proteomic analysis: Allergens, venom-like proteins, antibacterial properties, Royal Jelly Proteins, serine proteases, and their inhibitors. Journal of Natural Products. 2019;**82**(5):1217-1226. DOI: 10.1021/acs.jnatprod.8b00968

[12] Bogdanov S, Jurendic T, Sieber R, Gallmann P. Honey for nutrition and health: A review. Journal of Apicultural Research. 2013;**27**(6):677-689. DOI: 10.1080/07315724.2008.10719745

[13] Kuropatnicki AK, Kłósek M, Kucharzewski M. Honey as medicine: Historical perspectives. Journal of Apicultural Research. 2018;**57**(1):113-118. DOI: 10.1080/00218839.2017.1411182

[14] Nikhat S, Fazil M. History, phytochemistry, experimental pharmacology and clinical uses of honey: A comprehensive review with special reference to Unani medicine. Journal of

Ethnopharmacology. 2022;**57**(1):113-118. DOI: 10.1080/00218839.2017.1411182

[15] Zhang XH, Qing XD, Mu ST, Wang D, Zheng JJ, Zhou SJ, et al. Authentication of honey of different nectar sources and antioxidant property evaluation by phenolic composition analysis with chemometrics. Food Control. 2021;**124**:107900. DOI: 10.1016/J.FOODCONT.2021.107900

[16] Silva LR, Videira R, Monteiro AP, Valentão P, Andrade PB. Honey from Luso region (Portugal): Physicochemical characteristics and mineral contents. Microchemical Journal. 2009;**93**(1):73-77. DOI: 10.1016/J.MICROC.2009.05.005

[17] Machado De-Melo AA, Almeida-Muradian LB d, Sancho MT, Pascual-Maté A. Composition and properties of *Apis mellifera* honey: A review. Journal of Apicultural Research. 2017;**57**(1):5-37. DOI: 10.1080/00218839.2017.1338444

[18] Boussaid A, Chouaibi M, Rezig L, Hellal R, Donsì F, Ferrari G, et al. Physicochemical and bioactive properties of six honey samples from various floral origins from Tunisia. Arabian Journal of Chemistry. 2018;**11**(2):265-274. DOI: 10.1016/J.ARABJC.2014.08.011

[19] Maione C, Barbosa F, Barbosa RM. Predicting the botanical and geographical origin of honey with multivariate data analysis and machine learning techniques: A review. Computers and Electronics in Agriculture. 2019;**157**:436-446. DOI: 10.1016/J.COMPAG.2019.01.020

[20] Nemo R, Bacha K. Microbial quality, physicochemical characteristics, proximate analysis, and antimicrobial activities of honey from Anfilo district. Food Bioscience. 2021;**42**:101132. DOI: 10.1016/J.FBIO.2021.101132

[21] Osés SM, Cantero L, Puertas G, Fernández-Muiño MÁ, Sancho MT. Antioxidant, antimicrobial and anti-inflammatory activities of ling-heather honey powder obtained by different methods with several carriers. LWT. 2022;**159**:113235. DOI: 10.1016/J.LWT.2022.113235

[22] European Parliament [Internet]. 2018. Available from: https://www.europarl.europa.eu/news/en/headlines/economy/20180222STO98435/key-facts-about-europe-s-honey-market-infographic [Accessed: January 18, 2022]

[23] Anklam E. A review of the analytical methods to determine the geographical and botanical origin of honey. Food Chemistry. 1998;**63**(4):549-562. DOI: 10.1016/S0308-8146(98)00057-0

[24] Kaskonien V, Petras VR. Floral markers in honey of various botanical and geographic origins: A review. Comprehensive Reviews in Food Science and Food Safety. 2010;**9**:14. DOI: 10.1111 / j.1541-4337.2010.00130.x

[25] Molan PC. Authenticity of honey. In: Food Authentication. Norwich, UK: Springer, CSL Food Science Laboratory; 1996. pp. 259-303. DOI: 10.1007/978-1-4613-1119-5_8

[26] Seraglio SKT, Silva B, Bergamo G, Brugnerotto P, Gonzaga LV, Fett R, et al. An overview of physicochemical characteristics and health-promoting properties of honeydew honey. Food Research International. 2019;**119**:44-66. DOI: 10.1016/J.FOODRES.2019.01.028

[27] Iglesias MT, Martín-Álvarez PJ, Polo MC, De Lorenzo C, González M, Pueyo E. Changes in the free amino acid contents of honeys during storage at ambient temperature. Journal of Agricultural and Food Chemistry.

2006;**54**(24):9099-9104. DOI: 10.1021/JF061712X

[28] Feás X, Pires J, Iglesias A, Estevinho ML. Characterization of artisanal honey produced on the northwest of Portugal by melissopalynological and physico-chemical data. Food and Chemical Toxicology. 2010;**48**(12):3462-3470. DOI: 10.1016/J.FCT.2010.09.024

[29] Silva PM d, Gauche C, Gonzaga LV, Costa ACO, Fett R. Honey: Chemical composition, stability and authenticity. Food Chemistry. 2016;**196**:309-323. DOI: 10.1016/J.FOODCHEM.2015.09.051

[30] Seraglio SKT, Schulz M, Brugnerotto P, Silva B, Gonzaga LV, Fett R, et al. Quality, composition and health-protective properties of citrus honey: A review. Food Research International. 2021;**143**:110268. DOI: 10.1016/J.FOODRES.2021.110268

[31] Bobis O, Moise AR, Ballesteros I, Reyes ES, Durán SS, Sánchez-Sánchez J, et al. Eucalyptus honey: Quality parameters, chemical composition and health-promoting properties. Food Chemistry. 2020;**325**:126870. DOI: 10.1016/j.foodchem.2020.126870

[32] Mureșan CI, Cornea-Cipcigan M, Suharoschi R, Erler S, Mărgăoan R. Honey botanical origin and honey-specific protein pattern: Characterization of some European honeys. LWT. 2021:112883. DOI: 10.1016/J.LWT.2021.112883

[33] FNAP FN dos A de P. Boletim Estatístico O Setor Apícola em números: Análise e avaliação da situação em Portugal na União Europeia e no Mercado Mundial de Mel Existências Mercado de Produtos Apícolas Sanidade. Lisboa, Portugal: Federação Nacional de Apicultores do Norte; 2017. pp. 1-30

[34] Geana EI, Ciucure CT. Establishing authenticity of honey via comprehensive Romanian honey analysis. Food Chemistry. 2020;**306**:125595. DOI: 10.1016/J.FOODCHEM.2019.125595

[35] Pauliuc D, Ciursă P, Ropciuc S, Dranca F, Oroian M. Physicochemical parameters prediction and authentication of different monofloral honeys based on FTIR spectra. Journal of Food Composition and Analysis. 2021;**102**:104021. DOI: 10.1016/J.JFCA.2021.104021

[36] Eyer M, Neumann P, Dietemann V. A look into the cell: Honey storage in honeybees, *Apis mellifera*. PLoS One. 2016;**11**(8):e0161059. DOI: 10.1371/journal.pone.0161059

[37] Laughton AM, Boots M, Siva-Jothy MT. The ontogeny of immunity in the honey bee, *Apis mellifera* L. following an immune challenge. Journal of Insect Physiology. 2011;**57**(7):1023-1032. DOI: 10.1016/j.jinsphys.2011.04.020

[38] Britannica, The Editors of Encyclopaedia. "honeybee". Encyclopedia Britannica; 23 Sep. 2024. Available from: https://www.britannica.com/animal/honeybee [Accessed: October 16, 2024]

[39] Domingos AT, Nóbrega MM, Silva RA. Biologia das abelhas *Apis mellifera*: Uma revisão bibliográfica. Acta Apicola Brasilica. 2016;**4**(2):08-12. DOI: 10.18378/aab.v3i2.4584

[40] Dolezal AG, Toth AL. Feedbacks between nutrition and disease in honey bee health. Current Opinion in Insect Science. 2018;**26**:114-119. DOI: 10.1016/J.COIS.2018.02.006

[41] Wright GA, Nicolson SW, Shafir S. Nutritional physiology

and ecology of honey bees. Annual Review of Entomology. 2018;**63**(1):327-344. DOI: 10.1146/annurev-ento-020117-043423

[42] Seeley TD. Social foraging in honey bees: How nectar foragers assess their colony's nutritional status. Behavioral Ecology and Sociobiology. 1989;**24**(3):181-199. DOI: 10.1007/BF00292101

[43] Cruz-Landim C. Abelhas: morfologia e função de sistemas. UNESP. Editora UNESP; 2009. p. 408. DOI: 10.7476/9788539304301

[44] Kadri SM, Zaluski R, Orsi R. Nutritional and mineral contents of honey extracted by centrifugation and pressed processes. Food Chemistry. 2017;**218**:237-241. DOI: 10.1016/J.FOODCHEM.2016.09.071

[45] Sammataro D. In: Press CU, editor. The Beekeeper's Handbook. 4th ed. 2011

[46] Wang J, Li QX. Chemical composition, characterization, and differentiation of honey botanical and geographical origins. In: Advances in Food and Nutrition Research. Academic Press; 2011. pp. 89-137. DOI: 10.1016/B978-0-12-385989-1.00003-X

[47] Conti ME, Canepari S, Finoia MG, Mele G, Astolfi ML. Characterization of Italian multifloral honeys on the basis of their mineral content and some typical quality parameters. Journal of Food Composition and Analysis. 2018;**74**:102-113. DOI: 10.1016/J.JFCA.2018.09.002

[48] Adgaba N, Al-Ghamdi AA, Getachew A, Tadesse Y, Belay A, Ansari MJ, et al. Characterization of honeys by their botanical and geographical origins based on physico-chemical properties and chemo-metrics analysis. Journal of Food Measurement and Characterization. 2017;**11**(3):1106-1117. DOI: 10.1007/s11694-017-9487-4

[49] Sakač MB, Jovanov PT, Marić AZ, Pezo LL, Kevrešan ŽS, Novaković AR, et al. Physicochemical properties and mineral content of honey samples from Vojvodina (Republic of Serbia). Food Chemistry. 2019 Mar;**15**(276):15-21. DOI: 10.1016/j.foodchem.2018.09.149

[50] Escuredo O, Dobre I, Fernández-González M, Seijo MC. Contribution of botanical origin and sugar composition of honeys on the crystallization phenomenon. Food Chemistry. 2014;**149**:84-90. DOI: 10.1016/J.FOODCHEM.2013.10.097

[51] Cheung Y, Meenu M, Yu X, Xu B. Phenolic acids and flavonoids profiles of commercial honey from different floral sources and geographic sources. International Journal of Food Properties. 2019;**22**(1):290-308. DOI: 10.1080/10942912.2019.1579835

[52] Hermosín I, Chicón RM, Cabezudo MD. Free amino acid composition and botanical origin of honey. Food Chemistry. 2003;**83**(2):263-268. DOI: 10.1016/S0308-8146(03)00089-X

[53] Escuredo O, González-Martín MI, Rodríguez-Flores MS, Seijo MC. Near infrared spectroscopy applied to the rapid prediction of the floral origin and mineral content of honeys. Food Chemistry. 2015;**170**:47-54. DOI: 10.1016/J.FOODCHEM.2014.08.061

[54] Louveaux J, Maurizio A, Vorwohl G. Methods of melissopalynology. Bee World. 1978;**59**(4):139-157. DOI: 10.1080/0005772X.1978.11097714

[55] Siddiqui AJ, Musharraf SG, Choudhary MI, Rahman A. Application of analytical methods

in authentication and adulteration of honey. Food Chemistry. 2017;**217**:687-698. DOI: 10.1016/J.FOODCHEM.2016.09.001

[56] Reyes ES, Sanchéz JS. Botanical Classification. Bee Products - Chemical and Biological Properties. Cham: Springer; 2017. pp. 3-19. DOI: 10.1007/978-3-319-59689-1_1

[57] Kasprzyk I, Depciuch J, Grabek-Lejko D, Parlinska-Wojtan M. FTIR-ATR spectroscopy of pollen and honey as a tool for unifloral honey authentication. The case study of rape honey. Food Control. 2018;**84**:33-40. DOI: 10.1016/J.FOODCONT.2017.07.015

[58] González-Miret ML, Terrab A, Hernanz D, Fernández-Recamales MÁ, Heredia FJ. Multivariate correlation between color and mineral composition of honeys and by their botanical origin. Journal of Agricultural and Food Chemistry. 2005;**53**(7):2574-2580. DOI: 10.1021/jf048207p

[59] Can Z, Yildiz O, Sahin H, Akyuz Turumtay E, Silici S, Kolayli S. An investigation of Turkish honeys: Their physico-chemical properties, antioxidant capacities and phenolic profiles. Food Chemistry. 2015;**180**:133-141. DOI: 10.1016/J.FOODCHEM.2015.02.024

[60] Cianciosi D, Forbes-Hernández T, Afrin S, Gasparrini M, Reboredo-Rodriguez P, Manna P, et al. Phenolic compounds in honey and their associated health benefits: A review. Molecules. 2018;**23**(9):2322. DOI: 10.3390/molecules23092322

[61] Ciulu M, Spano N, Pilo MI, Sanna G, Battino M, Niki E, et al. Recent advances in the analysis of phenolic compounds in unifloral honeys. Molecules. 2016;**21**(4):451. DOI: 10.3390/MOLECULES21040451

[62] Davies AMC. Amino acid analysis of honeys from eleven countries. Journal of Apicultural Research. 1975;**14**(1):29-39. Available from: https://www.tandfonline.com/doi/abs/10.1080/00218839.1975.11099798

[63] Davies AMC. The application of amino acid analysis to the determination of the geographical origin of honey. International Journal of Food Science and Technology. 1976;**11**(5):515-523. DOI: 10.1111/j.1365-2621.1976.tb00752.x

[64] Santos S, Maia M, Barros A, Gouvinhas I. Assessment of phenolic content, antioxidant and anti-aging activities of honey from *Pittosporum undulatum* vent. Naturalized in the Azores Archipelago (Portugal). Applied Sciences. 2023;**13**(3):1788. DOI: 10.3390/MOLECULES21040451

[65] MEL DOS AÇORES - DOP - MEL DOS AÇORES - DOP - Instituto de Alimentação e Mercados Agrícolas - Portal [Internet]. Available from: https://portal.azores.gov.pt/web/iama/-/mel-dos-acores-dop

[66] Alves A, Ramos A, Gonçalves MM, Bernardo M, Mendes B. Antioxidant activity, quality parameters and mineral content of Portuguese monofloral honeys. Journal of Food Composition and Analysis. 2013;**30**(2):130-138. DOI: 10.1016/j.jfca.2013.02.009

[67] Alqarni AS, Owayss AA, Mahmoud AA, Hannan MA. Mineral content and physical properties of local and imported honeys in Saudi Arabia. Journal of Saudi Chemical Society. 2014;**18**(5):618-625. DOI: 10.1016/j.jscs.2012.11.009

[68] Kamal MA, Klein P. Determination of sugars in honey by liquid chromatography. Saudi Journal of Biological Sciences. 2011;**18**(1):17-21. DOI: 10.1016/J.SJBS.2010.09.003

[69] Damodaran S, Parkin KL. Fennema's food chemistry. In: Fachinetto MR, editor. Techbooks. Taylor & Francis; 2010. 1104 p

[70] Smanalieva J, Senge B. Analytical and rheological investigations into selected unifloral German honey. European Food Research and Technology. 2009;**229**(1):107-113. DOI: 10.1007/s00217-009-1031-2

[71] Escuredo O, Míguez M, Fernández-González M, Carmen SM. Nutritional value and antioxidant activity of honeys produced in a European Atlantic area. Food Chemistry. 2013;**138**(2-3):851-856. DOI: 10.1016/j.foodchem.2012.11.015

[72] Won SR, Li CY, Kim JW, Rhee HI. Immunological characterization of honey major protein and its application. Food Chemistry. 2009;**113**(4):1334-1338. DOI: 10.1016/J.FOODCHEM.2008.08.082

[73] White JW, Kushnir I. The enzymes of honey: Examination by ionexchange chromatography, gel filtration, and starch-gel electrophoresis. Journal of Apicultural Research. 1967;**6**(2):69-89. DOI: 10.1080/00218839.1967.11100163

[74] Sak-Bosnar M, Sakač N. Direct potentiometric determination of diastase activity in honey. Food Chemistry. 2012;**135**(2):827-831. DOI: 10.1016/J.FOODCHEM.2012.05.006

[75] Nozal MJ, Bernal JL, Toribio L, Jiménez JJ, Martín MT. High-performance liquid chromatographic determination of methyl anthranilate, hydroxymethylfurfural and related compounds in honey. Journal of Chromatography. A. 2001;**917**(1-2):95-103. DOI: 10.1016/S0021-9673(01)00702-6

[76] Tornuk F, Karaman S, Ozturk I, Toker OS, Tastemur B, Sagdic O, et al. Quality characterization of artisanal and retail Turkish blossom honeys: Determination of physicochemical, microbiological, bioactive properties and aroma profile. Industrial Crops and Products. 2013;**46**:124-131. DOI: 10.1016/J.INDCROP.2012.12.042

[77] Grainger MNC, Owens A, Manley-Harris M, Lane JR, Field RJ. Kinetics of conversion of dihydroxyacetone to methylglyoxal in New Zealand mānuka honey: Part IV – Formation of HMF. Food Chemistry. 2017;**232**:648-655. DOI: 10.1016/J.FOODCHEM.2017.04.066

[78] Cherchi A, Spanedda L, Tuberoso C, Cabras P. Solid-phase extraction and high-performance liquid chromatographic determination of organic acids in honey. Journal of Chromatography. A. 1994;**669**(1-2):59-64. DOI: 10.1016/0021-9673(94)80336-6

[79] Desmouli A. Le miel: origine et composition., Actualités Pharmaceutiques. 2013;**52**:18-21. DOI: 10.1016/j.actpha.2013.10.004

[80] León-Ruiz V, Vera S, González-Porto AV, San Andrés MP. Analysis of water-soluble vitamins in honey by isocratic RP-HPLC. Food Analytical Methods. 2013;**6**(2):488-496. DOI: 10.1007/s12161-012-9477-4

[81] Đogo Mračević S, Krstić M, Lolić A, Ražić S. Comparative study of the chemical composition and biological potential of honey from different regions of Serbia. Microchemical Journal. 2020;**152**:104420. DOI: 10.1016/J.MICROC.2019.104420

[82] Pohl P, Stecka H, Greda K, Jamroz P. Bioaccessibility of Ca, Cu, Fe, Mg, Mn and Zn from commercial bee honeys. Food Chemistry. 2012;**134**(1):392-396. DOI: 10.1016/J.FOODCHEM.2012.02.065

[83] Lazarević KB, Jovetić MS, Tešić ŽL. Physicochemical parameters as a tool for the assessment of origin of honey. Journal of AOAC International. 2017;**100**(4):840-851. DOI: 10.5740/JAOACINT.17-0143

[84] Dżugan M, Wesołowska M, Zaguła G, Kaczmarski M, Czernicka M, Puchalski C. Honeybees (*Apis mellifera*) as a biological barrier for contamination of honey by environmental toxic metals. Environmental Monitoring and Assessment. 2018;**190**(2):101. DOI: 10.1007/S10661-018-6474-0

[85] Anastasiadi M, Pratsinis H, Kletsas D, Skaltsounis AL, Haroutounian SA. Grape stem extracts: Polyphenolic content and assessment of their in vitro antioxidant properties. LWT - Food Science and Technology. 2012;**48**(2):316-322. DOI: 10.1016/j.lwt.2012.04.006

[86] Gouvinhas I, Queiroz M, Rodrigues M, Barros AIRNA. Evaluation of the phytochemistry and biological activity of grape (*Vitis vinifera* L.) stems: Toward a sustainable winery industry. In: Polyphenols in Plants. Academic Press; 2019. pp. 381-394. DOI: 10.1016/B978-0-12-813768-0.00023-2

[87] Machado NFL, Domínguez-Perles R. Addressing facts and gaps in the phenolics chemistry of winery by-products. Molecules. 2017:48. DOI: 10.3390/molecules22020286

[88] Dai J, Mumper RJ. Plant phenolics: Extraction, analysis and their antioxidant and anticancer properties. Molecules. 2010:7313-7352. DOI: 10.3390/molecules15107313

[89] Burton GJ, Jauniaux E. Oxidative stress. Best Practice & Research. Clinical Obstetrics & Gynaecology. 2011;**25**(3):287-299. DOI: 10.1016/J.BPOBGYN.2010.10.016

[90] Bravo L. Polyphenols: Chemistry, dietary sources, metabolism, and nutritional significance. Nutrition Reviews. 1998;**56**(11):317-333. DOI: 10.1111/J.1753-4887.1998.TB01670.X

[91] Tsao R. Chemistry and biochemistry of dietary polyphenols. Nutrients. 2010;**2**(12):1231-1246. DOI: 10.3390/NU2121231

[92] Abraão AS, Fernandes N, Silva AM, Domínguez-Perles R, Barros A. *Prunus lusitanica* L. fruits as a novel source of bioactive compounds with antioxidant potential: Exploring the unknown. Antioxidants. 2022;**11**(9):1738. DOI: 10.3390/ANTIOX11091738

[93] Yao LH, Jiang YM, Shi J, Tomás-Barberán FA, Datta N, Singanusong R, et al. Flavonoids in food and their health benefits. Plant Foods for Human Nutrition. 2004;**59**(3):113-122. DOI: 10.1007/s11130-004-0049-7

[94] Pascual-Teresa S, Moreno DA, García-Viguera C. Flavanols and anthocyanins in cardiovascular health: A review of current evidence. International Journal of Molecular Sciences. 2010;**11**(4):1679-1703. DOI: 10.3390/IJMS11041679

[95] Scalbert A, Williamson G. Dietary intake and bioavailability of polyphenols. The Journal of Nutrition. 2000;**130**(8):2073S-2085S. DOI: 10.1093/jn/130.8.2073S

[96] Silici S, Sagdic O, Ekici L. Total phenolic content, antiradical, antioxidant and antimicrobial activities of rhododendron honeys. Food Chemistry. 2010;**121**(1):238-243. DOI: 10.1016/J.FOODCHEM.2009.11.078

[97] Trautvetter S, Koelling-Speer I, Speer K. Confirmation of phenolic acids and flavonoids in honeys by UPLC-MS.

Apidologie. 2009;**40**(2):140-150. DOI: 10.1051/apido/2008072

[98] Ronsisvalle S, Lissandrello E, Fuochi V, Petronio Petronio G, Straquadanio C, Crascì L, et al. Antioxidant and antimicrobial properties of *Casteanea sativa* Miller chestnut honey produced on Mount Etna (Sicily). Natural Product Research. 2019;**33**(6):843-850. DOI: 10.1080/14786419.2017.1413568

[99] Tomás-Barberán FA, Martos I, Ferreres F, Radovic BS, Anklam E. HPLC flavonoid profiles as markers for the botanical origin of European unifloral honeys. Journal of the Science of Food and Agriculture. 2001;**81**(5):485-496. DOI: 1002/jsfa.836

[100] Perna A, Intaglietta I, Simonetti A, Gambacorta E. A comparative study on phenolic profile, vitamin C content and antioxidant activity of Italian honeys of different botanical origin. International Journal of Food Science and Technology. 2013;**48**(9):1899-1908. DOI: 10.1111/ijfs.12169

[101] Di Marco G, Gismondi A, Panzanella L, Canuti L, Impei S, Leonardi D, et al. Botanical influence on phenolic profile and antioxidant level of Italian honeys. Journal of Food Science and Technology. 2018;**55**(10):4042-4050. DOI: 10.1007/s13197-018-3330-8

[102] Escuredo O, Silva LR, Valentão P, Seijo MC, Andrade PB. Assessing Rubus honey value: Pollen and phenolic compounds content and antibacterial capacity. Food Chemistry. 2012;**130**(3):671-678. DOI: 10.1016/J.FOODCHEM.2011.07.107

[103] Pyrzynska K, Biesaga M. Analysis of phenolic acids and flavonoids in honey. TrAC Trends in Analytical Chemistry. 2009;**28**(7):893-902. DOI: 10.1016/J.FOODCHEM.2011.07.107

[104] Liang Y, Cao W, Chen W, jun, Xiao X hong, Zheng J bin. Simultaneous determination of four phenolic components in citrus honey by high performance liquid chromatography using electrochemical detection. Food Chemistry. 2009;**114**(4):1537-1541. DOI: 10.1016/j.foodchem.2008.11.024

[105] Hossen MS, Ali MY, Jahurul MHA, Abdel-Daim MM, Gan SH, Khalil MI. Beneficial roles of honey polyphenols against some human degenerative diseases: A review. Pharmacological Reports. 2017;**69**(6):1194-1205. DOI: 10.1016/J.PHAREP.2017.07.002

[106] Ferreres F, Andrade P, Gil MI, Tomás-Barberán FA. Floral nectar phenolics as biochemical markers for the botanical origin of heather honey. Zeitschrift für Lebensmittel Untersuchung und Forschung. 1996;**202**(1):40-44. DOI: 10.1007/BF01229682

[107] Nolan VC, Harrison J, Cox JAG. Dissecting the antimicrobial composition of honey. Antibiotics. 2019;**8**(4):251. DOI: 10.3390/antibiotics8040251

[108] Estevinho L, Pereira AP, Moreira L, Dias LG, Pereira E. Antioxidant and antimicrobial effects of phenolic compounds extracts of Northeast Portugal honey. Food and Chemical Toxicology. 2008;**46**(12):3774-3779. DOI: 10.1016/J.FCT.2008.09.062

[109] Shen N, Wang T, Gan Q, Liu S, Wang L, Jin B. Plant flavonoids: Classification, distribution, biosynthesis, and antioxidant activity. Food Chemistry. 2022:132531. DOI: 10.1016/J.FOODCHEM.2022.132531

[110] Vuolo MM, Lima VS, Maróstica Junior MR. Phenolic Compounds: Structure, Classification, and Antioxidant

Power. Bioactive Compounds: Health Benefits and Potential Applications. Woodhead Publishing; 2019. pp. 33-50. DOI: 10.1016/B978-0-12-814774-0.00002-5

[111] Loizzo MR, Sicari V, Pellicanò T, Xiao J, Poiana M, Tundis R. Comparative analysis of chemical composition, antioxidant and anti-proliferative activities of Italian *Vitis vinifera* by-products for a sustainable agro-industry. Food and Chemical Toxicology. 2019;**127**:127-134. DOI: 10.1016/J.FCT.2019.03.007

[112] Habib HM, Kheadr E, Ibrahim WH. Inhibitory effects of honey from arid land on some enzymes and protein damage. Food Chemistry. 2021;**364**:130415. DOI: 10.1016/J.FOODCHEM.2021.130415

[113] Mareček V, Mikyška A, Hampel D, Čejka P, Neuwirthová J, Malachová A, et al. ABTS and DPPH methods as a tool for studying antioxidant capacity of spring barley and malt. Journal of Cereal Science. 2017;**73**:40-45. DOI: 10.1016/J.JCS.2016.11.004

[114] Gheldof N, Wang XH, Engeseth NJ. Buckwheat honey increases serum antioxidant capacity in humans. Journal of Agricultural and Food Chemistry. 2003;**51**(5):1500-1505. DOI: 10.1021/jf025897t

[115] Mahdi-Pour B, Jothy SL, Latha LY, Chen Y, Sasidharan S. Antioxidant activity of methanol extracts of different parts of *Lantana camara*. Asian Pacific Journal of Tropical Biomedicine. 2012;**2**(12):960-965. DOI: 10.1016/S2221-1691(13)60007-6

[116] Frankel EN. Nutritional benefits of flavonoids. In: Food Factors Cancer Prevention. Springer; 1997. pp. 613-616. DOI: 10.1007/978-4-431-67017-9_121

[117] Ben Sghaier M, Skandrani I, Nasr N, Franca MGD, Chekir-Ghedira L, Ghedira K. Flavonoids and sesquiterpenes from Tecurium ramosissimum promote antiproliferation of human cancer cells and enhance antioxidant activity: A structure–activity relationship study. Environmental Toxicology and Pharmacology. 2011;**32**(3):336-348. DOI: 10.1016/j.etap.2011.07.003

[118] Gouvinhas I, Machado N, Sobreira C, Domínguez-Perles R, Gomes S, Rosa E, et al. Critical review on the significance of olive phytochemicals in plant physiology and human health. Molecules. 2017;**22**(11):1-35. DOI: 10.3390/MOLECULES22111986

[119] Wang TY, Li Q, Bi KS. Bioactive flavonoids in medicinal plants: Structure, activity and biological fate. Asian Journal of Pharmaceutical Sciences. 2018;**13**(1):12-23. DOI: 10.1016/J.AJPS.2017.08.004

[120] Dossi N, Toniolo R, Impellizzieri F, Tubaro F, Bontempelli G, Terzi F, et al. A paper-based platform with a pencil-drawn dual amperometric detector for the rapid quantification of ortho-diphenols in extravirgin olive oil. Analytica Chimica Acta. 2017;**950**:41-48. DOI: 10.1016/J.ACA.2016.11.030

[121] Silva FAM, Borges F, Guimarães C, Lima JLFC, Matos C, Reis S. Phenolic acids and derivatives: Studies on the relationship among structure, radical scavenging activity, and physicochemical parameters. Journal of Agricultural and Food Chemistry. 2000;**48**(6):2122-2126. DOI: 10.1021/jf9913110

[122] Chmiel JA, Daisley BA, Pitek AP, Thompson GJ, Reid G. Understanding the effects of sublethal pesticide exposure on honey bees: A role for

probiotics as mediators of environmental stress. Frontiers in Ecology and Evolution. 2020;**8.** Article 22. DOI: 10.3389/fevo.2020.00022

[123] Das A, Datta S, Mukherjee S, Bose S, Ghosh S, Dhar P. Evaluation of antioxidative, antibacterial and probiotic growth stimulatory activities of *Sesamum indicum* honey containing phenolic compounds and lignans. LWT - Food Science and Technology. 2015;**61**(1):244-250. DOI: 10.1016/J.LWT.2014.11.044

[124] Castiglioni S, Stefano M, Astolfi P, Carloni P. Chemometric approach to the analysis of antioxidant properties and colour of typical Italian monofloral honeys. International Journal of Food Science and Technology. 2017;**52**(5):1138-1146. DOI: 10.1111/ijfs.13397

[125] Apraj V, Pandita N. Evaluation of skin anti-aging potential of *Citrus reticulata* Blanco peel. Pharmacognosy Research. 2016;**8**(3):160. DOI: 10.4103/0974-8490.182913

[126] Rodrigues F, Cádiz-Gurrea MDLL, Nunes MA, Pinto D, Vinha AF, Linares IB, et al. Cosmetics. In: Polyphenols: Properties, Recovery, and Applications. Elsevier; 2018. pp. 393-427. DOI: 10.1016/B978-0-12-813572-3.00012-9

[127] Di PA, Santos-Buelga C, Era B, González-Paramás A, Tuberoso IG, Medda R, et al. Sardinian honeys as sources of xanthine oxidase and tyrosinase inhibitors. Food Science and Biotechnology. 2017:27. DOI: 10.1007/s10068-017-0275-z

[128] Karaçelik AA, Sahin H. Determination of enzyme inhibition and antioxidant activity in some chestnut honeys. Foods and Raw Materials. 2018;**6**(1):210-218. DOI:10.21603/2308-4057-2018-1-210-218

[129] Ersoy E, Eroglu E, Boga M, Abdullah M. Anti-aging potential and anti-tyrosinase activity of three Hypericum species with focus on phytochemical composition by LC – MS/MS. Industrial Crops and Products. 2019;**141**:1-9. DOI: 10.1016/j.indcrop.2019.111735

[130] Aguilar-Toalá JE, Hernández-Mendoza A, González-Córdova AF, Vallejo-Cordoba B, Liceaga AM. Potential role of natural bioactive peptides for development of cosmeceutical skin products. Peptides. 2019;**122**:170170. DOI: 10.1016/j.peptides.2019.170170

[131] Aumeeruddy MZ, Aumeeruddy-Elalfi Z, Neetoo H, Zengin G, Blom A, Fibrich B, et al. Pharmacological activities, chemical profile, and physicochemical properties of raw and commercial honey. Biocatalysis and Agricultural Biotechnology. 2019;**18**:101005. DOI: 10.1016/j.bcab.2019.01.043

[132] Mathen C, Soman G, Peter S, Thergaonkar R, Teredesai M. Evaluation of anti-elastase and antioxidant activity in antiaging formulations containing terminalia extracts. International Journal of Herbal Medicine. 2014;**2**(2):95-99. ISSN: 2321-2187

[133] Leal C, Gouvinhas I, Santos RA, Rosa E, Silva AM, José M, et al. Potential application of grape (*Vitis vinifera* L.) stem extracts in the cosmetic and pharmaceutical industries: Valorization of a by-product. Industrial Crops and Products. 2020;**154**:112675. DOI: 10.1016/j.indcrop.2020.112675

Section 2

# Lipophilic Antioxidants and Micronutrients

Chapter 6

# Fat-Soluble Antioxidants: Role of Postprandial Lipoproteins

*Ngoc-Anh Le*


## Abstract

Many commonly known antioxidants, from probucol to vitamin E, are fat-soluble and have been shown to be most effective when administered with meals. Following meal consumption, these compounds are incorporated into intestinal lipoproteins, known as chylomicrons, and secreted into the circulation. These lipid-carrying particles are responsible for the transport of newly absorbed dietary fat for delivery to peripheral tissues. In the bloodstream, chylomicrons interact with heparin-releasable lipases common known as lipoprotein lipase and hepatic triglyceride lipase. Bothe lipases are anchored along the endothelial wall via heparan sulfate proteoglycans and have triglycerides as their preferred substrate. During this process, as dietary triglycerides are hydrolyzed and transported across the endothelium, we hypothesize that antioxidants carried in chylomicrons would be delivered directly to the arterial wall where they would be most effective in quenching reactive oxygen species generated by activated macrophages. Thus, the metabolism of postprandial lipoproteins is a key process in the defense against oxidative stress and may provide the path for effective antioxidant management. In this chapter, we will review the evidence in support of the hypothesis that postprandial lipoproteins may contribute to the delivery of fat-soluble antioxidants that are administered orally.


**Keywords:** lipoproteins, chylomicrons, postprandial, lipoprotein lipase, reactive oxygen species, vitamin E

## 1. Introduction

Atherosclerosis is a chronic condition associated with the accumulation of cholesterol along the wall of blood vessels leading to the occlusion of normal blood flow [1]. In his 1978 Lyman Duff lecture, Zilversmit reviewed his work on the development of atherosclerosis in the rabbit model with emphasis on the role of postprandial lipoproteins [2]. The data suggested a close interaction between the endothelial lipoprotein lipase and triglyceride-rich (TG-rich) lipoproteins [3]. While subsequent research implicated the role of oxidative stress in the development and progression of atherosclerotic disease [1, 4–6], the role of TG-rich lipoproteins in this process is not understood.

Oxidative stress, a state of imbalance between oxidants and antioxidants in favor of oxidants, is associated with disruption of redox signaling and metabolic functions [7] resulting in many chronic conditions, from type 2 diabetes mellitus [8], atherosclerosis [4] to neurodegenerative diseases [9]. Multiple lines of research have suggested a role for pharmacologic as well as nutritional antioxidants in the management of these conditions. While several observational studies with various agents have reported promising results, several case-control studies as well as large prospective studies have yielded conflicting results. Several explanations have been suggested, including differences in the mode and time of delivery, as well as differences in plasma levels versus concentrations at the site(s) of action.

In this section we will review the evidence that suggest a role of TG-rich lipoproteins, in general, and specifically postprandial lipoproteins as a double-edged sword in the propagation of oxidant radicals from a local site of inflammation to other tissues as well as in the delivery of fat-soluble antioxidants.

## 2. Metabolism of postprandial lipoproteins

For the routine medical examination, most diagnostic blood tests are based on measurement of the levels of metabolites in fasting plasma. Considering the fact that a typical individual would be consuming three main meals each day, 4 to 6 hours apart, with possible snacks in-between, one is constantly in a non-fasted state. Throughout the waking hours, there are postprandial lipoproteins carrying newly absorbed nutrients for delivery throughout the body. Dietary fats and fat-soluble vitamins are packaged into large spherical particles, ranging from 80 to 1200 nm in diameter, and are known as intestinal lipoproteins or chylomicrons. **Figure 1** illustrates the different organs that are involved in the postprandial response.

Plasma chylomicrons are constituted of triglycerides (TG, 85–92%), phospholipids (PL, 6–12%), cholesterol (CHOL, 1–3%) and protein (1–2%). The protein component serves as solubilizing coat to allow these lipid-rich particles to circulate in the aqueous environment and be delivered to the appropriate peripheral tissues [10]. These proteins are referred to as apolipoproteins and consist of apoAs, apoB-48, apoCs, and apoE. ApoAs serve primarily as part of the solubilizing coat and do not play a role in the metabolism of these intestinal lipoproteins. On the other hand, apoB-48 is the major structural protein and is required for the formation and secretion of chylomicrons. Genetic deficiency in apoB-48, known as abetalipoproteinemia, is associated with fat malabsorption [11]. ApoCs are responsible for the modulation of lipoprotein

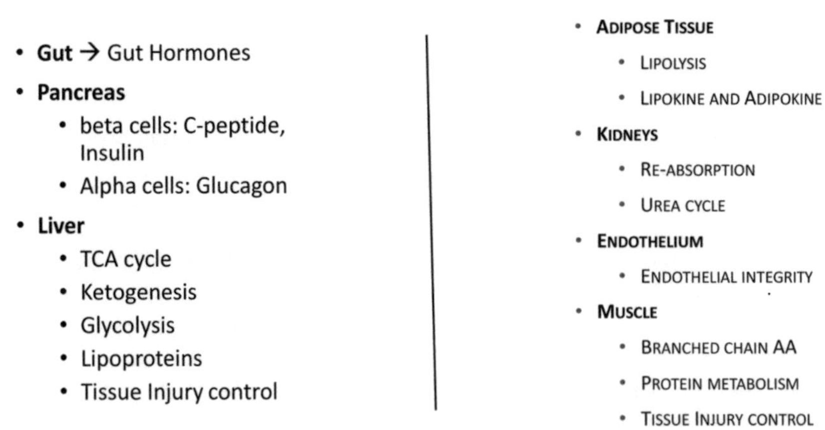


**Figure 1.**
*Organs involved in the postprandial response.*

lipase activity in the hydrolysis of triglycerides. ApoC-II that is present on the surface of chylomicrons is a required activator for lipoprotein lipase (LPL) [12]. Genetic deficiency in apoC-II is associated with significant elevations in plasma TG and the clinical condition known as type I hyperlipoproteinemia or hyperchylomicronemia. On the other hand, apoC-III, also present on the surface of chylomicrons, acts as an inhibitor of LPL activity [13]. ApoE serves as the ligand for the remnant receptor that is responsible for the uptake of the TG-depleted chylomicrons [14, 15].

Chylomicrons are assembled in enterocytes from dietary lipids and enter the systemic circulation via the lacteals. In the circulation they interact with lipoprotein lipase (LPL) that is attached to the endothelium throughout the body. LPL is responsible for the cleavage of TG into monoglycerides and fatty acids that can freely move into the cells for storage and energy utilization. Zilversmit and co-workers have previously demonstrated a link between the localization of lipoprotein lipase and accumulation of cholesterol transported in TG-rich chylomicrons [3]. **Figure 2A** schematizes the interactions between TG-rich lipoproteins and endothelial lipoprotein lipase. As the result of the opposite actions of apoC-II and apoC-III, partially delipidated chylomicrons are released back into the circulation and are available to interact with other lipoprotein lipase sites downstream.

The tight junction formed between TG-rich lipoproteins and the endothelium during TG hydrolysis are suitable for the flow of newly released monoglycerides and free fatty acids into the sub-endothelium. This flow of metabolites, however, could be bi-directional. We postulate that certain metabolites could also diffuse from the sub-endothelium onto the lipoproteins that are still in the circulation. In the presence of an inflamed arterial wall, excess reactive oxygen species (ROS) that are generated by resident activated macrophages could potentially seed fatty acids, preferentially highly oxidizable polyunsaturated fatty acids present on plasma TG-rich lipoproteins (**Figure 2B**). These particles could subsequently deliver oxidatively modified fatty acids to other tissues and thus propagating the oxidative process.

To examine this hypothesis, we carried out a number of studies looking at the meal-induced changes in oxidative markers. In a group of patients with documented coronary artery disease (CAD) we reported an acute and transient reduction in the levels of circulating autoantibodies (AAb) against malondialdehyde (MDA)-modified low-density lipoproteins (LDL), MDA-LDL following meal consumption [16]. This meal-induced reduction in AAb levels was not observed in a group a young healthy individuals with no history of heart disease [16]. In a group of patients with hypercholesterolemia, we subsequently reported that this transient response could not be demonstrated following oral challenges enriched either saturated or monounsaturated fatty acids but was specific for test meal containing polyunsaturated fatty acids that are highly susceptible to oxidative modification [17].

More recent unpublished data from our laboratory further support the significance of oxidative modification during the interactions between plasma lipoproteins and the arterial wall. In a group of 57 men and women we were able to analyze baseline plasma and plasma obtained at 15 minutes following an intravenous bolus injection of heparin (60 IU/kg body weight). Participants were enrolled in a clinical trial conducted at the University of Miami (courtesy of Dr. Ronald Goldberg and his team) and received the heparin to assess the activities of heparin-releasable lipolytic enzymes. Participants were classified as diabetes mellitus (DM), impaired glucose tolerance (IGT), and normal glucose tolerance (NGT) according to the response to the standard oral glucose tolerance test. Pre- and post-heparin plasma were immediately frozen at −80°C and shipped in dry ice to our laboratory at the end of the study.

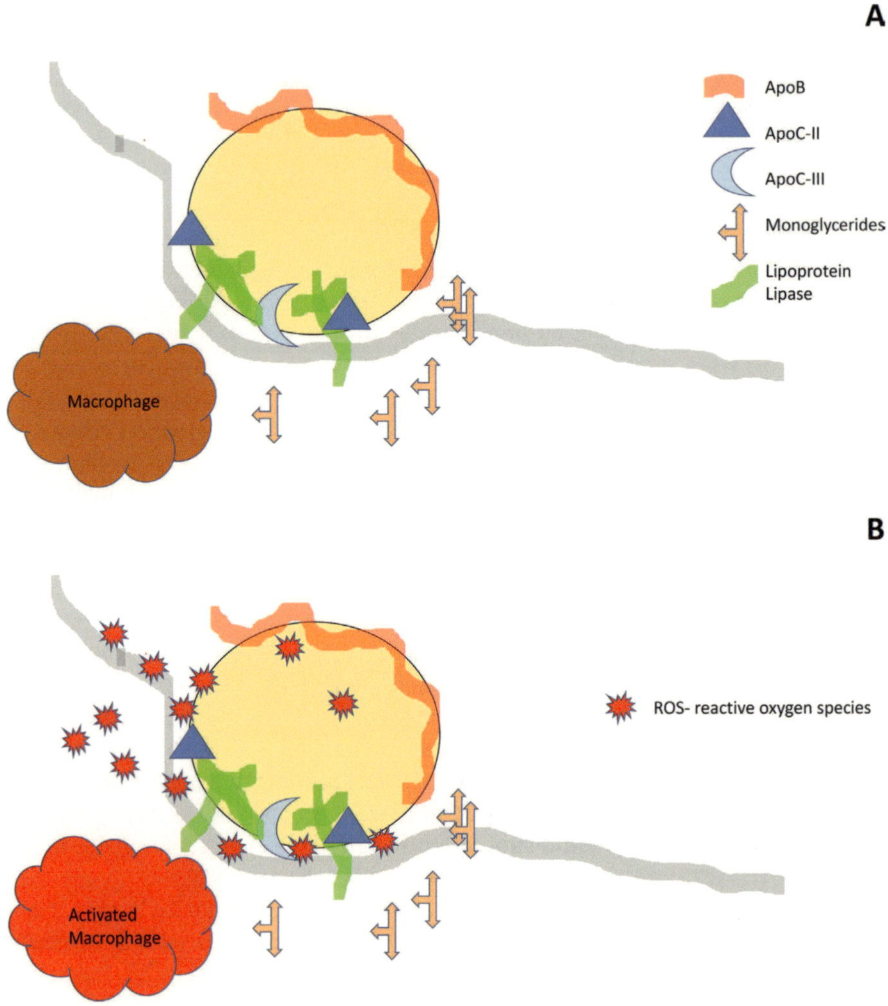


**Figure 2.**
*Interactions of TG-rich lipoproteins with endothelial lipoprotein lipase.*

Plasma levels of autoantibodies against MDA-LDL (AAb) and LDL-IgG immune complexes (IC) were measured by ELISA as previously described [16]. Plasma levels of oxLDL were determined by enzyme-linked immunoassay (ELISA) using a commercial kit (www.mercodia.com). **Figure 3** illustrates the percent increased (%D) in plasma levels of oxLDL, AAb, and IC following heparin injection. There was a 25% increase in plasma oxLDL for NGT as compared to 22% for DM + IGT (not significant). In fact, while the increase from pre-heparin levels were statistically significant ($p < 0.01$), diabetic status did not affect the changes in any measured parameter following heparin. This observation suggests that there are significant markers of oxidatively modified lipoproteins attached to the arterial wall that could be released by heparin.

Using the same ELISA assay for oxLDL (www.mercodia.com), we measured plasma levels of oxLDL and apoB in frozen samples previously collected from our postprandial studies [16, 17]. Plasma apoB was determined by immunoturbidometric method on the AU480 chemistry autoanalyzer (Beckman Diagnostics, LaBrea, CA)

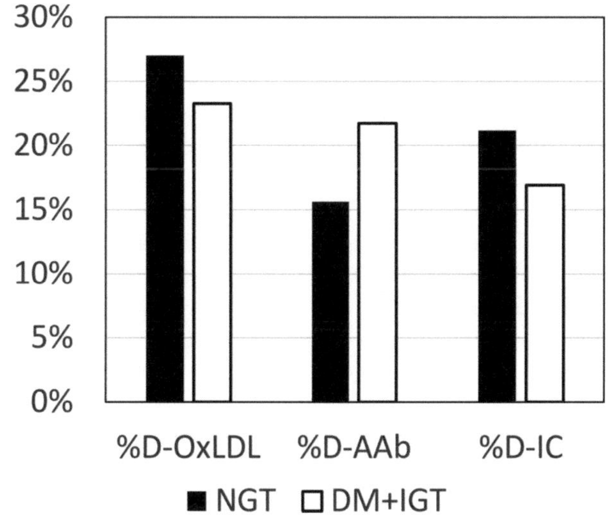


**Figure 3.**
*Increases in markers of lipoprotein oxidation following intravenous heparin.*

using reagents and calibrators from Sekisui Diagnostics (www.sekisui-dx.com). **Figure 4** illustrates the fold-increases in plasma apoB and the ratio of oxLDL/apoB following meal consumption. While there was minimal change in plasma apoB during the postprandial period for these two representative individuals, there was statistically significant increase in oxLDL/apoB at 4 hours following the meal. In fact, in individuals with documented CAD, it would appear that the oxLDL levels were elevated by 2 hours after the meal and remained elevated at 6 hours in some individuals. Additional studies are needed to confirm these increases in direct markers of lipoprotein oxidation following heparin and meal consumption.

As summarized in this section, the interactions of plasma lipoproteins, in particular TG-rich lipoproteins with endothelial lipase could potentially play a central role in atherogenesis.

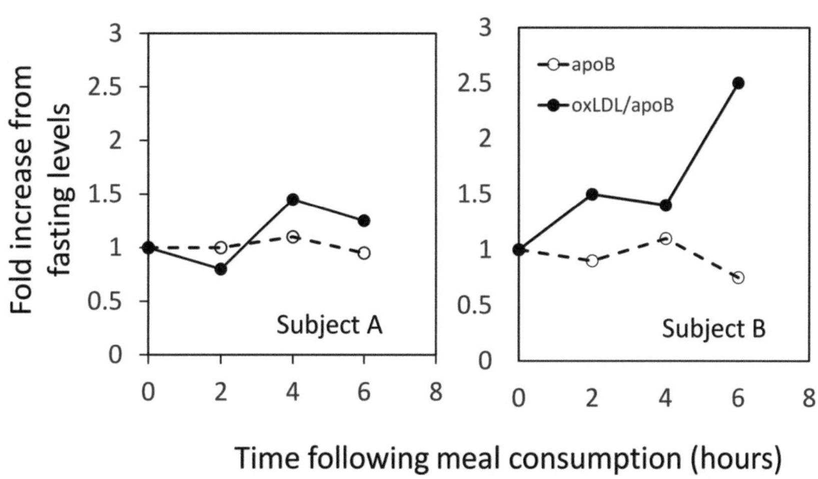


**Figure 4.**
*Increases in oxLDL following meal consumption.*

## 3. Probucol as a fat-soluble antioxidant

While hypolipidemic drugs, in particular bile acid resins [18] and 3-hydroxy-3-methylglutaryl coenzyme A (HMG CoA) reductase inhibitors [19], have been the focus of pharmaceutical approach to the management of atherosclerosis [20], recent studies would suggest that certain antioxidants may also play a role in the prevention of atherosclerosis.

Probucol is a diphenolic compound that has a modest cholesterol-lowering effect before the introduction of the statins [21]. In vitro experiments indicated that probucol can inhibit oxidative modification of LDL [22], inhibit lipid accumulation in THP-1 cells, and enhance the release of cholesterol from macrophages [23]. In the landmark study using the Watanabe heritable hyperlipidemic (WHHL) rabbit model, Carew et al. [24] demonstrated that in spite of negligeable reduction in plasma cholesterol, animals treated with probucol had significantly less arterial disease as compared to animals treated with the HMG CoA reductase inhibitor lovastatin. In view of the observation that probucol is a lipid-soluble compound and is found associated with plasma LDL, subsequent studies demonstrated that LDL isolated from patients treated with probucol was resistant to ex vivo oxidation [25]. The ability of probucol to protect LDL from oxidative modification appears to be concentration dependent [26]. Furthermore, lipoprotein oxidation does not proceed until available probucol has been degraded [26].

While the mechanism responsible for the in vivo oxidative modification of LDL is still poorly understood, several in vitro processes have been demonstrated, including exposure to lipid oxidation products [27], monocyte-macrophages [28], endothelial cells [29] and smooth muscle cells [30]. Damaged endothelium can generate reactive intermediates that result in the oxidation of LDL [4] which trigger the release of growth factors responsible for tissue proliferation. Several trials with probucol in the prevention of restenosis after coronary angioplasty highlighted the importance of the timing of antioxidant therapy in the management of atherosclerosis. In an earlier trial when probucol + lovastatin was started 48 hours prior to angioplasty, minimal protection from restenosis was noted [31]. When probucol was administered 30 days prior to angioplasty, significant improvement in all indices of restenosis was observed [32]. While plasma levels of both probucol and alpha-tocopherol were at plateau level at the time of angioplasty, treatment with multivitamins (beta carotene + vitamin E + vitamin C) did not have any effect on the rate of restenosis.

The key take-away points from these studies are (1) the antioxidant protection depends on the concentration of probucol on the LDL particles and (2) the efficacy of the antioxidant therapy depends on prior exposure of the arterial wall to lipoproteins that are protected from oxidative modification. It should also point out that the modest cholesterol-lowering effect (10–20%) achieved with probucol was accounted by reduction in both low-density lipoproteins (LDL) and high-density lipoproteins (HDL). The reduction in HDL-cholesterol was reported to be the result of reduced production of the primary proteins of HDL [21] and is considered undesirable as HDL is regarded as a negative risk factor for coronary heart disease [33].

## 4. AGI-1067 as a fat-soluble antioxidant

Succinobucol also known as AGI-1067 is a synthetic analogue of probucol with strong anti-oxidant properties [34, 35]. In the Canadian Antioxidant Restenosis Trial (CART-1), treatment with either AGI-1067 or probucol before and after percutaneous coronary intervention (PCI) resulted in increases in luminal area at the PCI site [36].

In a series of animal studies including cholesterol-fed cynomolgus monkeys, cholesterol-fed rabbits, LDL-receptor (LDLr) −/− and apoE −/− mice, lipoproteins fractionated from frozen plasma by fast-phase liquid chromatography [37, 38] were shown to contain AGI-1067 [34]. LDL fractions from animals treated with AGI-1067 had reduced reactivity against a monoclonal antibody specific to oxidatively modified LDL and were protected from ex vivo oxidation in the presence of $Cu_2SO_4$ [34].

Consistent with these observations, an investigator-initiated, nested case-control study with frozen plasma from participants in the CART-1 trial also noted that both very low density lipoproteins (VLDL) and LDL from individuals receiving AGI-1067 were protected from ex vivo oxidative modification. Frozen plasma obtained from 40 participants at baseline and at 6-month of treatment with either AGI-1067 or placebo were adjusted to a density of 1.21 gm/mL to isolate the lipoprotein-rich fraction by ultracentrifugation. The lipoprotein-rich supernatants were subsequently fractionated by fast-phase liquid chromatography [37, 38] and VLDL and LDL fractions were pooled based on the cholesterol profile as previously described. ApoB contents in the pooled fractions corresponding to VLDL and LDL was quantitated by immunoassay and used to determine the amount of $Cu_2SO_4$ required for the ex vivo oxidation assay [39]. The detailed methodology has been published previously [40]. **Figure 5** illustrates the decrease in oxidative susceptibility of LDL isolated from a participant receiving AGI-1067 (**Figure 5A**) as compared to the lack of change for LDL isolated from an individual receiving placebo (**Figure 5B**). **Figure 6** illustrates the decrease in oxidative susceptibility of VLDL after 6-month therapy with AGI-1067 (**Figure 6A**) as compared to VLDL from an individual treated with placebo (**Figure 6B**). It should be noted that the ratio of $Cu_2SO_4$ to apoB was the same for all incubations and differences in maximum OD (270 nm) reflected total lipid contents in the incubation mixture. The mean increase in lag times with AGI-1067 was 260% (±50%) for isolated LDL and 110% (±35%) for isolated VLDL (unpublished data).

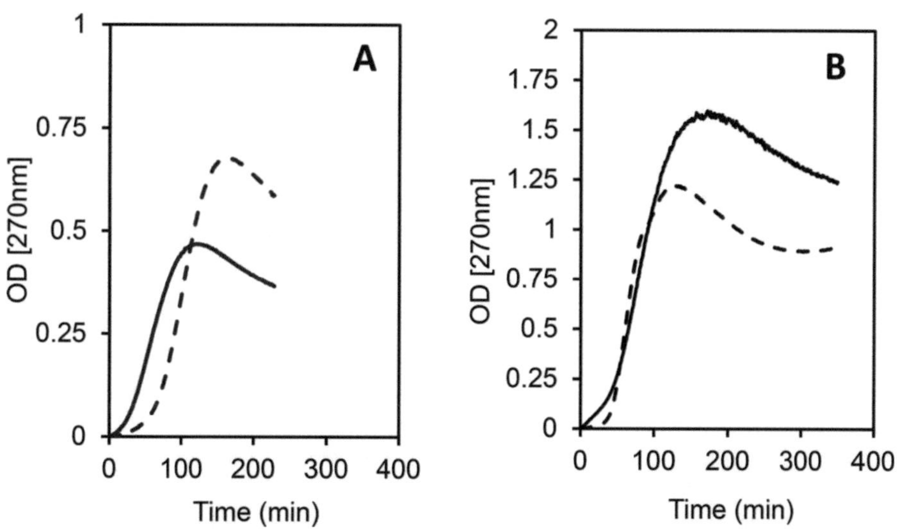


**Figure 5.**
*Ex vivo oxidative susceptibility of LDL with AGI-1067 (Panel A) v. placebo (Panel B). Baseline (solid line) post-treatment (dashed line).*

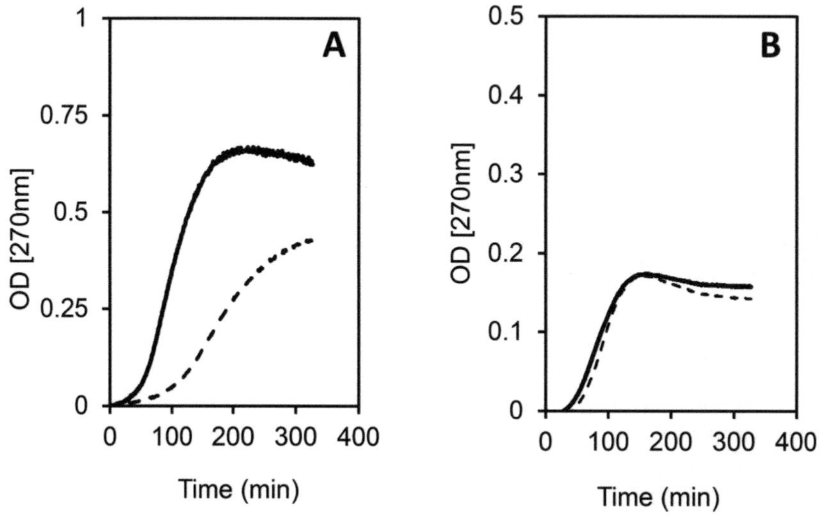


**Figure 6.**
*Ex vivo oxidative susceptibility of isolated VLDL with AGI-1067 (Panel A) v. placebo (Panel B). Baseline (solid line) post-treatment (dashed line).*

## 5. Vitamin E as a fat-soluble antioxidant

With this backdrop of beneficial effect of antioxidant agents on the progression of coronary artery atherosclerosis, there was considerable interest in trials with nutritional antioxidants [41, 42]. In the Basel Study [43] of healthy males monitored over a 7-year period, while there was no association between coronary death and vitamin E levels at baseline, risk of coronary death was higher in the lowest quartile of vitamin C and beta-carotene. Several studies have suggested that antioxidant vitamins, specifically vitamin E, in humans may decrease the susceptibility of LDL to oxidation ex vivo [44–46]. In a non-randomized trial with serial coronary angiography, Hodis et al. [47] reported that men with previously coronary artery bypass graft surgery demonstrated less lesion progression when colestipol-niacin therapy was supplemented with vitamin E and C. They also found no benefit of vitamin A and C supplementation in patients receiving placebo [47].

In a double-blind, placebo controlled trial with α-tocopherol or placebo, administered after successful percutaneous transluminal coronary angioplasty, the 31% reduction in incidence of restenosis with tocopherol did not reach statistical significance [48] in spite of a high dose of vitamin E supplementation (1200 IU/day). Possible explanations for the lack of significance ranged from small sample size to short-term follow-up, timing of the supplementation, and mode of administration. A substudy of the Heart Outcomes Prevention Evaluation (HOPE) trial in patients with known cardiovascular disease, supplementation with vitamin E at 400 IU/day had no effect on atherosclerosis progression as assessed by ultrasound measurements [49]. In the secondary prevention trial (CHAOS, Cambridge Heart Antioxidant Study) of patients with angiographically proven CAD, there was a 47% reduction in the primary end-point (cardiovascular death and nonfatal myocardial infarction) with vitamin E supplementation as compared to placebo [50]. In fact, this result was attributable to a reduction in the risk for nonfatal myocardial infarction with an 18% increase in cardiovascular death in the vitamin E group. Subsequent analysis suggested that only a small number of the deaths was actually compliant with the vitamin E regimen [51]. A number of studies also failed to demonstrate conclusive benefit of vitamin E supplementation, including the Gruppo Italiano per lo Studio Della Sopravvivvenza nell'Infarto Miocardico (GISSI)-Prevenzione Trial [52], the Heart Protection Study [53], and the Alpha-Tocopherol/Beta-Carotene Cancer (ATBC) prevention study [54].

A review of vitamin E metabolism may shed some light into these somewhat conflicting observations [55]. There are several natural forms of vitamin E typically classified as either tocopherols or tocotrienols, with α, β, γ, δ isoforms. The activity of these various forma depends on its absorption, transport, and delivery to the tissues. Dietary vitamin E are absorbed, secreted into the circulation as chylomicrons carrying newly absorbed TG. As chylomicrons are metabolized in the circulation, vitamin E are transferred to peripheral tissues. This association between chylomicron metabolism and vitamin E might explain the benefit of dietary but not supplemental vitamin E in heart disease outcome [56]. The importance of fat absorption and vitamin E bioavailability was further confirmed by a series of study by Traber and co-workers [57–59].

In the liver, α-tocopherol transfer protein (α-TTP) preferentially enriched hepatic very low density lipoproteins (VLDL) with α-tocopherol. This process explains why α-tocopherol is the most abundant form of vitamin E in plasma, but γ-tocopherol is the most abundant form in western diet. As chylomicrons are responsible for the delivery of dietary TG to peripheral tissues, VLDL are responsible for the delivery of newly synthesized TG to peripheral tissues. As VLDL are metabolized, α-tocopherol are transferred to other plasma lipoproteins, including low density lipoproteins (LDL) and high density lipoproteins (HDL). Impaired metabolism of VLDL would be expecting to affect the transfer of α-tocopherol to LDL and thus affect its resistance to oxidative modification.

In a recent review, Vardi et al. emphasized the importance of proper patient selection in the assessment of the role of vitamin E in the prevention of cardiovascular disease [60]. To ensure the optimal bioavailability of fat-soluble antioxidants, it is critical to have an adequate amount of dietary fats to enhance the absorption and incorporation of antioxidants in the TG-rich chylomicrons for secretion into the bloodstream.

## 6. Role of postprandial lipoproteins in the delivery of antioxidants

As in the case of both probucol and AGI-1067 which are fat soluble and are most effective when administered with meals, the bioavailability of vitamin E depends on fat absorption and incorporation of vitamin E in chylomicrons for delivery to the peripheral tissues. Fat-soluble antioxidants are secreted by the intestine as chylomicrons and are transferred to peripheral tissues as the TG-rich lipoproteins are hydrolyzed by lipoprotein lipase along the arterial wall. Thus, by delivering antioxidant agents directly to the peripheral tissues, it may be possible to affect the oxidative status of the arterial wall. This process may explain the increased benefit observed when treatment with probucol was initiated several weeks prior to angioplasty [31, 32] and allowed to continue for extended period after the procedure.

In view of the data from Traber and co-workers [58] on the bioavailability of vitamin E with different administration regimens, new clinical trials with specific instructions for participants to take the vitamin E supplement with meals would be required to demonstrate the benefits of vitamin E in cardiovascular disease prevention.

## 7. Conclusion

The bioactivity of dietary supplements and oral agents depends on several factors including absorption, transport and delivery. We have reviewed studies that support the hypothesis that meal consumption is associated with in vivo generation of lipoprotein-associated biomarkers of oxidative stress [16] as well as changes in ex vivo oxidative modification of lipoproteins [40]. We also presented preliminary data suggesting that subpopulation of lipoproteins that are retained along the arterial wall consists primarily of oxidatively modified particles. The in vivo generation of lipoprotein-associated lipoproteins reported in patients with inflamed endothelium suggests that the interactions of TG-rich lipoproteins during postprandial lipemia could allow the transfer of ROS from the sub-endothelium to plasma lipoproteins [16]. The observation that this postprandial response is transient and specific for meals enriched in polyunsaturated fatty acids [17] further suggest that the type of fatty acids may affect antioxidant property. Without the double bonds, monounsaturated fatty acids are not susceptible to oxidative modification from ROS transferred to plasma lipoproteins during and would not promote the propagation of ROS to other tissues. In conclusion, for optimal benefit from interventions with fat-soluble antioxidants, it is important that the antioxidant be administered with meal and the consideration for the type and quantity of fatty acids be taken into account to ensure the optimal synthesis and secretion of intestinal chylomicrons. Impaired metabolism of these intestinal lipoproteins would affect the delivery of antioxidants to the peripheral tissues and thus the beneficial effect of the treatment.

## Author details

Ngoc-Anh Le
Biomarker Core Laboratory, Foundation for Atlanta Veterans Education and Research (FAVER), Atlanta VAHCS, Decatur, Georgia, USA

*Address all correspondence to: anhleatcsl@gmail.com

## References

[1] Ross R. The pathogenesis of atherosclerosis: A perspective for the 1990's. Nature. 1993;**362**:801-809

[2] Zilversmit DB. Atherogenesis: A postprandial phenomenon. Circulation. 1979;**60**:473-485

[3] Zilversmit DB. A proposal linking atherogenesis to the interaction of endothelial lipase with TG-rich lipoproteins. Circulation Research. 1973;**33**:633-638

[4] Steinberg D, Parthasarathy S, Carew TE, Khoo JD, Witztum JL. Beyond cholesterol: Modifications of LDL that increase its atherogenicity. The New England Journal of Medicine. 1989;**320**:915-924

[5] Ross R. Atherosclerosis: An inflammatory disease. The New England Journal of Medicine. 2002;**340**:115-126

[6] Steinberg D, Witztum JL. Oxidized low-density lipoprotein and atherosclerosis. Arteriosclerosis, Thrombosis, and Vascular Biology. 2010;**30**(12):2311-2316

[7] Sies H. Oxidative stress: A concept in redox biology and medicine. Redox Biology. 2015;**4**:180-183

[8] Tousoulis D, Papageorgiou N, Andourlakis E, Siasos G, Latsios G, Tentolouris K, et al. Diabetes mellitus-associated vascular impairment. Journal of the American College of Cardiology. 2013;**62**:667-678

[9] Moreira P, Sayre L, Zhu X, Nunomura A, Smith MA, Perry G. Detection and localization of markers of oxidative stress by in situ methods: Application in the study of Alzheimer disease. Methods in Molecular Biology. 2010;**610**:419-434

[10] Dominiczak M. In: Rifai N, Warnick G, Domniczak M, editors. Handbook of Lipoprotein Testing. Washington, DC: Amer Assoc Clin Chem; 2000. pp. 1-29

[11] Schaefer E, McNamara J. In: Rifai N, Warnick G, Dominiczak M, editors. Handbook of Lipoprotein Testing. Washington, DC: AACC Press; 2000. pp. 77-101

[12] Brown WV, Levy RI, Fredrickson DS. Further characterization of apoproteins from the human plasma very low density lipoproteins. The Journal of Biological Chemistry. 1970;**245**:6588-6594

[13] Brown WV, Baginsky ML. Inhibition of lipoprotein lipase by an apoprotein of very-low density lipoproteins. Biochemical and Biophysical Research Communications. 1972;**46**:375-382

[14] Rensen PCN, Herijgers N, Netscher MH, et al. Particle size determines the specificity of apoE-containing TG-rich emulsions for the LDL receptor versus hepatic remnant receptor in vivo. Journal of Lipid Research. 1997;**38**:1070-1084

[15] Havel RJ. Receptor and non-receptor mediated uptake of chylomicron remnants by the liver. Atherosclerosis. 1998;**141**:S1-S7

[16] Le N-A, Li X, Kyung S, Brown WV. Evidence for the in vivo generation of oxidatively modified epitopes in patients with documented CAD. Metabolism. 2000;**49**:1271-1277

[17] Gradek Q, Harris M, Yahia N, Davis W, Le N-A, Brown W.

Polyunsaturated fatty acids acutely suppress antibodies to malondialdehyde-modified LDL in patients with vascular disease. The American Journal of Cardiology. 2004;**93**:881-885

[18] Group L-CW. The lipid research clinics coronary primary prevention trial results: I. Reduction in incidence of coronary heart disease. Journal of the American Medical Association. 1984;**251**:351-364

[19] Brown G, Albers JJ, Fisher LD, Schaefer SM, Lin JT, Kaplan C, et al. Regression of CAD as a result of intensive lipid-lowering therapy in men with high levels of apoB. The New England Journal of Medicine. 1990;**323**:1289-1298

[20] Dujovne C, Harris W. The pharmacological treatment of dyslipidemia. Annual Review of Pharmacology and Toxicology. 1989;**29**:265-288

[21] Atmeh R, Stewart J, Boag D, Packard C, Lorimer A, Shepherd J. The hypolipidemic action of probucol: A study of its effects on high and low density lipoproteins. Journal of Lipid Research. 1983;**24**:588-595

[22] Parthasarathy S, Young SG, Witztum JL. Probucol inhibits oxidative modification of LDL. The Journal of Clinical Investigation. 1986;**73**:641-644

[23] Yamamoto A, Hara H, Takaichi S, Wakasugi J, Tommikawa M. Effect of probucol on macrophages, leading to regression of xanthomas and atheromatous vascular lesions. The American Journal of Cardiology. 1988;**62**:31B-36B

[24] Carew T, Schwenke D, Steinberg D. Antiatherogenic effect of probucol unrelated to its hypocholesterolemic effects. Proceedings of the National Academy of Sciences of the United States of America. 1987;**84**:7725-7729

[25] Regnstrom J, Walldius G, Carlson L, Nilsson J. Effect of probucol treatment on the susceptibility of LDL isolated from hypercholesterolemic patients to become oxidatively modified in vitro. Atherosclerosis. 1990;**82**:43-51

[26] Barnhart RL, Busch SJ, Jackson RL. Concentration-dependent antioxidant activity of probucol in LDL in vitro: Probucol degradation precedes lipoprotein oxidation. Journal of Lipid Research. 1989;**30**:1703-1710

[27] Fogelman AM, Schechter I, Seager J. Malondialdehyde alteration of LDL leads to cholesteryl ester accumulation in human monocyte macrophages. Proceedings of the National Academy of Sciences of the United States of America. 1980;**77**:2214-2218

[28] Cathcart M, Morel D, Chilsom G. Monocytes and neutrophils oxidize LDL making it cytotoxic. Journal of Leukocyte Biology. 1985;**38**:341-350

[29] Steinbrecher U, Parsarathy S, Leake D, Witztum JL, Steinberg D. Modification of LDL by endothelial cells involves lipid peroxidation and degradation of LDL phospholipids. Proceedings of the National Academy of Sciences of the United States of America. 1984;**81**:3883-3887

[30] Heinecke J, Baker L, Rosen H, Chait A. Superoxide-mediated modification of LDL by arterial smooth muscle cells. The Journal of Clinical Investigation. 1986;**77**:757-761

[31] O'Keefe JJ, Stone G, McCallister BJ, Maddex C, Ligon R, Kacich R, et al. Lovastatin plus probucol for prevention of restenosis after percutaneous transluminal coronary angioplasty.

The American Journal of Cardiology. 1996;77:649-652

[32] Tardif J, Cote G, Lesperance J, Bourassa M, Lambert J, Doucet S, et al. Probucol and multivitamins in the prevention of restenosis after coronary angioplasty. New England Journal of Medicine. 1997;**337**:365-372

[33] Gordon T, Castelli WP, Hjortland MC, Kannel WB, Dawber TR. HDL as a protective factor against CHD. The American Journal of Medicine. 1977;**62**:707-714

[34] Sundell C, Somers P, Meng C, Hoong L, Suen K, Hill R, et al. AGI-1067: A multifunctional phenolic antioxidant, lipid modulator, anti-inflammatory and antiatherosclerotic agent. The Journal of Pharmacology and Experimental Therapeutics. 2003;**305**:1116-1123

[35] Meng C, Somers P, Hoong L, Zheng X, Ye Z, Worsencroft K, et al. Discovery of novel phenolic antioxidants as inhibitors of vascular adhesion molecule-1 expression for use in chronic inflammatory diseases. Journal of Medicinal Chemistry. 2004;**47**:6420-6432

[36] Tardif J, Gregoire J, Schwatrz L, Title L, Laramee L, Reeves F, et al. Effects of AGI-1067 and probucol after percutaneous coronary interventions. Circulation. 2003;**107**:552-558

[37] Innis-Whitehouse W, Li X, Brown WV, Le NA. An efficient chromatographic system for lipoprotein fractionation using whole plasma. Journal of Lipid Research. 1998;**40**:679-690

[38] Le NA, Innis-Whitehouse W, Li X, Bakker-Arkema R, Black D, Brown WV. Lipid and apolipoprotein levels and distribution in patients with hypertriglyceridemia: Effect of triglyceride reductions with atorvastatin. Metabolism. 2000;**49**(2):167-177

[39] Esterbauer H, Striegl G, Puhl H, Rotheneder M. Continuous monitoring of in vitro oxidation of human LDL. Free Radical Biology & Medicine. 1989;**6**:67-75

[40] Le N-A, Farkas-Epperson M, Sweeney ME, Wilson PWF, Brown WV. Effect of ABT-335 (fenofibric acid) on meal-induced oxidative stress in patients with metabolic syndrome. Atherosclerosis. 2013;**231**:268-273

[41] Chilsolm G. Antioxidants and atherosclerosis: A current assessment. Clinical Cardiology. 1991;**14**:25-30

[42] Harris W. The prevention of atherosclerosis with antioxidants. Clinical Cardiology. 1992;**15**:636-640

[43] Gey K, Stagelin H, Eichholzer M. Poor plasma status of carotene and vitamin C is associated with higher mortality from ischemic heart disease and stroke. Basel Prospective Study. The Clinical Investigator. 1993;**71**:3-6

[44] Dieber-Rotheneder M, Puhl H, Waeg G, Striegl G, Esterbauer H. Effect of oral supplementation with d-a-tocopherol on the vitamin E content of human LDL and its oxidation resistance. Journal of Lipid Research. 1991;**32**:1325-1332

[45] Princen H, van Poppel G, Vogelezang C, Buytenhek R, HJ K. Supplementation with vitamin E but not beta-carotene in vivo protects LDL from lipid peroxidation in vitro: Effect of cigarette smoking. Arteriosclerosis and Thrombosis. 1992;**12**:554-562

[46] Jialal I, Grundy S. Effect of dietary supplementation with alpha-tocopherol

on the oxidative modification of LDL. Journal of Lipid Research. 1992;**87**:597-601

[47] Hodis H, Mack W, LaBree L, Cashin-Hemphill L, Sevanian A, Johnson R, et al. Serial coronary angiographic evidence that antioxidant vitamin intake reduces progression of coronary atherosclerosis. Journal of the American Medical Association. 1995;**273**:1849-1854

[48] DeMaio S, King S, Lembo N, Roubin G, Hearn J, Bhagavan H, et al. Vitamin E supplementation, plasma lipids and incidence of restenosis after percutaneous transluminal coronary angioplasty. Journal of the American College of Nutrition. 1992;**11**:68-73

[49] Lonn E, Yusuf S, Dzavik V, Doris I, Yi Q, Smith S, et al. Effects of ramipril and vitamin E on atherosclerosis. Circulation. 2001;**103**:919-925

[50] Stephens N, Parsons A, Brown M, Schofield P, Kelly F, Cheeseman K, et al. Randomised controlled trial of vitamin E in patients with coronary disease: Cambridge Heart Antioxidant Study (CHAOS). Lancet. 1996;**347**:781-786

[51] Mitchinson M, Stephens N, Parsons A, Bligh E, Schofield P, Brown M. Mortality in the CHAOS trial. Lancet. 1999;**353**:381-382

[52] Anonymous. Dietary supplementation with n-3 polyunsaturated fatty acids and vitamin E after MI: Results of the GISSI-Prevenzione trial. Lancet. 1999;**354**:447-455

[53] MRC/BHF. MRC/BHF Heart Protection Study of cholesterol-lowering therapy and of antioxidant vitamin supplementation in a wide range of patients at increased risk of coronary heart disease: Early safety and efficacy experience. European Heart Journal. 1999;**20**:725-741

[54] Virtamo J, Rapola J, Ripatti S, Heinonen O, Taylor P, Albanes D, et al. Effect of vitamin E and beta carotene on the incidence of primary nonfatal myocardial infarction and fatal coronary heart disease. Archives of Internal Medicine. 1998;**158**:668-675

[55] Schmolz L, Birringer M, Lorkowski S, Wallert M. Complexity of vitamin E metabolism. World Journal of Biological Chemistry. 2016;**7**:14-43

[56] Kushi L, Folsom A, Prineas R, Mink P, Wu Y, Bostick R. Dietary antioxidant vitamins and death from coronary heart disease in postmenopausal women. The New England Journal of Medicine. 1996;**334**:1156-1162

[57] Traber M, Goldberg I, Davidson E, Lagmay N, Kayden HJ. Vitamin E uptake by human intestinal cells during lipolysis in vitro. Gastroenterology. 1990;**98**:96-103

[58] Leonard S, Good C, Gugger E, Traber M. Vitamin E bioavailability from fortified breakfast cereal is greater than that from encapsulated supplements. The American Journal of Clinical Nutrition. 2004;**79**:86-92

[59] Traber M, Leonard S, Ebenuwa I, Violet P, Niyyati M, Padayatty S, et al. Vitamin E catabolism in women as modulated by food and by fat studied using 2 deuterium-labeled alpha-tocopherols in a 3-phase nonrandomized crossover study. The American Journal of Clinical Nutrition. 2021;**113**:92-103

[60] Vardi M, Levy N, Levy A. Vitamin E in the prevention of cardiovascular disease: The importance of proper patient selection. Journal of Lipid Research. 2013;**54**:2307-2314

Chapter 7

# The Importance of Vitamin E in Cardiovascular Health

*Sofia Kimi Uehara, Andrea Ramalho and Claudia Teresa Bento*


## Abstract

Vitamin E (VE) is the main antioxidant of the cell membrane. Research has extensively explored the use of VE in preventing or treating diseases associated with oxidative stress in humans, with a particular emphasis on cardiovascular diseases (CVD). Studies suggest a range of effects beyond its well-established antioxidant action, including anti-inflammatory, immunomodulatory, lipid-lowering, anticoagulant, antidiabetic, and neuroprotective properties. Regarding the effects of VE on the prevention or treatment of CVD and cardiovascular outcomes, the studies showed conflicting results. The benefits of VE were suggested as: (1) the inverse association between increased intake of VE and the risk of stroke, (2) reduced CVD mortality rate, (3) increased glutathione and reduced C-reactive protein, malondialdehyde, LDL, and blood pressure, (4) lower risk of CVD and metabolic syndrome with higher intake of VE, and (5) lower systolic blood pressure. However, a positive association between VE supplementation and diastolic blood pressure was observed. Other adverse effects of VE supplementation have been suggested, such as an increased risk of hemorrhagic stroke and a small but significant increase in the risk of hospitalization in patients with heart failure. Therefore, it is not possible to recommend VE supplementation for the prevention and treatment of CVD.


**Keywords:** vitamin E, antioxidant, oxidative stress, atherosclerosis, cardiovascular diseases

## 1. Introduction

Vitamin E, the primary fat-soluble vitamin in plasma, exists in four isoforms (alpha, beta, gamma, and delta-tocopherol), with alpha-tocopherol standing out as the most active and extensively studied biological form. This positioning within the lipid portion of cell membranes enables vitamin E to safeguard unsaturated phospholipids from lipid oxidation. Its protective mechanism involves the ability to convert free radicals into non-harmful metabolites by donating hydrogen [1].

The antioxidant properties of vitamin E have led to extensive research in various areas, including inflammation [1], immunomodulation [2], lipid profile [3], anticoagulant [4], antidiabetic [5], neuroprotective [6], and cardiovascular diseases [7], a leading global cause of death [8]. Atherosclerosis, an inflammatory disease caused by damage to the intimal layer resulting in chronic arterial lumen obstruction, forms the foundation for several cardiovascular diseases [9]. The oxidative process contributes to the progression of the inflammatory response. Reactive oxygen species [10] have also been linked to additional processes related to atherosclerosis, including increased vasoconstriction, angiogenesis, and apoptosis of endothelial cells.

In scientific literature, studies assessing the effects of vitamin E on the prevention or treatment of cardiovascular diseases and cardiovascular outcomes in humans have been carried out and the results suggest that this vitamin is associated with a lower stroke risk, could reduce the incidence of cardiovascular events and may improve cardiometabolic health in patients with diabetes [7, 11–19]. However, vitamin E was associated with significant increase in the risk of heart failure hospitalization [20]. In this chapter, the physiological aspects, food sources, nutritional deficiency and the effects of vitamin E on human health were addressed, with an emphasis on cardiovascular diseases.

## 1.1 Chemical composition, food sources, and nutritional deficiency

In recent years, important metabolic aspects of vitamin E have been identified, as well as antioxidant, anticholesterolemic, anti-inflammatory, anticancer, neuroprotective, and cardioprotective properties [21, 22].

Vitamin E exists in eight lipophilic natural forms, including α-, β-, γ-, and δ-tocopherol (αT, βT, γT, δT) and α-, β-, γ-, and δ-tocotrienol (αTE, βTE, γTE, δTE). Different isoforms of tocopherols and tocotrienols differ in the 5 or 7 position of the chromanol ring with an H or CH3 group. All tocopherols and tocotrienols are potent antioxidants with lipoperoxyl radical-scavenging activities (**Figure 1**) [1].

Tocopherols and tocotrienols are present in various vegetable oils (palm oil, rice bran oil, and annatto) and grains (wheat, oats, barley, rice, and rye) [1]. Palm oil contains a healthy balance of saturated and unsaturated fats, including monounsaturated oleic acid (40%), polyunsaturated linoleic acid (10%), saturated palmitic acid (44%), as well as stearic acid (6%) [23]. There is no indication that palm oil consumption has any particular impact on health compared to other fats rich in saturated fatty acids [24, 25]. Edible oils are the main sources of tocotrienols, such as palm oil, rice bran oil, grape seed oil, and cranberry seed oil. Food sources of tocotrienols are barley, wheat, and annatto [26, 27].

**Table 1** below shows the amount of vitamin E (in milligrams) per 100 grams or 100 mL of food, in descending order.

For more detail, Siger & Górnaś [27] checked the tocopherol (T) and tocotrienol (T3) profiles in four hundred and eighteen vegetable oil samples, of which fifteen species were dominated by tocotrienols, while sixty-seven by tocopherols. A relationship was observed between the taxonomic origin of the plant and the presence of the characteristic tocochromanol profile in the oils obtained from these plants [27].

Tocotrienol is generally highly lipophilic and to be absorbed by enterocytes requires lipolysis into micelles [29]. The presence of food has been shown to increase the absorption of tocotrienols by at least 1.5 times in fed human compared to those in a fasted state [30]. Alpha-tocotrienol recorded the highest absolute bioavailability, followed by γ- and δ-tocotrienol. In humans α-tocotrienol was the most abundant vitamin detected in plasma and lipoproteins following consumption of a single dose of tocotrienol-rich vitamin E from palm oil (70% tocotrienols, 30% tocopherol) [31].

Tocotrienols exhibit excellent biological activities. Tocopherols and tocotrienols differ by the presence of saturated and unsaturated isoprenoid side chains,

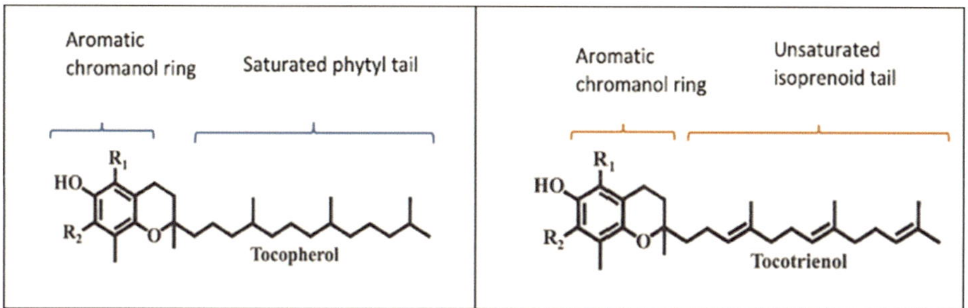


| | R1 | R2 |
|---|---|---|
| **α-tocopherol** | $CH_3$ | $CH_3$ |
| **β-tocopherol** | $CH_3$ | H |
| **γ-tocopherol** | H | $CH_3$ |
| **δ-tocopherol** | H | H |

| | R1 | R2 |
|---|---|---|
| **α-tocotrienol** | $CH_3$ | $CH_3$ |
| **β-tocotrienol** | $CH_3$ | H |
| **γ-tocotrienol** | H | $CH_3$ |
| **δ- tocotrienol** | H | H |

**Figure 1.**
*Chemical structure of vitamin E. Adapted from: Zainal et al., 2022 [6].*

| Food (100 g /100 mL) | Amount of vitamin E (mg) |
|---|---|
| Hazelnut oil | 47.2 |
| Sunflower oil | 41.08 |
| Almond oil | 39.2 |
| Roasted sunflower seeds | 38.33 |
| Roasted sesame seeds | 36.33 |
| Safflower oil | 34.1 |
| Almond | 23.75 |
| Hazelnut | 23 |
| Canola oil | 17.46 |
| Peanut oil | 15.69 |
| Corn oil | 14.3 |
| Olive oil | 11.42 |
| Soybean oil | 8.18 |
| Brazil nuts | 7.14 |
| Peanut / Groundnuts | 6.94 |
| Palm oil | 3.81 |
| Egg yolk | 2.58 |
| Avocado | 2.07 |
| Walnuts | 1.8 |
| Sesame oil | 1.4 |

| Food (100 g /100 mL) | Amount of vitamin E (mg) |
|---|---|
| Cashew nuts | 0.92 |
| Linseed oil | 0.71 |
| Coconut oil | 0.09 |

*Source: Food Composition Table. Adapted from: Departamento de Informática em Saúde. Escola Paulista de Medicina. Universidade Federal de São Paulo. [Internet]. http://tabnut.dis.epm.br/alimento [28].*

**Table 1.**
*Main food sources of vitamin E.*

respectively. To allow for greater fluidity and uniform distribution in the phospholipid bilayer. In addition, α-tocopherol is preferentially retained by body tissues via the α-tocopherol transfer protein (α-TTP), but tocotrienols are rapidly degraded into short-chain carboxychromanols to be more efficient in eliminating peroxyl radicals due to more effective interaction in membrane environments. However, clinical evidence on the effects of palm oil use on diseases such as cancer and cardiovascular diseases is insufficient [6].

The human body is unable to produce vitamin E, and its deficiency manifests obvious symptoms. Insufficiency of Vitamin E can be caused by factors such as a lack of fat absorption or protein malnutrition, among others [32].

### 1.2 Regulation of metabolism and excretion of vitamin E

In the intestine, tocopherols and tocotrienols from the diet appear to be absorbed in a similar way along with dietary fat and are secreted in chylomicron particles along with triacylglycerol, phospholipids, and cholesterol. Vitamin E bound to chylomicrons is transported via the lymphatic system to muscles, bone marrow, adipose tissue, skin, and brain. In these tissues, the vitamin E forms are taken up by a process mediated by lipoprotein receptors that is not well understood [33].

In the liver, αT is preferentially bound to the α-tocopherol transfer protein (α-TTP). Together with the ATP-binding cassette transporter ABCA1 (also known as the cholesterol efflux regulatory protein, an important regulator of cellular cholesterol and phospholipid homeostasis), α-TTP incorporates αT into lipoproteins that trans-transport to other tissues via the circulation. In contrast to the high affinity of αT (100%), α-TTP has a much lower affinity compared to 50, 10–30 or 1% affinity for βT, γT, and δT, respectively. Large portions of non-αT forms are catabolized in the liver via ω-hydroxylation initiated by cytochrome P450 (CYP4F2) and oxidation followed by β-oxidation of the phytyl chain to generate -hydroxychromanol (130-OH), various carboxychromanols and the terminal metabolite 30-carboxychromanol (30-COOH) or (20-carbox-ethyl)-6-hydroxychromans (CEHCs). Conjugation such as sulfation and glucuronidation of the phenolic on chromanol can occur in parallel with β-oxidation when there is a high intake of vitamin E forms [33].

In addition, protein control, absorption, and excretion of vitamin E may also play a role in its bioavailability. Cytochrome P450-mediated hydroxylation and oxidation of the side chain are responsible for initiating vitamin E catabolism. The activity of tocopherol-ω-hydroxylase is associated with CYP4F2, but not with other isoforms, including CYP3A. CYP4F2, a microsomal ω-hydroxylase, was initially described to metabolize leukotriene B4 via ω-hydroxylation [17]. ω-hydroxylase helps explain some aspects related to the bioavailability of vitamin E. For example, αT supplementation depletes plasma and tissue γT. This observation is probably partly explained by the preference of α-TTP affinity for αT, as increased αT can compete with γT in TTP binding and therefore leaves more γT for catabolism by hydroxylases. Meanwhile, increased αT intake can stimulate γT catabolism due to its effector activity towards CYP4F2 [2].

Lithocholic acid is oxidized at the C-3 position by human cytochrome P450 3A (CYP3A) to form 3-ketocholic acid, and this metabolite is cytotoxic in human intestinal and liver cells. α-Tocotrienol, γ-tocotrienol, δ-tocotrienol and a vitamin E mixture rich in tocotrienol, but not α-tocopherol, differentially inhibited the 3-oxidation of lithocholic acid catalyzed by liver and intestinal microsomes and recombinant CYP3A4 and CYP3A5 [2].

Tocotrienols can be found in the plasma 4 hours after use and have a reduced half-life in circulation, while those of α- and γ-Tocopherols were 57 and 13 h. However, tocotrienols are more effective despite lower bioavailability and absorption, and their deposition takes place in adipose tissue and the heart, in the metabolism in the liver, or the crossing of the blood-brain barrier. Tocotrienols have strong therapeutic properties but require accessibility to target organs [33].

The excretion of vitamin E and its metabolites also influences the tissue retention of these compounds. Short-chain carboxychromanols and their sulfated derivatives or glucuronidated homologs are excreted in the urine, while unconjugated carboxychromanols are found mainly in the feces. Bardowell et al. [34] estimated that up to 80% of the total metabolites were excreted in the feces, and unmetabolized tocopherols and tocotrienols are also eliminated by biliary excretion. Interestingly, fecal excretion of γT and δT is markedly increased in response to their supplementation, which probably prevents excessive accumulation of these tocopherols [34].

### 1.3 Antioxidant function

Vitamin E is one of the most important essential vitamins for supporting the regulation of oxidative stress in humans. All forms of vitamin E are potent antioxidants, as they eliminate peroxyl lipid radicals by donating hydrogen from the phenolic group on the chromanol ring. On the other hand, tocotrienols have been suggested to be better than αT at scavenging peroxyl radicals due to a more even distribution of tocotrienols in the phospholipid bilayer and more effective interaction with peroxyl radicals than tocopherols in membrane environments [35].

Inflammation, physical and mental stress, cigarette smoke, air pollutants, and chemical additives such as dyes and acidulants, among others, increase the generation of free radicals in the body [36]. Free radicals (FR) are considered destructive because they alter cell membranes, enzymes, and DNA. Free radicals and the intensity of chronic inflammation promote aging and various chronic diseases such as neurodegenerative and cardiovascular diseases [35].

Oxidative stress also increases under unbalanced levels of free radicals and antioxidants. It has been suggested that natural antioxidants may be beneficial to human health [37]. Among these, vitamin E stands out because it is a natural compound and has strong antioxidant activity, inhibiting lipid peroxidation and the generation of reactive oxygen species (ROS) [38].

The chromanol ring of tocotrienols plays a fundamental role in reducing methylation, highlighting the ease with which this molecule can be incorporated into cell membranes, and because of this tocotrienols are 40–60 times more effective than α-tocopherol in inhibiting the generation of pro-inflammatory cytokines. Among tocotrienols, γ tocotrienols and δ tocotrienols are supposedly more efficient at eliminating peroxyl radicals [39]. Tocotrienols are part of the vitamin E family. The potential of tocotrienols as a nutraceutical ingredient is largely underestimated due to low oral bioavailability, which is a common problem associated with fat-soluble bioactive compounds.

Fu et al. [39] showed that nanoencapsulation technology offers innovative solutions to improve the delivery mechanisms of these compounds and found that both nanovesicles (NV-T3) and solid lipid nanoparticles (NP-T3) showed increased accumulation in the kidneys and liver (5-fold) compared to the control group, while selectivity for α-tocotrienol was evident for NP-T3. In the brain and liver of rats that received NP-T3, α-tocotrienol emerged as the dominant congener (>80%). Acute oral administration of nanoencapsulated tocotrienols showed no signs of toxicity. The study concluded greater bioavailability and selective accumulation of tocotrienol congeners in tissues when delivered via nanoencapsulation.

### 1.4 Anti-inflammatory function

Inflammation is characterized by the production of a plethora of reactive oxygen and nitrogen species and pro-inflammatory mediators, including lipid mediators, notably prostaglandins and leukotrienes, and cytokines such as TNF-α and interleukin-6 (IL-6) which, in turn, aggravate inflammation and lead to excessive damage to host tissues. Prostaglandins and leukotrienes play important roles in mediating the inflammation response. For example, prostaglandin E2 (PGE2), which is produced from the oxidation of AA catalyzed by COX-1 and COX-2, causes pain and fever, besides activating the formation of cytokines. Leukotriene B4 (LTB4), another lipid mediator derived from AA through the reaction catalyzed by 5-lipoxygenase (5-LOX) in neutrophils, is one of the most potent chemotactic agents. Due to the central roles of PGE2 and LTB4 in inflammation, COXs, and 5-LOX have been recognized as key targets for drug therapy against chronic diseases [6].

Tocotrienols exhibit superior anti-inflammatory efficacy compared to tocopherols and a lack of adverse effects [6]. γTocopherol, δTocopherol, and γTocotrienol strongly inhibited A23187-stimulated LTB4 and LTC4-like human neutrophils isolated from peripheral blood, while αT was much less effective. Jiang et al. [40] showed that different forms of vitamin E partially inhibited COX-2-mediated PGD2 and PGE2 formation in lipopolysaccharide (LPS)-stimulated RAW264.7 macrophages and IL-1β-activated lung epithelial cells, respectively, with a potency ratio of δT AND γTE 4 γT ≫ αT. Although γT, δT, and γTE had no effect on endotoxin- or cytokine-stimulated COX-2 up-regulation, they suppressed COX-2 activity in cellular environments. On the other hand, none of these forms of vitamin E appeared to inhibit purified COX-2 activity. These results support the notion that forms of vitamin E are weak at inhibiting COX-2 tors, such as salicylic acid and paracetamol, both of which inhibit PGE2 in intact cells, but show weak inhibition of purified COXs (**Figure 2**) [40, 42–44].

Over the last 20 years, research into vitamin E has focused mainly on αT and its antioxidant effects, and much more recently (in the last 10 years), the focus of vitamin E research has been on investigating different tocopherols and tocotrienols, their metabolism, and their non-antioxidant activities. But αT supplementation has failed to provide consistent benefits for the prevention of chronic diseases, including cancer and cardiovascular diseases, in many large clinical intervention studies [45–48].

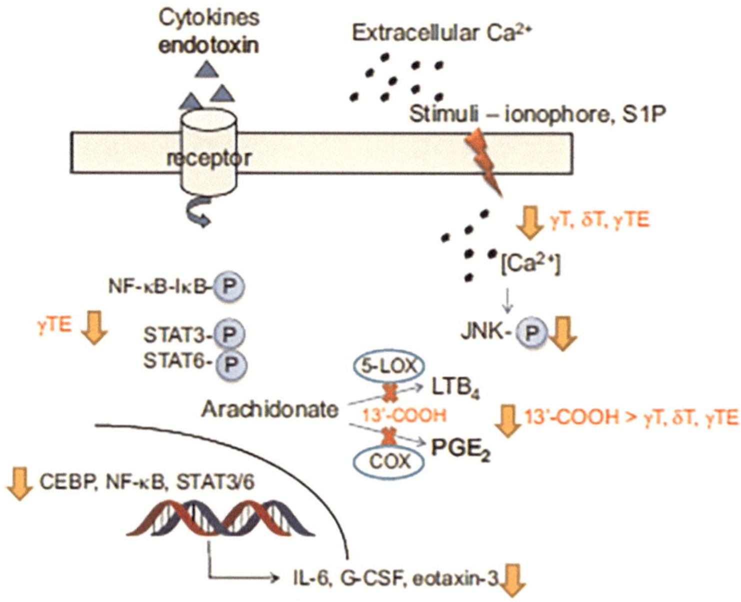


**Figure 2.**
*Anti-inflammatory activities of vitamin E. In epithelial cells, macrophages, and neutrophils, γT, δT, and γTE modestly inhibit PGE2 and LTB4 without inhibiting COX and 5-LOXactivity. 130-COOH spotently inhibit COX-1/COX-2 and 5-LOX enzyme activity (redcrosses). In neutrophils, vitamin forms suppress ionophore-orS1P (sphingosine1-phos-phate)-stimulated calcium influx and its downstream signaling. In macrophages and lung epithelial cells, γTE inhibits activation of NF-κB and STAT6/3 as well as their regulated genes, including cytokines and chemokines. Adapted from: Jiang, Q, 2014 [41].*

One explanation is that αT may be beneficial for individuals deficient in αT and/or other micronutrients, which may be caused by low dietary intake of this vitamin E or αT depletion due to pathologies, smoking, alcoholism, and malabsorption. In these subclinical conditions, αT supplementation is likely to be beneficial [48].

γ-tocopherol (γT) is the second most abundant form of vitamin E in blood and tissues, while α-tocopherol (αT) is the predominant vitamin E in tissues. Research has revealed that γT has unique antioxidant and anti-inflammatory activities relevant to disease prevention compared to αT. Although both compounds are potent lipophilic antioxidants, γT, but not αT, can capture reactive nitrogen species forming 5-nitro-γT, and appears to show superior protection of function in mitochondria. γT inhibits ionophore-stimulated leukotrienes by blocking the translocation of 5-lipoxygenase (5-LOX) in leukocytes, decreases prostaglandins catalyzed by cyclooxygenase-2 (COX-2) in macrophages and blocks the growth of cancer cells, but not healthy cells. For these activities, γT is stronger than αT. In addition, γT is more extensively metabolized than αT via side-chain oxidation initiated by cytochrome P-450 (CYP4F2), which leads to the formation of metabolites including 13′-carboxychromanol (13′-COOH) and carboxyethylhydroxychroman (γ-CEHC). 13’-COOH and γ-CEHC are the predominant metabolites found in feces and urine, respectively. Interestingly, γ-CEHC has natriuretic activity and 13′-COOH inhibits COX-1/−2 and 5-LOX activity. Consistent with these mechanistic findings of γT and metabolites, studies show that γT supplementation attenuates inflammation and disease symptoms in animal models with induced inflammation, asthma, and cancer. In addition, γT supplementation decreased inflammation markers in patients with kidney disease and mild asthma.

These observations support that γT may be useful against diseases associated with inflammation [41, 49, 50].

Findings have provided mechanistic insights into the role of γT in the prevention and management of diseases observed in animal and human studies, as seen in a recent review, in which Jiang et al. [51] that γT is shown to be superior to αT in detoxifying nitrogen dioxide ($NO_2$), peroxynitrite (ONOO-), and copper-zinc superoxide dismutase (SOD)/$H_2O_2$/$NO_2$-, via the formation of 5-NγT. Consistently, 5-NγT was elevated in zymosan-induced-peritonitis in rats and during $FeCl_3$ patch-induced occlusive thrombus formation in rats. Supplementation of pharmaceutical doses of γT (~1200 mg) temporarily increases this vitamin E in the plasma to up to 30–40 μM.

Below is a **Table 2** of the bioactivities of γT and metabolites, adapted from the above-mentioned review.

## 1.5 Immunomodulatory function

In the adaptive immune system, Th17 cells have pro-inflammatory properties, while Treg cells have negative immunoregulatory effects. The Th17/Treg balance plays an important role in the treatment of autoimmune diseases [52]. Th17 differentiation is regulated by the cytokine IL-6 via the JAK/STAT3 signaling pathway. In the IL-6/JAK/STAT3 signaling pathway, IL-6 binds to the membrane receptors IL-6Rα and gp130 (IL-6Rβ) to regulate JAK activation. JAK induces STAT3 that combines with another STAT3 protein to form a phosphorylated dimer. Finally, the dimer enters the nucleus and activates transcription factors involved in Th17 differentiation [53].

| **Activities in cells or animal models Mechanisms** | | |
|---|---|---|
| Activity of γT; Comparing γT and αT | • Scavenging NOx ↓nitrotyrosine, ↓nitrite<br>• γT but not αT shows these effects<br>• ↓LTB4, ↓cysteinyl leukotrienes: γT's IC50 ~ 5–25 μM, αT's IC50 ≥ 50 μM<br>• ↓PGE2, ↓PGD2: γT's IC50 ~ 4–7.5 μM αT's IC50 > 50 μM<br>• ↓TxB2 in platelets γT's IC50 ~ 8–25 μM αT's IC50 ≥ 50 μM<br>• ↓cancer cell growth: γT's IC50 ~ 25–50 μM αT's IC50 > 100 μM | • Form 5-NγT<br>• Block ionophore-stimulated Ca2+ influx and 5- LOX translocation<br>• Modest inhibition of COX-2 after chronic incubation with macrophages or generating 13′ -COOH that inhibits COX-2<br>• Inhibit human recombinant COX-1 activity<br>• Modulate sphingolipids including intracellular accumulation of dihydroceramides and dihydrosphingosine |
| 13′ -COOH | • ↓ LTB4<br>• ↓ PGE2, ↓ TxB2<br>• cancer cell growth | • Inhibit 5-LOX activity (IC50 0.04–1 μM)<br>• Inhibit COX-1 and COX-2 enzyme (IC50 ~ 4 μM)<br>• Modulate sphingolipids including intracellular accumulation of ceramides and dihydroceramides |
| γ-CEHC | • Natriuretic<br>• ↓PGE2, ↓PGD2<br>• ↓JNK-p, prevent IκBα degradation | • Inhibit 70 pS potassium channel in the kidney thick ascending limb cells<br>• Inhibit COX-2 activity (IC50 ≥ 30 μM)<br>• Not determined |

*Adapted from: Q. Jiang, Metabolism of natural forms of vitamin E and biological actions of vitamin E metabolites, Free Radic Biol Med 179 (2022) 375–387 [51].*

**Table 2.**
*Bioactivities of γT and metabolites.*

The role of γ-Tocotrienol in the differentiation and function of T helper 17 (Th17)/regulatory (Treg) cells is not fully understood. However, *in vitro*, γ-Tocotrienol suppressed Th17 differentiation by inhibiting the signal transducer and activator of transcription 3 (STAT3) phosphorylation in the interleukin 6 (IL-6)/Janus kinase (JAK)/STAT3 signaling pathway. γ-Tocotrienol also inhibited HIF1A expression in Th17 metabolism [54].

In addition, γ-Tocotrienol can inhibit nuclear factor kappa B (NF-κB) signaling via positively regulating A20 protein expression [17]. And γ-Tocotrienol inhibits lipopolysaccharide-induced IL-6, tumor necrosis factor alpha, granulocyte-macrophage colony-stimulating factor, and other pro-inflammatory cytokines [55].

### 1.6 Hypolipidemic effect

Tocotrienols have shown hypolipidemic, as well as atherogenic and nephroprotective effects since hyperlipidemia is intrinsically connected to oxidative stress leading to cardiovascular dysfunction, atherosclerosis, and progressive chronic renal dysfunction [3, 56, 57].

Tocotrienols can inhibit the action of hepatic hydroxy-methylglutaryl coenzyme A (HMG-CoA) reductase, the enzyme responsible for cholesterol biosynthesis, which has been attributed to their lipid-lowering properties that reduce cholesterol production. This activity has been demonstrated in experimental and clinical studies by several researchers [3, 58].

In experimental studies, atherosclerotic animals supplemented with tocotrienols observed a reduction in serum levels of vascular cell adhesion protein-1 (VCAM-1), IL-6, and matrix metalloproteinase-12 (MMP-12) and LDL cholesterol [59, 60].

### 1.7 Anti-coagulant function

T3 can act as an anticoagulant, thereby delaying the progression of atherosclerosis and reducing total plasma triglycerides (TG) and apolipoprotein-A1 (Apo-A1) [4]. Oxidative stress, as described in [61], induces inflammation and endothelial damage. T3 exerts beneficial effects on serum LDL, CRP, ApoB, and glycation end products. Simultaneously, it increases HDL levels, collectively reducing the risk of developing atherosclerosis.

### 1.8 Tocotrienols as an anti-diabetic agent

Diabetes is a chronic disease that impairs the body's ability to produce or use insulin properly [55]. After most food is converted into glucose and absorbed, insulin is released in response to an increase in blood glucose [62].

Clinical and experimental investigations have shown that T3 supplementation can help reduce the risk of diabetes and mitigate associated conditions such as brain health care, nerve pain relief, preventive eye care, and kidney well-being. Due to this, it is considered an antihyperglycemic agent [5, 63].

### 1.9 Vitamin E as a neuroprotector

Tocotrienols have been investigated as nutraceuticals, particularly in the context of metabolic diseases and degenerative conditions [6]. In recent years, clinical studies on the neuroprotective effect of tocotrienols have gained significant results.

Supplementation for up to 12 months, as reported in [source], has been shown to reduce white matter lesion volume and alleviate pain in diabetic patients with peripheral neuropathy [64, 65]. These clinical benefits have been attributed mainly to the antioxidant and anti-inflammatory mechanisms of tocotrienols [64, 65].

Alcoholism affects around 2 billion people worldwide. Withdrawal causes a neuroinflammatory response which increases anxiety. Villas Boas et al. [66] conducted a study with male Wistar rats divided into four groups and subjected to a forced and chronic rehabilitation procedure, involving a liquid diet containing 6 and 8% ethanol for 15 days, followed by an abrupt interruption of treatment. The results showed that α-tocopherol showed anxiolytic action and anti-inflammatory activity, attenuating the anxious behavior of abstinent animals and reducing neuroinflammation in amygdaloid nuclei (AN) and medial hypothalamic nucleus (mHN); and the intensity of the anxiolytic and anti-inflammatory effect of α-tocopherol is dose-dependent. [66].

Toxicological evaluation is critical and has been tested for safety in experimental tests. One of the known side effects of tocopherols is the modification of vitamin K's function in blood coagulation. However, *in vivo* and clinical investigations into tocotrienol have revealed no detectable toxicity [67].

According to Fu et al. [53], T3 supplementation at 400 mg/day for 14 days did not affect hematotoxicity or hepatotoxicity in individuals with metabolic diseases. In addition, the Institute of Medicine suggests taking 200–1000 mg of vitamin E per day which can only be obtained through supplementation.

The recommended dietary intake of vitamin E (α-Toc) is around 15 mg per day. In addition, the Food and Drug Administration (FDA) has recommended tocotrienol intake of 13 to 20 mg per day from food. Tocotrienol supplementation has proven to be safe and acceptable in >30 clinical trials with varying dosages of tocotrienol in various formulations and study demographics [68].

More clinical trials investigating the potential use of tocopherols, tocotrienols, and combinations must be carried out to ensure effective treatments from this important vitamin.

## 2. Vitamin E in cardiovascular diseases

Cardiovascular diseases (CVD) represent a major cause of death worldwide [8]. *Acute myocardial infarction (AMI) and stroke, as reported by Pan American Health Organization* [8], *account for the majority of deaths*. Smoking, a sedentary lifestyle, poor diet, obesity, systemic arterial hypertension (SAH), diabetes mellitus (DM), and dyslipidemia contribute to the onset of CVDs [8]. Atherosclerosis, the basis of various forms of CVD, is the most common alteration found in arteries due to the accumulation of cholesterol and the proliferation of smooth muscle cells in the tunica intima [9]. This process gives rise to a fibrous plaque that protrudes into the lumen, altering the tunica media and contributing to a series of circulatory complications due to the inflammatory response triggered in the vessel wall. Increased serum concentrations of low-density lipoprotein (LDL) are the main factor involved in atherosclerosis [9].

When serum LDL concentrations rise, this lipoprotein is deposited in the arteries, resulting in oxidation and the formation of oxidized LDL (LDL-ox), which attacks endothelial cells that start expressing various molecules (intercellular adhesion molecule, E-selectin, vascular adhesion molecule, and interleukin-8) that act to recruit and bring neutrophils into the tunica media of the vessels. LDL-ox itself is chemotactic for monocytes. This process increases the inflammatory response, promoting the migration of monocytes to the tunica intima and their differentiation into macrophages that phagocytize LDL-ox, transforming it into foam cells. These cells are components of atherosclerotic plaques which, in more advanced stages, develop into fibrous plaques that impair blood flow, resulting in necrosis [9].

LDL-ox phagocytosis is characterized by the increased production of digestive proteases by macrophages and an increase in reactive oxygen species [9]. Oxidative stress is one of the most important factors involved in atherosclerosis and its complications [69]. The oxidative process favors the progression of the inflammatory response, making the damage more complex and triggering clinical manifestations such as *angina pectoris*. Other processes related to atherosclerosis (increased vasoconstriction, angiogenesis, and apoptosis of endothelial cells) have also been associated with reactive oxygen species [10].

Controlling dyslipidemia and inflammation and protecting the vascular endothelium should be part of the treatment of atherosclerosis. Lifestyle changes are the main determinants of CVD prevention, including diet, which can influence atherogenesis directly or through changes in classic risk factors (blood pressure, blood glucose, and serum lipids) [70].

Considering the role of LDL-ox in the atherosclerotic process, the use of antioxidants such as vitamin E (VE) has aroused the interest of researchers. This interest is due to the anti-atherogenic mechanisms that have been attributed to VE (**Table 3**).

In an attempt to answer the question: "*Is there consistent scientific evidence to indicate that increased dietary intake of VE or that elevated serum concentrations of VE influence cardiovascular health*?", meta-analyses (MA) [11, 12], systematic reviews (SR) with [13, 14] or without MA [15–19], critical review of MA [20] and umbrella review of MA and SR with MA [7] were evaluated. The overall characteristics of the articles included in this chapter are shown in **Table 4**.

The literature surveyed shows conflicting results (benefits, adverse effects, and no effect) regarding the effects of VE on the prevention or treatment of CVD and cardiovascular outcomes. In 5 studies, the benefits of VE were suggested as: (1) the

| |
|---|
| Remove peroxyl radicals, interrupting the chain of lipidic peroxidation; |
| Protect the polyunsaturated fatty acids from damage by free radicals; |
| Protect circulating lipoproteins; |
| Act in the proper functioning of the cell membranes; |
| Increase the resistance of LDL to oxidation; |
| Reduce platelet adhesion and aggregation; |
| Inhibit vitamin K-dependent coagulation factors; |
| Stimulate endothelin production; |
| Attenuate the inhibition of nitric oxide production mediated by LDL-ox; |
| Inhibit the secretion of interleukin-1 by monocytes by inhibiting protein kinase C; |
| Influence the vasodilator function of the arteries. |

*Adapted from M. Batlouni, Hipótese oxidativa da aterosclerose e emprego dos antioxidantes na doença arterial coronária, Arq Bras Cardiol 68 (1) (1997) 55–63 [10].*

**Table 3.**
*VE antiatherogenic mechanisms.*

| Authors/ year | Objective | Total/Type of the study | Characteristics of the studies | Quality of the studies | Results | Limitations |
|---|---|---|---|---|---|---|
| Cheng et al. [11] | • Investigate the relationship between vitamin E (VE) intake and risk for stroke (MA) | • 9 (prospective cohort study or nested case–control study) | • Total participants (n = 220,371), age (40 to 99 years), gender: studies with men (n = 4), men and women (n = 4) and with women (n = 1);<br>• 3284 stroke events and follow-up period (6.1 to 16.5 years);<br>• Countries: United States (n = 2), Netherlands (n = 2), Finland (n = 2), Italy (n = 1), China (n = 1) and Japan (n = 1). | • > 7 stars (high quality; Newcastle-Ottawa scale) | • Sources of VE assessed by the studies were: food (n = 8), supplement (n = 3), and total [food plus supplement (n = 2)];<br>• Dietary intake of VE ranged from 6.4 to 39.45 mg/day;<br>• High dietary intake of VE was inversely associated with stroke risk (RR =0.83, 95% CI: 0.73, 0.94<br>• in follow-up studies of less than 10 years. | • Limited number of studies, heterogeneity between studies, inaccurate estimation of vitamin E intake, and the majority of studies were carried out in America and Europe and few in East Asia. |
| Machado et al. [15] | • To verify the association between dietary intake or serum concentrations of antioxidant micronutrients and CAC in individuals with no previous history of CVD.<br>• (SR) | • 9, with VA (n = 1; CSOS), β-carotene (n = 1; CSOS), VC (n = 4; 1 RCT and 3 CSOS), VE (n = 4; 1 RCT and 3 CSOS), Mg (n = 5; CSOS) and Zn (n = 1; CSOS). | • Total number of participants (n = 7349);<br>• Adults and the elderly (both sexes);<br>• Sample size (from 80 to 34,553);<br>• 1 study (VE) with adult HIV patients and 1 study (Mg) with adult dialysis patients;<br>• Total number of studies that assessed:<br>o dietary intake of VA (n = 1), β-carotene (n = 1), VC (n = 3), VE (n = 3), Mg | • No data | • Mg was inversely associated with CAC;<br>• VA, β-carotene, VC and Zn were not associated with CAC;<br>• Results of the association between CAC and VE are controversial. | • Use of different cut-off points for defining CAC, failure to carry out a meta-analysis, use of FFQ to assess dietary intake and failure to consider other dietary components with antioxidant action;<br>• Only studies in English were included. |

| Authors/ year | Objective | Total/Type of the study | Characteristics of the studies | Quality of the studies | Results | Limitations |
|---|---|---|---|---|---|---|
| | | | (n = 2) and Zn (n = 1);<br>o plasma concentrations of VE (n = 1) and Mg (n = 3);<br>o dietary intake and plasma concentrations of VC (n = 1). | | | |
| Han et al. [13] | • Evaluate the effectiveness of vitamin supplements (alone or in combination) in preventing CVD<br>• (SR with MA) | • 43 RCT, of which: 1 ($VB_6$), 2 (VA), 2 (VA + VE), 3 ($VB_9$ + $VB_{12}$), 4 (B9), 4 (VC + VE), 5 (VA + VC + VE), 5 (VD), 7 ($VB_9$ + $VB_{12}$ + $VB_6$) and 10 (VE). | • Total participants (n = 384,248), age (from 40 to 84), average sample size (4584);<br>• Countries: Great Britain (n = 1), Norway (n = 1), China (n = 1), the Netherlands (n = 1), Switzerland (n = 1), Canada (n = 1), Australia (n = 1), Scotland (n = 1), Israel (n = 2), France (n = 2), Italy (n = 2), Germany (n = 2), Finland (n = 2), the United Kingdom (n = 4) and the United States (n = 21); | • Acceptable quality (Cochrane Collaboration tool) | • VD seems to act in the prevention of cardiovascular events (OR = 0.84, 95% CI: 0.72, 0.98, P = 0.000);<br>• VE appears to reduce the cardiovascular mortality rate **(OR = 0.92, 95% CI: 0.86, 0.99, P = 0.020)**;<br>• B vitamins appear to reduce the incidence of stroke compared to placebo (OR = 0.81, 95% CI: 0.70, 0.94, P = 0.006):<br>• Combination of vitamins has no efficacy in preventing CVD (OR = 1.00, 95% CI: 0.96, 1.05). | • Differences regarding frequency and dose of supplementation;<br>• Results cannot be extrapolated to people with pre-existing diseases. |

| Authors/ year | Objective | Total/Type of the study | Characteristics of the studies | Quality of the studies | Results | Limitations |
|---|---|---|---|---|---|---|
| | | | • Amount of VE from 298 IU/day to 800 IU/ day;<br>• Duration of the studies: from 4 weeks to 1 year. | | | |
| Khan et al. [20] | • Evaluate the effectiveness of interventions (dietary interventions and nutritional supplements) on the incidence of HF and outcomes in patients with HF<br>• (Critical evaluation of MA) | • 14 meta-analyses of RCT | • Total participants (n = 176,097) | • Low quality of evidence | • Interventions included: 2 dietary interventions (salt reduction and Mediterranean diet) and 12 supplements (n3-series PUFA, n-6-series PUFA, PUFA, iron, coQ10, VE, VD, VD + calcium, thiamine, beet juice, Chinese medicinal herbs, and L-carnitine);<br>• CoQ10 (↓ all-cause mortality) [RR = 0.69 (95% CI 0.50–0.96);<br>• $I^2$ =0%; low certainty of evidence];<br>• Mediterranean diet (↓ hospitalization) **[RR = 0.45**<br>• (95% CI 0.26–0.79); $I^2$ =0%; low certainty of evidence];<br>• Chinese medicinal herbs [MD −143.19 | • Heterogeneity of the patients' baseline characteristics, wide variety in the follow-up period, lack of a dose–response test, the possible exposure of the control group to supplements (markets or in certain foods) and only articles in English were included. |

| Authors/ year | Objective | Total/Type of the study | Characteristics of the studies | Quality of the studies | Results | Limitations |
|---|---|---|---|---|---|---|
| | | | | | (95% CI − 218.90 to −67.48); $I^2$ =99%; low certainty]<br>• and n-3 PUFA [MD −131.11 (95% CI −165.30 to −96.92); $I^2$ =51%; moderate certainty] (↓ B-type natriuretic peptide);<br>• Iron [MD<br>• 3.55 (95% CI 1.20–5.90); $I^2$ = 82%; moderate certainty], vitamin D [MD 4.33 (95% CI 0.33–8.33); $I^2$ =90%; moderate certainty]<br>• L-carnitine [MD 0.49 (95% CI 0.24–0.74); $I^2$ =70%; moderate certainty],<br>• and thiamine [MD 3.59 (95% CI 0.91–6.27); $I^2$ =68%; moderate certainty]<br>• (↑ left ventricular ejection fraction);<br>• VD (↓ performance in the 6 min walk test) [MD −39.43 (95% CI −76.00 to −2.86); $I^2$ =5.5%; high certainty];<br>• VE supplementation (↑ hospitalization) | |

| Authors/ year | Objective | Total/Type of the study | Characteristics of the studies | Quality of the studies | Results | Limitations |
|---|---|---|---|---|---|---|
| | | | | | [RR = 1.21 (1.04–1.40), $I^2 = 0\%$; moderate certainty of evidence]. | |
| Shah et al. [16] | • Evaluate the effects of VE (dietary or serum) on the cardiovascular system<br>• (SR) | • 25, comprising CSOS (n = 3), case–control study (n = 5), RCT (n = 6) and cohort study (n = 11) | • Total participants (n = 360,840), 63.54% men and 36.46% women;<br>• Studies with women (n = 2), men (n = 4) and men and women (n = 19);<br>• Study countries: United States (n = 3), China (n = 3), Italy (n = 3), Japan (n = 3), Australia (n = 2) and the rest of the studies (n = 11) in 11 individual countries. | • Only high-quality studies (score > 7) with a low risk of bias were included. | • Conditions assessed: cardiovascular outcomes, CVD mortality, SAH, BP variation, vascular endothelial function, HF, AMI, coronary artery disease, recurrence of atrial fibrillation, stroke, LDL peroxidation;<br>• Two studies evaluated cardiovascular outcomes in patients with DM2 and two studies evaluated the cardiovascular outcomes in patients with CKD;<br>• VE has no role in CVD prevention and cardiovascular mortality;<br>• High serum concentrations of VE may increase the risk of AMI and stroke. | • Only studies in English were included, excluding articles that were not available in full and the non-availability of baseline VE values in several studies. |

| Authors/ year | Objective | Total/Type of the study | Characteristics of the studies | Quality of the studies | Results | Limitations |
|---|---|---|---|---|---|---|
| Hamedi-Kalajahi et al. [17] | • Evaluate the effect of alpha-tocopherol supplementation on ApoA1 and ApoB concentrations<br>• (RS with MA) | • 7 RCTs | • Total number of participants (n = 1284), sample size of studies (from 16 to 112 subjects), age (from 25.4 to 59 years);<br>• Studies with men and women (n = 6) and with men (n = 1);<br>• Studies with diabetic patients (n = 1), obese men (n = 1), on HD (n = 1), on dialysis (n = 1), healthy (n = 1), with hypercholesterolemia (n = 1) and with cardiac ischemia (n = 1);<br>• Study countries: United States (n = 1), Europe (n = 4) and Iran (n = 2);<br>• Dose of supplement (from 200 IU/d to 1200 IU/d);<br>• Duration of the study (from 0.85 to 34.2 weeks). | • Acceptable quality (Cochrane Collaboration tool) | • VE supplementation (alpha-tocopherol) did not influence ApoA1 (SMD = 0.22 IU/d; 95% CI: _0.38, 0.28; P = 0.481) and apoB100 (SMD = _0.62 IU/d; 95% CI: _1.94,<br>• 0.70; P = 0.360) concentrations | • Wide variety in sample size and failure to investigate the genetic component or factors associated with lifestyle |
| Li et al. [12] | • To investigate the influence of tocotrienols | • 17 ECR | • Total number of participants (n = 1354), study | • Jadad score ≥ 3 (n = 9; consistent) and < 3 (n = 8; inconsistent) | • ↓PAS (SMD: _0.616 mmHg, 95% CI: | • Small number of studies, heterogeneity of studies in terms of |

| Authors/ year | Objective | Total/Type of the study | Characteristics of the studies | Quality of the studies | Results | Limitations |
|---|---|---|---|---|---|---|
| | supplementation on anthropometry (BW, BMI, and WC) and cardiometabolic indices (BP, MDA, IL-6, HbA1C, ALT, AST, creatinine, TNF-α and glycemia) in individuals, regardless of health condition.<br>• (MA) | | sample size (18 to 240 subjects), age (18 to 80 years);<br>• History of current disease: DM2 (n = 6), dyslipidemia (n = 2), osteopenia (n = 1), hepatic steatosis (n = 1), breast cancer (n = 1), HD (n = 1), CVD (n = 1) and healthy individuals (n = 4);<br>• Studies with men only (n = 2) and women only (n = 2) and both sexes (n = 13) and study countries: Malaysia (n = 9), Pakistan (n = 1), Australia (n = 1), India (n = 1), Iran (n = 2), Germany (n = 1) and the United States (n = 2);<br>• Dose of supplement (from 27 to 860 mg/d);<br>• Duration of the study (from 60 days to 5 years). | | _1.123, _0.110, p = 0.017)<br>• and ↑PAD (SMD: 0.249 mmHg, 95% CI: 0.053, 0.446, p = 0.013);<br>• Did not influence glycemia (SMD: _0.045 mg/dl, 95% CI: _0.258, 0.168, p = 0.681), HbA1C (SMD: _0.282, 95% CI: _0.676, 0.111, p = 0.160),<br>• CRP (SMD: _0.077 mg/l, 95% CI: _0.309, 0.156, p = 0.519), MDA (SMD: _0.447 nmol/ml, 95% CI: _1.061, 0.167, p = 0.153),<br>• IL-6 (SMD: _0.031 pg./ml, 95% CI: _0.380, 0.319, p = 0.864),<br>• creatinine (SMD: _0.133 mg/dl, 95% CI: _0.606, 0.339, p = 0.580),<br>• AST (SMD: _0.195 U/L, 95% CI: _0.460, 0.071, p = 0.151), ALT (SMD: _0.183 U/L, 95% CI: _0.534, 0.169, p = 0.309), | clinical and demographic data of participants and dose of tocotrienols supplementation |

| Authors/ year | Objective | Total/Type of the study | Characteristics of the studies | Quality of the studies | Results | Limitations |
|---|---|---|---|---|---|---|
| | | | | | • weight (SMD: 0.063 kg, 95% CI: _0.200, 0.327, $p = 0.637$),<br>• BMI (SMD: _0.065 kg/$m^2$, 95% CI: _0.258, 0.129, $p = 0.512$)<br>• and WC (SMD: _0.337 cm, 95% CI: _0.696, 0.022, $p = 0.066$) | |
| O'Connor et al. [18] | • To evaluate the effects of supplementation (alone or in combination) of vitamins and minerals on the prevention of CVD and cancer in healthy adults without micronutrient deficiencies.<br>• (SR) | • 84, of which 78 were RCTs and 6 were cohort studies | • Total participants ($n = 739{,}803$), average age 61 years, 65.1% women and the majority of participants (80.4%) were white. | • From reasonable to good quality | • Supplements evaluated (VA, β-carotene, $VB_9$, $VB_3$, $VB_6$, $VB_{12}$, VC, VD, VE, calcium, Mg and Se);<br>• Vitamin and mineral supplementation was associated with little or no benefit in preventing cancer, CVD and death;<br>• No statistically significant association between VE use and the composite outcome of any cardiovascular disease event;<br>• Little benefit in reducing cancer incidence with the use of multivitamin supplements;<br>• β-carotene was associated with an | • Participants in the studies were predominantly healthy with no known nutritional deficiencies. Therefore, the results cannot be extrapolated to individuals with physical symptoms, medical conditions or nutritional deficiencies. |

| Authors/ year | Objective | Total/Type of the study | Characteristics of the studies | Quality of the studies | Results | Limitations |
|---|---|---|---|---|---|---|
| | | | | | increased risk of lung cancer<br>• and cardiovascular mortality;<br>• Some supplements have been associated with an increased risk of hip fractures (VA), kidney stones (VC and calcium)<br>• and hemorrhagic stroke (VE) | |
| An et al. [14] | • Evaluate the effect of nutrient supplementation on cardiovascular risk factors[1]<br>• (SR with MA) | • 884 RCTs | • Total participants (n = 883,627) | • Of the 256 MAs included, the quality of evidence results was: high (n = 8; 3.13%), moderate (n = 191; 74.61%), low (n = 54; 21.09%) and very low (n = 3; 1.17%). | • 27 different types of nutrients were included[2];<br>• Moderate to high quality evidence that supplementation of n-3 and n-6 PUFA, L-arginine, L-citrulline, $VB_9$, VD, Mg, Zn, α-lipoic acid, coQ10, melatonin, catechin, curcumin, flavanol, genistein, and quercetin reduces coronary risk factors (CRFs);<br>• Supplementation of n-3 PUFA: ↓ cardiovascular mortality (RR = 0.93; 95% CI: 0.88–0.97), | • Some studies with short duration (e.g. < 1 month), low quality evidence (for >20% of MAs included), the effect of the combination of antioxidants was not evaluated and the average dose used by RCTs does not correspond to the optimal dose for clinical effects. |

| Authors/ year | Objective | Total/Type of the study | Characteristics of the studies | Quality of the studies | Results | Limitations |
|---|---|---|---|---|---|---|
| | | | | | ↓AMI (RR = 0.85; 95% CI: 0.78–0.92) and ↓ coronary heart disese events (RR = 0.86; 95% CI: 0.80–0.93);<br>• Folate supplementation: ↓ risk of stroke (RR = 0.84; 95% CI: 0.72–0.97);<br>• CoQ10 supplementation: ↓ all-cause mortality (RR = 0.68; 95% CI: 0.49–0.94);<br>• VC (RR = 1.07 [95% CI: 0.92 to 1.25), VD (RR = 1.00 [95% CI: 0.85 to 1.18), VE (RR = 1.00 [95% CI: 0.92 to 1.08]) and Se (RR = 0.97 [95% CI: 0.83 to 1.14) supplementation (no effect on CVD);<br>• β-carotene supplementation: ↑ cardiovascular mortality (RR: 1.12; 95% CI: 1.06–1.18) and ↑ risk of stroke (RR: 1.09; 95% CI: 1.01–1.17). | |

| Authors/ year | Objective | Total/Type of the study | Characteristics of the studies | Quality of the studies | Results | Limitations |
|---|---|---|---|---|---|---|
| Dludla et al. [19] | • To evaluate the effect of dietary supplements on plasma GSH concentrations in patients with DM (SR) | • 12 RCTs | • Total participants (n = 359), sample size (from 5 to 52 patients) and age (from 28 to 85 years);<br>• Countries: Iran (n = 7), Italy (n = 1), Serbia (n = 1), China (n = 1), Mexico (n = 1) and Brazil (n = 1). | • 9 RCTs with high quality evidence and 3 RCTs with moderate quality evidence | • Dietary supplements evaluated: thiopronine (n = 1), CoQ10 (n = 1), Se (n = 2), curcumin (n = 1), n-3 PUFA (n = 1), VC (n = 1), VE (n = 1), VD (n = 1), glutathione (n = 1), Brazil nuts (n = 1), melatonin (n = 1), cocoa (n = 1) and *Salvia miltiorrhiza* extract (n = 1);<br>• CoQ10, Se, curcumin, n-3 PUFA, VE and VD supplements (↑ GSH, ↓LDL, ↓PCR and ↓MDA concentrations and ↓PA). | • Small number of studies, no evaluation of the long-term effects of supplements and small sample size |
| Zhang et al. [7] | • Evaluate the effect of VE intake (supplemental – alpha tocopherol, dietary or total) on multiple health outcomes<br>• (Umbrella review of MA and SR with MA) | • 27 meta-analyses of observational (cohort and case–control) and intervention studies (controlled and uncontrolled RCTs) | • 29 health outcomes, of which 13 and 16 were related[3] and unrelated[4] to cancer, respectively | • Most of the evidence was low (47.1%) or very low (44.1%)<br>• (AMSTAR score) | • VE intake was associated with a lower risk of cancer: breast, lung (OR = 0.89, 95% CI: 0.81,0.97), esophagus (OR = 0.47, 95% CI: 0.36, 0.60), stomach (RR = 0.76, 95% CI: 0.67, 0.85), kidney (RR = 0.88, 95% CI 0.72,1.08), bladder (RR = 0.89; 95% CI: 0.78,1.00), and cervical (OR = 0.68; 95% CI: 0.49, 0.94), | • MA was based on retrospective studies and most of the evidence was low;<br>• Results of the studies have a large deviation, a combination of micronutrients was often used in the studies;<br>• The amount (minimum and maximum) of VE intake was not clearly specified. |

| Authors/ year | Objective | Total/Type of the study | Characteristics of the studies | Quality of the studies | Results | Limitations |
|---|---|---|---|---|---|---|
| | | | | | Parkinson's disease (RR = 0.84; 95% CI: 0.71–0.99; $I^2$ = 51.9%; p = 0.244), depression (SMD = −0.88; 95% CI: −1.54, −0.21; $I^2$ = 87.0%; p < 0.001), cataracts (RR = 0.90; 95% CI: 0.80, 1.00; $I^2$ = 25.4%; p = 0.244), fractures (RR = 0.66; 95% CI: 0.46–0.95; $I^2$ = 94.2%; p = 0.00), metabolic syndrome (RR = 0.92; 95% CI: 0.85, 1.00; $I^2$ = 67.1%; p < 0.001) and CVD (RR = 0.92; 95% CI: 0.46–0.95; $I^2$ = 55.5%; p = 0.244). | |

*VE (vitamin E), AVC (cerebrovascular accident, stroke), MA (meta-analysis), VA (vitamin A), CSOS (cross-sectional observational studies), RCTs (randomized controlled trials), CRF (coronary risk factor), Mg (magnesium), Zn (zinc), vitamin C (vitamin C), CAC (coronary artery calcification), FFQ (food frequency questionnaire), CVD (cardiovascular disease), SR (systematic review), $VB_6$ (vitamin $B_6$), $VB_9$ (vitamin $B_9$), $VB_{12}$ (vitamin $B_{12}$), VD (vitamin D), HF (heart failure), PUFA (polyunsaturated fatty acid), CoQ10 (coenzyme Q10), ↓ (reduction), SAH (systemic arterial hypertension), BP (blood pressure), AMI (acute myocardial infarction), CAD (coronary artery disease), LDL (low-density lipoprotein), DM2 (type 2 diabetes mellitus), CKD (chronic kidney disease), apo (apolipoprotein), HD (hemodialysis), BW (body weight), BMI (body mass index), WC (waist circumference), MDA (malondialdehyde), IL-6 (interleukin-6), HbA1C (glycated hemoglobin), ALT (alanine aminotransferase), AST (aspartate aminotransferase), TNF-α (tumor necrosis factor alpha), SBP (systolic blood pressure), DBP (diastolic blood pressure), $VB_3$ (vitamin $B_3$), Se (selenium), GSH (glutathione), CRP (C-reactive protein), RR (relative risk), CI (confidence interval), OR (odds ratio), MD (mean differences), SMD (standardized mean difference), ARD (absolute risk difference) and HR (hazard ratio).*
*[1]BP, lipid profile, glycemia, all-cause mortality, cardiovascular mortality, AMI, stroke, CAD, arrhythmia and DM2.[2]PUFA of the n-3 series, PUFA of the n-6 series, PUFA of the n-9 series, hesperidin, L-arginine, L-citrulline, $VB_9$, VC, VD, VE, Mg, Zn, α-lipoic acid, coQ10, lycopene, Se, Mg, Zn, anthocyanin, quercetin, resveratrol, isoflavones, flavonoids, melatonin, catechin, curcumin, flavanol and genistein.[3]Cancer (breast, lung, esophagus, stomach, pancreas, kidney, bladder, cervical, glioma, thyroid, colorectal, endometrium, ovary, non-Hodking lymphoma), total cancer and cancer mortality.[4]CVD, Parkinson's disease, depression, age-related cataracts, metabolic syndrome, fractures, stroke, anxiety, glaucoma, obesity and all-cause mortality.*

**Table 4.**
*Characteristics of the studies and main findings on the effect of vitamin E on cardiovascular health.*

inverse association between increased intake of VE (diet, supplement, or combination) and the risk of stroke [11], (2) reduced CVD mortality rate with VE supplementation [13], and (3) lower systolic blood pressure with VE supplementation [12]. A recent SR showed that VE may improve cardiometabolic healthy by increasing glutathione levels, reducing C-reactive protein, malondialdehyde and LDL levels and controlling blood pressure [19]. A recent umbrella review suggests a lower risk of CVD and metabolic syndrome with higher intake of VE (diet, supplementation or combination) [7].

It is worth noting that, in the MA by Li et al. [12], a positive association between VE supplementation and DBP was found by unknown mechanisms. Other adverse effects of VE supplementation have been suggested, such as an increased risk of hemorrhagic stroke [18] and a small but significant increase in the risk of hospitalization in patients with heart failure (HF) [20]. When offered in high doses, VE starts to act as a pro-oxidant, generating oxidative stress that worsens HF [71], which could explain the increased risk of hospitalization in HF patients.

In the SR by Shah et al. [16], Asian studies showed a significant positive association between serum delta-tocopherol [72] and gamma-tocopherol [73] and increased risk for AMI and increased mortality from hemorrhagic stroke (in women), respectively. In these studies, the authors showed no effect of serum alpha-tocopherol and gamma-tocopherol for AMI [72] and reduced mortality from hemorrhagic stroke [73] for alpha-tocopherol. Another interesting finding is that in the study by Nagao et al. [73], serum gamma-tocopherol was associated with lower stroke mortality in men. These results suggest the influence of VE isoform and gender on cardiovascular outcomes.

There was no effect of VE on CVD prevention and cardiovascular mortality [16, 18], anthropometry and cardiometabolic indices (except blood pressure) [12], CVD and DM2 [14] and serum concentrations of apo A-1 and apo B-100 [17]. Although it did not influence apo A-1 and apo B-100 concentrations, VE was associated with a reduction in LDL-ox concentrations in dialysis patients [74] and diabetics [75], contributing to a reduction in the atherogenic process. In their SR, Machado et al. [15] found that the relationship between VE and coronary artery calcification (CAC), a risk factor for atherosclerosis, coronary events, and cardiovascular mortality, is controversial, with reports of no association [76, 77] or a positive association between VE intake and CAC [78, 79].

Two SRs evaluated the effect of antioxidant supplements alone or in combination on the prevention of CVD [13, 18] and cancer [18]. The use of combined antioxidants was found to have little or no advantage over antioxidants alone. A small reduction in cancer incidence with combined antioxidants was observed in healthy adults without micronutrient deficiencies [18].

Several factors may explain the discrepancy between the studies. In the MA by Khan et al. [20], the quality of the evidence and studies was considered low. In the MA of Li et al. [12], 47.1% of the studies were considered inconsistent. In the umbrella review by Zhang et al. [7], most of the evidence was low (47.1%) or very low (44.1%). In addition, the VE isoform tested was not identified in most studies. Only 2 studies [7, 17] identified the isoform (alpha) of the supplement. Asian results suggest the influence of the VE isoform on cardiovascular outcomes [72, 73]. The characteristics of the participants should also be considered. For example, in the SR of O'Connor et al. [18], where VE supplementation was associated with an increased risk of hemorrhagic stroke, in 55.5% of studies, women were the majority, aged between 55 and 66 years and the frequency of smokers ranged from 2–100%. Smoking is one of the independent risk factors for stroke [80]. All studies had limitations, such as the small number of studies, sample size, language of the articles, follow-up period, and supplement dose. Thus, caution is needed in interpreting the results.

The amount of VE intake associated with a given effect was not clearly specified by the studies, except for Cheng et al. [11] who suggested a dietary VE intake between 15 to 39.45 mg/day for stroke prevention. Bell & Grochoski [81] suggested that healthy individuals should not use VE supplements in amounts greater than 1000 mg/day and complement it, stating that daily VE intake of up to 800 IU appears to be safe and beneficial. Daily VE intake between 400 and 800 IU appears to reduce the risk for several diseases, including CVDs, in healthy individuals [81].

It is worth noting that VE can promote the degradation of drugs used in the treatment of cancer, CVD, SAH, or DM [7]. In addition, it has been suggested that VE doses above 300 mg/day may interfere with the action of tamoxifen, cyclosporine, aspirin, or warfarin [82]. The drug-VE interaction should be considered when interpreting the results.

## 3. Conclusion

Therefore, recommending VE supplementation for CVD prevention and treatment is not feasible due to the lack of consensus on dosage, duration, patient profiles benefiting from supplementation, efficacy of combined antioxidant supplements, and potential adverse effects. Further studies, characterized by good methodological rigor and evidence quality, are necessary to provide clarity on these uncertainties. Hence, prioritizing a diet abundant in fruits, vegetables, whole grains, seeds, oilseeds, and moderate amounts of vegetable oils is recommended for obtaining antioxidant nutrients. This emphasis on dietary sources is preferred over relying on antioxidant supplements, whether isolated or combined.

## Author details

Sofia Kimi Uehara[1], Andrea Ramalho[2] and Claudia Teresa Bento[2*]

1 State University of Rio de Janeiro, Brazil

2 Federal University of Rio de Janeiro, Brazil

*Address all correspondence to: cbento@nutricao.ufrj.br

## References

[1] Jiang Q. Natural forms of vitamin E: Metabolism, antioxidant, and anti-inflammatory activities and their role in disease prevention and therapy. Free Radical Biology & Medicine. 2014;**72**: 76-90

[2] Wong SY, Teo JSM, Chai SF, Yeap SL, Lau AJ. Vitamin E analogues differentially inhibit human cytochrome P450 3A (CYP3A)-mediated oxidative metabolism of lithocholic acid: Impact of δ-tocotrienol on lithocholic acid cytotoxicity. Toxicology. 2019;**423**:62-74

[3] Cheng HS, Ton SH, Tan JBL, Kadir KA. The ameliorative effects of a tocotrienol-rich fraction on the AGE-RAGE axis and hypertension in high-fat-diet-fed rats with metabolic syndrome. Nutrients. 2017;**9**:984

[4] Prasad KN, Edwards-Prasad J. Vitamin E and cancer prevention: Recent advances and future potentials. Journal of the American College of Nutrition. 1992;**11**(5):487-500

[5] Shahidi F, de Costa AC. Tocopherols and tocotrienols in common and emerging dietary sources: Occurrence, applications, and health benefits. International Journal of Molecular Sciences. 2016;**17**(10):1745

[6] Zainal Z, Khaza'ai H, Radhakrishnan AK, Chang SK. Therapeutic potential of palm oil vitamin E-derived tocotrienols in inflammation and chronic diseases: Evidence from preclinical and clinical studies. Food Research International. 2022;**156**:111175

[7] Zhang T et al. Vitamin E intake and multiple health outcomes: An umbrella review. Frontiers in Public Health. 2023; **11**:1035674

[8] Organização Pan-Americana da Saúde, Doenças Cardiovasculares [Internet]. 2023. Available from: https://www.paho.org/pt/topicos/doencas-cardiovasculares [Accessed: November, 2023].

[9] Corrêa-Camacho CR, Dias-Melicio LA, Soares AMVC. Aterosclerose, uma resposta inflamatória. Arq Ciênc Saúde. 2007;**14**(1):41-48

[10] Batlouni M. Hipótese oxidativa da aterosclerose e emprego dos antioxidantes na doença arterial coronária. Arquivos Brasileiros de Cardiologia. 1997;**68**(1):55-63

[11] Cheng P et al. Vitamin E intake and risk of stroke: A meta-analysis. The British Journal of Nutrition. 2018;**120**: 1181-1188

[12] Li F, Xu B, Soltanieh S, Zanghelini F, Abu-Zaid A, Sun J. The effects of tocotrienols intake on obesity, blood pressure, inflammation, liver and glucose biomarkers: A meta-analysis of randomized controlled trials. Critical Reviews in Food Science and Nutrition. 2021;**62**(26):7154-7167

[13] Han J, Zhao C, Cai J, Liang Y. Comparative efficacy of vitamin supplements on prevention of major cardiovascular disease: Systematic review with network meta-analysis. Complementary Therapies in Clinical Practice. 2020;**39**:101142

[14] An P et al. Micronutrient supplementation to reduce cardiovascular risk. JACC. 2022;**80**(24): 2269-2285

[15] Machado AD, Andrade GRG, Levy J, Ferreira SS, Marchioni DM. Association between vitamins and minerals with antioxidant effects and coronary artery

calcification in adults and older adults: A systematic review. Current Pharmaceutical Design. 2019;**25**: 2474-2479

[16] Shah S et al. A systematic review of effects of vitamin E on the cardiovascular system. Cureus. 2021; **13**(6):e15616

[17] Hamedi-Kalajahi F et al. A systematic review and meta-analysis on the impact of oral vitamin E supplementation on apolipoproteins A1 and B100. Clinical Nutrition. 2021;**46**: 106-114

[18] O'Connor EA et al. Vitamin and mineral supplements for the primary prevention of cardiovascular disease and cancer. Update evidence report and systematic review for the US preventive services task force. JAMA. 2022;**327**(23): 2334-2347

[19] Dludla PV et al. Dietary supplements potentially target plasma glutathione levels to improve cardiometabolic health in patients with diabetes mellitus: A systematic review of randomized clinical trials. Nutrients. 2023;**15**(4):944

[20] Khan MS et al. Dietary interventions and nutritional supplements for heart failure: A systematic appraisal and evidence map. European Journal of Heart Failure. 2021;**23**:1468-1476

[21] Torquato P et al. Analytical strategies to assess the functional metabolome of vitamin E. Journal of Pharmaceutical and Biomedical Analysis. 2016;**124**:399-412

[22] Phang SCW, Ahmad B, Kadir KA, Palanisamy UDM. Effects of tocotrienol-rich fraction supplementation in patients with type 2 diabetes: A systematic review and meta-analysis of randomized controlled trials. Advances in Nutrition. 2023;**14**:1159-1169

[23] Absalome MA et al. Biochemical properties, nutritional values, health benefits and sustainability of palm oil. Biochimie. 2020;**178**:81-95

[24] Marangoni F et al. Palm oil and human health. Meeting report of NFI: Nutrition Foundation of Italy symposium. International Journal of Food Sciences and Nutrition. 2017;**68**:643-655

[25] Sen CK, Rink C, Khanna S. Palm oil-derived natural vitamin E alphatocotrienol in brain health and disease. Journal of the American College of Nutrition. 2010;**29**:314s-323s

[26] Shahidi F, Pinaffi-Langley ACC, Fuentes J, Speisky H, de Camargo AC. Vitamin E as an essential micronutrient for human health: Common, novel, and unexplored dietary sources. Free Radical Biology & Medicine. 2021;**176**: 312-321

[27] Siger A, Górnas P. Free tocopherols and tocotrienols in 82 plant species' oil: Chemotaxonomic relation as demonstrated by PCA and HCA. Food Research International. 2023;**164**:112386

[28] Universidade Federal de São Paulo. Tabela Composição Alimentos – UNIFESP. São Paulo: Departamento de Informática em Saúde. Escola Paulista de Medicina. Universidade Federal de São Paulo. [Internet]. UNICAMP. 2011. Available from: http://tabnut.dis.epm.br/alimento

[29] Torres LCR, Sartori AGO, Silva APS, Alencar SM. Bioaccessibility and uptake/epithelial transport of vitamin E: Discoveries and challenges of *in vitro* and *ex vivo* assays. Food Research International. 2022;**162**:112143

[30] Yap WN et al. Gamma-and delta-tocotrienols inhibit skin melanin

synthesis by suppressing constitutive and UV-induced tyrosinase activation. Pigment Cell & Melanoma Research. 2001;**23**:688-692

[31] Fairus S, Nor RM, Cheng HM, Sundram K. Alpha-tocotrienol is the most abundant tocotrienol isomer circulated in plasma and lipoproteins after postprandial tocotrienol-rich vitamin E supplementation. Nutrition Journal. 2012;**11**:5

[32] Rizvi S, Raza ST, Ahmed F, Ahmed A, Abbas S, Mahdi F. The role of vitamin e in human health and some diseases. Sultan Qaboos University Medical Journal. 2014;**14**(2):e157-e165

[33] Waniek S et al. Vitamin E (α-and γ-tocopherol) levels in the community: Distribution, clinical and biochemical correlates, and association with dietary patterns. Nutrients. 2018;**10**(1):3

[34] Bardowell SA, Ding X, Parker RS. Disruption of P450-mediated vitamin E hydroxylase activities alters vitamin E status in tocopherol supplemented mice and reveals extra-hepatic vitamin E metabolism. Journal of Lipid Research. 2012;**53**(12):2667-2676

[35] Soysal P, Arik F, Smith L, Jackson SE, Isik AT. Inflammation, frailty and cardiovascular disease. Advances in Experimental Medicine and Biology. 2020;**1216**:55-64

[36] Phaniendra A, Jestadi DB, Periyasam L. Free radicals: Properties, sources, their implication in various diseases. Indian Journal of Clinical Biochemistry. 2015;**30**(1):11-26

[37] Jideani AJO, Silungwe H, Takalani T, Omolola AO, Udeh HO, Anyasi TA. Antioxidant-rich natural fruit and vegetable products and human health. International Journal of Food Properties. 2021;**24**:41-67

[38] Khorami SAH, Mutalib MSA, Khazai'ai H. Comparative effect of vitamin E compounds on HSP70 expression in response to acute redox imbalance in Chang liver cells Research Square. April 21st, 2020. DOI: q0.21203/rs.3.rs-22336/vl

[39] Trela-Makowej A, Kruk J, Zadlo A, Szymanska R. Antioxidant activity of vitamin E homologues in model systems, [Internet]. 2021. DOI: 10.21203/rs.3.rs-199353/v1

[40] Jiang Z, Yin X, Jiang Q. Natural forms of vitamin E and 130-carboxychromanol, a long-chain vitamin E metabolite, inhibit leukotriene generation from stimulated neutrophils by blocking calcium influx and suppressing 5-lipoxygenase activity, respectively. Journal of Immunology. 2011;**186**:1173-1179

[41] Jiang Q. Natural forms of vitamin E: Metabolism, antioxidant, and anti-inflammatory activities and their role in disease prevention and therapy. Review article. Free Radical Biology & Medicine. 2014;**72**:76-90

[42] Boutaud O, Aronoff DM, Richardson JH, Marnett LJ, Oates JA. Determinants of the cellular specificity of acetaminophen as an inhibitor of prostaglandin H(2)synthases. Proceedings of the National Academy of Sciences of the United States of America. 2002;**99**:7130-7135

[43] Jiang Q, Elson-Schwab I, Courtemanche C, Ames BN. Gamma-tocopherol and its major metabolite, in contrast to alpha-tocopherol, inhibit cyclooxygenase activity in macrophages and epithelial cells. Proceedings of the National Academy of Sciences of the

United States of America. 2000;**97**(21): 11494-11499

[44] Jiang Q, Yin X, Lill MA, Danielson ML, Freiser H, Huang J. Long-chain carboxychromanols, metabolites of vitamin E, are potent inhibitors of cyclooxygenases. Proceedings of the National Academy of Sciences of the United States of America. 2008;**105**: 20464-20469

[45] Papaioannou D, Cooper KL, Carroll C, Hind D, Squires H, Tappenden P, et al. Antioxidants in the chemoprevention of colorectal cancer and colorectal adenomas in the general population: A systematic review and meta-analysis. Colorectal Disease. 2011; **13**(10):1085-1099

[46] Moya-Camarena SY, Jiang Q. The role of vitamin E forms in cancer prevention and therapy—Studies in human intervention trials and animal models. In: Sarkar FH, editor. Nutraceuticals and Cancer. New York: Springer; 2012. pp. 323-354. (Chap.15)

[47] Dolara P, Bigagli E. Collins A, antioxidant vitamins and mineral supplementation, life span expansion and cancer incidence: Acritical commentary. European Journal of Nutrition. 2013;**51**:769-781

[48] Myung SK et al. Efficacy of vitamin and antioxidant supplements in prevention of cardiovascular disease: Systematic review and meta-analysis of randomised controlled trials. BMJ. 2013; **346**(f10). DOI: 10.1136/bmj.f10

[49] Dovizio M, Alberti S, Guillem-Llobat P, Patrignani P. Role of platelets in inflammation and cancer: Novel therapeutic strategies. Basic & Clinical Pharmacology & Toxicology. 2014; **114**(1):118-127

[50] Debbabi M et al. Protective effects of alpha-tocopherol, gamma-tocopherol and oleic acid, three compounds of olive oils, and no effect of trolox, on 7-ketocholesterol-induced mitochondrial and peroxisomal dysfunction in microglial BV-2 cells. International Journal of Molecular Sciences. 2016; **17**(12):1973

[51] Jiang Q. Metabolism of natural forms of vitamin E and biological actions of vitamin E metabolites. Free Radical Biology & Medicine. 2022;**179**:375-387

[52] Menn GL, Jablo'nska A, Chen Z. The effects of post-translational modifications on Th17/Treg cell differentiation. Biochimica et Biophysica Acta - Molecular Cell Research. 2022;**6**: 119223

[53] Fu JY, Meganathan P, Gunasegaran N, Tan DMY. Effect of nano-delivery systems on the bioavailability and tissue biodistribution of vitamin E tocotrienols. Food Research International. 2023;**171**:113048

[54] Zhao H et al. γ-Tocotrienol inhibits T helper 17 cell differentiation via the IL-6/JAK/STAT3 signaling pathway. Molecular Immunology. 2022;**151**: 126-133

[55] Kim Y, Gromovsky AD, Brown JM, Chung S. Gamma-tocotrienol attenuates the aberrant lipid mediator production in NLRP inflammasome-stimulated macrophages. The Journal of Nutritional Biochemistry. 2018;**58**:169-177

[56] Grimm MOW, Mett J, Hartmann T. The impact of vitamin E and other fat-soluble vitamins on Alzheimer s disease. International Journal of Molecular Sciences. 2016;**17**(11):1785

[57] Ismail M et al. Safety and neuroprotective efficacy of palm oil and

tocotrienol-rich fraction from palm oil: A systematic review. Nutrients. 2020; **12**(2):521

[58] Malavolta M et al. Pleiotropic effects of tocotrienols and quercetin on cellular senescence: Introducing the perspective of senolytic effects of phytochemicals. Current Drug Targets. 2016;**17**:447-459

[59] Rahman TA, Hassim NF, Zulkafli N, Muid S, Kornain NK, Nawawi H. Atheroprotective effects of pure tocotrienol supplementation in the treatment of rabbits with experimentally induced early and established atherosclerosis. Food & Nutrition Research. 2016;**60**:31525

[60] Das S et al. Cardioprotection with palm tocotrienol: Antioxidant activity of tocotrienol is linked with its ability to stabilize proteasomes. American Journal of Physiology. Heart and Circulatory Physiology. 2012;**302**(11):H2447

[61] Kosmas CE et al. High-density lipoprotein (HDL) functionality and its relevance to atherosclerotic cardiovascular disease. Drugs Context. 2018;**7**:212525

[62] American Diabetes Association. Standards of medical care in diabetes. Diabetes Care. 2009;**32**(suppl 1):S13-S61

[63] Skyler JS et al. Differentiation of diabetes by pathophysiology. Natural History, and Prognosis, Diabetes. 2017; **66**(2):241-255

[64] Birringer M, Lorkowski S. Vitamin E: Regulatory role of metabolites. IUBMB Life. 2019;**71**(4):479-486

[65] Wells SR, Jennings MH, Rome C, Hadjivassiliou V, Papas KA, Alexander JS. α-, γ- and δ-tocopherols reduce inflammatory angiogenesis in human microvascular endothelial cells. The Journal of Nutritional Biochemistry. 2010;**21**(7):589-597

[66] Villas Boas GR, Paes MM, Cunha MS, Ponsoni LF, da Silveira APS, Oesterreich SA. Evaluation of the effect of alpha-tocopherol on anxiety and the neuroinflammatory process during alcohol withdrawal in a model of forced and chronic self-administration of liquid diet containing ethanol: Behavioral and neurochemical evidence. Alcohol. 2022; **104**:31-44

[67] Yang C, Jiang Q. Vitamin E δ-tocotrienol inhibits TNF-α-stimulated NF-κB activation by up-regulation of anti-inflammatory A20 via modulation of sphingolipid including elevation of intracellular dihydroceramides. The Journal of Nutritional Biochemistry. 2019;**64**:101-109

[68] Dhodapkar MM, Ramachandran XSR, Wallach JD. Characterization and corroboration of safety signals identified from the US Food and Drug Administration adverse event reporting system, 2008-19: Cross sectional study. BMJ. 2022;**379**:e071752

[69] Violi F, Nocella C, Lofredo L, Carnevale R, Pignatelli P. Interventional study with vitamin E in cardiovascular disease and meta-analysis. Free Radical Biology & Medicine. 2022;**178**:26-41

[70] Hu FB, Willett WC. Optimal diets for prevention of coronary heart disease. JAMA. 2002;**288**(20):2569-2578

[71] Bowry VW, Ingold KU, Stocker R. Vitamin E in human low-density lipoprotein. When and how this antioxidant becomes a pro-oxidant. The Biochemical Journal. 1992;**288**(Pt 2): 341-344

[72] Naidoo N et al. Plasma vitamin E and coenzyme Q10 are not associated with a

lower risk of acute myocardial infarction in Singapore Chinese adults. The Journal of Nutrition. 2012;**142**:1046-1052

[73] Nagao M, Moriyama Y, Yamagishi K, Iso H, Tamakoshi A. Relation of serum α- and γ-tocopherol levels to cardiovascular disease-related mortality among Japanese men and women. Journal of Epidemiology. 2012;**22**:402-410

[74] Diepeveen S et al. Effects of atorvastatin and vitamin E on lipoproteins and oxidative stress in dialysis patients: A randomised-controlled trial. Journal of Internal Medicine. 2005;**257**(5):438-445

[75] Keihan GS, Gharib MH, Momeni A, Hemati Z, Sedighin R. A comparison between the effect of cuminum and vitamin E on the level of leptin, paraoxonase 1, HbA1c and oxidized LDL in diabetic patients. International Journal of Molecular and Cellular Medicine. 2016;**5**(4):229-235

[76] Arad Y, Spadaro LA, Roth M, Newstein D, Guerci AD. Treatment of asymptomatic adults with elevated coronary calcium scores with atorvastatin, vitamin C, and vitamin E: The St. Francis heart study randomized clinical trial. Journal of the American College of Cardiology. 2005;**46**(1): 166-172

[77] de Oliveira Otto MC et al. Dietary micronutrient intakes are associated with markers of inflammation but not with markers of subclinical atherosclerosis. The Journal of Nutrition. 2011;**141**(8):1508-1515

[78] Hatzigeorgiou C, Taylor AJ, Feuerstein IM, Bautista L, O'Malley PG. Antioxidant vitamin intake and subclinical coronary atherosclerosis. Preventive Cardiology. 2006;**9**(2): 75-81

[79] Falcone EL et al. Micronutrient concentrations and subclinical atherosclerosis in adults with HIV. The American Journal of Clinical Nutrition. 2010;**91**(5):1213-1219

[80] de Souza DP, Waters C. Perfil epidemiológico dos pacientes com acidente vascular cerebral: pesquisa bibliográfica. Brazilian Journal of Health Review. 2023;**6**(1):1466-1478

[81] Bell SJ, Grochoski GT. How safe is vitamin E supplementation? Critical Reviews in Food Science and Nutrition. 2008;**48**:760-774

[82] Miller ER, Pastor-Barriuso R, Dalal D, Riemersma RA, Appel LJ, Guallar E. Metaanalysis: High-dosage vitamin E supplementation may increase all-cause mortality. Annals of Internal Medicine. 2005;**142**(1):37-46

Chapter 8

# Carotenoids and Their Antioxidant Power

*Nuriye Arslansoy and Ozkan Fidan*


## Abstract

Carotenoids are natural products found in photosynthetic organisms such as plants, algae, and some bacteria species. Humans and animals cannot synthesize carotenoids, and they obtain these molecules through their diet. The common structure of carotenoids contains conjugated double bonds that provide color formation in the visible spectrum, at 400–500 nm. In photosynthetic organisms, carotenoids contribute to color formation for various purposes, such as sex selection, protection from predators, and light-harvesting to increase the spectral range of photosynthesis. The conjugated double bonds not only provide color formation but also provide antioxidant properties to carotenoid molecules. Studies have shown that carotenoids are capable of scavenging free radicals and reactive oxygen species, as well as quenching singlet oxygen molecules. The antioxidant power of carotenoids results in several health benefits. These include anticancer, neuroprotective, and anti-atherosclerotic activities. This chapter aims to review the antioxidant activities and health benefits of major carotenoids, beginning with their structure and synthesis, and also discussing their natural sources.


**Keywords:** carotenoids, antioxidant, natural products, pigment, health benefits

## 1. Introduction

Carotenoids are a natural group of lipid-soluble pigments [1]. They consist of polyunsaturated isoprene units ($C5H8$) and have 30–50 carbons in their structure with conjugated double bonds [2]. These conjugated double bonds provide the color formation in carotenoid molecules [3]. The most naturally abundant carotenoids are C40 carotenoids and the functional groups at the end of the chain diversify the carotenoids, leading to the formation of more than 750 carotenoid molecules [2, 3]. Depending on oxygen presence, carotenoids are divided into two groups: carotenes consisting of only carbon and hydrogen, and xanthophylls containing oxygen in their structure [3]. Carotenoids are mainly biosynthesized in photosynthetic organisms such as plants, algae, some bacteria, and fungi species [3]. However, animals and humans cannot synthesize carotenoids; thus they must obtain them from dietary sources [4].

Carotenoids play a significant role in photosynthesis due to two main functions: as photoprotective agents and as light harvesting pigments [5]. In addition, carotenoids have exhibited various biological activities such as antioxidant, anticancer, anti-atherosclerotic properties [6]. Also, it was shown that increased carotenoid consumption correlates with a lower risk of chronic diseases [4]. Another important aspect of carotenoids is their role as provitamin A. Vitamin A is involved in various processes, including vision, reproduction, cell differentiation and proliferation, and immunity. Vitamin A deficiency can cause several symptoms, such as xerophthalmia, increased susceptibility to severe infections, increased mortality, and detrimental effects on growth and fetal development [7]. β-carotene, α-carotene, and β-cryptoxanthin exhibit provitamin-A activity and thus, are useful to combat vitamin A deficiency [7]. Apart from all these health benefits, a study revealed that long-term supplementation of β-carotene has increased the cognitive abilities of the men subjects [8]. Overall, carotenoids play a significant role in human health due to their biological activities.

## 2. Biosynthesis of carotenoids

Carotenoids are mainly found in photosynthetic organisms and synthesized from plastids [9]. The most abundant carotenoids are the C40 carotenoids, and their synthesis starts from 5-carbon isoprenoid units as isopentenyl diphosphate (IPP) and dimethylallyl diphosphate (DMAPP) [9]. These two precursors are synthesized from one of the two non-homologous pathways: the mevalonate (MVA) pathway and the 2-*C*-methyl-D-erythritol 4-phosphate (MEP) pathway [2, 10]. In the first step of MVA pathway, three acetyl-CoA molecules are condensed to yield 3-hydroxy-3-methylglutaryl-CoA (HMG-CoA), which is subsequently reduced to MVA via the catalysis of HMG-CoA reductase. The phosphorylation of MVA twice and subsequent decarboxylation result in the formation of IPP, and through the catalysis of IPP isomerase, IPP is converted to DMAPP. On the other hand, the starting point of MEP pathway is the synthesis of 1-deoxy-D-xylulose 5-phosphate (DXP) from pyruvate and D-glyceraldehyde-3-phosphate (G3P) by head-to-head condensation of these two molecules, catalyzed via DXP synthase. Intramolecular rearrangement of DXP results in formation of 2-*C*-methyl-D-erythritol (ME) and the subsequent reduction of ME generates MEP. Finally, MEP is converted to IPP and DMAPP via four intermediates [11].

The complete biosynthesis of certain carotenoids is shown in **Figure 1**. The two precursors, IPP and DMAPP, first form the C10 molecule geranyl pyrophosphate (GPP) via head-to-tail condensation. GPP and IPP yields C15 molecule farnesyl pyrophosphate (FPP) by addition, and further addition of one more IPP molecule to FPP results in the formation of C20 intermediate precursor geranylgeranyl pyrophosphate (GGPP). The head-to-head condensation of two GGPP molecules leads to prephytoene diphosphate (PPPP), which forms phytoene via the removal of the diphosphate group and stereospecific proton abstraction. Phytoene is a colorless molecule with three conjugated double bonds and provides the C40 skeleton of carotenoids. It is the primary precursor to synthesize C40 carotenoids. For the formation of phytoene from IPP, there are four enzymes to catalyze the steps. These enzymes are IPP isomerase for isomerization of IPP to DMAPP, GPP synthase to convert IPP and DMAPP into GPP, GGPP synthase to catalyze conversion of IPP and DMAPP to GGPP, and lastly, phytoene synthase that catalyzes the formation of phytoene from GGPP [9].

Conjugated carbon-carbon double bonds provide the required chromophore structure for carotenoid coloration. Desaturation of phytoene by phytoene desaturase (PDS) in higher plants and carotene isomerase (CrtI) in bacteria and fungi species

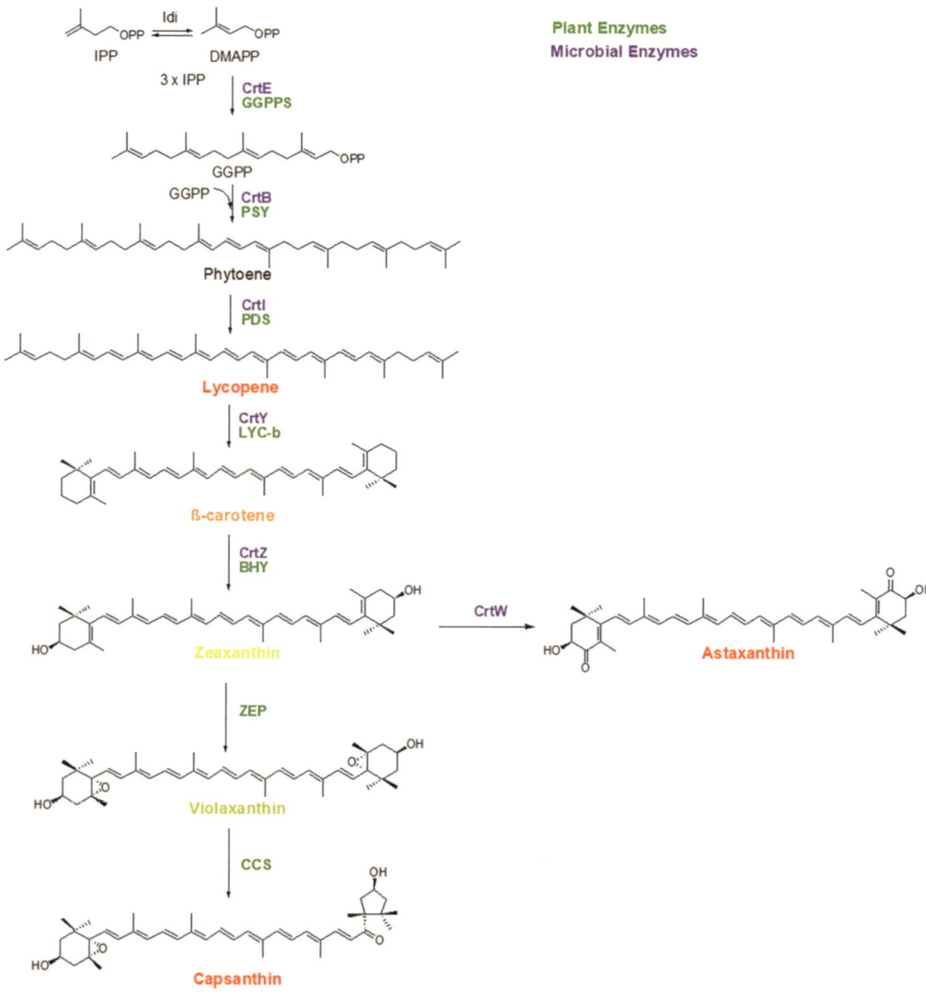


**Figure 1.**
*Biosynthesis pathway of carotenoids.*

leads to formation of lycopene at four steps by introducing one desaturation at each step [9]. Increased number of conjugated carbon-carbon double bonds results in red colored lycopene [2, 9]. Phytoene is found as 15-*cis* isomer in higher plants while lycopene is observed as all-*trans*, which indicates the isomerization that is facilitated by light and catalyzed by plant desaturases at some point [9].

Lycopene is the precursor for the synthesis of cyclic and bicyclic carotenoids, which is catalyzed by lycopene cyclases [2]. At first, cyclization of lycopene by lycopene-β-cyclase (LYC-b) by the introduction of β-rings and lycopene-ε-cyclase (LYC-e) by the introduction of ε-rings yields β-carotene and α-carotene. Further modifications of carotenes via β-carotene ketolase and β-carotene hydroxylase generate several C40 carotenoids that belong to the xanthophylls group [2]. The first xanthophylls formed from β-carotene are β-cryptoxanthin and zeaxanthin, which are obtained via hydroxylation on C-3 and C-3′ positions of the β-rings by β-carotene hydroxylase. Zeaxanthin epoxidase (ZEP) modifies 3-hydroxy β-rings of zeaxanthin by introducing 5,6-epoxy groups and yields antheraxanthin and violaxanthin.

Finally, capsanthin/capsorubin synthase (CCS) forms cyclopentane rings (κ-rings) from 3-hydroxy-5,6-epoxy β-rings of antheraxanthin and violaxanthin, and results in capsanthin and capsorubin, which are the ketolated carotenoids [9]. Astaxanthin is synthesized by β-carotene ketolase, catalyzing the conversion of zeaxanthin into astaxanthin [12]. C50 and C30 carotenoids are rare, and C50 carotenoids are synthesized through the addition of two DMAPP molecules on C40 carotenoids [13], while C30 carotenoids are synthesized by two ways: by the condensation of FPP molecules or from the oxidative cleavage of C40 carotenoids [2, 14].

## 3. Dietary sources of carotenoids

Carotenoids cannot be synthesized by humans and must be obtained through daily dietary intake. Most important sources of carotenoids in the human diet are fruits and vegetables, which provide almost 50 different carotenoids. Various carotenoids can be detected in human blood plasma, including zeaxanthin, β-cryptoxanthin, lutein, lycopene, β-carotene, and α-carotene [15]. A study revealed that the highest carotenoid-containing fruits and vegetables are red peppers, carrots, apricots, nectarines, plums, peaches, and the lowest in cherries [16]. **Table 1** summarizes some of the fruits and vegetables and their carotenoid contents as μ/100 g vegetable or fruit.

Due to its provitamin A activity, β-carotene is considered one of the most important carotenoids. Mainly, apricot, mango, and pumpkin species are rich in β-carotene and also contain a portion of α-carotene [18]. Sixty to seventy percent of total carotenoids of apricot is determined as β-carotene [19]. Orange vegetables such as carrots, and dark green leafy vegetables like lettuce and spinach, are other sources rich in β-carotene [20, 21]. According to the database (https://fdc.nal.usda.gov/fdc-app.html#/?component=1122), lycopene is found in tomato (46 mg/100 g FW (fresh weight)), watermelon (1.6–3.5 mg/100 g FW), papaya (1.8–4.2 mg/100 g FW), and guava (3.2–7.0 mg/100 g FW) [17]. Eighty to ninety percent of the total pigments in tomatoes are determined to be lycopene, making tomatoes and tomato-based products such as ketchup and sauce significant sources of lycopene [22, 23]. Gac fruit aril is not included in the database, yet. However, gac fruit aril is determined as the highest β-carotene (20.5 mg/100 g FW) and lycopene (146.6 mg/100 g FW) containing fruit [24].

Corn seeds and egg yolks are good sources of lutein and zeaxanthin due to their high content of these carotenoids [23]. In egg yolk, 21.8 μg/g FW lutein and 13.4 μg/g FW zeaxanthin were determined [25]. The lutein and zeaxanthin content of egg yolk is coming from corn seeds; hence chicken diet is primarily depending on corn seeds [23]. Total lutein and zeaxanthin content from four different corn cultivars was determined between 947 and 2758 μg/100 g [26]. Spinach (11.93 mg/100 g) and kale (39.55 mg/100 g) are found as high-level lutein and zeaxanthin containing vegetables [27]. Carotenoids can exist in esterified form among plants. For instance, the high level of zeaxanthin dipalmitate (35.7 mg/g FW) was determined in goji berry at fully ripe stage, while 5% of total carotenoids was estimated as β-cryptoxanthin monopalmitate (2.2 mg/g FW) [28]. β-Cryptoxanthin is provitamin A xanthophyll and the major carotenoid found in mandarins and oranges [15]. β-Cryptoxanthin was determined in both skin (283–1254 μg/g FW) and pulp (76.5–287 μg/g FW) of the persimmon with different levels depending on the cultivar [29].

In addition to plants, carotenoids are also synthesized via various microorganisms, including certain algae, bacteria, and fungi species. Algae are

| Carotenoid | Description | Carotenoid per 100 g (μg) |
|---|---|---|
| Lycopene | Tomato powder | 46,300 |
| | Tomato, sun-dried | 45,900 |
| | Catsup | 12,100 |
| | Rose Hips, wild (Northern Plains India) | 6800 |
| | Guavas, common, raw | 5200 |
| β-Carotene | Peppers, sweet, red, freeze-dried | 42,900 |
| | Carrot, dehydrated | 34,000 |
| | Grape leaves, raw | 16,200 |
| | Sweet potato, frozen, cooked, baked, without salt | 12,500 |
| | Tomato powder | 10,300 |
| α-Carotene | Carrot, dehydrated | 14,300 |
| | Peppers, sweet, red, freeze-dried | 6930 |
| | Pumpkin, raw | 4020 |
| β-Cryptoxanthin | Species, pepper, red or cayenne | 6250 |
| | Species, paprika | 6190 |
| | Papayas, raw | 589 |
| | Peppers, sweet, red, raw | 490 |
| Zeaxanthin | Eggs, Grade A, large, egg yolk | 546 |
| | Spinach, mature | 466 |
| | Spinach, baby | 191 |
| Lutein | Spinach, mature | 7450 |
| | Spinach, baby | 5830 |
| | Eggs, Grade A, large, egg yolk | 612 |

**Table 1.**
*Carotenoid contents of different fruits and vegetables according to information from [17].*

pigment-producing photosynthetic organisms and perceived as a good source of bioactive compounds including carotenoids [30]. Astaxanthin is a red carotenoid with an antioxidant activity 10 times higher than other carotenoids [31]. Main microalgae species producing astaxanthin are *Haematococcus pluvialis*, *Chlorella zofingiensis*, and *Chlorococcum* sp. [32]. Richest source of astaxanthin is *H. pluvialis* with 80% (relative to biomass) of the total carotenoids and *H. pluvialis* is the primary industrial source for the production of astaxanthin [30, 32]. Halotolerant green algae *Dunaliella salina* synthesizes carotenoids such as β-carotene, α-carotene, lutein, and lycopene [33]. Relatively, 86% of total carotenoids from *D. salina* was determined as β-carotene that makes *D. salina* the richest source of β-carotene [33]. Lutein and zeaxanthin are synthesized via algal species such as *Scenedesmus* spp., *Chlorella* spp., *Rhodophyta* spp., or *Spirulina* spp. *Chlorella* genus is the best source for industrial level lutein production. β-carotene, β-cryptoxanthin, and zeaxanthin are synthesized by *Spirulina platensis* [30]. Generally, algae species produce various carotenoids, and they are the most suitable organisms for industrial scale-up processes to synthesize carotenoids.

Despite the high yield of microalgal carotenoid biosynthesis, the cost of carotenoid biosynthesis via microalgae in an industrial process is higher than that of bacteria since the microalgae require longer cultivation time and depend on light to produce carotenoids [34]. Some bacterial species are also natively capable of biosynthesizing carotenoids and gain attention for carotenoid biosynthesis due to shorter cultivation time [34]. In literature, *Brevibacterium linens* was reported as high β-cryptoxanthin (0.3 mg/mL) producing bacteria [35], while *Paracoccus zeaxanthinifaciens* was reported as high zeaxanthin (13.8 mg/L) producer [36] and *Paracoccus carotinfaciens* (E-396) was reported as high astaxanthin producer (19.9 mg/L) [37]. Besides, some endophytic bacteria species are identified as native carotenoid producers. For instance, *Pseudomonas* sp. 102,515 was isolated from the leaves of *Taxus chinensis* and capable of producing an antioxidant natural product, zeaxanthin diglucoside (380 ± 12 mg/L) [38].

## 4. Antioxidant potential and health benefits of carotenoids

### 4.1 Functional roles of different carotenoid molecules

Carotenoids have various functions in nature, including light-harvesting, photoprotection, coloration for sexual purposes, and protection in animals, as well as provitamin A activity in vertebrates [39]. All carotenoids possess light-harvesting and photoprotection properties due to conjugated double bonds, which can exhibit $\pi \rightarrow \pi^*$ transition following light-absorption and subsequent high-energy excitation [39]. This light-harvesting property of carotenoids increases the spectral range of photosynthesis in plants [40]. Besides, light absorption and excitation result in color formation and carotenoids exhibit yellow- orange or red color, which are in the visible spectrum with wavelengths 400–500 nm [39].

In the presence of at least 9 conjugated double bonds, molecules exhibit singlet oxygen quenching activity and radical scavenging activity that provides carotenoids' antioxidant properties. In addition, the linear system of conjugated C-C double bonds renders carotenoids potent antioxidants via high reducing potential in lipid formation via oxidation. When carotenoid molecules interact with membranes or dissolve in lipid structures, they can protect these structures from oxidative damage caused by aggressive radical species, thereby preventing irreversible destruction [40]. The effective neutralization of reactive oxygen species and other free radicals by carotenoids in both photosynthetic and non-photosynthetic organisms provides protection from oxidative damage. This makes carotenoids excellent candidates as natural antioxidants [41].

### 4.2 Role of carotenoids in reducing oxidative stress via antioxidant properties

Reactive oxygen species (ROS) are formed by cells for normal cellular functions, such as intracellular signaling and redox regulation. ROS molecules are derived from oxygen and can be extremely reactive in different forms such as hydroxyl radical, or less reactive like superoxide and hydrogen peroxide. Intracellular free radicals, which have unpaired electrons, are often considered as ROS. The excessive formation of ROS is called oxidative stress and causes severe damage in cellular structures [42, 43]. ROS and free radicals start a chain reaction by interacting with biomolecules like lipids, proteins, and DNA, and form free radicals readily [42, 43]. Newly formed free radicals further react with another free radical to eliminate the unpaired electron or with a free radical scavenger, a primary antioxidant, to break the chain [42]. Antioxidant is defined as "any substance that delays, prevents or removes oxidative damage to a target molecule" [44]. Antioxidants are investigated under two groups as enzymatic antioxidants and non-enzymatic antioxidants [43] or as preventative and chain-breaking antioxidants [45].

Carotenoids are non-enzymatic, chain-breaking, and lipid soluble antioxidants due to their hydrophobic nature [45, 46]. Hydrophobic scavengers are located on the cellular membrane or found in lipoproteins, preventing lipid peroxidation by interrupting chain reactions of ROS or free radicals [47]. Chain-breaking antioxidants interfere with the chain reaction via trapping the chain-carrying radicals [45]. Chain-breaking antioxidants, such as α-tocopherol, tend to donate hydrogen atom to trap free radicals [48]. However, a study investigated the radical scavenging activity of four carotenoids, astaxanthin, β-carotene, canthaxanthin, and zeaxanthin, to reveal the radical scavenging mechanism of carotenoids and concluded that carotenoids demonstrated the antioxidant activity via the addition of radicals to conjugated polyene chain in their structure unlike the other natural antioxidants. The conjugated polyene chain stabilizes the free radicals by the resonance of conjugated double bonds under low oxygen pressure similar to physiological tissue environment [48].

Carotenoids are secondary free radical scavengers and physical singlet oxygen quenchers [47]. Singlet oxygen ($^1O_2$) is an excited state of oxygen where an electron changes spin and yields two unpaired electrons with opposite spins. It is responsible for $^1O_2$ dependent lipid peroxidation and DNA damage [49, 50]. Vitamin-E (α-tocopherol) is the most efficient peroxyl radicals' scavenger in cellular membrane phospholipid bilayer structure [47]. Importantly, the inhibition of $^1O_2$ dependent lipid peroxidation was also achieved by carotenoids including β-carotene, astaxanthin, and lycopene, due to their $^1O_2$ scavenging activity as great as that of α-tocopherol [49]. An *in vitro* study to show the protection of cells from lipid peroxidation via singlet oxygen quenching activity of carotenoids concluded that the highest protection was provided by lycopene, followed by astaxanthin and then, β-carotene [51]. Basically, carotenoids prevent lipid peroxidation and cellular damage through radical scavenging or singlet oxygen quenching activities, rendering carotenoids effective natural antioxidants and important dietary supplements for health.

### 4.3 Potential health benefits associated with carotenoid consumption

In human plasma, 12 carotenoids have been detected, including lutein, zeaxanthin, lycopene, α-carotene, β-carotene, and β-cryptoxanthin [52]. It has been hypothesized that dietary carotenoids could reduce cancer rate, and several studies have been conducted to test this hypothesis. These studies considered different types of cancer, and revealed that a high consumption of carotenoid-rich vegetables and fruits was associated with a decreased risk of cancer [53]. Epidemiological studies investigated the mechanism behind reduced risk of cancer associated with carotenoids and showed that the supplementation of vitamin A does not decrease the risk of cancer, unlike carotenoids. This suggests that carotenoids may exert their anticancer effects through mechanisms other than those related to vitamin A activity [54]. The antioxidant activity of carotenoids is well-known for their ability to scavenge radicals and quench singlet oxygen. Both radicals and singlet oxygen can damage biomolecules in cells through mechanisms such as lipid peroxidation or DNA damage. Since the genetic changes caused by DNA damage can led to cancer formation, carotenoids can decrease the risk of cancer through their antioxidant activities [34, 53]. For instance, a study showed that there was an inverse relationship between developing prostate cancer and the consumption of lycopene rich tomato and tomato-based products. This suggests that lycopene has a significant impact to reduce the prostate cancer development risk [55]. In an in vitro study on breast cancer, it was demonstrated that carotenoids, including astaxanthin, β-carotene, and lutein, exhibited synergistic effects with anticancer drug doxorubicin. These carotenoids increased the ROS-mediated apoptosis of cancer cells while showing no toxicity to healthy cells [56]. Overall, various carotenoids exhibit anticancer activity attributed to their radical scavenging and singlet oxygen quenching capabilities. The supplementation of carotenoids holds promise for the treatment and prevention of several cancer types.

Oxidative stress may contribute to the pathogenesis of Alzheimer's disease. Astaxanthin was reported with its antioxidant activity as well as *in vivo* and *in vitro* neuroprotective effects. In a study investigating the neuroprotective effect of astaxanthin for Alzheimer's disease using HT22 cells, it was found that astaxanthin could be beneficial for the treatment of neurological diseases such as Alzheimer's disease, since it significantly suppressed the accumulation of ROS in HT22 cells and was involved in the regulation of other biomolecules implicated in the progression of the disease [57]. In the case of subarachnoid hemorrhage, neuronal apoptosis plays a significant role in the development of pathogenesis. The administration of astaxanthin after subarachnoid hemorrhage was shown to reduce the neuronal apoptosis and mitigate secondary brain injury, thereby alleviating brain dysfunction. This neuroprotective effect of astaxanthin involves modulation of the Akt/Bad pathway [58]. In addition, oxidative stress is also involved in aging and reduces cognitive abilities such as memory and language [59]. In 2013, a study revealed that serum lutein levels were significantly associated with better cognitive abilities in the subjects over 80 years old. Additionally, serum levels of lutein, zeaxanthin, and β-carotene were all correlated with better cognitive abilities in subjects over 100 years old. Besides, it was found that brain lutein and β-carotene levels are associated with better cognitive abilities [59]. It may be concluded that carotenoids are also beneficial for improving cognitive abilities and preventing neurodegenerative diseases.

Macula is a part of the eye that contains high density of cone cells and provides vision. It is exposed to light and oxidative stress on a regular basis. Age-related macular degeneration (AMD) is a chronic disease that progresses to blindness and oxidative stress aids the manifestation of AMD. Zeaxanthin and lutein are accumulating on the human macula to protect the macula through scavenging light-induced radicals and reducing oxidative stress. Previous studies indicated that the antioxidant activity of carotenoids in macula reduces the ROS levels and prevents damage caused from oxidative stress, which ultimately helps to reduce the risk of AMD. It was concluded that the supplementation of lutein and zeaxanthin is a preventative treatment against AMD [60]. Skin is the largest organ in the human body and is exposed continuously to light and solar radiation, like eyes. The solar radiation and light exposure cause the generation of wrinkles, pigmentation, and premature aging as well as ROS formation [34]. Lycopene and β-carotene supplementation was reported to provide protection against UV-induced skin damage while elevated dietary intake of these carotenoids protected the skin from UV-induced erythema in humans [61].

Lutein is mainly suggested for eye-related disorders, but it has a higher antioxidant potential than β-carotene and lycopene, which makes it a suitable target for the prevention and treatment of cardiovascular diseases. It has been showed that lutein exhibits anti-atherosclerotic activity. Lutein is involved in the regulation of certain pathways related to antioxidant enzyme production and reduces ROS via both direct antioxidant activity and expression of antioxidant enzymes. Low density lipoproteins (LDL) play a key role in the development of atherosclerosis, and lutein reduces the distribution of LDL particles in plasma and accumulation of LDL particles in the aorta. There are several other mechanisms that lutein involves and prevents cardiovascular disorders and thus, it is a good candidate to prevent or delay atherosclerosis [62]. Several studies demonstrated that lycopene exhibits similar mechanisms of action to other carotenoids, including the modulation of oxidative stress through the regulation of different pathways. Additionally, various cleavage products of lycopene were shown to interact with transcription factors and regulate the overexpression of antioxidants [63]. Lycopene was also considered as a promising molecule for the treatment and prevention of cardiovascular diseases, neurodegenerative diseases, and various cancer types due to its high antioxidant activity [63].

## 5. Conclusion

Carotenoids are secondary metabolites synthesized mainly by photosynthetic organisms such as plants, algae, and certain bacterial species. They are synthesized from IPP and DMAPP through the catalysis of several enzymes and have a conjugated polyene chain in their structure. The conjugated polyene chain provides carotenoids with antioxidant activities to scavenge free radicals and ROS and quench singlet oxygen. Since animals and humans cannot synthesize carotenoids, they must obtain them through their diet. Fruits and vegetables contain different types of carotenoids, some of which have been detected in human blood plasma, indicating the absorption and distribution of dietary carotenoids in the human body. Carotenoids exhibit various biological activities such as anticancer, neuroprotective, and anti-atherosclerotic properties due primarily to their antioxidant activity. Therefore, carotenoids are beneficial to human health for both the prevention and treatment of various diseases. However, the supply chain of carotenoids depends on extraction from plants, which is not sufficient for industry, is affected by climate change, and also requires longer time for plant growth. Currently, biotechnological processes are gaining attention for the microbial synthesis of carotenoids. There are various microbial species that can synthesize carotenoids. Carotenoid producing microorganisms can be used for industrial scale-up production as well as genetically modified microorganisms. In the industrial process of microbial carotenoid biosynthesis, synthetic biology and metabolic engineering approaches are useful to increase the yield. However, high amounts of production might be toxic to microorganisms. Further studies should investigate alternative methods for the high-yield and low-cost production of carotenoids.

## Conflict of interest

The authors declare no conflict of interest.

## Author details

Nuriye Arslansoy[1] and Ozkan Fidan[2*]

1 Department of Bioengineering, Graduate School of Science and Engineering, Abdullah Gül University, Kayseri, Turkey

2 Faculty of Natural and Life Sciences, Department of Bioengineering, Abdullah Gül University, Kayseri, Turkey

*Address all correspondence to: ozkan.fidan@agu.edu.tr

## References

[1] Böhm V. Carotenoids. Antioxidants. 2019;**8**(11):516

[2] Fernandes AS, TC do N, Jacob-Lopes E, De RVV, Zepka LQ. Introductory chapter: Carotenoids—A brief overview on its structure, biosynthesis, synthesis, and applications. In: Progress in Carotenoid Research. Rijeka: IntechOpen; 2018

[3] Westphal A, Böhm V. Carotenoids. Properties, distribution, bioavailability, metabolism and health effects. Ernahrungs Umschau. 2015;**62**(11):196-207

[4] Rao AV, Rao LG. Carotenoids and human health. Pharmacological Research. 2007;**55**:207-216

[5] Cogdell RJ, Gardiner AT. [18] Functions of carotenoids in photosynthesis. Methods in Enzymology. 1993;**214**(C):185-193

[6] Russell RM, Paiva SAR. β-Carotene and other carotenoids as antioxidants. Journal of the American College of Nutrition. 1999;**18**(5):426-433

[7] Meléndez-Martínez AJ. An overview of carotenoids, apocarotenoids, and vitamin A in agro-food, nutrition, health, and disease. Molecular Nutrition and Food Research. 2019;**63**(15):e1801045

[8] Grodstein F, Kang JH, Glynn RJ, Cook NR, Gaziano JM. A randomized trial of beta carotene supplementation and cognitive function in men: The physicians' health study II. Archives of Internal Medicine. 2007;**167**(20):2184-2190

[9] Fraser PD, Bramley PM. The biosynthesis and nutritional uses of carotenoids. Progress in Lipid Research. 2004;**43**(3):228-265

[10] Rico J, Duquesne K, Petit JL, Mariage A, Darii E, Peruch F, et al. Exploring natural biodiversity to expand access to microbial terpene synthesis. Microbial Cell Factories. 2019;**18**(1):23

[11] Kuzuyama T, Seto H. Two distinct pathways for essential metabolic precursors for isoprenoid biosynthesis. Proceedings of the Japan Academy Series B: Physical and Biological Sciences. 2012;**88**(3):41-52

[12] Tao L, Wilczek J, Odom JM, Cheng Q. Engineering a β-carotene ketolase for astaxanthin production. Metabolic Engineering. 2006;**8**(6):523-531

[13] Yang Y, Yatsunami R, Ando A, Miyoko N, Fukui T, Takaichi S, et al. Complete biosynthetic pathway of the C50 carotenoid bacterioruberin from lycopene in the extremely halophilic archaeon Haloarcula japonica. Journal of Bacteriology. 2015;**197**(9):1614-1623

[14] Mo XH, Sun YM, Bi YX, Zhao Y, Yu GH, Tan LL, et al. Characterization of C30 carotenoid and identification of its biosynthetic gene cluster in methylobacterium extorquens AM1. Synthetic and Systems Biotechnology. 2023;**8**(3):527-535

[15] Meléndez-Martínez AJ, Mandić AI, Bantis F, Böhm V, Borge GIA, Brnčić M, et al. A comprehensive review on carotenoids in foods and feeds: Status quo, applications, patents, and research needs. Critical Reviews in Food Science and Nutrition. 2022;**62**(8):1999-2049

[16] Leong SY, Oey I. Effects of processing on anthocyanins, carotenoids

and vitamin C in summer fruits and vegetables. Food Chemistry. 2012;**133**(4):1577-1587

[17] Dias MG, Olmedilla-Alonso B, Hornero-Méndez D, Mercadante AZ, Osorio C, Vargas-Murga L, et al. Comprehensive database of carotenoid contents in Ibero-American foods. A valuable tool in the context of functional foods and the establishment of recommended intakes of bioactives. Journal of Agricultural and Food Chemistry. 2018;**66**(20):5055-5107

[18] Britton G, Khachik F. Carotenoids in food. In: Carotenoids. Basel: Birkhäuser; 2009. pp. 45-66

[19] Sass-Kiss A, Kiss J, Milotay P, Kerek MM, Toth-Markus M. Differences in anthocyanin and carotenoid content of fruits and vegetables. Food Research International. 2005;**38**(8-9):1023-1029

[20] López A, Javier GA, Fenoll J, Hellín P, Flores P. Chemical composition and antioxidant capacity of lettuce: Comparative study of regular-sized (Romaine) and baby-sized (little gem and mini Romaine) types. Journal of Food Composition and Analysis. 2014;**33**(1):39-48

[21] Beltrán B, Estévez R, Cuadrado C, Jiménez S, Olmedilla AB. Carotenoid data base to assess dietary intake of carotenes, xanthophyls and vitamin A; its use in a comparative study of vitamin A nutritional status in young adults. Nutrición Hospitalaria. 2012;**27**(4):1334-1343

[22] Shi J, Le Maguer M. Lycopene in tomatoes: Chemical and physical properties affected by food processing. Critical Reviews in Food Science and Nutrition. 2000;**40**(1):1-42

[23] Saini RK, Prasad P, Lokesh V, Shang X, Shin J, Keum YS, et al. Carotenoids: Dietary sources, extraction, encapsulation, bioavailability, and health benefits—A review of recent advancements. Antioxidants. 2022;**11**(4):795

[24] Müller-Maatsch J, Sprenger J, Hempel J, Kreiser F, Carle R, Schweiggert RM. Carotenoids from gac fruit aril (*Momordica cochinchinensis* [Lour.] Spreng.) are more bioaccessible than those from carrot root and tomato fruit. Food Research International. 2017;**99**(2):928-935

[25] Ko EY, Saini RK, Keum YS, An BK. Age of laying hens significantly influences the content of nutritionally vital lipophilic compounds in eggs. Food. 2021;**10**(1):22

[26] Aguillón-Páez YJ, Díaz GJ. Lutein and zeaxanthin content in corn imported from three countries of the American continent and in corn cultivated in Colombian territory. Arquivo Brasileiro de Medicina Veterinária e Zootecnia. 2023;**75**(3):500-510

[27] Mrowicka M, Mrowicki J, Kucharska E, Majsterek I. Lutein and zeaxanthin and their roles in age-related macular degeneration—Neurodegenerative disease. Nutrients. 2022;**14**(4):827

[28] Hempel J, Schädle CN, Sprenger J, Heller A, Carle R, Schweiggert RM. Ultrastructural deposition forms and bioaccessibility of carotenoids and carotenoid esters from goji berries (*Lycium barbarum L.*). Food Chemistry. 2017;**218**:525-533

[29] Veberic R, Jurhar J, Mikulic-Petkovsek M, Stampar F, Schmitzer V. Comparative study of primary and secondary metabolites

in 11 cultivars of persimmon fruit (*Diospyros kaki L.*). Food Chemistry. 2010;**119**(2):477-483

[30] Pereira AG, Otero P, Echave J, Carreira-Casais A, Chamorro F, Collazo N, et al. Xanthophylls from the sea: Algae as source of bioactive carotenoids. Marine Drugs. 2021;**19**(4):188

[31] Guerin M, Huntley ME, Olaizola M. Haematococcus astaxanthin: Applications for human health and nutrition. Trends in Biotechnology. 2003;**21**(5):210-216

[32] Camacho F, Macedo A, Malcata F. Potential industrial applications and commercialization of microalgae in the functional food and feed industries: A short review. Marine Drugs. 2019;**17**(6):312

[33] Murthy KNC, Vanitha A, Rajesha J, Swamy MM, Sowmya PR, Ravishankar GA. In vivo antioxidant activity of carotenoids from *Dunaliella salina*—A green microalga. Life Sciences. 2005;**76**(12):1381-1390

[34] López GD, Álvarez-Rivera G, Carazzone C, Ibáñez E, Leidy C, Cifuentes A. Bacterial carotenoids: Extraction, characterization, and applications. Critical Reviews in Analytical Chemistry. 2023;**53**(6):1239-1262

[35] Asker D, Awad TS, Beppu T, Ueda K. Screening, isolation, and identification of zeaxanthin-producing bacteria. In: Methods in Molecular Biology. New York, USA: Humana Press; 2018

[36] Joshi C, Singhal RS. Modelling and optimization of zeaxanthin production by *Paracoccus zeaxanthinifaciens* ATCC 21588 using hybrid genetic algorithm techniques. Biocatalysis and Agricultural Biotechnology. 2016;**8**:228-235

[37] Hirasawa K, Tsubokura A. Method for separating carotenoid. US 8853460B2. 2014

[38] Fidan O, Zhan J. Discovery and engineering of an endophytic Pseudomonas strain from *Taxus chinensis* for efficient production of zeaxanthin diglucoside. Journal of Biological Engineering. 2019;**13**(1):66

[39] Britton G. Structure and properties of carotenoids in relation to function. The FASEB Journal. 1995;**9**(15):1551-1558

[40] Vershinin A. Biological functions of carotenoids—Diversity and evolution. BioFactors. 1999;**10**:99-104

[41] Cho KS, Shin M, Kim S, Lee SB. Recent advances in studies on the therapeutic potential of dietary carotenoids in neurodegenerative diseases. Oxidative Medicine and Cellular Longevity. 2018;**2018**:4120458

[42] Nordberg J, Arnér ESJ. Reactive oxygen species, antioxidants, and the mammalian thioredoxin system. Free Radical Biology and Medicine. 2001;**31**(11):1287-1312

[43] Nogueira CCR, Paixão ICNDP, Teixeira VL. Antioxidant activity of natural products isolated from red seaweeds. Natural Product Communications. 2014;**9**(7):1031-1036

[44] Halliwell B. Biochemistry of oxidative stress. Biochemical Society Transactions. 2007;**35**(5):1147-1150

[45] Burton GW, Ingold KU. β-Carotene: An unusual type of lipid antioxidant. Science. 1984;**224**(4649):569-573

[46] Cooper DA, Eldridge AL, Peters JC. Dietary carotenoids and certain cancers, heart disease, and age-related macular degeneration: A review of

recent research. Nutrition Reviews. 1999;**57**(7):201-214

[47] Chaudière J, Ferrari-Iliou R. Intracellular antioxidants: From chemical to biochemical mechanisms. Food and Chemical Toxicology. 1999;**37**(9-10):949-962

[48] Jørgensen K, Skibsted LH. Carotenoid scavenging of radicals. Z Lebensm Unters Forsch. 1993;**196**(5):423-429

[49] Fukuzawa K, Inokami Y, Tokumura A, Terao J, Suzuki A. Rate constants for quenching singlet oxygen and activities for inhibiting lipid peroxidation of carotenoids and α-tocopherol in liposomes. Lipids. 1998;**33**(8):751-756

[50] Edge R, McGarvey DJ, Truscott TG. The carotenoids as anti-oxidants—A review. Journal of Photochemistry and Photobiology B: Biology. 1997;**41**(3):189-200

[51] Tinkler JH, Böhm F, Schalch W, Truscott TG. Dietary carotenoids protect human cells from damage. Journal of Photochemistry and Photobiology. B. 1994;**26**(3):283-285

[52] Khachik F. Distribution and metabolism of dietary carotenoids in humans as a criterion for development of nutritional supplements. Pure and Applied Chemistry. 2006;**78**(8):1551-1557

[53] van Poppel G. Carotenoids and cancer: An update with emphasis on human intervention studies. European Journal of Cancer. 1993;**29**(9):1335-1344

[54] Ziegler RG. A review of epidemiologic evidence that carotenoids reduce the risk of cancer. Journal of Nutrition. 1989;**119**(1):116-122

[55] Rowles JL, Ranard KM, Applegate CC, Jeon S, An R, Erdman JW. Processed and raw tomato consumption and risk of prostate cancer: A systematic review and dose–response meta-analysis. Prostate Cancer and Prostatic Diseases. 2018;**21**:319-336

[56] Vijay K, Sowmya PRR, Arathi BP, Shilpa S, Shwetha HJ, Raju M, et al. Low-dose doxorubicin with carotenoids selectively alters redox status and upregulates oxidative stress-mediated apoptosis in breast cancer cells. Food and Chemical Toxicology. 2018;**118**:675-690

[57] Wen X, Huang A, Hu J, Zhong Z, Liu Y, Li Z, et al. Neuroprotective effect of astaxanthin against glutamate-induced cytotoxicity in HT22 cells: Involvement of the Akt/GSK-3β pathway. Neuroscience. 2015;**303**:558-568

[58] Zhang XS, Zhang X, Wu Q, Li W, Zhang QR, Wang CX, et al. Astaxanthin alleviates early brain injury following subarachnoid hemorrhage in rats: Possible involvement of Akt/bad signaling. Marine Drugs. 2014;**12**(8):4291-4310

[59] Johnson EJ, Vishwanathan R, Johnson MA, Hausman DB, Davey A, Scott TM, et al. Relationship between serum and brain carotenoids, α -tocopherol, and retinol concentrations and cognitive performance in the oldest old from the Georgia centenarian study. Journal of Aging Research. 2013;**2013**:951786

[60] Nguyen D, Thrimawithana T, Piva TJ, Grando D, Huynh T. Benefits of plant carotenoids against age-related macular degeneration. Journal of Functional Foods. 2023;**106**:105597

[61] Stahl W, Sies H. β-Carotene and other carotenoids in protection from

sunlight. American Journal of Clinical Nutrition. 2012;**96**(5):1179S-1184S

[62] Hajizadeh-Sharafabad F, Ghoreishi Z, Maleki V, Tarighat-Esfanjani A. Mechanistic insights into the effect of lutein on atherosclerosis, vascular dysfunction, and related risk factors: A systematic review of in vivo, ex vivo and in vitro studies. Pharmacological Research. 2019;**149**:104477

[63] Saini RK, Rengasamy KRR, Mahomoodally FM, Keum YS. Protective effects of lycopene in cancer, cardiovascular, and neurodegenerative diseases: An update on epidemiological and mechanistic perspectives. Pharmacological Research. 2020;**155**:104730

Chapter 9

# Selenium in Plants: Biofortification, Toxicity, and Redox State Balance

*Flávio José Rodrigues Cruz*

## Abstract

Selenium is a beneficial nutrient for plants because it improves some physiological indicators and increases plant growth and production in low concentrations. In general, in several plant species, selenium levels below 8 μM have a biofortifying effect. However, under high concentrations, selenium metabolization causes depletion of cellular levels of glutathione, protein misfolding, redox imbalance, and, consequently, reduced plant growth. This chapter proposes a systematic review of the biofortifying and toxic mechanisms triggered by selenium, based on physiological, biochemical, and plant growth responses. Understanding these responses will contribute to improving the understanding of the basic mechanisms of biofortification and selenium toxicity in higher plants, particularly in plants native to the Amazon. In these plants, selenium can modulate the greater synthesis of secondary metabolism compounds, which have been proven to have nutraceutical and pharmacological importance, such as spilanthol.

**Keywords:** beneficial nutrient, absorption, antioxidant, pro-oxidant, misfold protein, glutathione

## 1. Introduction

Micronutrient deficiency or hidden hunger represents an important public health problem, especially in developing countries. Hidden hunger occurs when food is deficient in vitamins or minerals and does not meet human nutritional needs. Iron, zinc, iodine and vitamin A are the micronutrients that are often insufficient in energy-rich but nutrient-poor diets. According to the World Health Organization, around two billion people are affected by chronic micronutrient and vitamin deficiency in the world [1, 2].

In this scenario, several strategies have been adopted to overcome this problem, such as diet diversification (increased consumption of foods of animal origin) [3], supplementation (consumption of concentrated sources of vitamins and minerals) [4], fortification (enrichment of foods during its processing with micronutrients) [4] and agronomic biofortification of plants (increasing the concentration of nutrients in edible plant organs through fertilization via soil or leaves) [5].

Dietary diversification comprises variation in food consumption to better meet macro and micronutrient intake needs. Nutrient diversification is related to the introduction or more frequent consumption of non-basic foods in existing diets. However, there are challenges linked to implementing dietary diversification as a single strategy. This is explained by the fact that foods naturally rich in micronutrients are often inaccessible to the poorest social segments of societies and locally produced foods may not be able to satisfy nutritional needs for some population groups in some environments.

Supplementation consists of providing concentrated sources of nutrients or other substances with nutritional or physiological effects, which supplement the normal diet. It offers a direct solution to micronutrient deficiencies in specific contexts, where they can be diagnosed and managed effectively. However, at a population level, supplementation to combat hidden hunger is not practical because it is expensive, dependent on the ability to reach those most at risk and dependent on population adherence. For the individual, supplementation may provide a short-term solution, however, it does not solve the long-term problem of a nutrient-poor diet.

Food fortification is defined as the practice of adding vitamins and minerals to normally consumed foods during processing to increase their nutritional value. It is a proven, safe and cost-effective strategy for improving diets and for preventing and controlling micronutrient deficiencies that can result in disease. However, this strategy was unsuccessful in some regions. For example, in Brazil, a flour biofortification program with iron to combat anemia in six-year-old children obtained results contrary to those expected, where there was an increase in anemia among the children evaluated.

Crop biofortification is another strategy, in addition to the strategies mentioned above, that can be implemented to combat micronutrient deficiencies in the human diet. By fertilizing the soil or plant leaves, it is possible to increase the content of micronutrients such as selenium in cereals consumed by humans and/or animals, leafy vegetables and forage plants intended for animal feed. Its implementation is simple and its costs are low when selenium is supplied via fertilizers with a positive effect on combating diseases related to selenium deficiency.

Agronomic biofortification with micronutrients is a simple and viable alternative because of its ease and low cost. In this context, several studies have shown an increase in the concentration of micronutrients in plants, as in the case of selenium (Se) [6–8].

Selenium is considered a beneficial element for plants [9, 10] because in small concentrations it attenuates the effects of biotic and abiotic stresses [11, 12]. However, in plants treated with concentrations above a tolerated limit, Se becomes toxic because it affects plant metabolism and consequently reduces plant growth [13–15].

In humans and animals, Se is considered essential [16] because it is part of the constitution of around 25 selenoproteins that act in antioxidant metabolism, hormonal balance, immune system, male fertility, resistance to virus infections and cancer prevention [17–22].

Soils have a positive influence on human health because they provide essential nutrients for the functioning of the human body. About 78% of the world's average calorie consumption per person originates from growing crops directly in the soil, and approximately 20% comes from land-based food sources that depend indirectly on the soil [23]. However, the interface between plant nutrition and human health depends on the type of soil and its properties, as well as the magnitude of absorption and accumulation of nutrients in the edible parts of plants. Therefore, ideal agricultural production for human health must provide foods rich in essential mineral elements, but with a lower content of toxic elements [24].

This review aimed to address the topic of biofortification, toxicity and redox status of plants grown in the presence of selenium, detailing the biochemical and physiological aspects of higher plants related to the mechanism of selenium absorption, antioxidant metabolism, photosynthesis and plant growth.

## 2. Crop biofortification and its importance

Crop biofortification or agronomic biofortification is the deliberate use of fertilizers to increase the concentration of mineral nutrients in edible parts of plants and to increase human intake of foods of plant or animal origin containing a target nutrient [5, 25, 26].

The nutrients used in agronomic biofortification can be applied to the soil, nutrient solution, or leaves. Foliar application makes Se readily available to plants for use in their metabolism. In this case, attention must be paid to the source and dose, because selenite is more toxic to plants compared to selenate, considering the same dose for each source. On the other hand, application to the soil may imply greater adsorption of Se in mineral colloids, in particular selenite. This fact may imply less availability of Se and less impact of biofortification of plants with Se, considering the physical and chemical properties of the soil such as pH, texture and organic matter content. After Se application, the nutrient is absorbed and accumulated in grains [27], tubers [28], vegetable fruits [29], vegetable leaves [30], tree fruits [31], or forage [32]. The biofortified crop can be consumed directly by people (cereals, fruits, or leaves) or fed to ruminant or non-ruminant animals (forage or cereals), which will provide food in the form of meat, milk, or eggs containing higher levels of the target nutrient for consumption human.

For a few decades, the presence of Se in the human diet has received greater attention worldwide due to some regions having soils poor in Se [33]. Soils containing total Se levels below 0.5 mg $kg^{-1}$ are considered deficient [34].

Se deficiency has been reported in various regions of the world such as China, the United Kingdom, Australia, New Zealand, Africa, central Siberia, Bangladesh, part of India and eastern Europe [19, 35, 36]. This scenario tends to worsen because global climate changes may contribute to increasing Se deficiency in soils in agricultural areas [37], because rising temperatures and water scarcity can affect water availability, nutrient absorption and, consequently, the crop cycle.

Furthermore, populations living in regions where soils are poor in Se and who eat foods with low Se concentrations may suffer from Se deficiency disorders due to the recommended daily intake (RDI) of Se being lower than that recommended for adults (55–200 μg) for the correct maintenance of metabolism and expression of selenoproteins [38, 39].

Biofortification of crops with Se is important in regions with soils poor in Se, because this nutrient is essential for humans. Se enters the chemical composition of selenoproteins, specifically in one or more selenoaminoacid residues. These proteins are involved in diverse metabolic and physiological functions, antioxidant defenses, fertility, muscle function and development, thyroid hormone metabolism and the immune system.

## 3. Selenium biofortification and antioxidant metabolism in plants

The boundary between toxicity and biofortification of plants with Se is narrow. In this aspect, Se concentrations below or above a limit tolerated by plants have beneficial or deleterious effects, respectively, on the physiological and biochemical aspects of plants. For example, concentrations equal to or lower than 8 μM of Se have a positive impact on the photosynthetic machinery and plant growth, reducing oxidative stress [40–43].

Se supplied in low concentrations to plants has a biofortifying effect because it increases its levels in plant tissues and elicits the biosynthesis of secondary metabolites or nutraceutical compounds beneficial to human health [44]. For example, some studies report an increase in the concentration of total phenolic compounds [45–48], phenolic acids [30, 47, 49], monomethylated forms of Se such as Se-methyl selenocysteine [50–52] and antioxidant activity [30, 48]. In plants from the Brassicaceae family, Se supplied through fertilization is incorporated into secondary metabolites via the selenoglucosinolate synthesis route. In this secondary metabolite, Se may be present in the form of methylselenoakyl, a group that probably derives from selenomethionine.

The mechanism of biofortifying action of Se in plants arises from its direct or indirect antioxidant effect [53, 54], because it minimizes oxidative stress [55, 56], protects and improves the functioning of the photosynthetic machinery [43], increases plant growth and production [55, 57].

## 4. Biofortification mechanism via leaves and roots

Fertilization methods can be divided into root or foliar fertilization, depending on the route by which the nutrient is absorbed [58]. Soil fertilization is affected by a range of factors such as soil temperature, salinity, and microbiota [59]. On the other hand, foliar fertilization as a supplementary fertilization strategy can deliver the nutrient directly to the target through the aerial part of the plant, reducing the negative impacts that originate from fertilization on the soil [60, 61]. If absorbed by leaves it does not follow the same route when supplied to the roots (transport from root to shoot). Thus, the assimilation and accumulation of Se is increased in the aerial part compared to application in the soil, where Se can be lost due to it being adsorbed or undergoing chemical transformation [62].

In tropical regions where weathered soils predominate, caution must be taken regarding the form of application and source of selenium, because weathered soils tend to have a clay fraction with high levels of iron and aluminum oxides. These two oxides adsorb more Se in the form of selenite compared to selenate. This fact indicates that foliar fertilization is the most appropriate and efficient in biofortification programs because the availability of Se to plants is greater.

In plants, selenium ($SeO_3^{2-}$ and $SeO_4^{2-}$) and sulfur ($SO_3^{2-}$ and $SO_4^{-2}$) share the same absorption, assimilation and translocation route because they have chemical similarities. Among the forms of Se, $SeO_4^{2-}$ is absorbed through the $SO_4^{-2}$ membrane transporter. Therefore, the exogenous supply of $SeO_4^{2-}$ reduces the absorption of $SO_4^{-2}$ [12, 62]. Furthermore, plants can absorb Se in the form of organic compounds such as selenocysteine and selenomethionine [63, 64].

The translocation order of Se species in wheat plants is $SeO_4^{2-}$ > Se-Met > $SeO_3^{2-}$/Se-Cys [65]. Studies with sulfate and phosphate transporters in *Arabdopisis thaliana* suggest that the SULTR1;1 and SULTR1;2 transporters facilitate the absorption of $SeO_4^{2}$. On the other hand, $SeO_3^{2-}$ absorption can occur through phosphate transporters [59, 66, 67].

The low-affinity transporters SULTR2;1 and SULTR3;5 are expressed in the pericycle and xylem parenchyma, cooperating in the loading of sulfate into the xylem [68, 69]. The SULTR2;1 transporter can be expressed when the Se concentration is high in the plant growth substrate [69]. In wheat, selenate supplementation induced greater sulfate absorption because selenate modulates the expression of sulfate transporters to facilitate its absorption [27].

Regarding crop biofortification, it is important to consider the chemical form and method of application of Se because both $SeO_4^{2-}$ and $SeO_3^{2-}$ are forms absorbed and accumulated by plants. However, $SeO_3^{2-}$ interacts strongly with the clay fraction of several types of soils in tropical and subtropical regions, making it less available to plants compared to $SeO_4^{2-}$ [59, 70, 71].

After being absorbed by the roots, selenite is readily incorporated into low-mobility organic compounds in the xylem. Selenate has greater mobility in the xylem and is therefore accumulated in greater quantities in the aerial part of plants compared to selenite. However, when selenite is applied foliarly, it is accumulated in greater quantities compared to selenate [59, 70]. Selenate, when supplied via soil or nutrient solution to plants, presents greater biofortification efficiency compared to selenite in potato [72], lettuce [73] and rice plants [27].

In plants, the Se content in plant tissues varies in response to the form of application and source of Se. In beet plants, the Se content in the aerial part is higher when selenate is supplied via the soil. But when it is supplied foliarly, selenite promotes a higher Se content in the aerial part. In the root of beet plants, the Se content is higher when selenate is supplied via the root or leaf [28]. Regarding the production of lentils for food, foliar application of Se was more effective than soil application in terms of production and accumulation of Se in grains [33]. In wheat (cultivar 202w17), foliar application of fertilizer enriched with selenate promoted greater production and Se content in grains compared to soil application [74].

The chemical similarity between selenium and sulfur, the shared absorption, assimilation and translocation of these two nutrients, as well as the existence of organic and inorganic sources of Se absorbable by plants, are factors that optimize plant biofortification programs with Se.

## 5. Selenium toxicity and deficiency in plants and redox mechanism

Due to Se being a beneficial element for plants, there are no records of plant growth disorders resulting from its deficiency. However, Se supplementation above a limit tolerable by plants imposes toxicity risks, which result in reduced plant growth and production. In this sense, plants can be classified according to Se accumulation as non-accumulators (accumulate less than 100 μg Se $g^{-1}$ DM—forages and other plants), secondary accumulators (accumulate between 100 and 1000 μg Se $g^{-1}$ DM—Broccoli, Brassica, Helianthus and Aster) and primary accumulators or hyperaccumulators (accumulate more than 1000 μg Se $g^{-1}$ DM—Stanleya, Neptunia, Astragulus and Canopis) [12].

Risks of toxicity and reduction in plant growth or production depend on the plant species, source, and dose/concentration of Se. For example, Zhang et al. [57] obtained a reduction in the production of rice grains cultivated with 100 g selenite $ha^{-1}$. Matraszek and Hawrylak-Nowak [75] found a reduction in the productivity of lettuce grown under 20 μM selenite.

Selenium toxicity depends on the source, dose/concentration, species, and age of the plant. The sensitivity of seedlings is greater compared to that of mature plants. Furthermore, plants are more sensitive to selenite compared to selenate [12, 76]. On a biochemical level, Se toxicity in plants is characterized by the incorporation of Se into the amino acids cysteine and methionine, replacing S in these amino acids (**Figure 1**). In a subsequent step, the Se-amino acids selenocysteine and selenomethionine replace the amino acids cysteine and methionine, respectively in the constitution of selenoproteins or non-specific selenoproteins [14].

At toxic levels of Se, the substitution of amino acids (cysteine or methionine) for Se-amino acids (selenocysteine or selenomethionine) affects the interaction between amino acid residues through intramolecular forces established by disulfide bonds, so that malformation or folding occurs. Proteins and, consequently, the formation of amorphous protein aggregates inside cells. Malformed proteins resulting from the replacement of amino acids (cysteine or methionine) with Se-amino acids (selenocysteine or selenomethionine) harm the activity of chloroplasts and mitochondria, resulting in a reduction in the activity of these organelles (**Figure 1**). Disorders in mitochondrial and chloroplast activity occur because under Se toxicity there is replacement of clusters of Fe-S centers by Fe-Se in the electron transport chain of mitochondria and chloroplasts [14]. Due to its larger size, the Fe-Se group does not fit properly into proproteins [12].

In another aspect, after its absorption, Se ($SeO_4^{2-}$ or $SeO_3^{2-}$) is subjected to reduction steps. In the conversion of $SeO_3^{2-}$ to $Se^{2-}$ followed by its incorporation into the amino acid cysteine, a compound is formed, selenodiglutathione. At this stage, the participation of reduced glutathione (GSH) is crucial. Under conditions of Se toxicity, cellular depletion of GSH occurs, which results in the imbalance of redox reactions and, consequently, the generation of reactive oxygen species [14], which interact with cellular structures, generating oxidative stress and reducing plant growth [77]. This

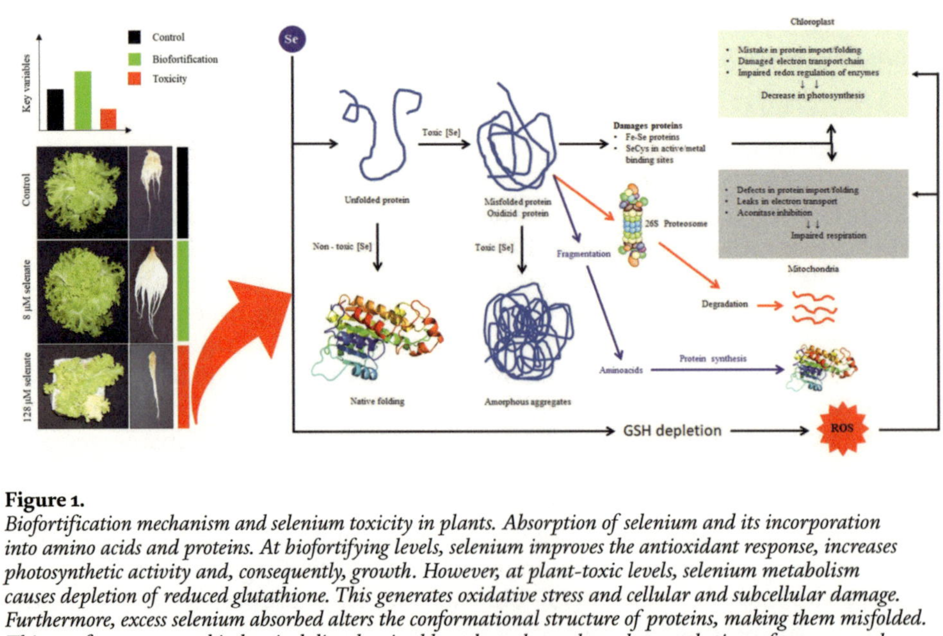


**Figure 1.**
*Biofortification mechanism and selenium toxicity in plants. Absorption of selenium and its incorporation into amino acids and proteins. At biofortifying levels, selenium improves the antioxidant response, increases photosynthetic activity and, consequently, growth. However, at plant-toxic levels, selenium metabolism causes depletion of reduced glutathione. This generates oxidative stress and cellular and subcellular damage. Furthermore, excess selenium absorbed alters the conformational structure of proteins, making them misfolded. This set of events causes biochemical disorders in chloroplasts that reduce photosynthetic performance and, ultimately, plant growth.*

imbalance in the availability of antioxidant substances generates a redox imbalance that can initiate the production of free radicals that are harmful to cells. However, plants under non-toxic doses of Se show improvement in redox status because of the pro-oxidant role of Se that induces an increase in the concentration of reduced glutathione, ascorbate, tocopherol and phenolic compounds in plants [78–81].

Regarding accumulator plants, despite accumulating high levels of Se, they do not develop symptoms of toxicity because in their chloroplasts the enzyme selenocysteine methyltransferase, which promotes the methylation of selenocysteine, transforming it into a volatile compound, dimethyl diselenide. This biochemical strategy prevents the excessive incorporation of Se into proteins, preventing the formation of proteins with structural changes [14].

On the other hand, in non-accumulator plants there is a differential response to Se source doses and cultivation medium. When growing plants in the presence of Se and in a nutrient solution, Se is readily available to the plants, considering that the factors pH, aeration and chemical composition of the nutrient solution are in ideal conditions for plant growth. Therefore, in these cultivation conditions, the expression of Se toxicity symptoms occurs in lower doses of selenite or selenate, in addition to the fact that the Se supply can be carried out according to the phenological stages of the plants, avoiding the loss of Se salts [82, 83].

On the other hand, when growing plants in the presence of Se and in soil, the total availability of Se to the plants is compromised by the physical and chemical characteristics of the soil, with the possibility of adsorption of Se by the clay in the soil or even leaching of Se in the case of sandier soils [84]. Therefore, this set of elements should be considered in Se biofortification programs in edible plants [85].

In studies with plants grown in the presence of Se, no toxic effects were reported given the aforementioned soil elements. However, the biofortifying effect and increase in growth and production has been reported [86].

Regarding cultures in nutrient solution, several studies report toxic effects of doses and sources of Se. For example, in *Cucumis sativus* (Cucurbitaceae) Se toxicity manifests itself in doses above 20 μM (selenite) and 80 μM (selenate) as they affect plant growth [87]. In *Zea maize* (Poaceae) and *Helianthus annuus* (Asteraceae) grown in nutrient solution, doses of 1 mg/L (selenite) and 3 mg/L (selenate) were toxic as they drastically reduced shoot growth [88]. In another study with *Cucumis sativus* (Curcubitaceae), a dose of 2 mg/L of selenite reduced shoot and root growth [88]. In *Pisum sativum* plants, (Fabaceaea), the fresh mass and length of the shoot and root were reduced by doses above 10 μM of selenite [89].

In plants, the boundary between toxicity and biofortification is narrow and dose-source dependent. Under biofortifying doses, Se improves plant growth by positively modulating antioxidant metabolism. This fact allows for better photosynthetic activity, even under abiotic stress, with a positive impact on plant growth and production.

Although the literature on the topic of selenium is quite profuse in studies on plants of agricultural and forestry interest, little is reported on native Amazonian vegetables of pharmaceutical, nutraceutical or food interest. These plants are widely consumed by populations in urban centers and cities further away from urban centers in the Amazon, but whose soils have low Se levels. Furthermore, the biofortifying and toxic effects of Se, the accumulation of Se in shoot organs and roots, the profile of secondary metabolites, photosynthetic and nitrogen metabolism in plants treated with Se, for example, are still unknown in these vegetables. This set of elements suggests new studies in the future involving the use of selenium and cultivation of these vegetables aiming at food security for local populations located in the Amazon.

## 6. Conclusions and future perspectives

In this review, mechanisms of absorption, translocation, and metabolism of Se in plants were shown. Furthermore, beneficial or toxic effects on the physiological and biochemical aspects of plants, as well as their biofortifying aspects. Se in low concentrations has a beneficial effect on plant growth and development. In high concentrations, Se has been linked to an advantageous effect on the growth of accumulating and non-accumulating plants. However, it is not considered an essential element for plants.

The maximum effects of exogenous Se application depend on the source, concentration, forms of application, and plant species. Brassicas are more efficient in absorbing and accumulating Se. They have a biofortifying action because they improve the nutritional status of plants by accumulating in edible plant structures. Furthermore, it stimulates the synthesis of nutraceutical substances beneficial to human health.

Se toxicity is related to the formation of nonspecific selenoproteins and the production of reactive free radical species, which inhibit physiological processes. On the other hand, low concentrations of Se are sufficient to modulate antioxidant activity in plants. The identification of sources, forms of application and doses/concentrations of Se allow gains in the growth, production and increments of this element. Furthermore, in other plant species not yet studied, nutraceutical substances should be the subject of experimental studies concerning the application of selenium. In this sense, there is little information regarding the biofortifying use of selenium in native Amazonian plants with nutraceutical and pharmaceutical potential. For example, in *Acmella oleracea* (L.) R. K. Jansen plants, the exogenous effect of selenium on the production of spilanthol, a substance with great pharmaceutical potential, is unknown. This suggests the need for future research to study which sources and doses of selenium can modulate increases in the production of spilanthol in leaves and/or inflorescences of *Acmella oleracea*.

## Author details

Flávio José Rodrigues Cruz
Federal Institute of Education, Science and Technology of Amapá,
Laranjal do Jari, AP, Brazil

*Address all correspondence to: fjrc@bol.com.br

## References

[1] WHO. World Health Report 2002: Reducing Risks, Promoting Healthy Life: Overview. Geneva: World Health Organization. p. 2002

[2] WHO. Global Prevalence of Vitamin A Deficiency in Populations at Risk 1995-2005. WHO Global Database on Vitamin A Deficiency. Geneva: World Health Organization; 2009

[3] Nair MK, Augustine LF, Konapur A. Food-based interventions to modify diet quality and diversity to address multiple micronutrient deficiency. Frontiers in Public Health. 2016;**3**:277

[4] Mannar MGV, Wesley AS. Food fortification. In: International Encyclopedia of Public Health. Amsterdam, The Netherlands: Elsevier; 2016

[5] Broadley MR, Alcock J, Alford J, Cartwright P, Foot I, Fairweather-Tait SJ, et al. Selenium biofortification of high-yielding winter wheat (*Triticum aestivum* L.) by liquid or granular Se fertilization. Plant and Soil. 2010;**332**:5-18

[6] Ramkissoon C, Degryse F, da Silva FC, Baird R, Young SD, Bailey EH, et al. Improving the efficacy of selenium fertilizers for wheat biofortification. Scientific Reports. 2019;**9**:19520. DOI: 10.1038/s41598-019-55914-0

[7] Rahim FP, Rocio CG, Adalberto BM, Rosaura SCL, Maginot NH. Agronomic biofortification with selenium in tomato crops (*Solanum lycopersicon* L. mill). Agriculture. 2020;**10**:486. DOI: 10.3390/agriculture10100486

[8] Lessa JHL, Raymundo JF, Corguinha APB, Martins FAD, Araujo AM, Santiago FEM, et al. Strategies for applying selenium for biofortification of rice in tropical soils and their effect on element accumulation and distribution in grains. Journal of Cereal Science. 2020;**96**:103125

[9] Epstein E. Silicon. Annual Review of Plant Biology. 1999;**50**:641-664

[10] Marschner H. Mineral Nutrition of Higher Plants. 3rd ed. London: Elsevier; 2012. 643 p

[11] Bakhat HF, Bibi N, Zia Z, Abbas S, Hammad HM, Fahad S, et al. Silicon mitigates biotic stresses in crop plants: A review. Crop Protection. 2018;**104**:21-34

[12] Hasanuzzaman M, Bhuyan MB, Raza A, Hawrylak-Nowak B, Matraszek-Gawron R, Al Mahmud J, et al. Selenium in plants: Boon or bane? Environmental and Experimental Botany. 2020;**178**:104170

[13] Terry N, Zayed AM, de Souza MP, Tarun AS. Selenium in higher plant. Annual Review of Plant Physiology and Plant Molecular Biology. 2000;**51**:401-432

[14] Van Hoewyk D. A tale of two toxicities: Malformed selenoproteins and oxidative stress both contribute to selenium stress in plants. Annals of Botany. 2013;**2013**(112):965-972

[15] Ribeiro DM, Júnior DDS, Cardoso FB, Martins AO, Silva WA, Nascimento VL, et al. Growth inhibition by selenium is associated with changes in primary metabolism and nutrient levels in *Arabidopsis thaliana*. Plant, Cell and Environment. 2016;**39**:2235-2246

[16] Schwarz K, Foltz CM. Selenium as an integral part of factor 3 against dietary

necrotic liver degeneration. Journal of the American Chemical Society. 1957;**79**:3292-3293

[17] Harthill M. Review: Micronutrient selenium deficiency influences evolution of some viral infectious diseases. Biological Trace Element Research. 2011;**3**:1325-1336. DOI: 10.1007/s12011-011-8977-1

[18] Fairweather-Tait SJ, Bao Y, Broadley MR, Collings R, Ford D, Hesketh JE, et al. Selenium in human health and disease. Antioxidants and Redox Signaling. 2011;7:1337-1383. DOI: 10.1089/ars.2010.3275

[19] Rayman MP. Selenium and human health. Lancet. 2012;**9822**:1256-1268

[20] Jian'an T, Daxian Z, Shaofan H, Wenyu Z, Ribang L. The relation of Keshan disease to the natural environment and the background of selenium nutrition. Acta Nutrition. 1982;**114**:175-182 In Chinese

[21] Valea A, Georgescu CE. Selenoproteins in human body: Focus on thyroid pathophysiology. Hormones (Athens, Greece). 2018;**2**:183-196. DOI: 10.1007/s42000-018-0033-5

[22] Rayman MP. Selenium intake, status, and health: A complex relationship. Hormones (Athens, Greece). 2020;**1**:9-14. DOI: 10.1007/s42000-019-00125-5

[23] Brevik EC. Soils and human health: An overview. In: Brevik EC, Burgess LC, editors. Soils and Human Health. Boca Raton: CRC Press; 2013. pp. 29-56

[24] Huang S, Wang P, Yamaji N, Ma JF. Plant nutrition for human nutrition: Hints from rice research and future perspectives. Molecular Plant. 2020;**13**:825-835

[25] White PJ, Broadley MR. Biofortification of crops with seven mineral elements often lacking in human diets - iron, zinc, copper, calcium, magnesium, selenium and iodine. The New Phytologist. 2009;**182**:49-84. DOI: 10.1111/j.1469-8137.2008.02738.x

[26] Adu MO, Asare PA, Yawson DO, Nyarko MA, Osei-Agyeman K. Agronomic biofortification of selected underutilized solanaceae vegetables for improved dietary intake of potassium (K) in Ghana. Heliyon. 2018;**4**:e00750. DOI: 0.1016/j.heliyon.2018.e00750

[27] Boldrin PF, de Figueiredo MA, Yang Y, Luo H, Giri S, Hart JJ, et al. Selenium promotes sulfur accumulation and plant growth in wheat (*Triticum aestivum*). Physiologia Plantarum. 2016;**1**:80-91. DOI: 10.1111/ppl.12465

[28] Oliveira VC, Faquin V, Guimarães KC, Andrade FR, Pereira J, Guilherme LRG. Agronomic biofortification of carrot with selenium. Ciência e Agrotecnologia. 2018;**2**:138-147. DOI: 10.1590/1413-70542018422031217

[29] Schiavon M, Dall'Acqua S, Mietto A, Pilon-Smits EA, Sambo P, Masi A, et al. Selenium fertilization alters the chemical composition and antioxidant constituents of tomato (*Solanum lycopersicon* L.). Journal of Agricultural and Food Chemistry. 2013;**61**(44):10542-10554. DOI: 10.1021/jf4031822

[30] Bachiega P, Salgado JM, Carvalho JE, Ruiz ALTG, Schwarz K, Tezotto T, et al. Antioxidant and antiproliferative activities in different maturation stages of broccoli (*Brassica oleracea Italica*) biofortified with selenium. Food Chemistry. 2016;**190**:771-776. DOI: 10.1016/j.foodchem.2015.06.024

[31] Groth S, Budke C, Neugart S, Ackermann S, Kappenstein FS,

Daum D, et al. Influence of a selenium biofortification on antioxidant properties and phenolic compounds of apples (*Malus domestica*). Antioxidants. 2020, 2020;**9**:187. DOI: 10.3390/antiox9020187

[32] Reynolds-MarzalMD,Rivera-MartínAM, RodrigoSM,SantamariaO,PoblacionesMJ. Biofortification of forage peas with combined application of selenium and zinc under Mediterranean conditions. Journal of Soil Science and Plant Nutrition. 2020;**21**:286-300. DOI: 10.1007/s42729-020-00360-5

[33] Thavarajah D, Thavarajah P, Vial E, Gebhardt M, Lacher C, Kumar S, et al. Will selenium increase lentil (Lens culinaris Medik) yield and seed quality? Frontiers in Plant Science. 2015;**6**:356. DOI: 10.3389/fpls.2015.00356

[34] Mirlean N, Seus- Arrache E, Vlasova O. Selenium deficiency in subtropical littoral pampas: Environmental and dietary aspects. Environmental Geochemistry and Health. 2017;**40**:1e14

[35] Chen L, Yang F, Xu J, Yun H, Hu Q, Zhang Y, et al. Determination of selenium concentration of rice in China and effect of fertilization of selenite and selenate on Se content of rice. Journal of Agricultural and Food Chemistry. 2002;**50**:5128-5130

[36] Broadley MR, White PJ, Bryson RJ, Meacham MC, Bowen HC, Johnson SE, et al. Biofortification of U.K. food crops with selenium (Se). Proceedings of the Nutrition Society. 2006;**65**:169-181

[37] Jones GD, Droz B, Greve P, Gottschalk P, Poffet D, McGrath SP, et al. Selenium deficiency risk predicted to increase under future climate change. National Academy of Sciences of the United States of America. 2017;**114**:201611576

[38] USDA-ARS. USDA National Nutrient Database for Standard Reference, Release 25. Washington, DC: USDA–ARS; 2012. Available from: https://www.nal.usda.gov/fnic/selenium

[39] WHO. Global Health Risks: Mortality and Burden of Disease Attributable to Selected Major Risks. 2009

[40] Jiang C, Zu C, Shen J, Shao F, Li T. Effects of selenium on the growth and photosynthetic characteristics of flue-cured tobacco (*Nicotiana tabacum* L.). Acta Societatis Botanicorum Poloniae. 2015;**1**:71-77. DOI: 10.5586/asbp.2015.006

[41] Bo C, Hai-feng L, Ying-ying L, Xin-hui Y, Ya-li S, Xiu-dong S, et al. Effects of selenium and sulfur on antioxidants and physiological parameters of garlic plants during senescence. Journal of Integrative Agriculture. 2016;**3**:566-572

[42] Prado RM, Cruz FJR, Ferreira RLC. Selenium biofortification and the problem of its safety. In: Waisundara V, Shiomi N, editors. Superfood and Functional Food - An Overview of Their Processing and Utilization. London, UK, London: IntechOpen; 2017. pp. 229-238

[43] Ferreira RLC, de Mello PR, de Souza Junior JP, Gratão PL, Tezotto T, Cruz FJR. Oxidative stress, nutritional disorders, and gas exchange in lettuce plants subjected to two selenium sources. Journal of Soil Science and Plant Nutrition. 2020;**20**:1215-1228. DOI: 10.1007/s42729-020-00206-0

[44] D'Amato R, Proietti P, Onofri A, Regni L, Esposto S, Servili M, et al. Biofortification (Se): Does it increase the content of phenolic compounds in virgin olive oil (VOO)? PLoS One. 2017;**4**:e0176580. DOI: 10.1371/journal.pone.0176580

[45] Saffaryazdi A, Lahouti M, Ganjeali A, Bayat H. Impact of selenium

supplementation on growth and selenium accumulation on spinach (*Spinacia oleracea* L.) Plants. Notulae Scientia Biologicae. 2012;**4**:95-100

[46] Goicoechea N, Garmendia I, Fabbrin EG, Bettoni MM, Palop JA, Sanmartín C. Selenium fertilization and mycorrhizal technology may interfere in enhancing bioactive compounds in edible tissues of lettuce. Scientia Horticulturae. 2015;**195**:163-172. DOI: 10.1016/j.scienta.2015.09.007

[47] Preciado-Rangel P, Hernández-Montiel LG, Valdez-Cepeda RD, de la Cruz-Lázaro E, Lara-Capistrán L, Morales-Morales B, et al. Biofortification with selenium increases bioactive compounds and antioxidant capacity in tomato fruits. Terra Latinoamericana. 2021;**39**:e979

[48] Skrypnik L, Novikova A, Tokupova E. Improvement of phenolic compounds, essential oil content and antioxidant properties of sweet basil (*Ocimum basilicum* L.) depending on type and concentration of selenium application. Plants. 2019;**8**:458. DOI: 10.3390/plants8110458

[49] Elguera JCT, Barrientos EY, Wrobel K, Wrobel K. Effect of cadmium (Cd(II)), selenium (Se(IV)) and their mixtures on phenolic compounds and antioxidant capacity in *Lepidium sativum*. Acta Physiologiae Plantarum. 2013;**35**:431-441. DOI: 10.1007/s11738-012-1086-8

[50] Shah M, Kannamkumarath SS, Wuilloud JCA, Wuilloud RG, Caruso JA. Identification and characterization of selenium species in enriched green onion (*Allium fistulosum*) by HPLC-ICP-MS and ESI-ITMS. Journal of Analytical Atomic Spectrometry. 2004;**19**:381-386

[51] Tsuneyoshi T, Yoshida J, Sasaoka T. Hydroponic cultivation offers a practical means of producing selenium-enriched garlic. The Journal of Nutrition. 2006;**3**:870S-872S. DOI: 10.1093/jn/136.3.870S

[52] Ávila FW, Yang Y, Faquin V, Ramos SJ, Guilherme LRG, Thannhauser TW, et al. Impact of selenium supply on Se-methylselenocysteine and glucosinolate accumulation in selenium-biofortified brassica sprouts. Food Chemistry. 2014;**165**:578-586. DOI: 10.1016/j.foodchem.2014.05.134

[53] Pílon-Smiths EAH, Quinn CF, Tapken W, Malagoli M, Schiavon M. Physiological functions of beneficial elements. Current Opinion in Plant Biology. 2009;**12**:267-274

[54] Natasha, Shahid M, Niazi NK, Khalid S, Murtaza B, Bibi I, et al. A critical review of selenium biogeochemical behavior in soil-plant system with an inference to human health. Environmental Pollution. 2018;**234**:915-934. DOI: 10.1016/j.envpol.2017.12.019

[55] Dai Z, Imtiaz M, Rizwan M, Yuan Y, Huang H, Shuxin TS. Dynamics of selenium uptake, speciation, and antioxidant response in rice at different panicle initiation stages. The Science of the Total Environment. 2019;**691**:827-834. DOI: 10.1016/j.scitotenv.2019.07.186

[56] Semida WM, El-Mageed TAA, Abdelkhalik A, Hemida KA, Abdurrahman HA, Howladar SM, et al. Selenium modulates antioxidant activity, osmoprotectants, and photosynthetic efficiency of onion under saline soil conditions. Agronomy. 2021;**11**:855. DOI: 10.3390/agronomy11050855

[57] Zhang M, Tang S, Huang X, Zhang F, Pang Y, Huang Q, et al. Selenium uptake, dynamic changes in selenium content and its influence on photosynthesis

and chlorophyll fluorescence in rice (*Oryza sativa* L.). Environmental and Experimental Botany. 2014;**107**:39-45

[58] Niu J, Liu C, Huang M, Liu K, Yan D. Effects of foliar fertilization: A review of current status and future perspectives. Journal of Soil Science and Plant Nutrition. 2020;**21**:104-118. DOI: 10.1007/s42729-020-00346-3

[59] Li HF, McGrath SP, Zhao FJ. Selenium uptake, translocation and speciation in wheat supplied with selenate or selenite. The New Phytologist. 2008;**178**:92-102

[60] Fernández V, Eichert T. Uptake of hydrophilic solutes through plant leaves: Current state of knowledge and perspectives of foliar fertilization. Critical Reviews in Plant Sciences. 2009;**28**:36-68. DOI: 10.1080/07352680902743069

[61] Bindraban PS, Dimkpa C, Nagarajan L, Roy A, Rabbinge R. Revisiting fertilisers and fertilisation strategies for improved nutrient uptake by plants. Biology and Fertility of Soils. 2015;**51**:897-911. DOI: 10.1007/s00374-015-1039-7

[62] Gupta M, Gupta S. An overview of selenium uptake, metabolism, and toxicity in plants. Frontiers in Plant Science. 2017;**7**:2074. DOI: 10.3389/fpls.2016.02074

[63] White PJ, Bowen HC, Parmaguru P, Fritz M, Spracklen WP, Spiby RE, et al. Interactions between selenium and sulphur nutrition in *Arabidopsis thaliana*. Journal of Experimental Botany. 2004;**55**:1927-1937

[64] White PJ, Bowen HC, Marshall B, Broadley MR. Extraordinarily high leaf selenium to sulphur ratios define 'Se accumulator' plants. Annals of Botany. 2007;**100**:111-118

[65] Kikkert J, Berkelaar E. Plant uptake and translocation of inorganic and organic forms of selenium. Archives of Environmental Contamination and Toxicology. 2013;**65**:458-465

[66] Hopper JL, Parker DR. Plant availability of selenite and selenate as influenced by the competing ions phosphate and sulfate. Plant and Soil. 1999;**210**:199-207

[67] Banning H, Stelling M, Konig S, Schoenberg R, Neumann T. Preparation and purification of organic samples for selenium isotope studies. PLoS One. 2018;**3**:e0193826

[68] Kataoka T, Hayashi N, Yamaya T, Takahashi H. Root-to-shoot transport of sulfate in Arabidopsis. Evidence for the role of SULTR3;5 as a component of low-affinity sulfate transport system in the root vasculature. Plant Physiology. 2004;**136**:4198-4204. DOI: 10.1104/pp.104.045625

[69] Takahashi H, Watanabe-Takahashi A, Smith FW, Blake-Kalff M, Hawkesford MJ, Saito K. The roles of three functional sulphate transporters involved in uptake and translocation of sulphate in *Arabidopsis thaliana*. The Plant Journal. 2000;**23**:171-182. DOI: 10.1046/j.1365-313X.2000.00768.x

[70] Yläranta T. Raising the selenium content of spring wheat and barley using selenite and selenate. Annales Agriculturae Fenniae. 1984;**1**:75-84

[71] Martínez M, Giménez J, dew Pablo J, Rovira M, Duro L. Sorption of selenium (IV) and selenium (VI) onto magnetite. Applied Surface Science. 2006;**1**:3767-3773

[72] Turakainen M, Hartikainen H, Seppnen MM. Effects of selenium treatments on potato (*Solanum tuberosum*

L.) growth and concentrations of soluble sugars and starch. Journal of Agricultural and Food Chemistry. 2004;**52**:5378-5382

[73] Ramos SJ, Faquin V, Guilherme LRG, Castro EM, Ávila FW, Carvalho GS, et al. Selenium biofortification and antioxidant activity in lettuce plants fed with selenate and selenite. Plant, Soil and Environment. 2010;**12**:584-588

[74] Xia Q, Yang Z, Shui Y, Liu X, Chen J, Khan S, et al. Methods of selenium application differentially modulate plant growth, selenium accumulation and speciation, protein, anthocyanins and concentrations of mineral elements in purple-grained wheat. Frontiers in Plant Science. 2020;**11**:1114. DOI: 10.3389/fpls.2020.01114

[75] Matraszek R, Hawrylak-Nowak B. Macronutrients accumulation in useable parts of lettuce as a_ected by nickel and selenium concentrations in nutrient solution. Fresenius Environmental Bulletin. 2009;**18**:1059-1065

[76] Kabata-Pendias A. Trace Elements in Soil and Plants. 3rd ed. Boca Raton: CRC Press; 2010. 315 p

[77] Mittler R. ROS are good. Trends in Plant Science. 2017;**1**:11-19. DOI: 10.1016/j.tplants.2016.08.002

[78] Hajiboland R, Rahmat S, Aliasgharzad N, Hartikainen H. Selenium-induced enhancement in carbohydrate metabolism in nodulated alfalfa (*Medicago sativa* L.) as related to the glutathione redox state. Soil Science and Plant Nutrition. 2015;**61**:676-687

[79] Ghosh S, Biswas AK. Selenium modulates growth and thiol metabolism in wheat (*Triticum aestivum* L.) during arsenic stress. American Journal of Plant Sciences. 2017;**8**:363-389

[80] Dawood MG, Sadak MS, Bakry BA, Kheder HH. Effect of glutathione and/or selenium levels on growth, yield, and some biochemical constituents of some wheat cultivars grown under sandy soil conditions. Bulletin of the National Research Centre. 2020;**44**:158

[81] Ishtiaq M, Mazhar MW, Maqbool M, Hussain T, Hussain AS, Casini R, et al. Seed priming with the selenium nanoparticles maintains the redox status in the water stressed tomato plants by modulating the antioxidant defense enzymes. Plants. 2023;**12**:1556

[82] Marschner H. Mineral Nutrition of Higher Plants. 2nd ed. London: Academic Press; 1995. 889 p

[83] Puccinelli M, Malorgio F, Pezzarossa B. Selenium enrichment of horticultural crops. Molecules. 2017;**22**:933. DOI: 10.3390/molecules22060933

[84] Abreu LB, Carvalho GS, Curi N, Guilherme LRG, Melo Marques JJGS. Sorção de selênio em solos do bioma Cerrado. Revista Brasileira de Ciências do Solo. 2011;**35**:1995-2003

[85] Boldrin PF, Faquin V, Ramos SJ, Guilherme LRG, Bastos CEA, Carvalho GS, et al. Selenato e selenito na produção e biofortificação agronômica com selênio em arroz. Pesquisa Agropecuária Brasileira. 2012;**47**:831-837

[86] Hawrylak-Nowak B, Matraszek R, Pogorzelec M. The dual effects of two inorganic selenium forms on the growth, selected physiological parameters and macronutrients accumulation in cucumber plants. Acta Physiologiae Plantarum. 2015;**37**:41

[87] Garousi F, Veres S, Kovács B. Comparison of selenium toxicity in sunflower and maize seedlings grown

in hydroponic cultures. Bulletin of Environmental Contamination and Toxicology. 2016;**97**:709-713

[88] Haghighi M, Sheibanirad A, Pessarakli M. Effects of selenium as a beneficial element on growth and photosynthetic attributes of greenhouse cucumber. Journal of Plant Nutrition. 2016;**39**(10):1493-1498

[89] Lehotai N, Lyubenova L, Schröder P, Feigl G, Ördög A, Szilágyi K, et al. Nitro-oxidative stress contributes to selenite toxicity in pea (*Pisum sativum* L). Plant and Soil. 2016;**400**:107-122

Section 3

# Bioactive Properties and Role in Oxidative Stress

Chapter 10

# Probiotics against Oxidative Stress

*Parisa Naji, Farshid Parvini and Mohammad Ali Farazi Fard*


## Abstract

Probiotics, as host-compatible microorganisms, possess antioxidant properties capable of inhibiting the growth of harmful bacteria, improving digestion, and altering the balance of intestinal microflora. Lactic acid bacteria (LAB) are among the most beneficial probiotics. These microbiotas not only play a vital role in nutrient uptake and production but also modulate gastrointestinal tract maturation, immune system function, infection resistance, and host metabolism. Antioxidants are components that elicit oxidation-reduction homeostasis. The term "oxidative stress" refers to the physiological imbalance in homeostasis, resulting in intracellular hyperproduction of reactive oxygen species (ROS). Retrospectively, the antioxidant capacity of LAB has been confirmed. LAB consumption has been demonstrated to be healthfully efficient in managing ROS-related disorders that affect the nervous, cardiovascular, respiratory, gastrointestinal, urogenital, reproductive, and metabolic systems, as well as mental and immune system disorders. Furthermore, various cancers progress due to ROS triggers. This chapter reviews LAB capability against oxidative stress and clinical trials with LAB usage.


**Keywords:** probiotics, *lactobacillus*, oxidative stress, antioxidants, ROS-related diseases

## 1. Introduction

Elie Metchnikoff, a Russian zoologist and considered the grandfather of modern probiotics, published "The Prolongation of Life," advocating the daily consumption of health-beneficial lactic acid bacteria (LAB) by humans. Henry Tissier, a French pediatrician, independently suggested that bifid bacteria could be a potentially curative option for pediatric patients with diarrhea [1]. Sixty years later, two biologists, D. M. Lilly and R. H. Stillwell, coined the term "probiotics" in 1965 to refer to viable non- or low-pathogenic microorganisms that produce beneficial substances promoting the growth of their host. The concept has evolved to the current definition that bacteria must be sufficiently ingested to play a central role in their host's health by the World Health Organization (WHO)/the Food and Agriculture Organization (FAO) in 2001 [2–4].

A diverse population, including bacteriophages [5], archaea [6], viruses, and bacteria—comprising more than 40% operational taxonomic units—inhabit various parts of the body such as the skin, oral mucosa, respiratory tract, gastrointestinal tract (GIT), and urogenital tract [7]. The GIT microbiota colonization starts in utero and continues after birth with two microbic sources, the maternal skin, milk, and vagina, besides the environment. The process is dependent on the GIT chemical environment, the degree of oxygenation, the immune system intervention, and the nutritional intake. The low pH resistance of the GIT and the absence of antibiotic resistance genes make probiotics efficient, resulting in higher concentrations of microorganisms in the colon, distal ileum, and jejunum, with counts reaching $10^{12}$, $10^{9}$, and $10^{4}$ colony-forming units/gram (CFU/g), respectively. The population in the stomach and duodenum is relatively low due to the presence of gastric acid, bile, and pancreatic juices. The mechanisms of action by which probiotics of $10^{14}$ microorganisms exert their effects are not only multidimensional but also strain-specific due to the existence of at least 1000 different species. The main mechanisms are

- the epithelial barrier development,
- the intestinal mucosa adhesion progression,
- the immune system modulation,
- the antibacterial substances production,
- the bacterial adhesion simultaneous suppression, and
- the pathogen competitive exclusion [7].

The prenatal human-microbiota symbiotic relationship not only contributes to nutrient uptake and production but also plays a crucial role in the GIT and immune system maturation, infection resistance, and host metabolism, primarily through fermentation. Furthermore, the relationship embodies a powerful tool to re-establish the GIT homeostasis and promote its health. As a result, probiotics have a cumulative effect on alleviating and eliminating GIT-related diseases, among others [7–11].

The most beneficial LAB species are *Lactobacillus* (*L.*), *Lactococcus*, *Propionibacterium*, *Bacillus*, *Bifidobacterium*, *Enterococcus*, and *Streptococcus*. Long-ago LAB strains usage has represented it as one of the safest bioresources probiotics in the food and beverage industry and the pharmaceutical one [7, 8, 12].

## 2. What is *Lactobacillus*?

The genus *Lactobacillus* belongs to the phylum *Firmicutes*, the class *Bacilli*, order *Lactobacillales*, and family *Lactobacillaceae*. This rod-shaped, nonspore-forming bacterium is a gram-positive and facultative anaerobe. The optimal growth temperature, sugar utilization, and metabolite production variations; the phenotypic traits; were primarily applied for the lactobacilli taxonomy. Far ahead, the chemical structure of the peptidoglycan, the mol% G + C content, DNA–DNA hybridization, and more recently gene's average nucleotide identity values and the whole bacterial genome sequence were classification criteria. Scientists have reclassified the nutritiously fastidious bacterium into 260 species in 25 genera, which subcategorizes into two emended, *L.* (*L. delbrueckii* and *Paralactobacillus*) and 23 novel genera (*Acetilactobacillus*, *Agrilactobacillus*, *Amylolactobacillus*, *Apilactobacillus*, *Bombilactobacillus*, *Companilactobacillus*, *Dellaglioa*, *Fructilactobacillus*,

*Furfurilactobacillus*, *Holzapfelia*, *Lacticaseibacillus*, *Lactiplantibacillus*, *Lapidilactobacillus*, *Latilactobacillus*, *Lentilactobacillus*, *Levilactobacillus*, *Ligilactobacillus*, *Limosilactobacillus*, *Liquorilactobacillus*, *Loigolactobacilus*, *Paucilactobacillus*, *Schleiferilactobacillus*, and *Secundilactobacillus*), that are phylogenetically, ecologically, and metabolically diverse [7, 13].

The nomenclature of LAB is based on their ability to produce lactic acid through fermentation metabolism, fermenting carbohydrates into lactic acid without requiring oxygen. Three subcategories of obligate homofermentative (*L. acidophilus* and *L. salivarius*), obligately heterofermentative (*L. reuteri* and *L. fermentum*), and facultatively heterofermentative (*L. casei* and *L. plantarum*) are described for *L.* It is noteworthy to mention that the L. reclassification results in simpler metabolic activities subcategories of homofermentative and heterofermentative LABs that metabolize hexoses through the Embden–Meyerhoff and the Phosphoketolase pathways, respectively to pyruvate [13, 14]. Approximately, 6.5% of the human heterogeneous probiotic ecosystem is constituted by the genus. Host characteristics such as genotype, mode of delivery, sex, age, weight, race, geographical location, and disease status, as well as environmental factors, including nutrition, stress, and medication, influence the abundance of LAB in the GIT [15]. For instance, numerical dominance of *L.* is observed in infants' feces, with counts ranging from $10^3$ to $10^7$ CFU/g, and in the ileum and oral cavity of a healthy adult, with counts of approximately $10^3$ to $10^4$ CFU/g [14, 16]. Main probiotic *L.* species include *L. acidophilus*, *L. brevis*, *L. casei*, *L. clausii*, *L. crispatus*, *L. delbrueckii subsp. bulgaricus*, *L. fermentum*, *L. gasseri*, *L. helveticus*, *L. johnsonii*, L. kalixensis, *L. paracasei subsp. paracasei*, *L. paracasei subsp. tolerans*, *L. plantarum*, *L. reuteri*, *L. rhamnosus*, *L. sakei*, and *L. ultunensis*. These species not only improve the GIT health and immune function but also synthesize antimicrobial agents, decrease cholesterol and unfavorable metabolite levels, and prevent colonization. Additionally, they exhibit antimutagenic and antitumoral effects [13, 14, 17].

## 3. The oxidative stress

The oxidation-reduction (redox) homeostasis is crucial as it pervades a wide range of fundamental biological processes, including bioenergetics, metabolism of biological macromolecules, and vital functions such as differentiation, immunity, and apoptosis [18]. The dynamic network of electron donors and acceptors is responsible for the intracellular steadiness of the redox. Not only do sufficient levels of two byproducts, reactive oxygen species (ROS) and reactive nitrogen species, contribute to redox maintenance but also multiple antioxidants, such as small molecules, vitamins, proteins, enzymes, transcription factors, and modulators [19]. The term "oxidative stress" was coined by Helmut Sies, defining it as the physiological homeostasis imbalance resulting in intracellular ROS hyperproduction containing hydrogen peroxide ($H_2O_2$), superoxide radicals ($O_2^{\bullet -}$), hydroxyl radicals (•OH), and singlet oxygen ($_1O^2$) [20, 21]. Many studies have not only confirmed antioxidants' mechanism and kinetics but also will identify the new potential ones through cell-based approaches, cell-free biological systems, and *in vivo* models. Nevertheless, byproduct reduction in *in vitro* and *in vivo* models does not necessarily act as an antioxidant in the body. The primary and secondary types of antioxidants prevent or detect a chain of oxidative propagation helping to reduce oxidative damage. The mechanisms by which antioxidants function and their roles in cellular processes are

- scavenging free radicals,
- scavenging superoxide, hydroxyl radical, stable radical, and other ROS,
- chelating metal ion ($Fe^{2+}$, $Fe^{3+}$, $Cu^{2+}$, and $Cu^{+}$),
- prevention of lipid peroxidation,
- controlling of ROS generation enzymes,
- activation of ubiquitous peroxiredoxins,
- prevention of protein modification,
- prevention of DNA damage, and
- participating in DNA repair and apoptosis mechanisms [22].

The crucial small nonprotein molecule, glutathione (GSH), which comprises a constitutive tripeptide of cysteine, glycine, and glutamate, was discovered in 1921. Functionally vital as the GSH has a role in DNA synthesis, translation, amino acid transportation system, and disulfide reduction, it is a determinant defender against oxidative stress. Cytosol contains the highest concentration of 1–10 micromolar (μM) GSH, accounting for 80–85%, followed by mitochondria and endoplasmic reticulum, 10–15% and <5%, respectively. Water-soluble vitamins, such as ascorbic acid, ascorbate, or vitamin C, besides two fat-soluble vitamins A and E, act as antioxidants. The concentration function of vitamin C has proved its anti- and pro-oxidant roles, the physiological (30–100 mg/kg body weight) and higher (1000 mg/kg body weight) concentration, respectively. Furthermore, peroxynitrite, intracellular ROS hyperproduction, and lipid peroxidation are scavenged and inhibited by vitamin A only at <20 kPa oxygen pressure. Moreover, vitamin E breaks the lipid radical chain reaction and scavenges $O_2^{\bullet-}$ and •OH with the help of its distinct structure. It is noteworthy to mention heme's oxidative break-down product, bilirubin, albumin, uric acid, and carotenoids are cellular antioxidants. On the other hand, thioredoxin (Trx), an inflammation-inducible small oxidoreductase protein, catalyzes disulfides into thiols through the oxidation of cysteines in a conserved CXXC motif. The Trx system is closely collaborated with the ubiquitous peroxiredoxin family. Three-subcategory ubiquitous peroxiredoxins, superoxide dismutase (SOD), glutathione peroxidase (GPx), and tetrameric catalase (CAT) are considered antioxidant enzymes. Additionally, transcription factors, such as nuclear factor E2-related factor 2, which controls over 200 genes, as well as microRNAs, are believed to modulate and regulate redox homeostasis, DNA repair, detoxification, proteasomal degradation, and apoptosis mechanisms [19, 23].

Large quantities of intracellular ROS hyperproduction are believed to participate in a diverse scope of either nonmalignant or malignant diseases that affect the nervous, cardiovascular, and respiratory tract disorders, including chronic obstructive pulmonary disease, GIT (e.g., inflammatory bowel disease and irritable bowel syndrome), urogenital, reproductive, and metabolic systems [20, 24], besides mental and immune system disorders [25]. Abundant studies, going into detail, have proved a significant correlation between either oxidative stress or an antioxidant deficiency,

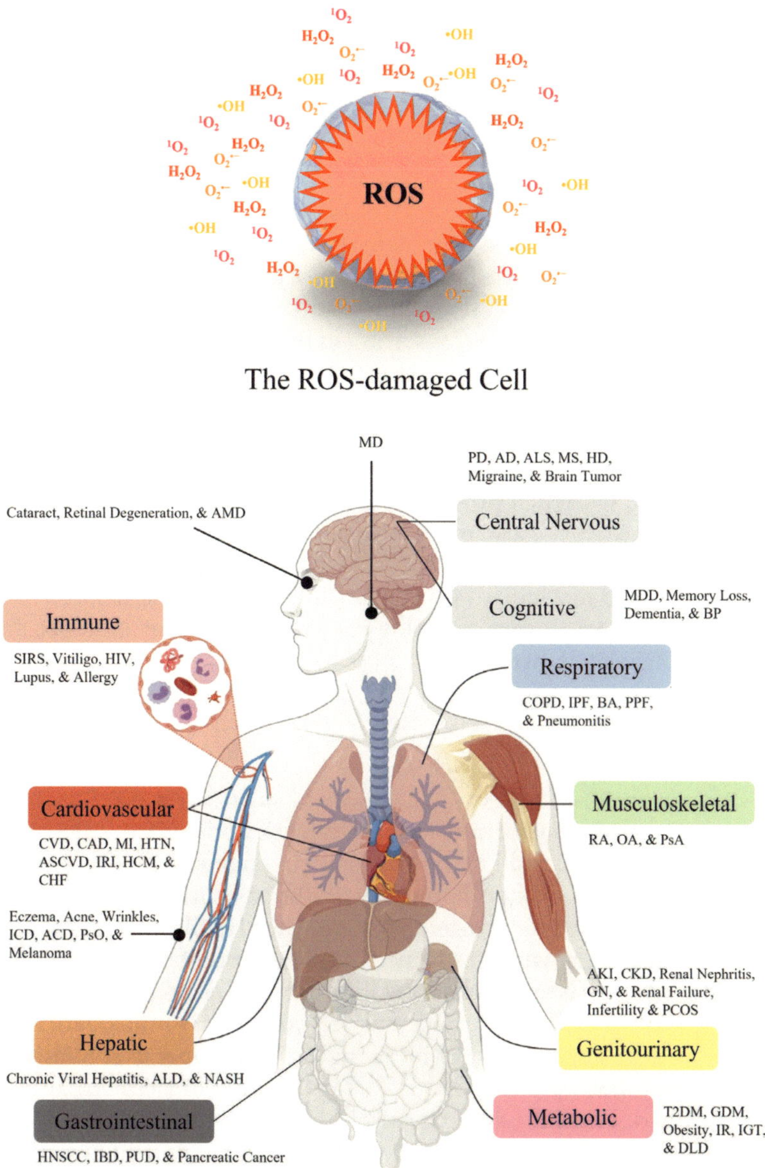


**Figure 1.**
*ROS-related diseases. The ROS-damaged cell can be either the primary pathological etiology of some diseases, such as AD, COPD, and RA, or a secondary contributor that triggers the disease progression, such as different cancers affecting all human systems. ROS: Reactive oxygen species; MD: Meniere's disease; PD: Parkinson's disease; AD: Alzheimer's disease; ALS: Amyotrophic lateral sclerosis; MS: Multiple sclerosis; HD: Huntington's disease; MDD: Major depressive disorder; BP: Bipolar disease; COPD: Chronic obstructive pulmonary disease; IPF: Idiopathic pulmonary fibrosis; BA: Bronchila asthma; PPF: Progressive pulmonary fibrosis; RA: Rheumatoid arthritis; OA: Osteoarthritis; PsA: Psoriatic arthritis; AKI: Acute kidney injury; CKD: Chronic kidney disease; GN: Glomerulonephritis; PCOS: Polycystic ovary syndrome; T2DM: Type 2 diabetes mellitus; GDM: Gestational diabetes mellitus; IR: Insulin resistance; IGT: Impaired glucose tolerance; DLD: Dyslipidemia; HNSCC: Head and neck squamous cell carcinoma; IBD: Inflammatory bowel disease; PUD: Peptic ulcer disease; ALD: Alcoholic liver disease; NASH: Nonalcoholic Steatohepatitis; ICD: Irritant contact dermatitis; ACD: Allergic contact dermatitis; PsO: Psoriasis; CVD: Cardiovascular disease; CAD: Coronary artery disease; MI: Myocardial infarction; HTN: Hypertension; ASCVD: Atherosclerotic cardiovascular disease; IRI: Ischemia–reperfusion injury; HCM: Hypertrophic cardiomyopathy; CHF: Congestive heart failure; SIRS: Systematic Infammatory response syndrome; HIV: Human immunodeficiency virus; AMD: Age-related muscular degeneration.*

for instance, the CAT mutation, and several nervous system diseases, including Parkinson's, Alzheimer's, amyotrophic lateral sclerosis, multiple sclerosis, and mental disorders (e.g., depression and dementia) [19, 20, 24, 26–28]. Furthermore, not only the *in vivo* research data point out the role of oxidative stress in cardiovascular diseases, including atherosclerosis, hypertension, ischemia, hypertrophic cardiomyopathy, and congestive heart failure but also *ex vivo* ones confirm it [29]. Moreover, chronic obstructive pulmonary disease, resulting from inflammation *via* macrophage and epithelial cell activation by environmental oxidants and different kinase activations, idiopathic pulmonary fibrosis, and bronchial asthma are described as redox imbalance diseases. Furthermore, many studies described oxidative stress as a cancer hallmark by ample mechanisms [30] resulting in the usage of countless antioxidants as chemotherapeutic drugs in various cancers such as brain tumors [31, 32], head and neck squamous cell carcinoma [33, 34], breast [35, 36], ovarian, testicular [37], colon [38], pancreatic [39–41], and liver cancer [42], as well as melanoma [43] and hematologic malignancies [44–46]. However, it is noteworthy to mention that the usage of LAB strains, with their protective roles against oxidative stress, not only alleviates various diseases but also modulates post-treatment complications, such as graft-versus-host disease and infection, as two causes of hematopoietic stem cell transplantation-related morbidity [47–51]. As shown in **Figure 1**, oxidative stress can be either the primary cause of pathology or a secondary contributor to disease progression in ROS-related diseases.

## 4. Antioxidant effects of probiotics

Although both native nonenzymatic and enzymatic antioxidant systems such as GSH, vitamins A, C, and E, and Trx, besides peroxiredoxins, SOD, GPx, and CAT, are protectively efficient, they cannot thoroughly prevent oxidative damage. Furthermore, the GIT environments, high- or low-temperature, low pH, bile salts, oxygen, and limited nutrition induce stress, which influences probiotics proliferation, function, and survival. Oxidative stress is one of the most life-threatening stressors that probiotic strains, such as LAB, must have a strong defense against. LAB antioxidant activities are of significant importance, as indicated by an increasing number of studies focusing on them. As a result, these products are replacing the usage of synthetic antioxidants such as ascorbic acid, butylated hydroxyanisole, butylated hydroxytoluene, gallic acid esters, and tert-butyl hydroquinone due to their carcinogenicity and non-ecofriendly characters [52–54]. LAB oxidative stress tolerance has been confirmed; however, mechanisms are blurred. These modes of action can be categorized into nonenzymatic and enzymatic mechanisms varying in different strains, following

- chelating metal ion ($Fe^{2+}$, $Fe^{3+}$, $Cu^{2+}$, and $Cu^{+}$),
- scavenging ROS through antioxidant metabolites and enzyme production,
- downregulating ROS generation enzyme activities,
- upregulating antioxidant enzyme activities, and
- regulating antioxidant signaling pathways [55].

L. strains represent chelator properties, the first nonenzymatic approach in redox modulation. Metal ions ($Fe^{2+}$ and $^{3+}$) oxidation prevention and capture are said to be managed by 4,7-diphenyl-1,10-phenanthroline disulfonic acid, deferoxamine, ethylenediamine tetra acetic acid, and penicillamine, which are administrated as chelators. $Fe^{2+}$ and $Fe^{3+}$ transition inhibits enzyme-catalyzed phosphate ester displacement reactions, resulting in hydroperoxides decomposition and peroxyl and alkoxyl radicals' production. In retrospective, *L. acidophilus* B, E, N1, 4356, *LA-1*, and *Farr*; *L. bulgaricus 12,278*, *448*, *449*, *Lb*, *1006*, and *11,842*. *L. casei* KCTC 3260, *L. helveticus CD6* were proved to be a $Fe^{2+}$ chelator. Lin et al. (1999) mentioned that LAB chelator properties may originate from the intracellular cell-free extracts [56]. The second mechanism is the LABs' metabolites, such as folic acid (folate or vitamin B9) and GSH. The L. *helveticus* CD6 verified antioxidant capacities by folic acid in the intracellular cell-free extract reported by Ahire et al., an Indian group in 2013 [57]. The GSH metabolites production of *L. fermentum E-3*, *E-18*, and *ME-3* was confirmed by Kullisaar et al. Furthermore, LAB is responsible for the regulation of the GIT microbiota composition. For instance, antimicrobial compounds of *L. rhamnosus GG* conquer diverse bacteria proliferation, resulting in harmful one abundance reduction [52, 58].

The first enzymatic approaches are the evolutionary-based Fe- and Mn-SOD employment by LAB strains such as *L. fermentum E-3*, *E-18*, and *L. casei BL23*. Another native antioxidant possibility is the LAB's CAT production. Different studies have demonstrated that the CAT-induced *L. slactis* and *L. casei BL23* prevent or alleviate the GIT-related diseases caused by ROS. Moreover, *L. plantarum CAI6*, *FC225*, and *SC4* strains, besides *L. rhamnosus GG* and *GG-CM* strains, are proven to have antioxidant signaling pathways, including Nrf2-Keap1-ARE, mitogen-activated protein kinases, and protein kinase C mediators in *in vitro* studies, respectively. Besides some LAB strains, including *L. fermentum CECT5716*, *L. coryniformis CECT5711* (K8), *L. gasseri CECT5714* (LC9), *L. acidophilus*, and *L. casei* play their antioxidant role by the regulation of ROS-produces enzymes [52, 53, 58].

## 5. Clinical trials of different L. strains following *in vitro* and preclinical studies

Since 1999, exceeding 75 *in vitro* studies have been conducted on different LAB strains (n = 19), *L. plantarum* (n = 19), *L. casei* (n = 11), *L. paracasei* (n = 6), *L. acidophilus* (n = 4), *L. helveticus* (n = 4), and *L. johnsonii* (n = 3), besides two studies on *L. brevis*, *L.* gasseri, *L. reuteri*, *L. rhamnosus* apiece. Few did test-tube experiment on *L. delbrueckii*, *L. fermentum*, *L. panis*, *L. pentosus*, and *L. sanfranciscensis* report. Many evaluations of LABs' features, including

- colonization,
- epithelial cells' adhesion,
- viability in the GIT,
- production of short-chain fatty acids,
- modulation of the GIT microbiota,

- immunomodulatory,
- immune response effects,
- anti-inflammatory properties, and
- antimutagenic and antitumoral characteristics.

are assessed in these *in vitro* studies. They not only assessed various features of LABs but also focused on elucidating the mechanisms by which LABs exert their antioxidant effects, providing insights into the molecular mechanisms underlying the beneficial effects in mitigating oxidative stress associated with ROS-related diseases.

Although the safety and efficacy therapeutic criteria of LAB strains have been verified by 150 preclinical studies, the quality is somewhat unclear. There are many parameters, which preclinical studies focus on to understand how LABs may lessen oxidative stress and related conditions by exerting antioxidant, anti-inflammatory, and tissue-protective effects, including

- energy metabolism,
- mitochondrial function,
- antioxidant activity assessments,
- ROS scavenging ability,
- oxidative damage reduction evaluations,
- lipid peroxidation levels measurement,
- inflammatory markers analysis,
- tissue damage, and
- repair investigation in ROS-affected tissues.

The supplementary file 1 contains a list of preclinical studies on LAB strains to date. Randomized controlled trials (RCT) simultaneously focused on LAB strains' usage with redox modulatory characteristics in healthy individuals in the early 2000s.

T. Kullisaar et al. (2003 and 2005) studied the antioxidative activity of *L. fermentum ME-3* in a pilot study encompassing 21 participants (16 women and five men) with a mean age of 50 [35 – 65] years in two consequent studies. The intervention was to receive 150 g of goats' milk without or with $3 \times 10^{11}$ *L. fermentum ME-3* CFU in five placebo and 16 study subjects for 3 weeks. The results have confirmed that not only does *L. fermentum ME-3* increase LAB population and variety, for instance, the isolation of *L. acidophilus* and *L. salivarius* of study subjects but also statistically associates with decreased lipoprotein fraction and oxidized low-density lipoprotein (LDL) (*P-value* = *0.003* and *0.05*, respectively), urine 8-isoprostanes (*P-value* = *0.005*), and improves serum total antioxidative activity and status [59, 60].

Eighteen healthy young men without a previous regular exercise regimen had participated in the Japanese study by W. Aoi et al. (2013). Three trials of rest with placebo intake, exercise with placebo intake, and exercise with *L. helveticus*-fermented milk intake were designed. The muscle soreness was reported to be suppressed by consuming fermented milk (*P-value* < *0.05*). Exercise significantly decreased the respiratory quotient and the oxygen radical absorbance capacity (*P-value* < *0.05*), although these reductions were negated by taking *L. helveticus*–fermented milk (*P-value* < 0.05). They suggested a probable association of muscle soreness and glucose metabolism improvement with LABs' antioxidant capacity [61].

NCT01069445 and NCT01179789 clinical trials were carried out under the RISTOMED project, evaluating probiotic supplementation consumption impact on nutritional parameters, intestinal microbiota, and inflammation in healthy older people [60]. *L. Valentini* et al. selected 69 healthy individuals (age range 65–85 years) with a body mass index of 22–30 $kg/m^2$ and a performance status scale of 0–2 based on the eastern cooperative oncology group (ECOG). The intervention improved homocysteine (*P-value* < 0.001), vitamin B12 (*P-value* = 0.001), and folate (*P-value* = 0.007) while reducing ESR (*P-value* = 0.05). Few data show significant effects on inflammation [62].

In 2006, an RCT in Poland cohort of 32 healthy subjects with a female-to-male ratio of 1:1 was carried out to assess *L. casei's* influence on plasma oxidative stress parameters, including $H_2O_2$, CAT, SOD, GSH, GPx, malondialdehyde, and free sulfhydryl groups concentrations and ferric reducing ability. Capsules containing lyophilized *L. casei* $4 \times 10^8$ CFU beside 400 mg of prebiotic Inulin were administered for 7 weeks. Dietary supplementation capsules not only increase the ferric reducing ability (*P-value* = 0.00008), glutathione S-transferase (*P-value* < 0.001), GSH (*P-value* < 0.01), CAT (*P-value* = 0.02), and free sulfhydryl groups (*P-value* < 0.05) but also decrease $H_2O_2$ (*P-value* < 0.01), malondialdehyde (*P-value* < 0.01), and oxidized glutathione (*P-value* < 0.05) concentrations. Positively efficient does *L. casei* strain on some parameters have, and no statistically significant increase is reported in GPx and SOD [63, 64].

The open-label pilot study of 45 clinically asymptomatic participants was designed by Rehema et al. in Estonia. The unique composition of innovative food supplement Reg'Activ Cholesterol (RAC) containing $6 \times 10^9$ CFU/g of *L. fermentum ME-3* with antioxidative and antiatherogenic properties was administered for 4 weeks. The RAC consumption trial was registered as ISRCTN55339917 at the University of Tartu. RAC was said to improve the cardiovascular and diabetes risk profile, body mass index, hemoglobin A1c, and low-density lipoprotein, besides influence positive oxidative stress and inflammatory profile [65].

Three RCTs conducted in 2019 by Carlo Tenore et al. [66], Park et al. [67], and Ebrahimi et al. [68], as well as studies by Liang et al. in 2021 [69], Ch. Ch. Hsu in 2021 [70] and NCT05346432 in 2022 [71] on healthy individuals have been proved the LAB strains' beneficial impacts on oxidative stress and probable therapeutic effects with or without any other substances on various ROS-related diseases. **Table 1** summarizes 27 RCTs utilizing LABs' probiotics in different GIT- and non-GIT diseases.

| # | Population Intervention | Dose (X × 10 (n)) CFU) | Participants (Placebo: Study) | FU Duration (Week) | Results | Author, Year (20-); Reference |
|---|---|---|---|---|---|---|
| 1 | Pregnant Women with Singleton Primigravida, age range [18 – 30] in their 3rd Trimester<br>Probiotic Yogurt of *S. thermophilus* + *L. bulgaricus* + Probiotic culture of L. *acidophilus LA5* + *B. lactis BB12* | 1 × [7] | 70 (33:37) | 9 | ↑ GSH, and GR (*P = 0.01*), GPx, and 8-oxoG (*P = 0.04*) | Sh. Jazayeri, 12; [72] |
| 2 | MetS Adult Pts<br>YAKULT light of *L. casei shirota* | 3 × 6.5 (9) | 38 (10:28) | 12 | ↑ Gut permeability, hsCRP, and LBP levels after 3 mths | H. Sourij, 12; [73] |
| 3 | T2DM Pts with a BMI < 35 kg/m$^2$, age range [30 – 60]<br>Probiotic Yogurt of *L. acidophilus La5* + *B. lactis Bb12* | 1.85, 1.79 × [6] | 64 (32:32) | 6 | ↑ SOD and GPx activities, and TAS (*P < 0.05*)<br>↓ FBS (*P < 0.01*), HbA1c, and MDA (*P < 0.05*) | A. Homayouni-Rad, 12; [74] |
| 4 | Diabetic Pts, age range [35 – 70]<br>Probiotic Supplement of Viable and Freeze-dried *L. acidophilus* + *L. casei* + *L. rhamnosus* +*L. bulgaricus* + *B. breve* + *B. longum* + *S. thermophilus* + Fructo-oligosaccharide + Lactose (100 mg) | 2, 7, 1.5 × [9], 2 × [8], 2 × [10], 7, 1.5 × [9] | 54 (27:27) | 8 | ↑ GSH (*P = 0.03*)<br>↓ FPG (*P = 0.01*) | A.Esmaillzadeh, 13; [75] |
| 5 | Pregnant Women with Singleton Primigravida, age range [18 – 35] in their 3rd Trimester<br>Symbiotic Food of a Viable and Heat-resistant *L. sporogenes* + Inulin (HPX) (0.04 g) | 1 × [7] | 52 (26:26) | 9 | ↑ GSH (*P = 0.004*)<br>↓ TG and VLDL (*P = 0.04*) | Z. Asemi, 14; [76] |
| 6 | T2DM Pts<br>Symbiotic Food of Viable and Heat-resistant *L. sporogenes* + Inulin (HPX) (0.04 g) | 1 × [7] | 70 (35:35) | 6 | ↑ GSH (*P < 0.001*) and UA (*P = 0.04*)<br>↓ Insulin (*P = 0.03*) and hs-CRP (*P = 0.01*) | A.Esmaillzadeh, 14; [77] |
| 7 | Diabetic Pts, age range [35 – 70]<br>Symbiotic Bread of *L. sporogenes* + (Inulin (0.07 g)) | 1 × [8] | 81 (27:54) | 8 | ↑ NO (*P < 0.001*)<br>↓ MDA (*P < 0.001*) | Z. Asemi, 15; [78] |
| 8 | Nonsmoking Lean (BMI: 19–25 kg/m$^2$) and obese (BMI: 30–45 kg/m$^2$) Participants without GIT Disease, age range [40 – 65]<br>BIO-tract Controlled-release Caplets of *L. reuteri SD5865* | 2 × [10] | 21 (10:11) | 4 | ↑ Glucose-stimulated GLP-1 and GLP-2 (*P < 0.01*), Insulin, and C-peptide (*P < 0.05*) | M. Roden and N. C. Schloot, 15; [79] |

| # | Population Intervention | Dose (X × 10 (n)) CFU) | Participants (Placebo: Study) | FU Duration (Week) | Results | Author, Year (20-); Reference |
|---|---|---|---|---|---|---|
| 9 | Inactive to Moderate Level of RA Women with a BMI < 40 kg/m², age range [20 – 80], Stable Medication for at least the prior 3 mths<br>Capsule of *L. casei 01* | 1 × [8] | 60 (30:30) | 8 | ↓ SOD ($P < 0.05$) | B. Alipour, 15; [80] |
| 10 | MDD Pts, age range [20–55]<br>Capsule of Viable and Freeze-dried *L. acidophilus, L. casei* + *B. bifidum* | 2 × [9] | 40 (20:20) | 8 | ↑ GSH ($P = 0.02$)<br>↓ BDI ($P = 0.001$), Insulin, HOMA-IR, and hs-CRP ($P = 0.03$) | Z. Asemi, 15; [81] |
| 11 | T2DM Pts with a BMI < 35 kg/m², age range [35 – 60], Confirmed Dx. for at least 1 yr.<br>Fermented Milk L. *acidophilus La-5* + *B. animalis subsp lactis BB-12* | 1 × [9] | 50 (25:25) | 6 | ↓ Fructosamine ($P = 0.04$), Mean values of HbA1c ($P = 0.02$), TC ($P = 0.04$), and LDL ($P = 0.03$) | L. B. Tonucci, 15; [82] |
| 12 | T2DM Pts<br>Symbiotic of *L. sporogenes* + Inulin (0.1 g) + Beta-carotene (0.05 g) | 1 × [7] | 102 (51:51) | 6 | ↑ NO and GSH ($P = 0.01$)<br>↓ Insulin, HOMA-IR ($P = 0.002$), HOMA-B ($P = 0.01$), TG, VLDL ($P = 0.02$) and TC:HDL ratio ($P = 0.001$) | A.Esmaillzadeh, 15; [83] |
| 13 | Pregnant Women, age range [18 – 37]<br>Probiotic Supplement of *L. acidophilus* + *L. casei* + *B. bifidum* | 2 × [9] | 60 (30:30) | 12 | ↑ QUICKI ($P = 0.03$)<br>↓ Insulin, HOMA-B ($P = 0.03$), HOMA-IR ($P = 0.04$)<br>Changes in TG ($P = 0.002$), GSH, hs-CRP ($P = 0.004$), NO, and TAC ($P < 0.001$) | Z. Asemi, 16; [84] |
| 14 | GDM Women who were not on Oral Hypoglycemic Agents, age range [18 – 40]<br>Symbiotic Capsule of *L. acidophilus T16* + *L. casei T2* + *B. bifidum T1* + Inulin (HPX) (800 mg) | 2 × [9] | 60 (30:30) | 6 | ↑ TAC ($P = 0.009$), and GSH ($P = 0.02$)<br>↓ hs-CRP ($P = 0.004$), MDA ($P = 0.02$), CS rate ($P = 0.04$), hyperbilirubinemia newborns, and newborns' hospitalization ($P = 0.006$) | Z. Asemi, 16; [85] |
| 15 | GDM Women who were not on Oral Hypoglycemic Agents<br>Probiotic Capsule of *L. acidophilus* + *L. casei* + *B. bifidum* | 2 × [9] | 60 (30:30) | 6 | ↑ TAC ($P = 0.002$)<br>↓ FPG ($P = 0.01$), hs-CRP ($P < 0.001$), MDA ($P = 0.03$), and MDA/TAC ratio ($P = 0.004$) | Z. Asemi and M. Jamilian, 17; [86] |

| # | Population Intervention | Dose (X × 10 (n)) CFU) | Participants (Placebo: Study) | FU Duration (Week) | Results | Author, Year (20-); Reference |
|---|---|---|---|---|---|---|
| 16 | Moderate to Severe Level of RA Women with a BMI < 40 kg/m$^2$, age range [25 – 70], Confirmed Dx. for at least 6 mths ago<br>Synbiotic Capsule of *L. acidophilus* + *L. casei* + *B. bifidum* + Inulin (HPX) (800 mg) | 2 × [9] | 54 (27:27) | 8 | ↑ DAS-28 ($P < 0.001$), VAS pain ($P < 0.001$), NO ($P = 0.008$), and GSH ($P = 0.005$)<br>↓ hs-CRP ($P = 0.001$), Insulin, HOMA-B ($P = 0.01$), and HOMA-IR ($P = 0.03$) | Z. Asemi, 17; [87] |
| 17 | Diabetic Pts undergoing HD for at least 1 yr., age range [18 – 80]<br>Capsule of *L. acidophilus* + *L. casei* + *B. bifidum* | 2 × [9] | 60 (30:30) | 12 | ↑ QUICKI and TAC<br>↓ FPG, Insulin, HOMA-IR, HOMA-B, HbA1c, hsCRP, MDA, SGA, and TIBC | Z. Asemi, 17; [88] |
| 18 | Diabetic Pts undergoing HD, age range [18 – 80]<br>Synbiotic Capsule of *L. acidophilus* + *L. casei* + *B. bifidum* + Inulin (HPX) (0.8 g) | 2 × [9] | 60 (30:30) | 12 | ↑ QUICKI ($P < 0.001$), TAC and GSH ($P < 0.001$)<br>↓ FPG ($P = 0.01$), Insulin, HOMA-IR, hsCRP ($P < 0.001$), and MDA ($P = 0.003$) | Z. Asemi, 18; [89] |
| 19 | AD Pts<br>*L. acidophilus* + *B. bifidum* + *B. longum* + Se (200 μg) | 2 × [9] | 78 (26:52) | 12 | Se supplementation:<br>↑ GSH ($P = 0.001$) and QUICKI ($P = 0.01$)<br>↓ hs-CRP ($P < 0.001$), Insulin ($P = 0.001$), HOMA-IR ($P = 0.002$), LDL ($P = 0.04$), and TC:HDL ratio ($P = 0.004$)<br>Probiotic + Se co-supplementation:<br>↑ MMSE ($P < 0.001$), TAC, GSH ($P = 0.001$), and QUICKI ($P < 0.006$)<br>↓ hs-CRP ($P < 0.001$), Insulin, HOMA-IR ($P < 0.001$), TG, VLDL ($P = 0.02$), LDL ($P = 0.04$), and TC:HDL ratio ($P = 0.005$) | Z. Asemi, 18; [90] |

| # | Population Intervention | Dose (X × 10 (n)) CFU) | Participants (Placebo: Study) | FU Duration (Week) | Results | Author, Year (20-); Reference |
|---|---|---|---|---|---|---|
| 20 | Nulliparous GDM Woman, Confirmed Dx. of (24–28-week) gestation, a pre-pregnancy BMI < 18.5 kg/m$^2$, no history of impaired glucose tolerance, T2DM, chronic disease, and smoking or alcohol consumption in early pregnancy, no use of fermented foods and Abs within 2 and 4 weeks of the intervention, respectively. Lack of acute GIT problems and no use of GCs or IST for 1 mth before intervention, age range [18–45]<br>Probiotic Capsule of *L. acidophilus LA-5* + *L. delbrueckii bulgaricus LBY-27* + *B. BB-12* + *S. thermophilus STY-31* | 4 × [9] | 52 (26:26) | 8 | ↑ hsCRP, TNF-α, MDA, GR, and GPx | N. Dolatkhah, 18; [91] |
| 21 | MI Pts with a recent Dx. who underwent Percutaneous Coronary Intervention<br>Capsule of *L. rhamnosus* | 1.6 (10) | 44 (22:22) | 12 | ↑ Mean QoL (*P* = *0.023*) and TAC (*P* = *0.009*)<br>↓ BDI (*P* = *0.045*), MDA (*P* = *0.033*), and hsCRP (*P* = *0.040*) | J. Moludi, 19; [92] |
| 22 | HD Pts<br>Symbiotic Tablet of *L. coagulans* + Fructo-oligosaccharides | U. K. | 50 (25:25) | 8 | ↓ Mean values of hs-CRP and MDA (*P* = *0.01*), TC, and LDL (*P* = *0.001*) | T. Tofighiyan, 19; [93] |
| 23 | CAD Pts, age range [18 – 85]<br>*L. rhamnosus G* + Inulin (HPX) (15 g) | 1.9 × [9] | 96 (24:72) | 8 | ↓ BDI (*P* = *0.001*), STAI-state (*P* = *0.021*), and STAI-trait, hs-CRP (*P* = *0.02*), LPS (*P* = *0.047*), and TNF-α (*P* = *0.032*) | J. Moludi, 21; [94] |
| 24 | PsO Pts, age range [18–50]<br>Probiotic Drink of *L.* strains | U.K. | 50 (25:25) | 8 | ↑ BDI (*P* = *0.017*), DLQI (*P* = *0.045*), TAC (*P* = *0.03*)<br>↓ PASI (*P* = *0.049*), PSS (*P* = *0.047*), hs-CRP (*P* = *0.015*), IL-6 (*P* = *0.05*), and MDA (*P* = *0.013*) | H. Razmi, 21; [95] |
| 25 | RA Pts<br>Sachet of Freeze L. *acidophilus La-14* + *L. casei Lc-11* + *L. lactis Ll-23* + *B. lactis Bl-04* + *B. bifidum Bb-06* | 1 × [9] | 42 (21:21) | 8 | ↑ Sulfhydryl group (*P* = *0.028*) and TRAP (*P* = *0.019*)<br>↓ WBC (*P* = *0.012*), TNF-α, NO (*P* = *0.004*), and IL-6 (P = 0.039) | A. N. Colado Simão, 21; [96] |

| # | Population Intervention | Dose (X × 10 (n)) CFU) | Participants (Placebo: Study) | FU Duration (Week) | Results | Author, Year (20-); Reference |
|---|---|---|---|---|---|---|
| 26 | MetS Pts, age range [20–65]<br>Probiotic Yogurt of *L. acidophilus La5* + *B. lactis Bb12* | 4.41 and 3.55 × [6] | 44 (22:22) | 8 | ↑ TAC (*P* < *0.05*)<br>↓ UA (*P* < *0.05*) | B. Pourghassem Gargari, 21; [97] |
| 27 | Mild to Moderate Dementia Pts, age range [50 – 90]<br>*L. plantarum PL-02* + *B. longum subsp. infantis BLI-02* + *B. breve Bv-889* + *B. animalis subsp. lactis CP-9* + *B. bifidum VDD088* | 5 × [7], 1 × [10] | 40 (20:20) | 12 | ↑ BDNF (P = 0.005) and SOD (P = 0.012)<br>↓ IL-1β (P = 0.041) | y. Sun, 23; [98] |

*CFU: Colony-forming Unit; FU: Follow-up; GSH: Glutathione; GR: Glutathione Reductase; GPx: Glutathione Peroxidase; 8-oxoG: 8-oxo-7,8-Dihydroguanine; MetS: Metabolic Syndrome; Pt: Patient; hs-CRP: High-sensitivity C-reactive Protein; LBP: Lipopolysaccharide-binding Protein; T2DM: Type 2 Diabetes Mellitus; BMI: Body Mass Index; SOD: Superoxide Dismutase; TAS: Total Antioxidant Status; FBS: Fasting Blood Glucose; HbA1c: Hemoglobin A1c; MDA: Malondialdehyde; FPG: Fasting Plasma Glucose; TG: Triglyceride; VLDL: Very-low-density Lipoprotein; UA: Uric Acid; NO: Nitric Oxide; Ab: Antibiotic; GIT: Gastrointestinal Tract; GLP: Glucagon-like Peptide; RA: Rheumatoid Arthritis; MDD: Major Depressive Disorder; BDI: Beck Depression Inventory; HOMA-IR: Homeostasis Model of Assessment-estimated Insulin Resistance; Dx: Diagnosis; TC: Total Cholesterol; LDL: Low-density Lipoprotein; HOMA-B: Homeostasis Model of Assessment-estimated B-cell function; HDL: High-density Lipoprotein; QUICKI: Quantitative Insulin Sensitivity Check Index; TAC: Total Antioxidant Capacity; GDM: Gestational Diabetes Mellitus; CS: Cesarean Section; DAS-28: Disease Activity Score-28 joints; VAS: Visual Analog Scale; HD: Hemodialysis; SGA: Subjective Global Assessment; TIBC: Total Iron-binding Capacity; AD: Alzheimer's Disease; Se: Selenium; MMSE: Mini-mental State Examination; GCs: Glucocorticoids; IST: Immunosuppressive Therapy; TNF-α: Tumor Necrosis Factor-α; MI: Myocardial Infarction; QoL: Quality of Life; CAD: Coronary Artery Disease; STAI: State–trait Anxiety Inventory; LPS: Lanterman-Petris-Short act; PsO: Psoriasis; DLQI: Dermatology Life Quality Index; PASI: Psoriasis Area and Severity Index; PSS: Progressive Systemic Sclerosis; IL: Interleukin; TRAP: Total Radical-trapping Antioxidant Parameter; WBC: White Blood Count; BDNF: Brain-derived Neurotrophic Factor. X × 10 (n), ↑, ↓, and P- indicate X × 10* $^{n}$ *CFU, increase, decrease, and P-value. The age range is shown in years.*

**Table 1.**
*Twenty-seven RCTs of LABs' probiotics usage in the treatment of various ROS-related diseases.*

## 6. Conclusion and prospects

In retrospect, LAB strains are recognized as "generally recognized as safe" by the U.S. Food and Drug Administration (FDA) and as having a "qualified presumption of safety" status by the European Food Safety Authority (EFSA), similar to gram-negative bacteria. As a result, nutrigenetics and nutrigenomics, personalized nutrition, besides precision medicine using genetics and microbiology technological advances, will push back the frontiers. Future research will focus on designing preferred probiotics, next-generation probiotics, and different sources of probiotics, such as environmental sources and fecal bacteriotherapy [7, 99, 100]. Not only do *in vitro* and preclinical research confirm the LAB's beneficial effects on nutrition uptake and production, the GIT and immune system maturations, infection resistance, and host metabolism, but they also have antioxidant features. So valuable are the characteristics that many RCTs focus on. Based on numerous studies confirming the possibilities and capabilities of LAB strains in a diverse scope of either nonmalignant or malignant diseases, including metabolic syndrome, type 2 and gestational diabetes mellitus, rheumatoid arthritis, Alzheimer's disease, coronary artery disease, and psoriasis, it concludes that these bacteria have safe therapeutic characteristics. Nevertheless, the LAB strains' clinical usage faces several challenges and limitations, including

- the difficulty or inability of live LAB strains to reach some organs, for instance, the GIT,
- the insufficient number of live LAB strains at the target organ,
- the technical difficulties, and
- the sensitivity of LAB strains to famine, oxygen, and stress.

The new emerging trends provide potential applications of LAB strains overcoming these challenges and limitations [15].

## Supplementary information

Additional supplementary file 1 may be found in the Supplementary information tab for this chapter: https://docs.google.com/document/d/1kqSG7tHS3jUug9AVWE35QzQJfWgxOJmE/edit?usp=sharing&ouid=107504097666027977930&rtpof=true&sd=true

## Acknowledgements

The authors express sincere gratitude to Semnan University for its facilities and cooperation.

## Conflict of interest

The authors declare no conflict of interest.

## Notes/thanks/other declarations

None.

## Author details

Parisa Naji[1], Farshid Parvini[1*] and Mohammad Ali Farazi Fard[2]

1 Faculty of Basic Sciences, Department of Biology, Semnan University, Semnan, Iran

2 Department of Medical Genetics, Shiraz University of Medical Sciences, Shiraz, Fars, Iran

*Address all correspondence to: f.parvini@semnan.ac.ir

## References

[1] Anukam KC, Reid G. Probiotics: 100 years (1907-2007) after Elie Metchnikoff's observation. Communicating Current Research and Educational Topics and Trends in Applied Microbiology. 2007;**1**:466-474

[2] Lilly DM, Stillwell RH. Probiotics: Growth-promoting factors produced by microorganisms. Science. 1965;**147**(3659):747-748

[3] Suez J, Zmora N, Segal E, Elinav E. The pros, cons, and many unknowns of probiotics. Nature Medicine. 2019;**25**(5):716-729

[4] Yan F, Polk DB. Probiotics and probiotic-derived functional factors—Mechanistic insights into applications for intestinal homeostasis. Frontiers in Immunology. 2020;**11**:1428

[5] Kiani AK, Anpilogov K, Dautaj A, Marceddu G, Sonna WN, Percio M, et al. Bacteriophages in food supplements obtained from natural sources. Acta Bio Medica: Atenei Parmensis. 2020;**91**(Suppl 13):e2020025

[6] Hania WB, Ballet N, Vandeckerkove P, Ollivier B, O'Toole PW, Brugère J-F. Archaebiotics: Archaea as pharmabiotics for treating chronic disease in humans. In: Archaea—New Biocatalysts, Novel Pharmaceuticals and Various Biotechnological Applications. London, UK, Rijeka, Croatia: IntechOpen; 2017. pp. 41-62

[7] Maftei N-M, Raileanu CR, Balta AA, Ambrose L, Boev M, Marin DB, et al. The potential impact of probiotics on human health: An update on their health-promoting properties. Microorganisms. 2024;**12**(2):234

[8] La Fata G, Weber P, Mohajeri MH. Probiotics and the gut immune system: Indirect regulation. Probiotics and Antimicrobial Proteins. 2018;**10**:11-21

[9] Santacroce L, Charitos IA, Bottalico L. A successful history: Probiotics and their potential as antimicrobials. Expert Review of Anti-infective Therapy. 2019;**17**(8):635-645

[10] Wieërs G, Belkhir L, Enaud R, Leclercq S, Philippart de Foy J-M, Dequenne I, et al. How probiotics affect the microbiota. Frontiers in Cellular and Infection Microbiology. 2020;**9**:454

[11] Zhang H, Duan Y, Cai F, Cao D, Wang L, Qiao Z, et al. Next-generation probiotics: Microflora intervention to human diseases. BioMed Research International. 2022;**2022**:5633403

[12] Yadav MK, Kumari I, Singh B, Sharma KK, Tiwari SK. Probiotics, prebiotics and synbiotics: Safe options for next-generation therapeutics. Applied Microbiology and Biotechnology. 2022;**106**(2):505-521

[13] Zheng J, Wittouck S, Salvetti E, Franz CM, Harris HM, Mattarelli P, et al. A taxonomic note on the genus lactobacillus: Description of 23 novel genera, emended description of the genus lactobacillus Beijerinck 1901, and union of Lactobacillaceae and Leuconostocaceae. International Journal of Systematic and Evolutionary Microbiology. 2020;**70**(4):2782-2858

[14] Un-Nisa A, Khan A, Zakria M, Siraj S, Ullah S, Tipu MK, et al. Updates on the role of probiotics against different health issues: Focus on lactobacillus. International Journal of Molecular Sciences. 2022;**24**(1):142

[15] Huang R, Wu F, Zhou Q, Wei W, Yue J, Xiao B, et al. Lactobacillus and intestinal diseases: Mechanisms of action and clinical applications. Microbiological Research. 2022;**260**:127019

[16] Rastogi S, Singh A. Gut microbiome and human health: Exploring how the probiotic genus lactobacillus modulate immune responses. Frontiers in Pharmacology. 2022;**13**:1042189

[17] Szajewska H, Hojsak I. Health benefits of lactobacillus rhamnosus GG and Bifidobacterium animalis subspecies lactis BB-12 in children. Postgraduate Medicine. 2020;**132**(5):441-451

[18] Juan CA, Pérez de la Lastra JM, Plou FJ, Pérez-Lebeña E. The chemistry of reactive oxygen species (ROS) revisited: Outlining their role in biological macromolecules (DNA, lipids and proteins) and induced pathologies. International Journal of Molecular Sciences. 2021;**22**(9):4642

[19] Tretter V, Hochreiter B, Zach ML, Krenn K, Klein KU. Understanding cellular redox homeostasis: A challenge for precision medicine. International Journal of Molecular Sciences. 2021;**23**(1):106

[20] Pizzino G, Irrera N, Cucinotta M, Pallio G, Mannino F, Arcoraci V, et al. Oxidative stress: Harms and benefits for human health. Oxidative Medicine and Cellular Longevity. 2017;**2017**:8416763

[21] Sies H. Oxidative stress: Introductory remarks. In: Sies H, editor. Oxidative Stress. London, UK: Academic Press; 1985

[22] Santos-Sánchez NF, Salas-Coronado R, Villanueva-Cañongo C, Hernández-Carlos B. Antioxidant compounds and their antioxidant mechanism. Antioxidants. 2019;**10**:1-29

[23] Kullisaar T, Songisepp E, Zilmer M. Probiotics and oxidative stress. In: Lushchak V, editor. Oxidative Stress-Environmental Induction and Dietary Antioxidants. London, UK: IntechOpen; 2012. pp. 203-222

[24] Forman HJ, Zhang H. Targeting oxidative stress in disease: Promise and limitations of antioxidant therapy. Nature Reviews Drug Discovery. 2021;**20**(9):689-709

[25] Hao Y, Zhu Y-j, Zou S, Zhou P, Hu Y-w, Zhao Q-x, et al. Metabolic syndrome and psoriasis: Mechanisms and future directions. Frontiers in Immunology. 2021;**12**:711060

[26] Liguori I, Russo G, Curcio F, Bulli G, Aran L, Della-Morte D, et al. Oxidative stress, aging, and diseases. Clinical Interventions in Aging. 2018;**13**:757-772

[27] Chen L, Liu B. Relationships between stress granules, oxidative stress, and neurodegenerative diseases. Oxidative Medicine and Cellular Longevity. 2017;**2017**:1809592

[28] Dhapola R, Beura SK, Sharma P, Singh SK, HariKrishnaReddy D. Oxidative stress in Alzheimer's disease: Current knowledge of signaling pathways and therapeutics. Molecular Biology Reports. 2024;**51**(1):48

[29] Chandrasekaran A, Idelchik M, Melendez J. Redox control of senescence and age-related disease. Redox Biology. 2017;**11**:91-102

[30] Arfin S, Jha NK, Jha SK, Kesari KK, Ruokolainen J, Roychoudhury S, et al. Oxidative stress in cancer cell metabolism. Antioxidants. 2021;**10**(5):642

[31] Lin C-J, Lee C-C, Shih Y-L, Lin T-Y, Wang S-H, Lin Y-F, et al. Resveratrol

enhances the therapeutic effect of temozolomide against malignant glioma in vitro and in vivo by inhibiting autophagy. Free Radical Biology and Medicine. 2012;**52**(2):377-391

[32] Fu J, Shao C-J, Chen F-R, Ng H-K, Chen Z-P. Autophagy induced by valproic acid is associated with oxidative stress in glioma cell lines. Neuro-Oncology. 2010;**12**(4):328-340

[33] Mirzaei S, Mohammadi AT, Gholami MH, Hashemi F, Zarrabi A, Zabolian A, et al. Nrf2 signaling pathway in cisplatin chemotherapy: Potential involvement in organ protection and chemoresistance. Pharmacological Research. 2021;**167**:105575

[34] Yen C-J, Hung C-H, Tsai W-M, Cheng H-C, Yang H-L, Lu Y-J, et al. Effect of exercise training on exercise tolerance and level of oxidative stress for head and neck cancer patients following chemotherapy. Frontiers in Oncology. 2020;**10**:1536

[35] Ahmed KM, Cao N, Li JJ. HER-2 and NF-κB as the targets for therapy-resistant breast cancer. Anticancer Research. 2006;**26**(6B):4235-4243

[36] Mizutani H, Shiga C, Imai M, Ikemura K, Kitamura Y, Ohta K, et al. Idarubicin, an Anthracycline, induces oxidative DNA damage in the presence of copper (II). Anticancer Research. 2020;**40**(10):5399-5404

[37] Perri M, Pingitore A, Cione E, Vilardi E, Perrone V, Genchi G. Proliferative and anti-proliferative effects of retinoic acid at doses similar to endogenous levels in Leydig MLTC-1/R2C/TM-3 cells. Biochimica et Biophysica Acta (BBA)-General Subjects. 2010;**1800**(9):993-1001

[38] Miki H, Uehara N, Kimura A, Sasaki T, Yuri T, Yoshizawa K, et al. Resveratrol induces apoptosis via ROS-triggered autophagy in human colon cancer cells. International Journal of Oncology. 2012;**40**(4):1020-1028

[39] Asensio-López MC, Soler F, Pascual-Figal D, Fernández-Belda F, Lax A. Doxorubicin-induced oxidative stress: The protective effect of nicorandil on HL-1 cardiomyocytes. PLoS One. 2017;**12**(2):e0172803

[40] Zhang L, Li J, Zong L, Chen X, Chen K, Jiang Z, et al. Reactive oxygen species and targeted therapy for pancreatic cancer. Oxidative Medicine and Cellular Longevity. 2016;**2016**:1616781

[41] Mogavero A, Maiorana MV, Zanutto S, Varinelli L, Bozzi F, Belfiore A, et al. Metformin transiently inhibits colorectal cancer cell proliferation as a result of either AMPK activation or increased ROS production. Scientific Reports. 2017;7(1):15992

[42] Wang W, Dong X, Liu Y, Ni B, Sai N, You L, et al. Itraconazole exerts anti-liver cancer potential through the Wnt, PI3K/AKT/mTOR, and ROS pathways. Biomedicine & Pharmacotherapy. 2020;**131**:110661

[43] Marino M, Fais S, Djavaheri-Mergny M, Villa A, Meschini S, Lozupone F, et al. Proton pump inhibition induces autophagy as a survival mechanism following oxidative stress in human melanoma cells. Cell Death & Disease. 2010;**1**(10):e87

[44] Jazirehi AR. Regulation of apoptosis-associated genes by histone deacetylase inhibitors: Implications in cancer therapy. Anti-Cancer Drugs. 2010;**21**(9):805-813

[45] Li J, Liu R, Lei Y, Wang K, Lau QC, Xie N, et al. Proteomic analysis revealed

association of aberrant ROS signaling with suberoylanilide hydroxamic acid-induced autophagy in Jurkat T-leukemia cells. Autophagy. 2010;**6**(6):711-724

[46] Miller CM, O'Sullivan EC, McCarthy FO. Novel 11-substituted ellipticines as potent anticancer agents with divergent activity against cancer cells. Pharmaceuticals. 2019;**12**(2):90

[47] Robin F, Paillard C, Marchandin H, Demeocq F, Bonnet R, Hennequin C. Lactobacillus rhamnosus meningitis following recurrent episodes of bacteremia in a child undergoing allogeneic hematopoietic stem cell transplantation. Journal of Clinical Microbiology. 2010;**48**(11):4317-4319

[48] Cohen SA, Woodfield MC, Boyle N, Stednick Z, Boeckh M, Pergam SA. Incidence and outcomes of bloodstream infections among hematopoietic cell transplant recipients from species commonly reported to be in over-the-counter probiotic formulations. Transplant Infectious Disease. 2016;**18**(5):699-705

[49] Sharma A, Tilak T, Bakhshi S, Raina V, Kumar L, Chaudhary S, et al. Lactobacillus brevis CD2 lozenges prevent oral mucositis in patients undergoing high dose chemotherapy followed by haematopoietic stem cell transplantation. ESMO Open. 2016;**1**(6):e000138

[50] Gorshein E, Wei C, Ambrosy S, Budney S, Vivas J, Shenkerman A, et al. *Lactobacillus rhamnosus* GG probiotic enteric regimen does not appreciably alter the gut microbiome or provide protection against GVHD after allogeneic hematopoietic stem cell transplantation. Clinical Transplantation. May 2017;**31**(5): e12947

[51] Koyama S, Fujita H, Shimosato T, Kamijo A, Ishiyama Y, Yamamoto E, et al. Septicemia from *lactobacillus rhamnosus* GG, from a probiotic enriched yogurt, in a patient with autologous stem cell transplantation. Probiotics and Antimicrobial Proteins. 2019;**11**(1):295-298

[52] Wang Y, Wu Y, Wang Y, Xu H, Mei X, Yu D, et al. Antioxidant properties of probiotic bacteria. Nutrients. 2017;**9**(5):521

[53] Feng T, Wang J. Oxidative stress tolerance and antioxidant capacity of lactic acid bacteria as probiotic: A systematic review. Gut Microbes. 2020;**12**(1):1801944

[54] Stoia M, Oancea S. Low-molecular-weight synthetic antioxidants: Classification, pharmacological profile, effectiveness and trends. Antioxidants. 2022;**11**(4):638

[55] Mu G, Gao Y, Tuo Y, Li H, Zhang Y, Qian F, et al. Assessing and comparing antioxidant activities of lactobacilli strains by using different chemical and cellular antioxidant methods. Journal of Dairy Science. 2018;**101**(12):10792-10806

[56] Lin M-Y, Yen C-L. Antioxidative ability of lactic acid bacteria. Journal of Agricultural and Food Chemistry. 1999;**47**(4):1460-1466

[57] Ahire JJ, Mokashe NU, Patil HJ, Chaudhari BL. Antioxidative potential of folate producing probiotic lactobacillus helveticus CD6. Journal of Food Science and Technology. 2013;**50**:26-34

[58] Heydari A, Parvini F, Fard NA. Functional foods and antioxidant effects: Emphasizing the role of probiotics. In: Current Topics in Functional Food. London, UK: IntechOpen; 11 Apr 2022

[59] Kullisaar T, Songisepp E, Mikelsaar M, Zilmer K, Vihalemm T,

Zilmer M. Antioxidative probiotic fermented goats' milk decreases oxidative stress-mediated atherogenicity in human subjects. British Journal of Nutrition. 2003;**90**(2):449-456

[60] Songisepp E, Kals J, Kullisaar T, Mändar R, Hütt P, Zilmer M, et al. Evaluation of the functional efficacy of an antioxidative probiotic in healthy volunteers. Nutrition Journal. 2005;**4**(1):1-10

[61] Iwasa M, Aoi W, Mune K, Yamauchi H, Furuta K, Sasaki S, et al. Fermented milk improves glucose metabolism in exercise-induced muscle damage in young healthy men. Nutrition Journal. 2013;**12**(1):1-7

[62] Valentini L, Pinto A, Bourdel-Marchasson I, Ostan R, Brigidi P, Turroni S, et al. Impact of personalized diet and probiotic supplementation on inflammation, nutritional parameters and intestinal microbiota–the "RISTOMED project": Randomized controlled trial in healthy older people. Clinical Nutrition. 2015;**34**(4):593-602

[63] Kleniewska P, Hoffmann A, Pniewska E, Pawliczak R. The influence of probiotic lactobacillus casei in combination with prebiotic inulin on the antioxidant capacity of human plasma. Oxidative Medicine and Cellular Longevity. 2016;**2016**:1340903

[64] Kleniewska P, Pawliczak R. Influence of synbiotics on selected oxidative stress parameters. Oxidative Medicine and Cellular Longevity. 2017;**2017**:9315375

[65] Kullisaar T, Zilmer K, Salum T, Rehema A, Zilmer M. The use of probiotic L. fermentum ME-3 containing Reg'Activ cholesterol supplement for 4 weeks has a positive influence on blood lipoprotein profiles and inflammatory cytokines: An open-label preliminary study. Nutrition Journal. 2016;**15**:1-6

[66] Tenore GC, Caruso D, Buonomo G, D'Avino M, Ciampaglia R, Maisto M, et al. Lactofermented annurca apple puree as a functional food indicated for the control of plasma lipid and oxidative amine levels: Results from a randomised clinical trial. Nutrients. 2019;**11**(1):122

[67] Chong H-X, Yusoff NAA, Hor Y-Y, Lew L-C, Jaafar MH, Choi S-B, et al. Lactobacillus plantarum DR7 improved upper respiratory tract infections via enhancing immune and inflammatory parameters: A randomized, double-blind, placebo-controlled study. Journal of Dairy Science. 2019;**102**(6):4783-4797

[68] Noorifard M, Moghaddam AD, Asemi Z, Farahani RH, Jazayeri SMM, Ebrahimi E. Effect of probiotic supplementation on oxidative stress enzymes and mental health of athletes. Annals of Military and Health Sciences Research. 2019;**17**(1):e84922

[69] Mei L-H, Zheng W-X, Zhao Z-T, Meng N, Zhang Q-R, Zhu W-J, et al. A pilot study of the effect of lactobacillus casei obtained from long-lived elderly on blood biochemical, oxidative, and inflammatory markers, and on gut microbiota in young volunteers. Nutrients. 2021;**13**(11):3891

[70] Fu S-K, Tseng W-C, Tseng K-W, Lai C-C, Tsai Y-C, Tai H-L, et al. Effect of daily oral lactobacillus plantarum PS128 on exercise capacity recovery after a half-marathon. Nutrients. 2021;**13**(11):4023

[71] Lee M-C, Ho C-S, Hsu Y-J, Huang C-C. Live and heat-killed probiotic lactobacillus paracasei PS23 accelerated the improvement and recovery of strength and damage biomarkers after exercise-induced muscle damage. Nutrients. 2022;**14**(21):4563

[72] Asemi Z, Jazayeri S, Najafi M, Samimi M, Mofid V, Shidfar F, et al.

Effect of daily consumption of probiotic yogurt on oxidative stress in pregnant women: A randomized controlled clinical trial. Annals of Nutrition and Metabolism. 2012;**60**(1):62-68

[73] Leber B, Tripolt N, Blattl D, Eder M, Wascher T, Pieber T, et al. The influence of probiotic supplementation on gut permeability in patients with metabolic syndrome: An open label, randomized pilot study. European Journal of Clinical Nutrition. 2012;**66**(10):1110-1115

[74] Ejtahed HS, Mohtadi-Nia J, Homayouni-Rad A, Niafar M, Asghari-Jafarabadi M, Mofid V. Probiotic yogurt improves antioxidant status in type 2 diabetic patients. Nutrition. 2012;**28**(5):539-543

[75] Asemi Z, Zare Z, Shakeri H, Sabihi S-s, Esmaillzadeh A. Effect of multispecies probiotic supplements on metabolic profiles, hs-CRP, and oxidative stress in patients with type 2 diabetes. Annals of Nutrition and Metabolism. 2013;**63**(1-2):1-9

[76] Taghizadeh M, Hashemi T, Shakeri H, Abedi F, Sabihi S-S, Alizadeh S-A, et al. Synbiotic food consumption reduces levels of triacylglycerols and VLDL, but not cholesterol, LDL, or HDL in plasma from pregnant women. Lipids. 2014;**49**:155-161

[77] Asemi Z, Khorrami-Rad A, Alizadeh S-A, Shakeri H, Esmaillzadeh A. Effects of synbiotic food consumption on metabolic status of diabetic patients: A double-blind randomized cross-over controlled clinical trial. Clinical Nutrition. 2014;**33**(2):198-203

[78] Bahmani F, Tajadadi-Ebrahimi M, Kolahdooz F, Mazouchi M, Hadaegh H, Jamal A-S, et al. The consumption of synbiotic bread containing lactobacillus sporogenes and inulin affects nitric oxide and malondialdehyde in patients with type 2 diabetes mellitus: Randomized, double-blind, placebo-controlled trial. Journal of the American College of Nutrition. 2016;**35**(6):506-513

[79] Simon M-C, Strassburger K, Nowotny B, Kolb H, Nowotny P, Burkart V, et al. Intake of lactobacillus reuteri improves incretin and insulin secretion in glucose-tolerant humans: A proof of concept. Diabetes Care. 2015;**38**(10):1827-1834

[80] Vaghef-Mehrabany E, Homayouni-Rad A, Alipour B, Sharif S-K, Vaghef-Mehrabany L, Alipour-Ajiry S. Effects of probiotic supplementation on oxidative stress indices in women with rheumatoid arthritis: A randomized double-blind clinical trial. Journal of the American College of Nutrition. 2016;**35**(4):291-299

[81] Akkasheh G, Kashani-Poor Z, Tajabadi-Ebrahimi M, Jafari P, Akbari H, Taghizadeh M, et al. Clinical and metabolic response to probiotic administration in patients with major depressive disorder: A randomized, double-blind, placebo-controlled trial. Nutrition. 2016;**32**(3):315-320

[82] Tonucci LB, Dos Santos KMO, de Oliveira LL, Ribeiro SMR, Martino HSD. Clinical application of probiotics in type 2 diabetes mellitus: A randomized, double-blind, placebo-controlled study. Clinical Nutrition. 2017;**36**(1):85-92

[83] Asemi Z, Alizadeh S-A, Ahmad K, Goli M, Esmaillzadeh A. Effects of beta-carotene fortified synbiotic food on metabolic control of patients with type 2 diabetes mellitus: A double-blind randomized cross-over controlled clinical trial. Clinical Nutrition. 2016;**35**(4):819-825

[84] Jamilian M, Bahmani F, Vahedpoor Z, Salmani A,

Tajabadi-Ebrahimi M, Jafari P, et al. Effects of probiotic supplementation on metabolic status in pregnant women: A randomized, double-blind, placebo-controlled trial. Archives of Iranian Medicine. 2016;**19**(10):687-682

[85] Karamali M, Nasiri N, Taghavi Shavazi N, Jamilian M, Bahmani F, Tajabadi-Ebrahimi M, et al. The effects of synbiotic supplementation on pregnancy outcomes in gestational diabetes. Probiotics and Antimicrobial Proteins. 2018;**10**:496-503

[86] Badehnoosh B, Karamali M, Zarrati M, Jamilian M, Bahmani F, Tajabadi-Ebrahimi M, et al. The effects of probiotic supplementation on biomarkers of inflammation, oxidative stress and pregnancy outcomes in gestational diabetes. The Journal of Maternal-Fetal & Neonatal Medicine. 2018;**31**(9):1128-1136

[87] Zamani B, Farshbaf S, Golkar HR, Bahmani F, Asemi Z. Synbiotic supplementation and the effects on clinical and metabolic responses in patients with rheumatoid arthritis: A randomised, double-blind, placebo-controlled trial. British Journal of Nutrition. 2017;**117**(8):1095-1102

[88] Soleimani A, Mojarrad MZ, Bahmani F, Taghizadeh M, Ramezani M, Tajabadi-Ebrahimi M, et al. Probiotic supplementation in diabetic hemodialysis patients has beneficial metabolic effects. Kidney International. 2017;**91**(2):435-442

[89] Soleimani A, Motamedzadeh A, Zarrati Mojarrad M, Bahmani F, Amirani E, Ostadmohammadi V, et al. The effects of synbiotic supplementation on metabolic status in diabetic patients undergoing hemodialysis: A randomized, double-blinded, placebo-controlled trial. Probiotics and Antimicrobial Proteins. 2019;**11**:1248-1256

[90] Tamtaji OR, Heidari-Soureshjani R, Mirhosseini N, Kouchaki E, Bahmani F, Aghadavod E, et al. Probiotic and selenium co-supplementation, and the effects on clinical, metabolic and genetic status in Alzheimer's disease: A randomized, double-blind, controlled trial. Clinical Nutrition. 2019;**38**(6):2569-2575

[91] Hajifaraji M, Jahanjou F, Abbasalizadeh F, Aghamohammadzadeh N, Abbasi MM, Dolatkhah N. Effect of probiotic supplements in women with gestational diabetes mellitus on inflammation and oxidative stress biomarkers: A randomized clinical trial. Asia Pacific Journal of Clinical Nutrition. 2018;**27**(3):581-591

[92] Moludi J, Alizadeh M, Mohammadzad MHS, Davari M. The effect of probiotic supplementation on depressive symptoms and quality of life in patients after myocardial infarction: Results of a preliminary double-blind clinical trial. Psychosomatic Medicine. 2019;**81**(9):770-777

[93] Kooshki A, Tofighiyan T, Miri M. A synbiotic supplement for inflammation and oxidative stress and lipid abnormalities in hemodialysis patients. Hemodialysis International. 2019;**23**(2):254-260

[94] Moludi J, Khedmatgozar H, Nachvak SM, Abdollahzad H, Moradinazar M, Sadeghpour TA. The effects of co-administration of probiotics and prebiotics on chronic inflammation, and depression symptoms in patients with coronary artery diseases: A randomized clinical trial. Nutritional Neuroscience. 2022;**25**(8):1659-1668

[95] Moludi J, Khedmatgozar H, Saiedi S, Razmi H, Alizadeh M, Ebrahimi B. Probiotic supplementation improves clinical outcomes and quality of life indicators in patients with plaque psoriasis: A randomized double-blind clinical trial. Clinical Nutrition ESPEN. 2021;**46**:33-39

[96] Cannarella LAT, Mari NL, Alcântara CC, Iryioda TMV, Costa NT, Oliveira SR, et al. Mixture of probiotics reduces inflammatory biomarkers and improves the oxidative/nitrosative profile in people with rheumatoid arthritis. Nutrition. 2021;**89**:111282

[97] Rezazadeh L, Alipour B, Jafarabadi MA, Behrooz M, Gargari BP. Daily consumption effects of probiotic yogurt containing lactobacillus acidophilus La5 and Bifidobacterium lactis Bb12 on oxidative stress in metabolic syndrome patients. Clinical Nutrition ESPEN. 2021;**41**:136-142

[98] Hsu Y-C, Huang Y-Y, Tsai S-Y, Kuo Y-W, Lin J-H, Ho H-H, et al. Efficacy of probiotic supplements on brain-derived neurotrophic factor, inflammatory biomarkers, oxidative stress and cognitive function in patients with Alzheimer's dementia: A 12-week randomized, double-blind active-controlled study. Nutrients. 2023;**16**(1):16

[99] Dempsey E, Corr SC. Lactobacillus spp. for gastrointestinal health: Current and future perspectives. Frontiers in Immunology. 2022;**13**:840245

[100] Ajam-Hosseini M, Akhoondi F, Parvini F, Fahimi H. Gram-negative bacterial sRNAs encapsulated in OMVs: An emerging class of therapeutic targets in diseases. Frontiers in Cellular and Infection Microbiology. 2024;**13**:1305510

Chapter 11

# Cell Protection by Oxidative Stress Mitigation Using Substances with Bioactive Properties

*Miroslava Požgajová, Lucia Klongová, Marek Kovár and Alica Navrátilová*


## Abstract

Stress caused by challenging environmental conditions is often associated with the rapid production of reactive oxygen species (ROS) that dramatically alter cellular redox homeostasis. ROS generation is tightly connected with its utilization, as impaired equilibrium of these processes results in oxidative stress having profound consequences for cell physiology. The balanced action of both antioxidant enzymes and non-enzymatic antioxidants counterbalances the harmful effects of ROS. Despite the functional antioxidant system of the cell, excessive ROS leads to disruption of vital cellular processes which is associated with the development of various lifestyle diseases, mainly cardiovascular diseases, and cancer. Enhanced consumption of foodstuffs with high and balanced antioxidant bioactive compounds is linked with the positive effects of beneficial antioxidants impacting cellular protection resulting in health promotion. In this chapter, we present an outline of how bioactive compounds determine their health-promoting properties.


**Keywords:** oxidative stress, cellular protection, ROS, antioxidant defense, bioactive substances

## 1. Introduction

### 1.1 ROS formation and utilization

Under physiological conditions, over 90% of oxygen ($O_2$) the living organism receives is utilized in the electron transport chain as a terminal acceptor of electrons. Human beings require oxygen for energy production through the oxidative metabolism of the mitochondria. They generate adenosine triphosphate (ATP), from which the energy necessary for almost all vital processes is released. Remaining oxygen molecules undergo partial one-electron reduction through stepwise addition of electrons resulting in the formation of oxygen products called reactive oxygen species (ROS). The superoxide anion is formed by the univalent reduction of triplet-state molecular oxygen ($^3O_2$). The stepwise molecular oxygen reduction has the following progress [1]:

$$O_2 + H^+ + e \rightarrow HO_2^{\bullet} \quad (1)$$

$$HO_2^{\bullet} + H^+ + e \rightarrow H_2O_2 \quad (2)$$

$$H_2O_2 + H + e \rightarrow HO^{\bullet} + H_2O \quad (3)$$

$$HO^{\bullet} + H^+ + e \rightarrow H_2O \quad (4)$$

In general, ROS are formed as by-products of aerobic metabolism, however, several other sources are also involved. Cellular ROS levels can increase as a consequence of UV irradiation, ionizing radiation, exposure to toxins such as heavy metals, due to chemotherapy treatment, or neighboring inflammatory cells. In a healthy organism, ROS, which include superoxide radical ($O_2^{\bullet-}$), singlet oxygen ($^1O_2$), $H_2O_2$, perhydroxyl radical $O_2H^{\bullet}$, or hydroxyl radical ($OH^{\bullet}$), regulate various physiological processes such as differentiation, stress signaling, long-distance signaling, redox levels, immune response, cell death, etc. [2].

### 1.2 Oxidative stress

Formation of ROS as a deteriorative by-product of aerobic metabolism, can occur in any organelle of the cell consisting of high redox-potential molecules capable to excite the atmospheric oxygen. Transition metal ions such as iron (Fe), that reacts with $H_2O_2$ in the Fenton reaction, are involved in the conversion of oxygen intermediates into highly reactive forms (ROS or free radicals).

Consequently, due to the high reactivity of ROS, their levels are maintained under strict control to prevent unintentional oxidation of biological macromolecules. Formed ROS, that are not utilized for physiological processes, are removed, or detoxified by pathways of enzymatic defense systems or compounds with antioxidant properties to control ROS signaling reactions. The imbalance between ROS generation and their sequestration/utilization, with certain consequences for cell physiology, is called oxidative stress [3]. Since stress-free conditions are nearly unreachable, living organisms have to develop mechanisms to resist stress in order to survive. Even though the majority of chronic diseases of the modern age are multifactorial in origin, they share one common factor, oxidative stress. Oxidative stress, resulting in the oxidation of lipids, nucleic acids, or proteins, has been suggested to contribute to several pathologies like cancer, cardiovascular diseases, diabetes, neurological disorders (Alzheimer's, Parkinson's, and Huntington's diseases and amyotrophic lateral sclerosis), genetic disorders (such as Down or Klinefelter syndrome), psychiatric diseases (including depression, schizophrenia, and bipolar disorder), renal or lung diseases, and aging.

### 1.3 Antioxidants

The protective activity of molecules with antioxidant properties to reduce oxidative stress and thereby prevent negative consequences of its action, is ensured either by antioxidant enzymes or small molecular weight antioxidants, or by the combination of both [4, 5]. The most prominent and well characterized molecules with antioxidant properties are depicted in **Table 1**.

Antioxidants inhibit or delay the oxidation of a substrate. The protective mechanism of antioxidants is classified in accordance with the general criteria as enzymatic

| Antioxidant | Abbreviation | Subcellular localization |
|---|---|---|
| *Enzymatic system* | | |
| Superoxide dismutase | Cu/Zn-SOD | Cytosol, plastid, peroxisome |
| | Mn-SOD | Mitochondrion |
| | Fe-SOD | Plastid |
| Catalase | CAT | Peroxisome |
| Ascorbate peroxidase | APX | Cytosol, mitochondrion, plastid, peroxisome |
| Dehydroascorbate reductase | DHAR | Cytosol |
| Glutathione reductase | GR | Cytosol, mitochondrion, plastid |
| Glutathione peroxidase | GPX | Cytosol, mitochondrion, plastid |
| Monodehydroascorbate reductase | MDAR | Cytosol, mitochondrion, plastid |
| Peroxiredoxin | Prx | Chloroplast, mitochondrion |
| *Non-enzymatic system* | | |
| Ascorbate | AsA | Apoplast, cytosol, mitochondrion, plastid peroxisome, nucleus, vacuole |
| β-Carotene | CAR | Plastid |
| Glutathione reduced | GSH | Cytosol, mitochondrion, plastid, nucleus |
| Polyamines | | Cytosol, mitochondrion, plastid, nucleus |
| α-Tocopherol | | Membranes |
| Zeaxanthin | | Plastid |

**Table 1.**
*Subcellular localization of enzymatic and non-enzymatic antioxidant systems.*

or non-enzymatic, preventative or system-repairing, endogenous or exogenous, primary and secondary, hydrosoluble and liposoluble, and natural and synthetic. Antioxidant molecules capable of sequestering free radicals by donation of hydrogen (chain breakers) are called primary antioxidants. Secondary antioxidants belong to compounds capable of singlet oxygen quenching, peroxide decomposition, chelating of metal, inhibiting oxidative enzymes, or absorbing UV radiation.

Trace elements Cu, Se, Mn, and Zn represent the essential ions in the structure of antioxidant enzymes in the cytosol, mitochondria, or in plasma: Cu/Zn superoxide dismutase (SOD), Mn-SOD, catalase (Cu, Fe), and glutathione peroxidases (GPX:Se). The antioxidant processes are initiated by SODs that transform the superoxide anion into hydrogen peroxide, followed by catalase and GPX-mediated neutralization of the various peroxides at both intra- and extra-cellular levels [6].

In addition, the harmful effects of oxidative stress can be reduced via the ingestion of dietary antioxidants present in various foods. Nutritional antioxidants vary in their composition and act through various mechanisms and in different compartments, however, the predominant type of action is free radicals scavenging. This is realized either by the direct neutralization of ROS, reduction of the peroxide levels, and reconstruction of oxidized membranes, or by iron quenching to limit ROS formation, or via lipid metabolism, as short-chain free fatty acids and cholesteryl esters neutralize ROS [5].

## 2. Natural antioxidants

Generally, the function of antioxidants is to delay or prevent oxidation of chemical molecules in the cell. Initially, the studies on the antioxidant molecules have been aimed at investigating their ability to prevent the degradation of unsaturated fatty acids. The main source of natural substances with health-promoting properties is plants. Although the use of plants and their metabolites to promote health dates back to ancient times, nowadays they still belong to the constitutive and inseparable part of medicine and pharmacology. Most of the compounds with bioactive properties do not play an essential role in plant growth or development. Rather, they are important for organism protection against pathogens or toxic elements. Alternatively, they serve as signaling substances for various physiological processes. The synthesis of substances with bioactive properties is not common for all plant species. Mostly it is limited to a genus or, in some cases, even to individuals. In general, these natural compounds are not considered as stand-alone drugs. They are usually used as supportive substances that contribute to the reduction of side effects of the drug treatment or enhance the curative effect of the pharmaceutical. Natural compounds constitute a base for the design and development of new pharmaceuticals (**Figure 1**) [7, 8].

### 2.1 Natural antioxidants of plant origin

Plants as an invaluable source of natural substances with versatile medicinal properties have been used to treat various diseases and discussed for centuries. Multidirectional activities exhibited by natural substances and their ability to influence key signaling pathways in the cell, mainly related to cancer and other civilization

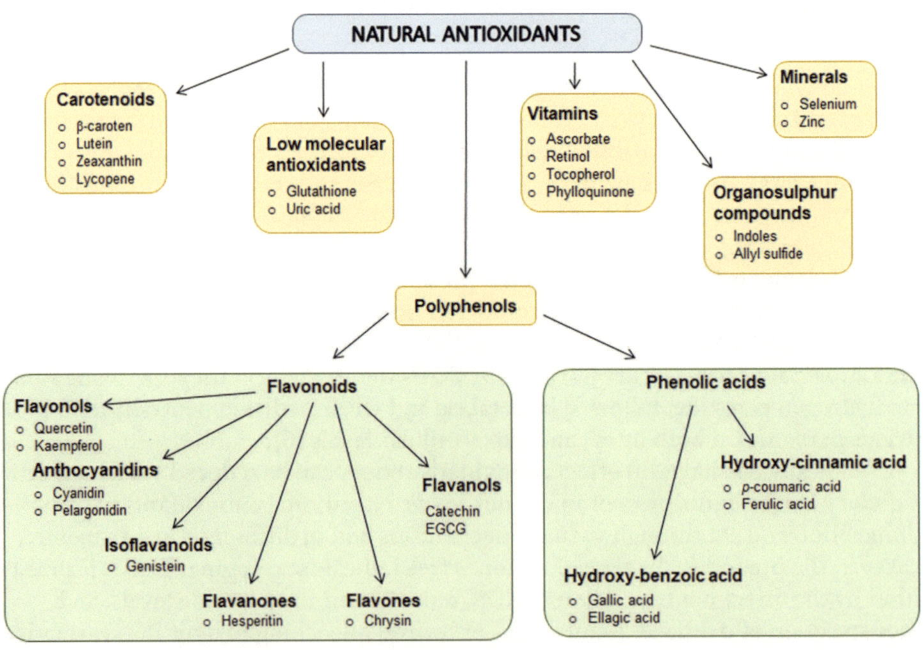


**Figure 1.**
*Schematic overview of the major natural antioxidants.*

diseases, drive these substances in an important research direction. Over the last decades, research has advanced the discovery and identification of plant components with beneficial properties and revealed their underlying mechanisms [7, 9].

### *2.1.1 Phenolic compounds*

The chemical composition of a specific group of secondary metabolites, phenolic compounds, comprises one or more aromatic rings triggering one or more hydroxyl groups capable of stabilizing and relocating unpaired electrons. In general, phenolic compounds are a heterogeneous group of phytocompounds that can be categorized as phenolic acids, flavonoids, coumarins, and tannins. The function of phenolic acids in plants has been associated with various physiological processes that include nutrient uptake, protein synthesis, enzyme activity, photosynthesis, or allelopathy. They exist in virtually all plants and are dispersed throughout the plant in seeds, leaves, roots, flowers, and stems [10, 11]. Except for their significant role in plant physiology, phenolic compounds included in the human diet have been connected to a variety of benefits associated with reduced risk of chronic diseases acting as an anti-allergenic, anti-atherogenic, anti-inflammatory, or antimicrobial medium. Most of the beneficial effects of phenolic compounds result from their antioxidant impact as they are capable to neutralize free radicals (i.e., superoxide anion, hydroxyl- radicals, and peroxy-radicals) thereby protecting a person from degenerative diseases such as cardiovascular disease, diabetes mellitus, neurodegenerative diseases, and cancer. They represent the largest known group of natural antioxidants [12–14]. The antioxidant activity of phenolic compounds is often determined by their ability to protect LDL lipids from oxidation. Several studies on various individuals of all ages under different health conditions such as overweight, obesity, metabolic syndrome, heavy smokers, patients with peripheral vascular disease, or subjects at high cardiovascular risk have been involved in clinical trials that involve the consumption of extracts obtained from various plant substances (e.g., different types of berries, grapes, olives, citrus fruits, various nuts, coffee, tea, or chocolate) the same way as particular phenolic compounds (e.g., gallic acid (GA), resveratrol, luteolinidin, coumaric acid, apigenidin, rosmarinic acid, caffeic acid, catechin, quercetin, ellagic acid, and rutin). These studies have been performed to confirm the strong antioxidant effect of dietary phenolics consumption at retarding LDL oxidation *in vivo* and thus promoting individuals' health [12]. The antioxidant activity of phenolic compounds that are beneficial to health is mainly achieved through (I.) collapse of oxidative chain reactions resulting in ROS elimination; (II.) pro-oxidants abolishment; (III.) recruitment of other antioxidants; (IV.) modification of enzymatic activity; and (V.) modulation of signal transduction pathways [15].

### *2.1.2 Phenolic acids*

Phenolic acids, a subclass of plant phenolics, possess a structure capable of the H-atom donation resulting in antioxidant activity through radical scavenging mechanism, radical quenching *via* electron donation, and singlet oxygen quenching. Phenolic acids have become the center of attention in the field of medicinal chemistry due to their significant health-promoting properties that include antimicrobial, anti-cancer, antiviral, anti-ulcer, anti-inflammatory, and anti-diabetic effects. Phenolic acids can be subcategorized to: hydroxybenzoic acids (e.g., pyrocatechuic acid, gallic acid (GA), gentisic acid, p-hydroxybenzoic acid, protocatechuic acid, or salicylic acid) and hydroxycinnamic acids (e.g., *p*-coumaric acid, ferulic acid, caffeic acid, isoferulic acid, rosmarinic acid, or chlorogenic acid) [16, 17].

Gallic acid (GA) is a naturally occurring organic compound that belongs to the family of phenolic acids. Its derivatives are widely found in fruits and plants. The esters of gallic acid possess a wide range of industrial applications, including their use as antioxidants in food, cosmetics, and pharmaceuticals. It has been demonstrated that gallic acid can mediate various health-promoting activities involved in anti-inflammation, anti-obesity, and anti-cancer effects. More recently, the anti-cancer activities of GA have been described to act via several biological pathways that include induction of programmed cell death, reticence of vasculature, tumor migration, metastasis, inflammation, cell cycle arrest, and oncogene expression. The activation of the NF-κB and Akt signaling pathways, as well as the activities of COX, ribonucleotide reductase, and glutathione reduced (GSH), are all inhibited by gallic acid, which has an impact on biochemical signaling pathways. Additionally, gallic acid inhibits the processes of carcinogenesis through activating ATM kinase signaling pathways. Furthermore, gallic acid is found to show synergism with other existing chemotherapeutic drugs [18–22].

Syringic acid is a phenolic molecule produced by plants through the shikimic acid pathway. It is commonly present in plant-based materials such as fruits, vegetables, and medicinal plants. It has been demonstrated by many studies that it can be used in a wide range of therapeutic applications in the prevention of diabetes, cardiovascular diseases, cancer, and cerebral ischemia due to its potent antioxidant, antimicrobial, anti-inflammatory, antiendotoxic, chemoprotective, anti-angiogenic, anti-glycating, anti-proliferative, anti-hyperglycaemic, neuro, and hepatoprotective activities. The therapeutic potential of syringing acid is largely attributed to the presence of methoxy groups onto the aromatic ring operating as an effective free radical scavenger leading to the oxidative stress alleviation [23]. In addition, syringic acid alters protein dynamics, the activity of enzymes, and a variety of transcription factors associated with angiogenesis, diabetes, inflammation, and cancer. In recent studies, it has been shown that treatment with syringic acid may help neurological dysfunction or behavioral impairments management [24]. Rats fed with syringic acid-enriched food displayed protective effects on diabetic cardiomyopathy by reducing lipid peroxidation and protein carbonylation related to the antioxidant activity of this phenolic acid. Furthermore, syringic acid administration has been linked to the amelioration of cardiac ischemia-reperfusion injury in the rat model, by blocking mitochondria-induced apoptosis that is controlled by the PI3K/Akt/GSK-3β signaling pathway [25, 26]. It is noteworthy that syringic acid effectively suppressed the frequency of UVB-induced skin tumors in mice by inhibiting the phosphorylation of the epidermal growth factor receptor (EGFR), Akt signaling pathways, and mitogen-activated protein kinases. Moreover, administration of syringic acid inhibited nicotinamide adenine dinucleotide phosphate (NADPH) oxidase activity resulting in intracellular ROS reduction and inhibited protein-tyrosine phosphatase-κ activity, a regulator of EGFR activation. Based on these findings, syringic acid may be a useful chemopreventive and therapeutic agent since it inhibits the Nox/PTP-κ/EGFR axis which is the primary mechanism by which it exerts strong chemopreventive activity in skin carcinogenesis [27].

Vanillic acid (VA) is a bioactive compound derived from natural sources such as vegetables, fruits, herbs, whole grains, juices, beers, and wines. It possesses various pharmacological activities such as antioxidants, and anti-proliferative, and it has been shown to block pro-inflammatory cytokines and suppress inflammatory cascades.

Due to its phenolic nature, vanillic acid exerts strong inhibitory effects on variable signal transduction pathways that include the Janus kinase (JAK)/signal transducer and activator of transcription (STAT), adenosine monophosphate-activated protein kinase (AMPK), Nod-like receptor family protein (NLRP), nuclear factor kappa B (NF-κB). Target of rapamycin (mTOR), and mitogen-activated signaling proteins (MAPK). These actions display diverse therapeutic potential, including cardioprotective and liver protective. Vanillic acid scavenges free radicals, boosts antioxidant status, lowers levels of lipid peroxidation, and improves mitochondrial function as a result of decreased ROS generation. As a result, it lowers cardiac dysfunction, especially that caused by cardiotoxicity (CTX) and myocardial ischaemia–reperfusion injury (MIRI) [28, 29]. Additionally, vanillic acid has been found to possess protective effects on the liver by reducing serum levels of transaminases, inflammatory cytokines, and collagen accumulation, thereby preventing liver fibrosis. Furthermore, practically every metabolic abnormality associated with non-alcoholic fatty acid disease (NAFLD), including hepatic steatosis, inflammation, and hepatic damage has been reported to be targeted by VA [30]. In addition, vanillic acid has been found to have a beneficial effect on diabetic nephropathic rats. This effect may be linked to its potent ability to sequester free radicals, as well as its ability to down-regulate NF-κB, TNF-α, and COX-2, while also up-regulating Nrf-2 proteins in renal tissue [31]. Moreover, through the reduction of ROS formation, stabilization of mitochondrial membrane potential, restriction of mPTP opening, regulation of caspase-3 activity, and eventually suppression of cardiomyocyte apoptosis, VA protects H9c2 cells against H/R-induced damage [32]. Interestingly, VA treatment has been associated with longevity due to its ability to increase thermotolerance, reduce protein aggregation, improve motility, and extend the life span by almost 50%, an extent that is hardly achieved with a natural compound. This reveals the therapeutical properties of VA as a novel geroprotector [33].

Protocatechuic acid (PCA) is one of the main metabolites of complex polyphenols such as anthocyanins, proanthocyanins, and procyanidins which widely exist in our daily diet as they are normally found in high concentrations in vegetables, fruits, and herbs. A growing body of evidence supports the concept that PCA can exert a variety of biological antioxidant, anti-inflammatory, antibacterial, antiviral as well as anti-hyperglycemic, neuroprotective antiosteoporotic, analgesia, and anti-aging effects by acting on different molecular targets. Moreover, as PCA suppresses *in vitro* chemical carcinogenesis and has pro-apoptotic and anti-proliferative effects on a variety of tissues. It appears to possess chemopreventive potential [34, 35]. Additionally, PCA has been shown to promote protection from metabolic syndrome; and preserve liver, kidney, and reproductive functions. As a result of PCA's ability to prevent oxidative stress via reduction of ROS generation, it is capable to attenuate cardiac hypertrophy as evidenced by the decreased heart-weight-to-body-weight ratio and the expression of Nppb. Thus, protocatechuic acid is supposed to prevent heart failure and is therefore suggested as a potential novel therapeutic agent against heart failure [36]. Moreover, PCA serves as a potential therapeutic to treat influenza as PCA administration to mice infected with mouse-adapted influenza strain A/Font Monmouth stopped mice from losing body weight and dying from an H1N1 infection. PCA modulated H1N1-induced TLR4/NF-κB activation and decreased lung index, virus titer, immune cells infiltration, and cytokine levels in the lung [37].

A phenolic plant hormone, salicylic acid (SA), is essential to defensive signaling, controlling growth, development, blooming, photosynthesis, and antioxidant enzymes in response to environmental changes. Challenging conditions of the environment caused by biotic and abiotic stress factors can mediate ROS overaccumulation that in turn results in oxidative stress. Therefore, a proper balance is essential to maintain redox homeostasis in plants, and salicylic acid is known as a chief regulator of ROS. The interplay of SA and ROS requires finetuning as, under some environmental circumstances, SA functions as an antioxidant (ROS scavenging) and pro-oxidant (ROS buildup), playing a dual role. Modulation of ROS metabolism by SA is tightly connected with its ability to directly bind both main $H_2O_2$-detoxifying enzymes the CAT and ascorbate peroxidase (APX) and thereby inhibit their activities. Similarly, pre-treatment with SA results in catalase 2 (CAT2) and glutathione S-transferases (GSTs) inhibition, the key players in plant defenses against pathogens. In addition, exogenously applied SA has been reported to affect antioxidant enzymes under different heavy metal stresses, like copper (Cu), lead (Pb), cadmium (Cd), and nickel (Ni) resulting in reduced lipid peroxidation that in turn leads to increased growth and photosynthesis. Furthermore, by regulating antioxidant enzymes such as GST, GPX, GR, MDHAR, DHAR, and GS as well as by boosting Assada and Halliwell to preserve ROS homeostasis, SA can reduce salt-derived oxidative stress and thereby increase biomass, chlorophyll content, and photosynthesis. Moreover, systemic acquired resistance (SAR) and the hypersensitive response (HR) are established by SA. This is supported by a strong and intricate network of SA that includes cascades of mitogen-activated protein kinases (MAPK), ROS, calcium ions ($Ca^{2+}$), nitric oxide (NO), and non-expressor of pathogenesis related protein-1 (NPR1) [38–41].

Ferulic and caffeic acid. Ferulic acid is highly abundant in a variety of foods of plant origin such as cereals, fruits including apples, oranges, and bananas, and vegetables including potatoes, tomatoes, and cabbage. Due to its antioxidant properties, ferulic acid protects macromolecules such as DNA and lipids from oxidative damage, which makes it a potential candidate for anti-cancer, anti-diabetic, anti-aging, and anti-inflammatory treatments. Studies on rats revealed the nephroprotective and antioxidant effects of ferulic acid and its treatment in diabetic rats inhibited peroxidation of lipids. Caffeic acid, abundantly found in a variety of mushrooms, tea, and coffee, has been shown to possess antiviral, anti-oxidant, anti-cancer, and anti-inflammatory properties. In addition, caffeic acid treatment due to its antioxidant activity, by direct free radical scavenging, protects against ischemia-reperfusion injury and shows hepatoprotective ability against oxytetracycline (OXT)-induced toxicity, anti-lipid peroxidation activity, and cardioprotective activity. Rats concomitantly treated with ferulic and caffeic acid had increased antioxidant activity and increased excretion of neutral and acidic sterols ($p < 0.05$). This resulted in lower hepatic and plasma cholesterol levels as well as increased excretion of fecal sterols and decreased absorption of dietary cholesterol. Furthermore, the iron chelating capacity of ferulic acid and caffeic acid in the brain tissues of iron-overloaded mice has been evaluated and revealed that such treatment significantly decreased iron content in both brain and serum samples. Moreover, the iron-induced decrease of the catalase activity was restored to that of the control group upon ferulic and caffeic acid administration. Thus, ferulic acid and caffeic acid might be possible natural iron chelators for brain iron overload therapy. Additionally, mice treatment with caffeic acid and ferulic acid attenuated the effects of iron overload on the rat serum biochemical parameters [42–44]. Strikingly, ferulic and caffeic acid supplementation have shown positive effects toward prevention of multiple aspects of the metabolic syndrome and liver steatosis in an obese mouse model [45], multifunctional properties against Alzheimer's disease (AD) [46], and the anti-cancer activity has been demonstrated by their ability to mediate cytotoxicity toward human breast cancer cell lines that in the future could be profitably applied for chemopreventive and/or chemotherapeutic purposes [47].

*P*-coumaric acid (*p*-CA) is found ubiquitously in plants and mushrooms either in the free form at low concentrations or in conjugated form at high concentrations. Fruits (such as grapes, apples, oranges, pears, and berries), vegetables (such as potatoes, onion, or beans), and cereals (such as wheat, maize, or oats) are rich sources of free or bound *p*-CA, and is thus often ingested in the regular diet of humans and animals. It has been demonstrated that particularly *p*-CA conjugates, exhibits a range of bioactivities, including the ability to mitigate oxidative cardiac damage, atherosclerosis, neuronal injury, gout, diabetes, anxiety, etc., as well as antimutagenic, anti-cancer, antioxidant, anti-ulcer, anti-inflammatory, and antiplatelet properties. The neuroprotective activity of *p*-CA can be connected to its potent antioxidant and anti-apoptotic abilities. The substance increased the synthesis of SOD, CAT, and GSH, while decreasing the levels of lipid peroxidation (LPO), and nitric oxide (NO). Its antioxidant activity is linked to the phenyl hydroxyl group in *p*-CA. However, except for hydroxyl radical, *p*-CA displays a much lower scavenging capacity for 2,2-diphenyl-1-picrylhydrazylhydrate (DPPH) superoxide anion radicals and $H_2O_2$ compared to its analogs (e.g., ferulic acid). An increase in the antioxidant activity of *p*-CA can be achieved upon conjugation with quinic acid, monosaccharides, and amine. Generally, bioavailability and the pathways via which *p*-CA metabolized change according to its chemical structure as it can be conjugated with other molecules such as organic acids, alcohols, mono- or oligosaccharides, lignin, or amine. Concerning its bioavailability, *p*-CA shows higher bioavailability than chlorogenic acid, rosmarinic acid, caffeic acid, and ferulic acid. The solubility of bound *p*-CA is affected by the molecules to which it is esterified. Water-soluble conjugates of *p*-CA are formed upon its binding to oligosaccharides, glycosides, amines, and alcohol, while *p*-CA esterified to long-chain alkyl alcohol, polysaccharides, and lignin, or etherified to lignin, produce water-insoluble conjugates. Absorption of *p*-CA by the gastrointestinal tract is much easier and quicker in its free form compared to its conjugates [48, 49].

*P*-coumaric acids' effect on increased resistance to high osmolarity has been demonstrated in *caenorhabditis elegans* as its administration led to a significant prolongation of the worms' survival under oxidative and osmosis stress. *p*-CA induced reduction in the intercellular ROS levels and increased oxidative resistance which was mediated by skn-1, an ortholog of the Nrf2 (nuclear factor erythroid 2-related factor 2) transcriptional factor. Downregulation of the osmosis regulatory gene, osr-1, might contribute to the increased tolerance to osmotic stress [50]. The protective effect of pCA against cisplatin (CDDP)-induced oxidative stress in ovarian tissue and diabetes-associated nephropathy in rats can be associated with its antioxidant properties as it improved changes in lipid peroxidation and activities of antioxidant enzymes: catalase, glutathione-S-transferase, and superoxide dismutase. It may thus be worthy to consider the usefulness of pCA as adjuvant therapy in cancer management and in the prevention of diabetic nephropathy [51, 52].

#### *2.1.3 Flavonoids*

Flavonoids belong to a class of naturally occurring polyphenolic compounds comprising the most common group of plant polyphenols. They are responsible for different colors of leaves, fruit, and flowers and provide much of the flavor to fruits and vegetables. To date, more than 5000 diverse types of flavonoids have been characterized. Six subclasses of flavonoids have been categorized to date that include flavones (e.g., tangeritin, apigenin, chrysin, luteolin, etc.), flavonols (e.g., quercetin, kaempferol, myricetin, etc.), flavanones (e.g., naringenin, eriodictyol, and hesperidin), catechins or flavanols (e.g., catechin, epicatechin, epicatechin gallate, gallocatechin, etc.), anthocyanidins (e.g., cyanidin, delphinidin, malvidin, pelargonidin, etc.), and isoflavones (e.g., glycitein, genistein, and daidzein). Structural variations have been described among the individual classes based on the arrangement and number of hydroxyl groups, and glycosylation or alkylation of these groups. Most of the flavonoids present in plants are attached to sugars (glycosides), although they might be occasionally found as aglycones, too. Ingestion of dietary polyphenolics leads to their digestion, resulting in their appearance in the circulatory system not as the parent compounds, but as phase II metabolites. In the gastrointestinal tract, the parent compounds and their metabolites are degraded giving rise principally to small phenolic acid and aromatic catabolites that are absorbed into the circulatory system [53, 54]. The potent health benefit of polyphenols and their metabolites and catabolites on human chronic diseases exerts from their striking antioxidant and free radical scavenging activities. The most prominent antioxidant mode of action of flavonoids is the direct scavenging of ROS through the breakdown of several free radical chain reactions. In general, flavonoids, due to their highly reactive hydroxyl group, are oxidized by radicals producing flavonoid radicals. This, in turn, results in a generation of more stable, less-reactive radicals. The antioxidant effects of flavonoids may be attributed to their ability to inhibit the chain reaction of lipid oxidation. Moreover, such pronounced antioxidant property of flavonoids leads to the anti-inflammatory, anticarcinogenic, anti-mutagenic, chemoprotective, and antimicrobial effects of those phytochemicals. In addition, flavonoids inhibit the synthesis and activities of pro-inflammatory mediators such as cytokines, eicosanoids, adhesion molecules, and C-reactive protein, exerting anti-inflammatory effects. Molecular activities of flavonoids include inhibiting transcription factors like NF-κB and activating protein-1 (AP-1). They also activate nuclear factor erythroid 2-related factor 2 (Nrf2) [55, 56]. These effects of nutritive and dietary polyphenols thus result in their dual role as cytoprotective and cytotoxic agents. They possess the ability to protect the normal healthy cells and cellular components, while they destroy or are toxic to the premalignant and neoplastic or malignant cells. Flavonoids can be found in a variety of plants. Some examples of beneficial compounds found in various fruits and vegetables include flavanol, flavonoid, and quercetin which are highly abundant in onion, broccoli, and apple. Catechin can be found in high amounts in different teas and fruits, while naringenin is considered the predominant flavanone in grapefruit. Glycitein, genistein, and daidzein are considered as the major isoflavanones in soybean. In addition, cyanogen glycosides are highly present in berries such as raspberry, blackberry, and black currant. It is noteworthy that the consumption of vegetables and fruits rich in antioxidant phytochemicals has proven to increase the antioxidant capacity of serum/plasma, thereby promoting the health-protective properties of those edibles. Altogether, the well characterized health-protection properties of flavonoids predetermine them to be utilized in future experiments or trials as attractive candidates for disease therapy [8, 57–59].

### *2.1.4 Coumarins*

Coumarins are fused six-membered oxygen-containing benzoheterocycles that join two synthetically useful rings: α-pyrone and benzene. Higher plants such as Rutaceae and Umbelliferae contain coumarins; however, coumarins can also be found highly abundand in essential oils such as lavender, cassia leaf, and cinnamon bark oil. Coumarin compounds are bioactive substances possessing remarkable biological properties, that include anti-cancer, antioxidant, antimicrobial, antibacterial, antifungal, anti-HIV, antihypertension, anti-coagulant, antiviral, anti-inflammatory, analgesics, anti-diabetic, or anti-depressive effects. The anti-cancer activity of coumarin and its derivatives has been described to be directed against numerous types of cancers such as prostate, renal, breast, laryngeal, lung, colon, CNS, leukemia, and malignant melanoma [60, 61]. Moreover, coumarins except for the cancer treatment support have the potential to neutralize the side effects associated with radiotherapy. The anti-cancer effect of coumarin-based agents like coumarin-triazole, coumarin-chalcone, coumarin-thiosemicarbazone derivatives, coumarin-metal complexes, furanocoumarin, pyranocoumarin, pyrone-substituted coumarin, and their derivatives have been found more potent than that of simple coumarin. Due to the great potential of coumarin scaffold it is extensively used in drug design and development. This scaffold is frequently used for designing small molecules with various biological activities. Coumarin nucleus possesses six positions available for substituents. The selection of functional groups suitable for anti-cancer activity is facilitated by three of those positions (C-3, C-4, and C-7), while the remaining positions (C-5, C-6, and C-8) are either never studied or are hardly ever used [62, 63]. Medicinal chemists and drug designers pay much attention to both natural and synthetic coumarin-based compounds for the sake of their diverse pharmacological and biological properties. Numerous therapeutic drugs containing coumarin scaffolds have become commercially used pharmacological agents extensively utilized in the clinic, such as Armillarisin (Antibiotic), Novobiocin (Antibiotic), Warfarin (Anti-Coagulant), Phenprocoumon (Anti-coagulant), or hymecromone (choleretic and antispasmodic). In addition, as coumarins can act on numerous receptors or signal transduction pathways that play a role as induction factors of cancer, they can serve as multi-targeted therapy against cancer development and progression which turns out to be more efficient as compared to the single-targeted therapy. Notably, it has been demonstrated that different coumarin subtypes and their derivatives can prevent the growth and spread of innumerable cancer cells by a variety of mechanisms such as cell cycle arrest, induction of apoptosis, carbonic anhydrase inhibition, inhibition of topoisomerase, kinase and Hsp90 inhibition, monocarboxylate transporter inhibition, sulfatase inhibition, inhibition of angiogenesis and DNA synthesis inhibition, inhibition of thioredoxin reductase, microtubule polymerization, ROS generation, regulation of STAT-3 and NF-κB factor, and expression of PI3- and MAP kinases [64].

### *2.1.5 Carotenoids*

Carotenoids, a group of nearly 600 compounds, constitute a ubiquitous group of isoprenoids, lipid-soluble pigments, produced in photoautotrophic plants, that are structurally similar to vitamin A. Provitamin A activity possesses approximately 50 of these compounds. It has been reported vitamin A, carotenoids, and provitamin A carotenoids can be effective antioxidants as they are very efficient in electron transfer reactions acting as physical quenchers of singlet oxygen, scavengers of other ROS, and stabilizers of peroxyl radicals by the hydrophobic chain of polyene units. In general, the longer the polyene chain, the greater the peroxyl radical stabilizing ability. It is quite clear that carotenoids can inhibit the propagation of radical-initiated lipid peroxidation, supporting their definition of antioxidants [65–67]. They interact synergistically with other antioxidants; mixtures of carotenoids are more effective than single compounds. The antioxidant potential of carotenoids is of particular significance to human health as an altered balance of antioxidants-ROS mediates oxidative stress, which is a critical factor in the pathogenesis of various chronic diseases. Regular intake of carotenoids has been associated with risk reduction for several chronic diseases, including the reduced incidence of obesity, type 2 diabetes, certain types of cancer, and even lower total mortality [68, 69]. Data coming from epidemiological studies and clinical trials have confirmed that carotenoids (β-carotene, α-carotene, lycopene, β-cryptoxanthin, zeaxanthin, crocin/crocetin, curcumin, and lutein) present in human serum were able to act as anticarcinogenic agents. Interestingly, carotenoids and retinoids have been shown to exert similar biological activities resulting in the inhibition of malignant tumor growth and the induction of apoptosis. Cancer chemoprevention derived by dietary carotenoids acts through PI3K/AKT/mTOR, and ERK-MAPK pathways and induces apoptotic processes through PPARs, IFNs, and RAR/RXR, which are the p53 intermediates in cancer signaling and play a role in the immune system and cellular differentiation. Moreover, carotenoids can stimulate immune system by supporting the B- and T-lymphocytes proliferation, activating the macrophages and cytotoxic T-cells, promoting the effector T-cell function, and the production of cytokines. Supplementation with carotenoids can thus influence a large number of downstream targets that can affect cell growth and modulate gene expression and immune responses [70–72]. Additionally, human studies have shown that carotenoids can protect the skin against photooxidative damage, and lutein, and zeaxanthin may be protective in eye diseases due to their ability to absorb damaging blue light that enters the eye. Epidemiological studies have demonstrated a positive correlation between an effective carotenoid intake with a breast, cervical, ovarian, colorectal cancers reduction and cardiovascular and eye diseases mitigation, indicating that carotenoids act in a dose- and time-dependent manner. The primary sources of carotenoids include a variety of fruits and vegetables, such for example the major sources of lycopene are tomato and tomato products [73–75]. Interestingly, astaxanthin, a natural carotenoid with superior antioxidant activity compared to other carotenoids, is known to exhibit a plethora of pharmacological effects. The protective effect of astaxanthin has been described on age-related diseases of multiple organs such as neurodegenerative diseases (such as Alzheimer's disease or stroke); bone-related diseases (like osteoarthritis or osteoporosis); cancers; cardiovascular disorders (such as myocardial infarction, hypertension, or atherosclerosis), complications associated with diabetes; eye, gastric, kidney, or liver disorders; and pulmonary, muscle, or skin disorders [76]. The specific structure of carotenoids, especially the conjugated double-bond system, determines many of the fundamental properties of these molecules. Carotenoid's structure affects their incorporation into biological membranes, as well as their interactions with ROS. In addition, the interaction of carotenoids with other co-antioxidants, primarily vitamins E and C significantly enhances their effectivity as antioxidants. However, carotenoids may lose their antioxidant effectivity at high concentrations or at high partial pressures of oxygen. Thus, the regulatory mechanisms of antioxidant and anti-inflammatory properties among carotenoids should be taken into consideration during the development of more effective therapeutic strategies to prevent chronic diseases [75, 77].

#### *2.1.6 Vitamins and minerals*

Micronutrients such as vitamins and minerals must meet their required daily intake as they act as immunomodulators and protect the host immune response and are thus essential for a healthy life. Vitamins are organic compounds, of which vitamin B1 was identified first, approximately 110 years ago, as an anti-beriberi factor. Thirteen groups of vitamins have been identified, nine water-soluble and four fat-soluble. Vitamin insufficiency is characterized by a specific deficiency symptom, such as peripheral neuropathy, dermatitis, blindness, rickets, and bleeding. The primary role of vitamins is to function as cofactors for functional proteins, such as enzymes and transcription factors. The importance of micronutrients is tightly connected to their involvement in various intracellular biological processes, biochemical signaling pathways, and synthesis of macromolecules, such as collagens, or small molecules that are necessary to maintain homeostasis (serotonin, adrenaline). Deficiencies in plasma concentrations of vitamins and minerals can lead to altered immune system function, contributing to unfavorable immunological conditions. The most prominent micronutrients are vitamins A, B, C, D, and E, riboflavin, β-carotene, selenium, zinc, and iron [78, 79]. Although beneficial effects of vitamins and minerals due to their rich bioactive properties are undoubtable, their over-supplementation may be associated with no protective effect or even harm [80–86]. However, numerous studies show health-promoting activities of vitamins and minerals exerting anti-cancer, anti-aging, immune-promoting effects, stroke prevention, preventing various medical conditions such as sepsis, ischemia, lifestyle-related diseases, etc. [78, 79, 87–91].

## 2.2 Natural antioxidants of animal origin

Meat and meat products are rich not only in essential nutrients but also possess a high nutritional value as they are rich in bioactive components that are either not found in other groups of food products or are found only in limited amounts. These compounds include primarily taurine, l-carnitine, choline, alpha-lipoic acid, conjugated linoleic acid, glutathione, creatine, coenzyme Q10, and bioactive peptides. Their antioxidant and health-promoting properties connected with their lipid-lowering, antihypertensive, anti-inflammatory, and immunomodulatory activities have been described. Bioactive substances (di- and tripeptides such as carnosine, anserine, glutathione, and ophidine) found in meat, poultry, and fish also exhibit a broad spectrum of health-promoting effects. They have been found to significantly improve lipid parameters, contribute to arterial blood pressure decrease, through inhibition of pro-inflammatory cytokines production reduce inflammatory conditions, promote detoxification, exhibit antibacterial and antimicrobial properties, reduce fat mass, and contribute to the organism protection from oxidative stress. Moreover, they participate in carbohydrate and lipid metabolism and are involved in energy generation in mitochondria [92].

Poultry eggs are considered one of the best sources of bioactive peptides producing pepsin, trypsin, chymotrypsin, and proteases produced peptides with antioxidant activities. Among the egg white proteins, ovalbumin, ovotransferrin, lysozyme, ovomucin, and ovomucoid are used to produce peptides with antioxidant activity. Structurally similar carnosine, anserine, and ophidine mediate free radicals scavenging as they inhibit lipid oxidation catalyzed by ionic iron, hydrogen peroxide-activated hemoglobin, or singlet oxygen. Glutathione is known to act as an antioxidant by neutralizing ROS and preventing oxidant-mediated cell death. However, these health-promoting activities depend on the dose of the bioactive compounds, and their amount present in meat depends on breed, age, gender, and breeding program [93, 94].

l-Carnosine (beta-alanyl-l-histidine) is a water-soluble endogenous dipeptide composed of β-alanine and l-histidine requiring carnosine synthetase and ATP for its biosynthesis. It can be found in the brain, kidneys, and skeletal muscles of fish, birds, and mammals. Studies have demonstrated that l-carnosine is capable of sequestering reactive oxygen and nitrogen species, thus exhibiting considerable antioxidant properties. Moreover, due to its ability to form complexes with metal ions (such as copper, iron, zinc, or cobalt), l-carnosine protects cells from injury. In addition, l-carnosine in combination with α-tocopherol exhibits a synergistic action [95].

l-Carnitine, a water-soluble quaternary amine, is synthesized in mammalian kidneys and the liver from lysine and methionine in the presence of vitamin C, iron, vitamin B6, and niacin. It is a conditionally essential nutrient for humans and animals as approximately only 25% of carnitine intake is synthesized in the body while the remaining 75% must be supplied with the diet. l-Carnitine plays a fundamental role in the generation of energy as it is involved in the activated long-chain fatty acids transport from the cytoplasm across the inner mitochondrial membrane, thus stimulating β-oxidation which plays an important role in human physiology. l-Carnitine through its antioxidant and anti-inflammatory properties plays a major part in protecting cellular membranes, preventing fatty acid accumulation and elimination of toxic metabolites [96].

Coenzyme Q10 also referred to as ubiquinone, is a fat-soluble vitamin-like compound. It can be found throughout the body, with its highest amount detected in muscles, the spleen, pancreas, heart, liver, kidneys, and the brain. The presence of vitamins B2, B6, B12, folic, and pantothenic acids lead to enhanced coenzyme production. The importance of ubiquinone is connected with the production of high-energy ATP molecules. Variable harmful conditions are associated with coenzyme Q10 deficiency, such as osteoporosis, fibromyalgia, cardiovascular diseases, neurodegenerative diseases, diabetes, periodontitis, nephropathies, or male infertility. Due to its antioxidant properties, it protects DNA against oxidative damage, while it also exhibits the capacity to regenerate other antioxidants, such as tocopherol or ascorbic acid, and prevents membrane lipids from peroxidation. It has been also suggested that ubiquinol (the reduced form of ubiquinone) participates in scavenging free radicals formed as a result of the metabolism of certain xenobiotics (e.g., anthracycline antibiotics) [97].

## 3. Synthetic antioxidants

One of the major disadvantage of using natural antioxidants as food additives is their insufficiency to cover the needs of the food industry due to extraction complications and limited stability [98]. The reason of increasing need of antioxidant supplementation in food arises from the oxidative damage caused by free radicals that lead to food spoilage, causing an unpleasant odor, loss of taste, and damaged tissues. Thus, as one of the most important aims of food technology is safety and taste, color, texture protection, appearance, nutritional value, and shelf life, synthetic antioxidants represent an advantageous alternative to natural antioxidants. Synthetic antioxidants cover a diverse range of raw materials, providing flexibility in their production, are cost-effective with minimal possible side effects [99]. Well-known and commonly used in the food industry are synthetic phenolic antioxidants such as butylated hydroxyanisole (BHA), butylated hydroxytoluene (BHT), tertiary butylhydroquinone (TBHQ), or propyl gallate (PG) [100]. The role of synthetic antioxidants in food stability is very important. They interact with peroxides generated by food during storage, and thereby help to protect food from spoilage under specific conditions, contributing to improved quality, stability and extended shelf life. When used correctly, they provide consumers with safer food and make economic aspects more efficient for producers [101]. However, the use of antioxidants in foods must be limited by the regulatory laws because consumption of synthetic antioxidants has to be controlled due to their toxic effects and carcinogenicity. Moreover, the inappropriate use of dietary supplements may result in antioxidative stress. Overdoses of antioxidant supplements cause negative effects as both, antioxidative and oxidative stresses lead to the imbalance of the antioxidant status of the organism causing damaging alterations to the cells that might result in carcinogenesis and aging [102, 103].

Antioxidant activity of butylated hydroxytoluene (BHT) is achieved by various mechanisms leading to the ROS generation suppression or to the free radicals scavenging. BHT is capable to scavenge reactive species that induce peroxidation, as it possesses metal chelating activity, particularly iron ions that catalyze oxidative processes leading to the release of hydroxyl radicals which in turn chelate hydroperoxides by Fenton reactions. Moreover, BHT scavenges $O_2^{\bullet-}$ mediated peroxides formation, assists in hindering the chain reactions of autoxidation, and reduces the local levels of $O_2$. By the EU law, BHT (E 321) is permissible to be administered in fats and oils, both alone or in combination with other antioxidants, at maximum levels of up to 100 mg/kg expressed as fat [104].

Butylated hydroxyanisole (BHA) is one of the phenolic antioxidants synthetized from the isomers 3-tert-butyl-4-hydroxyanisole (3-BHA) and 2-tert-butyl-4-hydroxyanisole (2-BHA) and can be used as food additive to prevent oxidation. Moreover, analyses performed with BHA have revealed that it can protect animals from chemically induced toxicity. In contrast to its beneficial effects, BHA has been shown to exhibit cytotoxic and carcinogenic effects. However, it has been demonstrated on a variety model system that BHA is not mutagenic in Salmonella with or without S9 metabolic activation, nor is BHA mutagenic in mammalian test systems in rat liver cells or Chinese hamster ovary cells. BHA belongs to many synthetic antioxidants widely used in different products including food, food packaging, pharmaceuticals, and cosmetics. Thus, due to its use as an antioxidant, exposure of BHA to human occurs not only from food but also in pharmaceuticals and cosmetic products [101, 105].

The maximum estimated daily intake of BHA from personal care products in Canada is 0.06–0.07 mg/kg/day [106].

Tertiary butylhydroquinone (TBHQ) is a synthetic antioxidant commonly used to enhance the effect of insulin *in vivo*. TBHQ noticeably enhances PI3-kinase-dependent Akt phosphorylation (p-Akt), which reflects insulin sensitivity in hepatocytes and nerve cells. Moreover, TBHQ mediated Nrf2 activation leads to upregulation of the expression of hemeoxygenase 1 (NO-1) and superoxide dismutase 2 (SOD2) resulting in GDM alleviation in mice. Numerous studies have shown that TBHQ possesses potent antioxidant activity and can exhibit neuroprotective, anti-Parkinson's disease, anticarcinogenic, and liver protective effects. In addition, the antioxidant activity of TBHQ prevents oxidative degradation and rancidity of foods rich in oils and fats upon food supplementation with TBHQ [107, 108].

## 4. Conclusions

Alterations to the redox balance of the organism result in oxidative stress leading to oxidative damage of biologically significant macromolecules that is a predominant cause of the development of chronic and age-related diseases. Substances possessing bioactive properties become the leading choice to treat or prevent such health-disturbing processes. The correlation between human health and the consumption of healthy and functional foods is confirmed by the development of medical sciences, dietetics, and nutrigenomics. With increasing awareness, consumers considerably prefer food with an added value containing either naturally occurring bioactive compounds, or enriched by natural bioactive ingredients, or enriched by synthetic antioxidants. Food fortification with biologically active substances becomes a powerful tool for increasing food quality. Although, due to concerns over the adverse health effects of some synthetic antioxidants, consumption of natural products is largely preferred. The use of plant extracts as antioxidants is becoming increasingly popular and is widely accepted by consumers. As food additives derived from natural raw materials are considered safe, plant extracts are a preferred source of antioxidants. Thus, growing interest in food with added value will prompt scientists to accelerate the research and technology to develop better ways of producing foodstuffs and food supplements containing higher amounts of health-promoting compounds. Enrichment of food products with antioxidants such as polyphenols will beneficially influence not only their oxidative stability but also, due to their enhanced intake to the human organism, they may reduce the incidence of degenerative diseases. Thus, the future research directions in the field of antioxidants and their applications will include characterization of novel substances with bioactive properties of either natural or synthetic origin that have great potential in health maintenance or improvement. In recent times, there has been a significant focus on thoroughly analyzing the combined effect of various types of antioxidants with the belief of synergistic positive action of tested molecules. Moreover, researchers are actively exploring the possibility of developing nano-antioxidants and nanoparticle carriers for antioxidants that can effectively increase their utilization by the body. By doing so, it is hoped that the potential health benefits of antioxidants can be effectively enhanced. In addition, the application of antioxidants into the food and medical healthcare products is also a matter of intensive science. This research area focuses on innovation and improvement of extraction, preservation, and processing technologies to maintain health-promoting properties of antioxidant molecules in the final products. These advancements are aimed to possess significant potential to improve the quality and effectiveness of healthcare products and contribute to better overall health outcomes.

## Acknowledgements

This work was financially supported by the Slovak Research and Development Agency under the contract number APVV-22-0294, the SUA grant agency project 12-GASPU-2021, and the project Vega 1/0583/23.

## Conflict of interest

The authors declare no conflict of interest.

*DOI: http://dx.doi.org/10.5772/intechopen.1004522*

## Author details

Miroslava Požgajová[1]*, Lucia Klongová[1], Marek Kovár[2] and Alica Navrátilová[3]

1 AgroBioTech Research Center, Slovak University of Agriculture in Nitra, Nitra, Slovakia

2 Faculty of Agrobiology and Food Resources, Institute of Plant and Environmental Sciences, Slovak University of Agriculture in Nitra, Nitra, Slovakia

3 Faculty of Agrobiology and Food Resources, Institute of Nutrition and Genomics, Slovak University of Agriculture in Nitra, Nitra, Slovakia

*Address all correspondence to: miroslava.pozgajova@uniag.sk

## References

[1] Pisoschi AM, Pop A. The role of antioxidants in the chemistry of oxidative stress: A review. European Journal of Medicinal Chemistry. 2015;**97**:55-74. DOI: 10.1016/j.ejmech.2015.04.040

[2] Ďúranová H, Šimora V, Ďurišová Ľ, Olexiková L, Kovár M, Požgajová M. Modifications in ultrastructural characteristics and redox status of plants under environmental stress: A review. Plants (Basel). 2023;**12**(8):1666. DOI: 10.3390/plants12081666

[3] Lushchak VI. Free radicals, reactive oxygen species, oxidative stress and its classification. Chemico-Biological Interactions. 2014;**224**:164-175. DOI: 10.1016/j.cbi.2014.10.016

[4] Jomova K, Raptova R, Alomar SY, et al. Reactive oxygen species, toxicity, oxidative stress, and antioxidants: Chronic diseases and aging. Archives of Toxicology. 2023;**97**(10):2499-2574. DOI: 10.1007/s00204-023-03562-9

[5] Berger MM. Can oxidative damage be treated nutritionally? Clinical Nutrition. 2005;**24**(2):172-183. DOI: 10.1016/j.clnu.2004.10.003

[6] Liang J, Gao Y, Feng Z, Zhang B, Na Z, Li D. Reactive oxygen species and ovarian diseases: Antioxidant strategies. Redox Biology. 2023;**62**:102659. DOI: 10.1016/j.redox.2023.102659

[7] Gielecińska A, Kciuk M, Mujwar S, et al. Substances of natural origin in medicine: Plants vs. cancer. Cells. 2023;**12**(7):1-34. DOI: 10.3390/cells12070986

[8] Chaudhary P, Janmeda P, Docea AO, et al. Oxidative stress, free radicals and antioxidants: Potential crosstalk in the pathophysiology of human diseases. Frontiers in Chemistry. 2023;**11**:1158198. DOI: 10.3389/fchem.2023.1158198

[9] Chagas M d SS, Behrens MD, Moragas-Tellis CJ, GXM P, Silva AR, Gonçalves-de-Albuquerque CF. Flavonols and flavones as potential anti-inflammatory, antioxidant, and antibacterial compounds. Oxidative Medicine and Cellular Longevity. 2022;**2022**:e9966750. DOI: 10.1155/2022/9966750

[10] Van Hung P. Phenolic compounds of cereals and their antioxidant capacity. Critical Reviews in Food Science and Nutrition. 2016;**56**(1):25-35. DOI: 10.1080/10408398.2012.708909

[11] Duan L, Zhang C, Zhao Y, Chang Y, Guo L. Comparison of bioactive phenolic compounds and antioxidant activities of different parts of *Taraxacum mongolicum*. Molecules. 2020;**25**(14):3260. DOI: 10.3390/molecules25143260

[12] Amarowicz R, Pegg RB. The potential protective effects of phenolic compounds against low-density lipoprotein oxidation. Current Pharmaceutical Design. 2017;**23**(19):2754-2766. DOI: 10.2174/1381612823666170329142936

[13] Saoudi MM, Bouajila J, Alouani K. Phenolic compounds of *Rumex roseus* L. extracts and their effect as antioxidant and cytotoxic activities. BioMed Research International. 2021;**2021**:e2029507. DOI: 10.1155/2021/2029507

[14] Gonçalves S, Moreira E, Grosso C, Andrade PB, Valentão P, Romano A. Phenolic profile, antioxidant activity and enzyme inhibitory activities of extracts from aromatic plants used in

Mediterranean diet. Journal of Food Science and Technology. 2017;**54**(1):219-227. DOI: 10.1007/s13197-016-2453-z

[15] Lopes M, Sanches-Silva A, Castilho M, Cavaleiro C, Ramos F. Halophytes as source of bioactive phenolic compounds and their potential applications. Critical Reviews in Food Science and Nutrition. 2023;**63**(8):1078-1101. DOI: 10.1080/10408398.2021.1959295

[16] Sehrawat R, Rathee P, Akkol EK, et al. Phenolic acids—Versatile natural moiety with numerous biological applications. Current Topics in Medicinal Chemistry;**22**(18):1472-1484

[17] Kumar N, Goel N. Phenolic acids: Natural versatile molecules with promising therapeutic applications. Biotechnology Reports. 2019;**24**:e00370. DOI: 10.1016/j.btre.2019.e00370

[18] Hassani S, Ghanbari F, Lotfi M, et al. How gallic acid regulates molecular signaling: Role in cancer drug resistance. Medical Oncology. 2023;**40**(11):308. DOI: 10.1007/s12032-023-02178-4

[19] Ow YY, Stupans I. Gallic acid and gallic acid derivatives: Effects on drug metabolizing enzymes. Current Drug Metabolism. 2003;**4**(3):241-248. DOI: 10.2174/1389200033489479

[20] Tuli HS, Mistry H, Kaur G, et al. Gallic acid: A dietary polyphenol that exhibits anti-neoplastic activities by modulating multiple oncogenic targets. Anti-Cancer Agents in Medicinal Chemistry. 2022;**22**(3):499-514. DOI: 10.2174/1871520621666211119085834

[21] Verma S, Singh A, Mishra A. Gallic acid: Molecular rival of cancer. Environmental Toxicology and Pharmacology. 2013;**35**(3):473-485. DOI: 10.1016/j.etap.2013.02.011

[22] Jiang Y, Pei J, Zheng Y, Miao YJ, Duan BZ, Huang LF. Gallic acid: A potential anti-cancer agent. Chinese Journal of Integrative Medicine. 2022;**28**(7):661-671. DOI: 10.1007/s11655-021-3345-2

[23] Srinivasulu C, Ramgopal M, Ramanjaneyulu G, Anuradha CM, Suresh KC. Syringic acid (SA)—A review of its occurrence, biosynthesis, pharmacological and industrial importance. Biomedicine & Pharmacotherapy. 2018;**108**:547-557. DOI: 10.1016/j.biopha.2018.09.069

[24] Ogut E, Armagan K, Gül Z. The role of syringic acid as a neuroprotective agent for neurodegenerative disorders and future expectations. Metabolic Brain Disease. 2022;**37**(4):859-880. DOI: 10.1007/s11011-022-00960-3

[25] Sabahi Z, Khoshnoud MJ, Hosseini S, Khoshraftar F, Rashedinia M. Syringic acid attenuates cardiomyopathy in Streptozotocin-induced diabetic rats. Advances in Pharmacological and Pharmaceutial Sciences. 2021;**2021**:5018092. DOI: 10.1155/2021/5018092

[26] Liu G, Zhang BF, Hu Q, Liu XP, Chen J. Syringic acid mitigates myocardial ischemia reperfusion injury by activating the PI3K/Akt/GSK-3β signaling pathway. Biochemical and Biophysical Research Communications. 2020;**531**(2):242-249. DOI: 10.1016/j.bbrc.2020.07.047

[27] Ha SJ, Lee J, Park J, et al. Syringic acid prevents skin carcinogenesis via regulation of NoX and EGFR signaling. Biochemical Pharmacology. 2018;**154**:435-445. DOI: 10.1016/j.bcp.2018.06.007

[28] Lashgari NA, Roudsari NM, Momtaz S, Abdolghaffari AH, Atkin SL,

Sahebkar A. Regulatory mechanisms of vanillic acid in cardiovascular diseases: A review. Current Medicinal Chemistry. 2023;**30**(22):2562-2576. DOI: 10.2174/0929867329666220831152608

[29] Yalameha B, Nejabati HR, Nouri M. Cardioprotective potential of vanillic acid. Clinical and Experimental Pharmacology & Physiology. 2023;**50**(3):193-204. DOI: 10.1111/1440-1681.13736

[30] Shekari S, Khonsha F, Rahmati-Yamchi M, Nejabati HR, Mota A. Vanillic acid and non-alcoholic fatty liver disease: A focus on AMPK in adipose and liver tissues. Current Pharmaceutical Design. 2021;**27**(46):4686-4692. DOI: 10.2174/1381612827666210701145438

[31] Singh B, Kumar A, Singh H, Kaur S, Arora S, Singh B. Protective effect of vanillic acid against diabetes and diabetic nephropathy by attenuating oxidative stress and upregulation of NF-κB, TNF-α and COX-2 proteins in rats. Phytotherapy Research. 2022;**36**(3):1338-1352. DOI: 10.1002/ptr.7392

[32] Yao X, Jiao S, Qin M, Hu W, Yi B, Liu D. Vanillic acid alleviates acute myocardial hypoxia/reoxygenation injury by inhibiting oxidative stress. Oxidative Medicine and Cellular Longevity. 2020;**2020**:8348035. DOI: 10.1155/2020/8348035

[33] Osorio-Paz I, Valle-Jiménez X, Brunauer R, Alavez S. Vanillic acid improves stress resistance and substantially extends life span in *Caenorhabditis elegans*. The Journals of Gerontology. Series A, Biological Sciences and Medical Sciences. 2023;**78**(7):1100-1107. DOI: 10.1093/gerona/glad086

[34] Masella R, Santangelo C, D'Archivio M, Li Volti G, Giovannini C, Galvano F. Protocatechuic acid and human disease prevention: Biological activities and molecular mechanisms. Current Medicinal Chemistry. 2012;**19**(18):2901-2917. DOI: 10.2174/092986712800672102

[35] Song J, He Y, Luo C, et al. New progress in the pharmacology of protocatechuic acid: A compound ingested in daily foods and herbs frequently and heavily. Pharmacological Research. 2020;**161**:105109. DOI: 10.1016/j.phrs.2020.105109

[36] Bai L, Han X, Kee HJ, et al. Protocatechuic acid prevents isoproterenol-induced heart failure in mice by downregulating kynurenine-3-monooxygenase. Journal of Cellular and Molecular Medicine. 2023;**27**(16):2290-2307. DOI: 10.1111/jcmm.17869

[37] Wang Q, Ren X, Wu J, et al. Protocatechuic acid protects mice from influenza A virus infection. European Journal of Clinical Microbiology & Infectious Diseases. 2022;**41**(4):589-596. DOI: 10.1007/s10096-022-04401-y

[38] Saleem M, Fariduddin Q, Castroverde CDM. Salicylic acid: A key regulator of redox signalling and plant immunity. Plant Physiology and Biochemistry. 2021;**168**:381-397. DOI: 10.1016/j.plaphy.2021.10.011

[39] Zaid A, Mohammad F, Wani SH, Siddique KMH. Salicylic acid enhances nickel stress tolerance by up-regulating antioxidant defense and glyoxalase systems in mustard plants. Ecotoxicology and Environmental Safety. 2019;**180**:575-587. DOI: 10.1016/j.ecoenv.2019.05.042

[40] Sharma A, Sidhu GPS, Araniti F, et al. The role of salicylic acid in plants exposed to heavy metals. Molecules. 2020;**25**(3):540. DOI: 10.3390/molecules25030540

[41] Guo B, Liu C, Liang Y, Li N, Fu Q. Salicylic acid signals plant defence against cadmium toxicity. International Journal of Molecular Sciences. 2019;**20**(12):2960. DOI: 10.3390/ijms20122960

[42] AAlikhani M, Khalili M, Jahanshahi M. The natural iron chelators' ferulic acid and caffeic acid rescue mice's brains from side effects of iron overload. Frontiers in Neurology. 2022;**13**:951725. DOI: 10.3389/fneur.2022.951725

[43] Yeh YH, Lee YT, Hsieh HS, Hwang DF. Dietary caffeic acid, ferulic acid and coumaric acid supplements on cholesterol metabolism and antioxidant activity in rats. Journal of Food and Drug Analysis. 2020;**17**(2):4. DOI: 10.38212/2224-6614.2292

[44] Figueredo KC, Guex CG, Graiczik J, et al. Caffeic acid and ferulic acid can improve toxicological damage caused by iron overload mediated by carbonic anhydrase inhibition. Drug and Chemical Toxicology. Published online November 29. 2022:144-155. DOI: 10.1080/01480545.2022.2152043

[45] Bocco BM, Fernandes GW, Lorena FB, et al. Combined treatment with caffeic and ferulic acid from *Baccharis uncinella* C. DC. (Asteraceae) protects against metabolic syndrome in mice. Brazilian Journal of Medical and Biological Research. 2016;**49**(3):e5003. DOI: 10.1590/1414-431X20155003

[46] He XX, Yang XH, Ou RY, et al. Synthesis and evaluation of multifunctional ferulic and caffeic acid dimers for Alzheimer's disease. Natural Product Research. 2017;**31**(6):734-737. DOI: 10.1080/14786419.2016.1219862

[47] Serafim TL, Carvalho FS, Marques MPM, et al. Lipophilic caffeic and ferulic acid derivatives presenting cytotoxicity against human breast cancer cells. Chemical Research in Toxicology. 2011;**24**(5):763-774. DOI: 10.1021/tx200126r

[48] Ferreira PS, Victorelli FD, Fonseca-Santos B, Chorilli M. A review of analytical methods for p-coumaric acid in plant-based products, beverages, and biological matrices. Critical Reviews in Analytical Chemistry. 2019;**49**(1):21-31. DOI: 10.1080/10408347.2018.1459173

[49] Pei K, Ou J, Huang J, Ou S. p-Coumaric acid and its conjugates: Dietary sources, pharmacokinetic properties and biological activities. Journal of the Science of Food and Agriculture. 2016;**96**(9):2952-2962. DOI: 10.1002/jsfa.7578

[50] Yue Y, Shen P, Xu Y, Park Y. p-Coumaric acid improves oxidative and osmosis stress responses in *Caenorhabditis elegans*. Journal of the Science of Food and Agriculture. 2019;**99**(3):1190-1197. DOI: 10.1002/jsfa.9288

[51] Mani A, Kushwaha K, Khurana N, Gupta J. p-Coumaric acid attenuates high-fat diet-induced oxidative stress and nephropathy in diabetic rats. Journal of Animal Physiology and Animal Nutrition (Berl). 2022;**106**(4):872-880. DOI: 10.1111/jpn.13645

[52] Ayazoglu Demir E, Mentese A, Kucuk H, Turkmen Alemdar N, Demir S. p-Coumaric acid alleviates cisplatin-induced ovarian toxicity in rats. The Journal of Obstetrics and Gynaecology Research. 2022;**48**(2):411-419. DOI: 10.1111/jog.15119

[53] Ross JA, Kasum CM. Dietary flavonoids: Bioavailability, metabolic effects, and safety. Annual Review of Nutrition.

2002;**22**(1):19-34. DOI: 10.1146/annurev.nutr.22.111401.144957

[54] Del Rio D, Rodriguez-Mateos A, Spencer JPE, Tognolini M, Borges G, Crozier A. Dietary (poly)phenolics in human health: Structures, bioavailability, and evidence of protective effects against chronic diseases. Antioxidants & Redox Signaling. 2013;**18**(14):1818-1892. DOI: 10.1089/ars.2012.4581

[55] George BP, Chandran R, Abrahamse H. Role of phytochemicals in cancer chemoprevention: Insights. Antioxidants. 2021;**10**(9):1455. DOI: 10.3390/antiox10091455

[56] Serafini M, Peluso I, Raguzzini A. Flavonoids as anti-inflammatory agents. Proceedings of the Nutrition Society. 2010;**69**(3):273-278. DOI: 10.1017/S002966511000162X

[57] Dias MC, Pinto DCGA, Silva AMS. Plant flavonoids: Chemical characteristics and biological activity. Molecules. 2021;**26**(17):5377. DOI: 10.3390/molecules26175377

[58] Galleano M, Verstraeten SV, Oteiza PI, Fraga CG. Antioxidant actions of flavonoids: Thermodynamic and kinetic analysis. Archives of Biochemistry and Biophysics. 2010;**501**(1):23-30. DOI: 10.1016/j.abb.2010.04.005

[59] Slika H, Mansour H, Wehbe N, et al. Therapeutic potential of flavonoids in cancer: ROS-mediated mechanisms. Biomedicine and Pharmacotherapy. 2022;**146**:112442. DOI: 10.1016/j.biopha.2021.112442

[60] Bhattarai N, Kumbhar AA, Pokharel YR, Yadav PN. Anticancer potential of coumarin and its derivatives. Mini Reviews in Medicinal Chemistry. 2021;**21**(19):2996-3029. DOI: 10.2174/1389557521666210405160323

[61] Xia D, Liu H, Cheng X, Maraswami M, Chen Y, Lv X. Recent developments of coumarin-based hybrids in drug discovery. Current Topics in Medicinal Chemistry. 2022;**22**(4):269-283. DOI: 10.2174/1568026622666220105105450

[62] Patil SA, Kandathil V, Sobha A, et al. Comprehensive review on medicinal applications of coumarin-derived imine-metal complexes. Molecules. 2022;**27**(16):5220. DOI: 10.3390/molecules27165220

[63] Kaur M, Kohli S, Sandhu S, Bansal Y, Bansal G. Coumarin: A promising scaffold for anticancer agents. Anti-Cancer Agents in Medicinal Chemistry. 2015;**15**(8):1032-1048. DOI: 10.2174/1871520615666150101125503

[64] Al-Warhi T, Sabt A, Elkaeed EB, Eldehna WM. Recent advancements of coumarin-based anticancer agents: An up-to-date review. Bioorganic Chemistry. 2020;**103**:104163. DOI: 10.1016/j.bioorg.2020.104163

[65] Bast A, Haenen GR, van den Berg R, van den Berg H. Antioxidant effects of carotenoids. International Journal for Vitamin and Nutrition Research. 1998;**68**(6):399-403

[66] Krinsky NI. The antioxidant and biological properties of the carotenoids. Annals of the New York Academy of Sciences. 1998;**854**:443-447. DOI: 10.1111/j.1749-6632.1998.tb09923.x

[67] Palace VP, Khaper N, Qin Q, Singal PK. Antioxidant potentials of vitamin A and carotenoids and their relevance to heart disease. Free Radical Biology & Medicine. 1999;**26**(5-6):746-761. DOI: 10.1016/s0891-5849(98)00266-4

[68] Vrdoljak N. Carotenoids and carcinogenesis: Exploring the antioxidant

and cell signaling roles of carotenoids in the prevention of cancer. Critical Reviews in Oncogenesis. 2022;**27**(3):1-13. DOI: 10.1615/CritRevOncog.2022045331

[69] Fiedor J, Burda K. Potential role of carotenoids as antioxidants in human health and disease. Nutrients. 2014;**6**(2):466-488. DOI: 10.3390/nu6020466

[70] Milani A, Basirnejad M, Shahbazi S, Bolhassani A. Carotenoids: Biochemistry, pharmacology and treatment. British Journal of Pharmacology. 2017;**174**(11):1290-1324. DOI: 10.1111/bph.13625

[71] Varghese R, Efferth T, Ramamoorthy S. Carotenoids for lung cancer chemoprevention and chemotherapy: Promises and controversies. Phytomedicine. 2023;**116**:154850. DOI: 10.1016/j.phymed.2023.154850

[72] Bohn T, Balbuena E, Ulus H, et al. Carotenoids in health as studied by omics-related endpoints. Advances in Nutrition. 2023;**14**(6):1538-1578. DOI: 10.1016/j.advnut.2023.09.002

[73] Krinsky NI, Johnson EJ. Carotenoid actions and their relation to health and disease. Molecular Aspects of Medicine. 2005;**26**(6):459-516. DOI: 10.1016/j.mam.2005.10.001

[74] Stahl W, Sies H. Antioxidant activity of carotenoids. Molecular Aspects of Medicine. 2003;**24**(6):345-351. DOI: 10.1016/s0098-2997(03)00030-x

[75] Ba W, Xu W, Deng Z, Zhang B, Zheng L, Li H. The antioxidant and anti-inflammatory effects of the main carotenoids from tomatoes via Nrf2 and NF-κB signaling pathways. Nutrients. 2023;**15**(21):4652. DOI: 10.3390/nu15214652

[76] Alugoju P, Krishna Swamy VKD, Anthikapalli NVA, Tencomnao T. Health benefits of astaxanthin against age-related diseases of multiple organs: A comprehensive review. Critical Reviews in Food Science and Nutrition. 2023;**63**(31):10709-10774. DOI: 10.1080/10408398.2022.2084600

[77] Young AJ, Lowe GM. Antioxidant and prooxidant properties of carotenoids. Archives of Biochemistry and Biophysics. 2001;**385**(1):20-27. DOI: 10.1006/abbi.2000.2149

[78] Shiozawa K, Agista AZ, Ohsaki Y, Shirakawa H. Emergence of novel functions of vitamins for the prevention of life-style related diseases. Journal of Nutritional Science and Vitaminology (Tokyo). 2022;**68**(Supplement):S8-S10. DOI: 10.3177/jnsv.68.S8

[79] Mitra S, Paul S, Roy S, et al. Exploring the immune-boosting functions of vitamins and minerals as nutritional food bioactive compounds: A comprehensive review. Molecules. 2022;**27**(2):555. DOI: 10.3390/molecules27020555

[80] US Preventive Services Task Force, Mangione CM, Barry MJ, et al. Vitamin, mineral, and multivitamin supplementation to prevent cardiovascular disease and cancer: US preventive services task force recommendation statement. Journal of the American Medical Association. 2022;**327**(23):2326-2333. DOI: 10.1001/jama.2022.8970

[81] Donkena KV, Karnes RJ, Young CYF. Vitamins and prostate cancer risk. Molecules. 2010;**15**(3):1762-1783. DOI: 10.3390/molecules15031762

[82] Manson JE, Cook NR, Lee IM, et al. Vitamin D supplements and prevention

of cancer and cardiovascular disease. The New England Journal of Medicine. 2019;**380**(1):33-44. DOI: 10.1056/NEJMoa1809944

[83] Aguilera-Méndez A, Boone-Villa D, Nieto-Aguilar R, Villafaña-Rauda S, Molina AS, Sobrevilla JV. Role of vitamins in the metabolic syndrome and cardiovascular disease. Pflügers Archiv. 2022;**474**(1):117-140. DOI: 10.1007/s00424-021-02619-x

[84] Koekkoek WACK, van Zanten ARH. Antioxidant vitamins and trace elements in critical illness. Nutrition in Clinical Practice. 2016;**31**(4):457-474. DOI: 10.1177/0884533616653832

[85] Asplund K. Antioxidant vitamins in the prevention of cardiovascular disease: A systematic review. Journal of Internal Medicine. 2002;**251**(5):372-392. DOI: 10.1046/j.1365-2796.2002.00973.x

[86] Hasanain B, Mooradian AD. Antioxidant vitamins and their influence in diabetes mellitus. Current Diabetes Reports. 2002;**2**(5):448-456. DOI: 10.1007/s11892-002-0110-6

[87] Bjørklund G, Peana M, Dadar M, et al. The role of B vitamins in stroke prevention. Critical Reviews in Food Science and Nutrition. 2022;**62**(20):5462-5475. DOI: 10.1080/10408398.2021.1885341

[88] Pedroza-García KA, Careaga-Cárdenas G, Díaz-Galindo C, Quintanar JL, Hernández-Jasso I, Ramírez-Orozco RE. Bioactive role of vitamins as a key modulator of oxidative stress, cellular damage and comorbidities associated with spinal cord injury (SCI). Nutritional Neuroscience. 2023;**26**(11):1120-1137. DOI: 10.1080/1028415X.2022.2133842

[89] Yan F, Zhao Q, Li Y, et al. The role of oxidative stress in ovarian aging: A review. Journal of Ovarian Research. 2022;**15**:100. DOI: 10.1186/s13048-022-01032-x

[90] Fagbohun OF, Gillies CR, Murphy KPJ, Rupasinghe HPV. Role of antioxidant vitamins and other micronutrients on regulations of specific genes and signaling pathways in the prevention and treatment of cancer. International Journal of Molecular Sciences. 2023;**24**(7):6092. DOI: 10.3390/ijms24076092

[91] Jain A, Tiwari A, Verma A, Jain SK. Vitamins for cancer prevention and treatment: An insight. Current Molecular Medicine;**17**(5):321-340

[92] Kulczyński B, Sidor A, Gramza-Michałowska A. Characteristics of selected antioxidative and bioactive compounds in meat and animal origin products. Antioxidants. 2019;**8**(9):335. DOI: 10.3390/antiox8090335

[93] Bhat ZF, Kumar S, Bhat HF. Bioactive peptides of animal origin: A review. Journal of Food Science and Technology. 2015;**52**(9):5377-5392. DOI: 10.1007/s13197-015-1731-5

[94] Abeyrathne EDNS, Nam K, Huang X, Ahn DU. Plant- and animal-based antioxidants' structure, efficacy, mechanisms, and applications: A review. Antioxidants (Basel). 2022;**11**(5):1025. DOI: 10.3390/antiox11051025

[95] Budzeń S, Rymaszewska J. The biological role of carnosine and its possible applications in medicine. Advances in Clinical and Experimental Medicine. 2013;**22**(5):739-744

[96] Flanagan JL, Simmons PA, Vehige J, Willcox MD, Garrett Q. Role of carnitine in disease. Nutrition and Metabolism. 2010;**7**(1):30. DOI: 10.1186/1743-7075-7-30

[97] Boreková M, Hojerová J, Koprda V, Bauerová K. Nourishing and health benefits of coenzyme Q10. Czech Journal of Food Sciences. 2008;**26**(4):229-241. DOI: 10.17221/1122-CJFS

[98] Saad B, Sing Y, Nawi M, et al. Determination of synthetic phenolic antioxidants in food items using reversed-phase HPLC. Food Chemistry. 2007;**105**(1):389-394. DOI: 10.1016/j.foodchem.2006.12.025

[99] Kim JM, Choi SH, Shin GH, et al. Method validation and measurement uncertainty for the simultaneous determination of synthetic phenolic antioxidants in edible oils commonly consumed in Korea. Food Chemistry. 2016;**213**:19-25. DOI: 10.1016/j.foodchem.2016.06.053

[100] Rodil R, Quintana JB, Cela R. Oxidation of synthetic phenolic antioxidants during water chlorination. Journal of Hazardous Materials. 2012;**199-200**:73-81. DOI: 10.1016/j.jhazmat.2011.10.058

[101] Xu X, Liu A, Hu S, et al. Synthetic phenolic antioxidants: Metabolism, hazards and mechanism of action. Food Chemistry. 2021;**353**:129488. DOI: 10.1016/j.foodchem.2021.129488

[102] Hassanpour SH, Doroudi A. Review of the antioxidant potential of flavonoids as a subgroup of polyphenols and partial substitute for synthetic antioxidants. Avicenna Journal of Phytomedicine. 2023;**13**(4):354-376. DOI: 10.22038/AJP.2023.21774

[103] Poljsak B, Šuput D, Milisav I. Achieving the balance between ROS and antioxidants: When to use the synthetic antioxidants. Oxidative Medicine and Cellular Longevity. 2013;**2013**:e956792. DOI: 10.1155/2013/956792

[104] Blundell R, Shah MA, Azzopardi JI, Iqbal S, Rasul A, Shah GM. Chapter 3.3—Butylated hydroxytoluene. In: Nabavi SM, Silva AS, editors. Antioxidants Effects in Health. United States: Elsevier Inc.; 2022. pp. 195-200. DOI: 10.1016/B978-0-12-819096-8.00033-1

[105] Felter SP, Zhang X, Thompson C. Butylated hydroxyanisole: Carcinogenic food additive to be avoided or harmless antioxidant important to protect food supply? Regulatory Toxicology and Pharmacology. 2021;**121**:104887. DOI: 10.1016/j.yrtph.2021.104887

[106] EFSA Panel on Food Additives and Nutrient Sources added to Food (ANS). Scientific opinion on the re-evaluation of butylated hydroxyanisole—BHA (E 320) as a food additive. EFSA Journal. 2011;**9**(10):2392. DOI: 10.2903/j.efsa.2011.2392

[107] Zhu TT, Zhu CN, Qiu Y, et al. Tertiary butylhydroquinone alleviated liver steatosis and increased cell survival via β-arrestin-2/PI3K/AKT pathway. Iranian Journal of Basic Medical Sciences. 2021;**24**(10):1428-1436. DOI: 10.22038/ijbms.2021.58156.12924

[108] Song H, Xu Y, Yang X, Rong X, Wang Y, Wei N. Tertiary butylhydroquinone alleviates gestational diabetes mellitus in C57BL/KsJ-Lep db/+ mice by suppression of oxidative stress. Journal of Cellular Biochemistry. 2019;**120**(9):15310-15319. DOI: 10.1002/jcb.28798

Chapter 12

# Oxidative Stress (OS) in Plants, Beneficial Interactions with Their Microbiome and Practical Implications for Agricultural Biotechnology

*Gustavo Alberto De la Riva, Juan Carlos Hernández González, Rolando Morán Valdivia and Rolando García González*

**Abstract**

Plants are aerobic, sessile, and autotrophic organisms that face a wide variety of climatic adversities and pathogen attacks. They have evolved to deal with such challenges, that is, the case of the antioxidant defense to avoid oxidative stress (OS) caused by the overproduction of reactive oxygen and nitrogen species (ROS/RONS). ROS/RONS can be by-products of many physiological functions and biochemical pathways, but particularly from the fundamental electronic transfer processes: photosynthesis and respiration. Photosynthesis is crucial for plant nutrition, trophic webs and maintenance of $O_2/CO_2$ balance in biosphere. Respiration is a source of energy for organisms. Both processes generate ROS and its overproduction can lead OS, modifying essential biomolecules and altering fundamental biochemical pathways and plant development. Antioxidant defense prevents such harmful accumulation of ROS. Plants interact with microbiota, a well-structured microbial community conferring adaptive and defense tools in both abiotic and biotic stressing conditions. We present the beneficial influence of the plant microbiome promotes the adaptability, resistance, and defense of plants using our results obtained in plants confronted drought. Microbiota can be used in agriculture in different ways, including adaptation to soil of micro-propagated plants, bioproducts for plant growth and pest control and processing and preservation of agriculture products.

**Keywords:** oxidative stress, antioxidant defense, plant microbiome, climate change, abiotic stress, biotic stress, biotechnology

## 1. Introduction

About 4.5 billion years ago our planet was formed from condensed cosmic debris, with primitive oceans and atmosphere, created by early volcanic activity, composed of such materials as gas methane ($CH_4$), ammonia ($NH_3$), and water vapor ($H_2O$)


IntechOpen

among others. Those hostile conditions, followed by the period of gradual cooling, climatic moderation and reducing conditions, generated new more complex organic molecules, including organic molecules, and made possible the first forms of anaerobic life [1]. In the atmosphere, oxygen ($O_2$) appeared as a result of water ($H_2O$) disruption by solar photolysis, shocking anaerobic life with a massive extinction. Survivors must adapt to using oxygen and cope with the potential dangers of oxygen and its metabolic by-products, ROS and RONS. Many of them are free radicals (hydroxyl radical (HO−), peroxyl (ROO−)) others are superoxide anions (($O_2^-$) hydrogen peroxide ($H_2O_2$), reactive aldehydes (ROCH) and nitric oxide (NO)). About 1–2% of total oxygen used in aerobic cells is diverted into ROS primarily in the mitochondria, chloroplasts, and peroxisomes [2, 3]. Survivors undergo evolutionary developed antioxidant defense capable of maintaining oxidative balance and maintaining ROS within certain values. The antioxidant defense includes antioxidant molecules (free amino acids, vitamins, proteins, polysaccharides, ions, and cell wall compounds), and antioxidant enzymes (including superoxide dismutase (SOD), glutathione peroxidase (GPO) and catalases (CAT) [4–6]. Plants are exposed to environmental conditions, including those that trigger and cause OS [7]. The complexity of plant biology and their living mode demands a fast adaptive response to environmental stressors for maintaining cellular homeostasis and essential physiological processes. Important processes like the disruption of seed dormancy, seed and spore germination, glyoxalase cycle, photosynthesis, respiration, and oxidation of amines, among others, are intrinsically linked with ROS generations [8, 9]. At low ROS levels, they function as fast-acting and signaling mediators, stimulating adaptive responses to abiotic and biotic stresses, keeping cellular homeostasis and plant development [10]. Increases in ROS trigger the activation of different adaptive and defense mechanisms against external abiotic (climate, salinity, extreme temperature, drought, and toxic soil components like heavy metals, agrochemicals, among others) and biotic stressors (phytopathogens, weeds) [9, 11–15]. Only 1–2% of total $O_2$ consumption in plant tissues becomes ROS [16, 17]. The antioxidant defense in plants displays a variety of antioxidant compounds and enzymes, with specific subcellular localization and particularly are presented in the most important ROS-generating sources. Climate change increases the presence of external stressors, representing a great challenge for agriculture and the conservation of natural ecosystems [18, 19]. In this chapter, we present an approach to OS in plants induced by biotic and abiotic stressors, their relationship and synergistic effect, as well as the importance of the microbiota in preventing and mitigating their adverse effects. We use some previously reported practical research as well as some data of our own experimental results, where the plant-microbe interactions are used to promote the adaptive response in plants to external stressors [19–22]. Such results encourage the development of microbial consortiums to improve plant development and induce adaptive responses in different wild and cropped plants [20–23].

## 2. Oxidative stress in plants

### 2.1 Brief overview of oxidative stress in plants

Plants are exposed to abiotic and biotic environmental stressors, that result in the overproduction of high ROS and RONS exceeding the capacity of the antioxidant defense and inducing OS. Specifically, ROS reacts and damage cellular structures, including cellular membranes, and biomolecules (proteins, lipids, carbohydrates and DNA), affecting homeostasis and important physiological functions.

Under normal conditions, ROS mainly derived from metabolic functions of chloroplasts, mitochondria, and peroxisomes [24–30]. Plant photosynthesis occurs in chloroplasts, this organelle displays a well-structured thylakoid membrane housing chlorophyll and the other components of photosynthetic systems (PS). Water is the final acceptor of electrons passing through the PS forming $O_2^-$. The balance between the production and the scavenging of ROS may be affected by various abiotic and biotic stressors. Those include higher levels of salinity, UV radiation, drought, heavy metals, temperature extremes, nutrient deficiency, air pollution, herbicides and pathogen attacks, lead adverse effects in plant cells, and activating the antioxidant defense to detoxify and prevent the damages [27, 28]. The most common ROS generated in plant cells are the free radicals $^{\bullet}OH$, $1O_2$, alkoxyl radical ($RO^{\bullet}$), and peroxyl radical ($ROO^{\bullet}$), while no radical species like $H_2O_2$, $O_2^{\bullet-}$, carbonate ($CO_3^{\bullet-}$), excited carbonyl group ($RO^*$), hypochlorous acid (HOCl), hypoiodous acid (HOI); hypobromous acid (HOBr) were found too [27, 28]. Peroxyl radical ($LOO^{\bullet}$), alkoxyl radical ($LO^{\bullet}$), hydroperoxyl radical ($HO_2^{\bullet}$) peroxynitrite ($HNO_3^-$), ozone ($O_3$), and trichloromethyl peroxyl radical ($C_{13}COO^{\bullet}$) could also be found in plant cells. The biological effects of ROS concentration vary up to the chemical reactivity, origin (pathway, subcellular compartment, or organelle), and diffusion rates [29–31].

Within certain concentration values, ROS act as specific signals activating antioxidant defense and promoting adaptive response to environmental conditions [32–34]. In chloroplast, ROS generation depends on the interaction of chlorophyll (chl) and light and the subsequent oxidation of chlI leading the transfer of an electron through ETC of PSI and PSII [34–36]. In PSI, increases in $H_2O_2$ activate photoreduction and the production of superoxide radicals (Mehler reaction), here oxygen can be reduced by ferredoxin to form superoxide ($O_2^{\bullet-}$), followed by reduction by SOD to $H_2O_2$, forming water and oxidized ascorbate by APO [37, 38]. In the presence of metal ions, such as $Fe^{2+}$, more highly reactive short-lived $HO^{\bullet}$ are formed from $O_2^{\bullet-}$ and $H_2O_2$ [38]. This reaction prevents over-reduction of the electron carriers and helps to keep a redox balance to protect the different components of the photosynthetic ETC, dissipating the excess of photons and the non-photochemical quenching-formation, and improving ATP/NADPH balance during photosynthesis [38]. Peroxisomal GOX plays a major role in ROS generation [39]. Here $O_2^{\bullet-}$ can be produced by XOD and NADPH oxidase in peroxisomal membrane [39]. The formed $O_2^{\bullet-}$ is dismutated into $H_2O_2$ by metalloenzymes SODs, (Cu-Zn-SOD and Mn-SOD). Other processes producing $H_2O_2$ in peroxisomes are: photorespiration GOX reaction, $O_2^{\bullet-}$ disproportionation, β-oxidation of fatty acids, flavin oxidase polyamine oxidase, sulfite oxidase, copper amine oxidase, and sarcosine oxidase activity [40].

Respiration occurs in mitochondria where ETC holds electrons generating a great amount of free energy, from which 0.2–2% of all transferred electrons interact with $O_2$ to form ROS [34–43]. In stressed plant, the complex I and III of mitochondrial ETC generate excessive ROS, particularly $O_2^{\bullet-}$ due to electron leakage, that later is catalyzed to $H_2O_2$ by Mn-SOD and Cu-Zn-SOD [40–42]. Mitochondria are the major ROS producers in the non-photosynthetic plant tissues but they are also produced in other plant cell compartments [42–44]. In plant cell wall ROS generation provoked a rigidity and growth reduction because of the polymerization of glycoproteins and phenolic compounds. NADH-dependent peroxidases produce $H_2O_2$ in the presence of NADH provided by malate dehydrogenase [42]. ROS can also be increased by the reduction of polyamines to quinine in the cell wall and the hydroperoxidation of polyunsaturated fatty acid driven by lipoxygenase [45]. Plasma membrane NADPH oxidase and quinine reductase mediate the $O_2^{\bullet-}$ generation when electron transfer from cytoplasmic NADPH and formed $O_2^{\bullet-}$ is transformed again to $H_2O_2$. In the endoplasmic reticulum $O_2^{\bullet-}$ is a by-product during reaction with an organic substrate to form an intermediate (Cyt P450-ROO$^-$), driven by Cyt P450 where NADPH acts as an electron donor [45]. In glyoxysome the fatty acid oxidation by GOX and urate oxidase generate both $O_2^{\bullet-}$ and $H_2O_2$ [46] as well as the XOD and AO are involved in ROS production in the cytosol [47].

## 2.2 Oxidative stress damage in plant cells

The antioxidant defense maintains the homeostasis in plant cells required levels. Daily and seasonal changes induce differences in the oxidative status of the plant are kept within harmless limits. External stressors can induce ROS overproduction disrupting the oxidative balance and provoke. Within specific values ROS are important intracellular and intercellular signaling during abiotic and biotic stress, sensing the integration of different environmental signals, activation of stress response, antioxidant defense and adaptive mechanisms of plant resilience [13]. OS can result IN extensive damage and cell death [28]. Excessive ROS cause lipid peroxidation (LPO) and the most susceptible target is the carbon atoms and the ester bond between fatty acid and glycerol. LPO affects polyunsaturated fatty acids of plasmatic membrane, as linoleic and linolenic acid, by the attack of $1O_2$ and $OH^\bullet$ in a multistep reaction, forming lipid dimmers, lipids-derived radicals with disruption of membranal functions, including PSI [28, 48]. PSI includes proteins, pigments, and co-factors joined in a complex responsible for the primary energy conversion during photosynthesis. During OS, this complex is the main target of $1O_2$ during OS, producing LOOH, lowering photosynthetic efficiency. Other harm to membranes includes oxidation of structural, receptors, and enzymatic complexes, causing loss of integrity and barrier functions, leading to cellular death [28, 48, 49]. Some LPO-resulted compounds, such as α, β-unsaturated hydroxyalkenal, and malondialdehyde (MDA), can react with proteins and majorly modify specific amino acids: arginine, lysine, proline, threonine and tryptophan. Those modified amino acids made these proteins more susceptible to proteolysis, and the loss of biological functions. That is the case of damaged D1 and D2 reaction center proteins via aminoacids oxidation of PhII resulted I in photoinhibition. Other of the most frequent protein targets are Rubisco and SOD promoting disulfide bond formation, oxidation, nitrosylation, glutathionylation, and sulfhydration [50, 51]. MDA is the most frequently used biomarker of oxidative stress in cells. ROS can oxidize deoxyribose sugar of DNA by $^\bullet OH$ attacks by taking $H^+$ from C—H bonds of 2-deoxyribose and methyl group, forming hydroxyl methyl urea, thymine glycol and lead to break locally the double-stranded DNA into single-strands [52]. Modifications of nucleotides and DNA-protein cross-linking can form some products, like 8-hydroxyquinine and dehydro-2′-deoxyguanosine [53]. Those damages, if not repaired, denaturation and DNA unfolding made DNA processing impossible [53]. DNA damages alter protein synthesis, photosynthesis, transcription, correct signal transduction, induce replicative errors, and generate genomic instability [31, 53].

The enzymes of glycolysis, Kreps' cycle (TCA) and pentose phosphate (PP) pathways are severely attacked by ROS, some of them, like glyceraldehyde 3-phosphate dehydrogenase and fructose-1,6-bisphosphate aldolase, can be completed inhibited, other can suffer the reduction in their activities. To compensate this situation, cells react by increasing the carbon flux to produce the necessary NADPH and the generation of in Ribose 5-P, and ribulose 5-P intermediates [53, 54]. In OS conditions, stressed plants reduce glycolysis and TCA activity and the abundance of citrate, isocitrate, fumarate, malate, succinate and 2-oxoglutarate have been observed [55].

### 2.3 Abiotic and biotic stresses in plants

The plant's adaptive response to climate change stands out as one of the most important issues to face by agriculture today. Those issues include both biotic and abiotic stress affecting the wealth and health of wild ecosystems and resulting in significant productive losses in cropped plants [54, 56]. Abiotic stress is closely related to affectation caused by climate change and anthropogenic activities, particularly economic projects without eco-friendly and sustainable technologies. Such factors as salinity, extreme temperature, drought, and toxic soil components like heavy metals, climatic factors agrochemicals and anthropogenic activities are listed as abiotic stressors. Texture, carbon, and organic matter content (C) content, essential plant nutrients, soil compaction, aeration, and moisture levels among others are considered as physicochemical factors. Biotic stressors are living organisms such as phytopathogens (virus, bacteria, fungi and nematodes, causing diseases), weeds (wild flora that competes for nutrients with cropped plants), and parasitic plants (plants take nutrients from cropped and timber plants). It is important to elucidate the plant adaptive response and the mechanisms of ROS production and OS to develop plant breeding programs and practical methods for the mitigation of adverse factors in agriculture and natural ecosystems.

#### *2.3.1 Abiotic stress*

Generation of ROS is one of the most common plant responses to different stresses, and it is the starting point at which various signaling pathways come together to modulate the plant response to environmental external changes. Integral Redox regulation is carried by interconnected pathway proteins that provides a rapid antioxidant response to protect cellular integrity. Its different components and mechanisms keep plant development and defense pathways. Each plant cellular compartment is equipped with its own ROS homeostasis control. ROS signaling is altered depending on the cell type, stage of development, and level of stress. Under different abiotic stress conditions production of ROS molecules in cells are recognized by various ROS sensors to produce stress-specific action for the development of acclimation and adaptive response. ROS signals are decoded through various redox reactions which induce specific gene expression modulating specific transcription factors that directly interact with the DNA.

*Drought stress* is one of the major limitations to global agricultural production. Drought induces stomal closure in leaves, and other photosynthetic tissues, reduces transpiration, $CO_2$ capture and fixation, and lowers photosynthesis rates. The imbalance between light capturing and its photosynthetic use causes ROS and lead to OS [22, 28, 46]. The over-exposition of chloroplasts to the light induces excess of photorespiration increasing the formation of $H_2O_2$ [28]. The reduction of ferredoxin and problems with the acceptance of electrons alter ETC and reduce regeneration of $NADP^+$ [57–59]. Drought increased the amount of $H_2O_2$ and MDA and other ROS [59, 60]. Those results were obtained when osmotic stress was induced by the addition of PEG 6000 (15–20%) in different plant species (*Brassica napus* L., *Oryza sativa* L., *Amaranthus*, *Triticum aestivum*, *Zea mays* L.) [61, 62]. In all case, tolerant cultivars showed reduced ROS production with reduced oxidative damage and better adaptive response, compared to the sensitive cultivar [61, 62]. In rice (*O. sativa* L.) subsp. japonica. cv. Nipponbare, drought stress (20% PEG) increased $O_2^{\bullet-}$ by 23%, enhanced $H_2O_2$ and increased MDA contents by 16%, compared to controls [63].

*Extreme temperature.* Temperature and water are the two most critical environmental factors that influence in plant development, from seed germination, and through all stages of its life cycle [60, 63]. Plants, as well as all the living organisms, have specific ranges of temperature represented as a minimum, maximum, and optimum surrounding temperature. The mechanisms of adaptation to high and low temperatures share some common responses, but distinct signaling components in high- and low-temperature response have differences, including the up- and down-regulation of specific transcription factors (TFs) such as heat shock factor (HSF) family. Temperature changes induce the activation of complex signaling networks resulting in decrease membrane thermostability; higher malondialdehyde (MDA) accumulation; and the generation of ROS such as hydrogen peroxide ($H_2O_2$), singlet oxygen ($1O_2$), superoxide ($O_2^-$), or hydroxyl radical [11, 52]. High temperatures can provoke oxidative stress, which harms plant cells and impairs their growth by releasing reactive oxygen species (ROS); and increased transpiration rates due to high temperatures and resulting also in water stress. High temperatures affect cell division, plant development and growth, pollen viability, fertilization, and fruit formation [60, 63, 64]. At low temperatures significant changes in the metabolism of ROS imply an in increase tyrosine nitration ($NO_2$) and LPO, ascorbate, glutathione, and the NADPH-generating dehydrogenases. Low temperature inhibits the activity of metabolic enzymes delaying many biochemical processes and plant growth and promoting the expression of cold-induced genes. Low-temperature stresses negatively impact plant growth and development by inhibiting the activity of metabolic enzymes [60, 63, 64].

*Salinity.* Salt excess is frequently accompanied by drought, both are caused by climatic effects combined with poor agricultural practices, such as inadequate irrigation and drainage, excessive water extraction from wells in coastal areas, extraction of peat, and poor drainage practices. The excess of salts causes osmotic stress, ion toxicity, genotoxicity, nutritional deficiency, and induces ROS overproduction, including MDA, and $H_2O_2$, in roots and mature leaves [62]. Direct DNA damage by ROS was observed in mung bean grown under severe salt stress by Comet Assay [52]. In salt-tolerant plants increases in $H_2O_2$ are correlated with lower cellular damage compared with the sensitive ones evidencing the signaling roles of ROS to stimulate the adaptive response [65, 66].

*Soil composition.* Soil is a complex and dynamic mixture composed of inorganic and organic substances, air, water, and living organisms interacting with plants and constitutes a dynamic, ecological and trophic network. Soil characteristics joined with climatic conditions determine their edaphological properties, ecology and productivity. The most common compounds present in soil contain elements like phosphorus, potassium and nitrogen, whereas the rest are presented in a smaller proportion. Some soils have an excessive content of heavy metalloids toxic to the majority of plants, causing morpho-physiological dysfunctions and OS [52, 67]. Increases in ROS production, LPO, MDA, and $H_2O_2$ were observed in *Pisum sativum* L. seedlings in substrate containing Ni [52, 67, 68] and similar results were obtained in other species as *A. thaliana* L [68], *V. radiata* L. [69], *B. napus* L. [70], *B. juncea* L. [71], and *Cucumis sativus* L. [72], evidencing that present toxicity induces OS [73, 74].

*UV radiation.* Another important stressor is UV (200–400 nm). About 95% of the UV reaching the planet is UV-A (315–400 nm) and the other 5% is UV-B (280–315 nm) and the majority of plants is sensible to type UV-B [75]. The UV excess directly attacks nucleic acids, nucleotides and proteins and also induces OS, damaging all cellular membranes and reducing all metabolic functions [76]. UV radiation affects photosynthesis, stomatal conductance, $CO_2$ assimilation, and electron transport chain. In mitochondria radiation interferes with electro transport chain (ETC) too, leading to increase production of $H_2O_2$, $O_2^{\bullet-}$, and MDA in a dose-dependent tendency [77–79]. Plants mitigate UV-B-induced damage by synthesizing more compounds like flavanol, anthocyanins, and proanthocyanidins [79].

*Ozone.* $O_3$ is a highly reactive molecule, an important pollutant and an external stressor for plants, whose atmospheric accumulation increases due the effects of climate change and anthropogenic activities. Ozone penetrates to the aqueous phase substomatal cavity, interfering with surface tissues and reaching mesophyll cells, where it is degraded to ROS in the apoplastic space of mesophyll cells, causing phytotoxic damages [80]. Plants adapted to higher $O_3$ stress can induce programmed cell death (PCD) by interaction with the synthesis of PCD mediators during stress conditions, including a cysteine proteinase vacuolar processing enzyme (VPE) [81]. The reaction of $O_3$ with membrane fatty acids promotes peroxidation (LPO) and showed increases in $H_2O_2$, $O_2^{\bullet-}$, $OH^{\bullet}$ and MDA levels [82].

*Soil pH.* Soil pH exerts great influence on the mobilization and uptake of different compounds and nutrients, cellular membrane functions, and internalization of phytotoxic compounds. The plasmatic membrane proton pumps try to combat the extreme pH by influx and efflux of $H^+$, generating ROS overaccumulation conducing to OS [83]. In some genotypes of wheat (*Triticum aestivum* L.) and alfalfa (*Medicago sativa* L.) under extreme acid conditions $H_2O_2$, LPO and LOX activities and membrane damage increased [84–87]. In alfalfa (*Medicago sativa* L.) under extreme alkaline conditions severe membrane damage was observed [84]. Wheat, in extreme pH conditions, increases generation of $H_2O_2$, MDA and LOX [85]. Acid pH facilitates the mobilization and absorption of toxic metals/metalloids (Fe, Cu, Mn, Zn, and Al) and restricts the assimilation of essential macronutrients (P, Mg, Ca, K, and Na) limiting plant nutrition and productivity in acid soils [86, 87]. In basic soils are common some deficiencies (P, Fe, Zn, Mn, Cu, and Mo), while in acid soils were found deficiencies in N, P, K, Ca, Mg, and S and excesses Fe, Mn, B, Zn, and Cu were found. [87]. The optimum pH for plant cells is considered around the cytoplasmic pH values (pH 7.0–7.5) [88–90].

Other ROS sources of ROS in plants are the toxicity of agrochemicals: herbicides, pesticides, fungicides, and vermicides, which affect plant cell membranes, lipids, photosynthetic pigments, and enzyme activities in a dose-effect manner tendency too [91, 92].

#### *2.3.2 Biotic stress*

Plant pathogens (viruses, bacteria, fungi, nematodes, and parasitic plants) are a cause of significant losses in crop productivity and ecosystem wealth. The infestation process follows a wide spectrum of physiological and biochemical reactions and symptoms accomplished by a decrease in photosynthesis, alterations in enzymatic patterns, changes in gene expression patterns, and the production and accumulation of ROS.

*Pathogen attack.* Plant–pathogen recognition is a process that includes different factors integrating in an overall result that determines the success of infestation [93].

When a potential plant host is unable to coordinate effective defense responses to prevent and reject pathogens from colonizing and proliferating within the host; it is considered that interaction is compatible [94]. In case of the incompatible interaction, plant defense is strong enough to prevent colonization and spread of the pathogen within host tissue [94, 95]. The plant immunity system is ROS-mediated induced and the plants generate first a primary ROS wave in response to preliminary infective contact by potential pathogens. In general, ROS synthesis occurs in two phases: I and II. Both phases differ in location, goal, magnitude, permanence in time, and magnitude. Phase I is a low-amplitude and transient action, like an alert signal. Phase II is more time prolonged and stronger with a typical increase of ROS production [95]. If a pathogen escapes recognition by the host, it fails to induce the second phase, and plants can not arm its integral defense system against the pathogen, suggesting a functional link between the ROS generation and the display of an effective immune response [96].

Plant pathogen recognition starts in the apoplast with the earliest oxidative reaction and the increment in ROS production $O_2^{\bullet-}$, and dismutated products of $H_2O_2$ [97]. The plasma membrane-localized pattern recognition receptors (PRRs) recognize the intruder and induce transient oxidative wave, mediated by respiratory burst oxidases (RBOH) and apoplastic peroxidases (aPO) [98]. RBOHs are signaling nodes in the ROS gene network of plants integrating a signal transduction pathway with ROS signaling. Plasma-membrane-localized PRRs are the main components of multiprotein complexes having additional transmembrane and cytosolic kinases. The pattern recognition receptors (PRRs) are a plasma membrane-localized class of proteins involved in the innate immune response of plants and located on the surface of plant cells, recognizing pathogen-associated molecular patterns (PAMPs). PRRs are considered the first line of defense against invading pathogens. In this process, some apoplastic enzymes are activated, as plasma membrane NADPH oxidases and cell wall peroxidases [13, 99]. Although the first oxidative wave is located in the apoplast, apoplastic ROS acts as mediator in cell signaling to induce ROS production in other cellular organelles and compartments (mitochondria and chloroplast) where ETC are placed. The ROS wave is followed by the accumulation of salicylic acid (SA) in its reduced form in the cytoplasm, accumulation of hormones and the deposition of callose at the cell wall and plasmodesmata, to prevent pathogen spread. Increases of ROS and salicylic acid after the pathogen recognition, induce redox-regulated transcriptional response mediated by NPR1, which is the central regulator of pathogens' attack response through a transcriptional gene expression cascade and mediated by salicylic acid and TGA transcription factors [100]. This process includes gene expression driven by redox signaling, while TGA transcription factors are involved in various cellular processes, connecting different hormonal and biochemical pathways, and regulatory intracellular and systemic elements. Usually, NPR1 is localized to the cytoplasm in oligomeric form with monomeric units bound by intermolecular bonds between Cys 82 and Cys 216. Monomeric NPR1 can be translocated into the nucleus for interacting with transcriptional factor TGA1 and subsequent regulatory defensive response. Translocation on monomeric NPR1 is antagonized by the presence of jasmonic acid, a plant hormone that promotes the S-nitrosylation of NPR1 on Cys156 leading oligomerization of monomeric NPR1, preventing its translocation to the cell nucleus. The nuclear localization of NPR1 is essential for the expression of PR genes [101]. Those transcriptional factors are essential to drive the gene expression including hormonal pathways, activation of genes encoded for pathogenesis related (PR) proteins, and transcriptional factors such as WRKY, a plant signaling system, modulating diverse biological functions, plant disease resistance and stress responses [24, 102–104]. Antioxidant enzymes allow the cells to resist the penetration of *Aphanomyces euteiches* and *Sclerotinia sclerotiorum* in plant tissues of lentil (*Lens culinaris*) [104]. In *Arabidopsis thaliana* L plants inoculated with the incompatible fungal pathogen *Alternaria brassicicola,* increments ROS generation and changes in the expression patterns of 35 different genes were observed from those of which 10 were down-regulated, and 25 were up-regulated. Among the genes that actived its expression (up-regulated) were those that code for the enzymes that drive the fatty acid ß-oxidation pathway [94]. The study of dynamic changes in enzymatic activities must be fundamental for understanding their role in resistance reactions and can be a useful tool for the preliminary selection of resistant cultivars. Enzymes like polyphenol oxidase (PPO) and pyranose peroxidase (POX) are important in the metabolism of ROS [105]. Experimental data show the role of oxidative enzymes in activating defense, like in the case of *F. oxysporum* in melon (*Cucumis melo*) [106] and in wheat cultivars assayed to a fungal pathogen, *Septoria tritici,* where resistant cultivar (cv. Stakado) generates higher magnitude ROS wave and susceptible cultivar (cv. Sevin) fails to generate the second Phase II of the ROS response [10, 107]. The face of *Septoria tritici* with wheat leaves infiltrated with catalase increases fungal penetration, colonization, and overall fungal biomass, suggesting that $H_2O_2$ is a critical factor in plant defense [107]. These results are consistent with those obtained in barley (*Hordeum vulgare*) *versus* two necrotrophic fungi, *Rhynchosporium secalis* and *Pyrenophora teres* [108]. In transgenic potato (*Solanum tuberosum*) transformed with a fungal gene encoding, glucose oxidase–conferred was observed resistance to bacterial soft rot disease and potato late blight [109]. In addition, $H_2O_2$ is also associated with the early events leading to the biosynthesis of phytoalexin [110]. ROS may be critical for establishing the hypersensitivity response (HR) of plants following infection and pathogen recognition to limit pathogen spread by initiating cell death at infected sites [111]. Low molecular weight elicitins proteins secreted by Phytophthora [112] induce an HR in many plant species [113].

Respiratory burst oxidase homolog (Rboh) genes have an important role in the pathogen-induced oxidative and hypersensitivity response (HR) [114]. In tobacco (*Nicotiana tabacum*), after elicitation induced with cryptogein, a 10-kD protein secreted by the oomycete *Phytophthora cryptogea*, tobacco cells transformed with antisense constructs of NtrbohD showed the same extracellular alkalinization as control plants, but they no longer produced ROS [10, 115]. In the tobacco benthi (*Nicotiana benthamiana*), the silencing of two Rboh cDNAs, NtrbohA and NtrbohB, lead to lower ROS levels and lower resistance to *Phytophthora infestans* showed that both genes are critical for $H_2O_2$ accumulation and for induction of resistance to this fungus [116]. In tomato (*Lycopersicon esculentum* M.), infection by *Botrytis cinerea* alters the action of the peroxisomal antioxidant system, and subsequently causes damage and enhances the pathogen-induced tissue senescence [13]. The capacity of plants to stand against sequentially or synergically stressors with harmless interference in their physiological functions depends on many aspects including nature, type and intensity of the stress and form of stress combination [24]. It was recently found that the increasing number and complexity of stressing combinations impact the plant, affecting the plant growth, declining survival dramatically [19, 117].

### *2.3.3 Antioxidant defense in plants*

Plants display an antioxidant defense based on nonenzymatic and enzymatic components. Antioxidant defenses are adapted and evolved according to the nature of plant cells. Some antioxidant components have been mentioned previously in this chapter when the oxidative damages were described. The antioxidant defense in plants is a complex, coordinated system displaying specificities in each organ, cellular organelles, and compartments, adapted to the plant living conditions. In steady stage conditions the antioxidant defense is able to maintain appropriate homeostasis and oxidative balance [13, 33, 45, 54]. The nature of the components in antioxidant defense has nonenzymatic or enzymatic natures. Nonenzymatic components include ascorbate (AsA), the major cellular redox buffers mostly located in cytoplasm, glutathione, (γ-glutamyl-cysteinyl-glycine, GSH), tocopherol, carotenoids, phenolic compounds, free aminoacids, vitamins, etc. Ascorbate (AsA) is a well-known and efficient antioxidant. It serves as a cofactor in large number of dioxygenases and other variety of biochemical reactions, scavenging primary ROS. It is mostly provided by the so-called Smirnoff-Wheeler pathway and from uronic acid intermediates [118, 119]. Part of AsA is translocated to the apoplast, where it acts as the first line of antioxidant defense [120]. Glutathione (GSH: γ-glutamyl cysteinyl-glycine) is one a crucial antioxidant located in cytosol, chloroplasts, endoplasmic reticulum, mitochondria, and other cell compartments. Although it is synthesized in the cytosol and chloroplast, where it can react with $O_2^{\bullet-}$, $^{\bullet}OH$, $H_2O_2$ and also has the capacity to detoxify cell from free radicals, protecting biopolymers from ROS and organic radicals [120]. The level of reduced form of GSH is an indicator of the oxidative status of the plant cells facing external stressors. Initially, external stressor stimulates GSH oxidation and increases in GSSH (glutathione oxidized form), followed by increment in GSH (glutathione reduced form), but sustained stressor strengths lead to OS [120].

Tocopherols (α, β, γ, and δ) are lipophilic molecules synthesized in photosynthetic tissues, the main form, α-tocopherol, is located in leaf chloroplasts, in the envelope, thylakoid membranes, and plastoglobuli. It scavenges ROS, usually $1O_2$ and OH from photosynthesis and protects lipids and thylakoid membranes from LPO, inducing tolerance to chilling, water deficit, and salinity in different plant species [121, 122]. Increases of alpha-tocopherol can help to mitigate plant stress tolerance, while low level of tocopherols indicates OS [122]. Carotenoids is another lipophilic antioxidant to $1O_2$ during photosynthesis. It is a pigment that absorbs light energy for further transfer to chlorophyll I (400 and 550 nm) harboring double bonds isoprene residues, allowing energic capture, caloric dissipation, and better adaptation of plants under drought/salinity stress [123, 124]. They are "de novo" synthesized in plant plastids of photosynthetic and sink organs. Carotenoids are important components of PhI and play a key role in photoprotection and adaptive response to stressors through cell signaling and phytohormone synthesis, as abscisic acid and strigolactone [125]. Other phenolic and polyphenolic compounds, such as flavonoids, tannins, hydroxycinnamate esters, and lignin, also have antioxidant properties [125].

The antioxidant defense also displays antioxidant enzymes, such as superoxide dismutase (SOD), catalase (CAT), guaiacol peroxidase (GPX), enzymes of ascorbate-glutathione (AsA-GSH), cycle ascorbate peroxidase (APX), monodehydroascorbate reductase (MDHAR), dehydroascorbate reductase (DHAR), and glutathione reductase (GR). Those enzymes are present in all aerobic and evolve together with organisms to respond properly against OS coordinately in different subcellular organelles and compartments [126]. Plant SOD belongs to the group of metalloenzymes catalyzing the dismutation of $O_2^{\bullet-}$ to $O_2$ and $H_2O_2$, and are present in three different isoenzymes differing in the associated metal they have: copper/zinc SOD (Cu/Zn-SOD), manganese SOD (Mn-SOD), and iron SOD (Fe-SOD) [126]. All genes encoding SOD isoforms are nuclear, and contain transit peptides for its translocation to the mitochondria (MnSOD, to chloroplasts (FeSOD), and to cytosol, chloroplast, peroxisome, and mitochondria (Cu/Zn-SOD isoforms). SOD overexpression is correlated with increased tolerance of the plant against environmental stresses and it could be used as an indirect selection criterion for screening drought-resistant plant materials [126, 127].

Other antioxidant enzymes are Catalases (CATs), which are highly active to $H_2O_2$ but much weaker against organic peroxides (R—O—O—R′) [127]. The major reservoirs of CAT in plant cells are the peroxisomes. Other enzymatic systems, like Xanthine Oxidase (XOD) are coupled to SOD [127, 128]. XOD are located in cytosol, chloroplast, and mitochondria in three isoforms. The guaiacol peroxidase (GPX) oxidizes aromatic electron found in compounds like guaiacol and pyragallo in vacuoles, cytosol and cell wall. Synthesis of ethylene and wound healing GPX increases its activity under the metal, chemical toxicity, and salinity stresses. In Safflower (*Carthamus tinctorius*), cultivars are tolerant to salinity by increasing of the GPX activity [128]. Ascorbate-glutathione cycle (AsA-GSH cycle) is the key enzyme in Halliwell-Asada pathway for the control of intracellular ROS levels [129]. The cycle involves ascorbate, glutathione, NADPH, and the related enzymes: ascorbate peroxidase (APX), monodehydroascorbate reductase (MDAR), dehydroascorbate reductase (DHAR), and glutathione reductase (GR) [129]. AsA and glutathione are linked through the ascorbate–glutathione cycle. In basal conditions, AsA is mostly found in its reduced form but under stress, ascorbic exists in oxidized form: dehydroascorbic acid (DHA). Alteration in the ratios AsA/DHA and GSH/GSSG are censoring parameters of cell oxidative status and of the correct work of antioxidant defense [28, 46]. In stressed conditions, ratio AsA/DHA decreases. The concentration of GSH is increased under stressed conditions showing the ability to deploy antioxidation activity so the ratio GSH/GSSG increases. Ascorbate peroxidase (APX) is another enzymatic component of antioxidant defense. It has different isoforms and it is reported to be an efficient antioxidant in response to abiotic stresses, such as drought, salinity, chilling, metal toxicity, and UV irradiation [46].

Antioxidant defense is a multifactorial interrelated system where whole antioxidant tools rather than a single detoxifying mechanism are necessary to keep oxidative balance. Consequently, the overexpression of one of its components probably will not change the antioxidant capacity of the whole pathway [130]. However, the overexpression in plants of one or various antioxidant enzymes can help to study their synergistic effect on stress tolerance [60, 67, 70].

## 3. Oxidative stress and plant microbiota

Plants live in a close relationship with soil microbiota. Microbiota is composed by microorganisms, such as bacteria, yeasts, fungi, protists and nematodes that form a complex, well-structured, interdependent trophic network. Symbiotic association are present in more than 70% of the vascular plants [131]. Microbiota displays multiple beneficial functions for plant development, soil productivity, biomass production, and conservation of the whole ecosystem, mitigating the negative effects of environmental stressors and OS damages [132]. It is crucial to make the nutrients accessible for plants and to improve water uptake. Plants and its microbiota produce ROS that coordinate the wealth and diversity of microbial populations, plant immune, and adaptive responses, allowing integrated regulation of ROS levels [8, 10, 133]. Microbiota is also known for their production of plant growth hormones, soil-borne, and systemic pathogen control [134]. The microbiota is distributed in different locations: surface (epiphytic), cavities, and vascular systems (endophytic), and plant vicinity, associated with the roots (rhizospheric) [131]. Some microorganisms can be endophytic and rhizospheric at the same time wild others are in free form in soils, but interact with rhizospheric microorganisms [22]. The endophytic bacteria, *Sphingomonas* sp. LK11, isolated from leaves of *Tephrosia apollinea*, mitigate salinity stress in wild-type tomato (*Solanum lycopersicum*) and tomato cultivar Got-3 by promoting root growth and expression of peroxiredoxin-glutathione S-transferase-, and glutaredoxin-related genes [133, 134]. Rice plants inoculated with the root endophytic fungus *Piriformospora indica*, showed improvement in roots and plant development due the proline accumulation [135]. The application of endophyte fungus *Metarhizium brunneum* strain CB15 in potato mitigated the nutrient deficiency by enhancing the photon capture capacity of PSII, better water uptakes and of N, P, and other nutrients [136]. Maize plants faced drought stress and inoculated with the endophytic fungus, *Piriformospora indica*, increase biomass, proline content, and enhanced up-regulation of antioxidants [6, 135–137]. In maize cultivars, tolerant to drought higher expression in drought-related genes were observed in plants inoculated with *Piriformospora indica* [137]. Wheat plants inoculated with six endophytic fungi enhanced the efficiency of photosynthesis increased plants development, grain weight, and germination rates under heat and drought [138].

Better adaptive response were obtained in japonica rice related to Paecilomyces *formosus* and faced heat stress. In this case, results evidenced the induction of downregulation of the stress-related signaling molecules, including ABA and jasmonic acid (JA) [139]. This fact can be explained by higher phytohormonal synthesis and expression of genes encoding for indole-3-acetamide hydrolase, aldehyde dehydrogenase for indoleacetic acid, and geranylgeranyl-diphosphate synthase, kaurene oxidase (P450-4), C13-oxidase (P450-3) for gibberellins synthesis [140]. Under drought conditions, chili pepper plants (*Capsicum annuum*) exhibited a significant increase in plant growth and higher activity in antioxidant enzymes: peroxidase, catalase and polyphenol oxidase, phenylalanine ammonia-lyase and Capsaicin synthase genes. Those activities induce phenylpropanoid biosynthesis, under the influence of endophytic fungus *Penicillium resedanum* LK6 [134]. The same results have been reported in wheat, maize, tomato, and sweet basil (*Ocimum basilicum*) plants inoculated with *Glomus fasciculatum*, *Cyclamen persicum*, *Rhizophagus irregularis*/*Variovorax paradoxus*, and *Glomus deserticola* [140–147].

Plant growth promotion bacteria (PGPR) synthesize phytohormones, including IAA, gibberellins, ethylene, ABA and cytokinins, promoting plants to become more tolerant of various abiotic environmental stresses, with increased plant dry biomass, flowering, and fruit production [148]. Strains of soil bacteria, such as *Azospirillum ssp.*, *Bacillus ssp.*, *Burkholderia ssp.*, *Erwinia ssp.*, *Pseudomonas ssp.*, *Rhizobium ssp.*, or *Serratia ssp.* among others, have been reported to be phosphate solubilizing, improver of nitrogen fixation, uptake of Fe, K, and Zn and other nutrients and production of plant hormones like IAA, an auxin involved in differentiation of vascular tissue, adventitious and lateral root, cell division, and shoot growth. [149]. They also stimulate the production of other enzymes like ACC-deaminase, chitinase, and glucanase, useful to control ethylene production and some phytopathogenic fungi [148]. Maize, sorghum, and soybean plants inoculated with PGPB better confront drought and salinity stresses by enhancing the concentration of ABA, regulating transcription of drought-related genes and increasing the production of salicylic acid, polyphenol, flavonoids, SOD, and 2,2-diphenyl-1-picryl-hydrazyl-hydrate (DPPH) scavenging activity [22, 133–139]. Secretion of ACC deaminase helps to improve tolerance to drought, keep homeostasis, reduction of ethylene production, and photosynthesis. That is the cases of *Bacillus aryabhattai* H19-1 and *Bacillus mesonae* H20-5 in tomato plants and *Variovorax paradoxus* in wheat, result in increasing proline, ABA, and improved antioxidant enzyme activities [150]. The absence of plant NADPH oxidase RBOHD (activator of ROS production) demonstrated that bacterial pathogen *Xanthomonas* sp. can alter phyllospheric and endospheric composition in microbiota [150]. Loss of RBOHD directly impacts the growth of individual bacteria promoting reconfiguration of microbe–microbe interactions, but the mechanism of this action remains unclear [150–153]. The dynamic of ROS-mediated plant microbiome interaction suggests that ROS also modulates plant immunity and microbiota homeostasis [153] and induces systemic tolerance (IST) [154]. Both plant and microbial communities produce ROS to coordinate the diversity and abundance of microbial populations of ROS gradients within plant organs (roots and shoots) and plant cell compartments (apoplast, plasma membrane, organelles) [155, 156]. The selection for ROS scavenging strains could explain compositional trends within plant microbiomes [155–157].

## 4. Characterization of beneficial bacterial strains to mitigate drought stress in maize and sorghum plants, a practical example

Microbiota plays an important role in plant development and the whole ecosystem [157]. We evaluate the plant growth and water stress resistance in maize and sorghum plants inoculated with of three selected bacterial isolates [22]. The selected strains were isolated from a natural ecosystem and preliminarily evaluated in activities considered important for deploying important biochemical activities to plant growth, nutrition and pest prevention, individually or in complementarity with each other, allowing them to be a part of a multifunctional microbial consortium. Those characteristics were: Gram-staining, IAA production, ACC deaminase activity in no stressed and stressed conditions, siderophores production, cellulase activity, chitinase activities, phosphate's solubilization, polyhydroxybutyrate (PHB) production, $H_2S$ production, and phospholipase activity. Each bacterium was classified by comparative analysis of 16 S amplicon sequences in each case and the results were compared to the data bank available and were classified as: *Pseudomonas ssp.* XiU1297, *Luteibacter ssp. 1292*, *Acinetobacter ssp*. XiU12138. Mutual compatibility was also assayed with positive results. Those three strains were subjected to several rounds of mutagenesis by culturing them in a minimal medium supplemented with guanidium thiocyanate and those capable of chelating heavy metals such as Al, Zn, and a mixture of Cu, Fe, and Ni were selected. Metal chelating putative mutants were individualized and analyzed to select those that, in addition to acquiring the ability to chelate metals, retained the original characteristics that we consider important for plants. Selected mutants of each bacterial species were used to form a bacterial consortium and designed as: *Pseudomonas ssp.* XiU1297-Nmut, *Luteibacter ssp.* XiU1292-Nmut, *Acinetobacter ssp.* XiU12138-Nmut (**Table 1**). The consortium contains about $5 \times 10^7$ colony-forming units of each strain and in drought-induced experiments in maize and sorghum plants in greenhouse conditions according with standard methodology during 6 week-period of drought stress [22].

Maize and sorghum plants were analyzed for the following parameters: root fresh weigh, root dry weight, leave-stem fresh weight, leave-stem dry weight. In case of maize, the best growth promotion was achieved by *Pseudomonas ssp.* XiU1297-Nmut, followed by *Luteibacter ssp.* XiU1292-NMut and *Acinetobacter ssp*. XiU12138-Nmut (**Figure 1**). For sorghum plants, the highest growth-promotions were obtained by *Luteibacter ssp.* XiU1292-Nmut followed by *Pseudomonas ssp.* XiU1297-NMut, and *Acinetobacter ssp*. XiU12138-Nmut (**Figure 2**).

| N° | Evaluated characteristics | Bacterial strain | | |
|---|---|---|---|---|
| | | ***Pseudomonas ssp.* XiU1297-Nmut** | ***Luteibacter ssp.* XiU1292-Nmut** | ***Acinetobacter ssp.* XiU12138-Nmut** |
| 1 | Gram staining | Gram Positive | Gram Positive | Gram Positive |
| 2 | Qualitative IAA | High | Medium | Medium |
| 3 | Quantitative IAA (μg/ml) | 25.34 ± 0.25 | 22.62 ± 0.19 | 23.09 ± 0.14 |
| 4 | ACC deaminase (no drought) (μM/mg α-ketobutyrate) | 5.35 ± 0.22 | 4.97 ± 0.31 | 5.15 ± 017 |
| 5 | ACC deaminase (drought) (μM/mg α-ketobutyrate) | 3.23 ± 0.16 | 3.18 ± 0.21 | 3.27 ± 0.13 |
| 6 | Siderophores production | Medium | Low | Medium |
| 7 | Cellulase activity | High | Medium | High |
| 8 | Chitinase activities | Medium | High | Medium |
| 9 | Solubilization of phosphates | High | High | High |
| 10 | Polyhydroxybutyrate (PHB) production | Medium | High | High |
| 11 | Production of H2S | Low | High | Medium |
| 12 | Phospholipase activity | High | High | High |
| 13 | Chelating Al | High | — | Medium |
| 14 | Chelating Zn | Medium | High | Low |
| 15 | Chelating Cu, Ni, Co | High | Medium | Medium |
| 16 | Intercompatibility | Compatible | Compatible | Compatible |
| 17 | Localization in host plants | Rhizospheric and endophytic | Rhizospheric and endophytic | Rhizospheric and endophytic |

*Nmut: Strain chemically mutagenized*

**Table 1.**
*Some characteristic and qualitative properties of individual bacterial strains were used as criteria for selection of Pseudomonas ssp. XiU1297-Nmut, and Luteibacter ssp. XiU1292-Nmut, Acinetobacter ssp. XiU12138-Mnut (De la Riva G.A. Unpublished data, [22]).*

We also explore some oxidative parameters after six weeks of hydric stress including: total antioxidant capacity, LPO, hydrogen peroxide assay, ascorbic acid assay, glutathione, assay, and MDA. We observed that treatments with selected bacterial strains promote better oxidative balance and better resistance to drought conditions in both maize and sorghum compared with negative control (no inoculated plants) and plants maintained in a normal irrigation regimen (**Table 2**). Relatively high levels of exogenous ROS production are common features of pathogenic microbes, whereas ROS scavenging (and/or low net ROS production) has been continually linked to beneficial microbes [14, 15, 148].

The short living period and nature of ROS (particularly $O_2^{\bullet-}$ and $^{\bullet}OH$) present serious methodological challenges for their detection and trajectory within complex systems, but established links and dynamic in microbiota, its wealth, structure, diversity, and concentration, seems to be in a function of these ROS gradients. The state-of the-arts in the knowledge on plant-microorganism interrelationships calls us to be cautelous when attempting to generalize microbial functions across varied environmental contexts, including the mechanisms of action [8–9, 157]. The relation

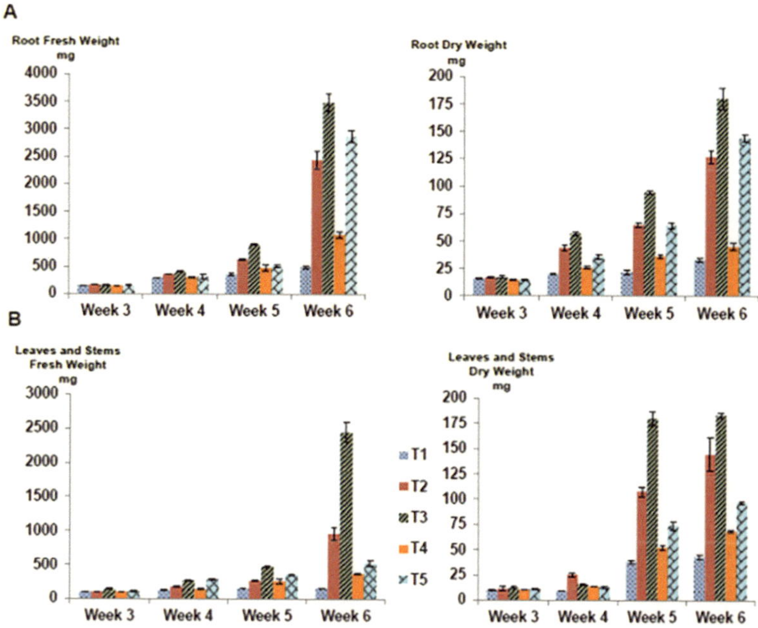


**Figure 1.**
*Effect of bacterial strains on maize´s roots (A), stems-leaves (B). The values correspond to fresh and dry weight. Treatments: T1-No bacterial treatment, T2-Luteibacter ssp. XiU1292-Nmut, T3-Pseudomonas ssp. XiU1297-Nmut, T4-Acinetobacter ssp. XiU12138-Nmut, T5-Daily irrigated plant without bacterial treatment. Water irrigation was discontinued at the end of week 3 and restarted at the end of week 4. Best results in maize were obtained by inoculation with Pseudomonas ssp. XiU1297-Nmut (T3). Nmut: Strain chemically mutagenized and selected [22].*

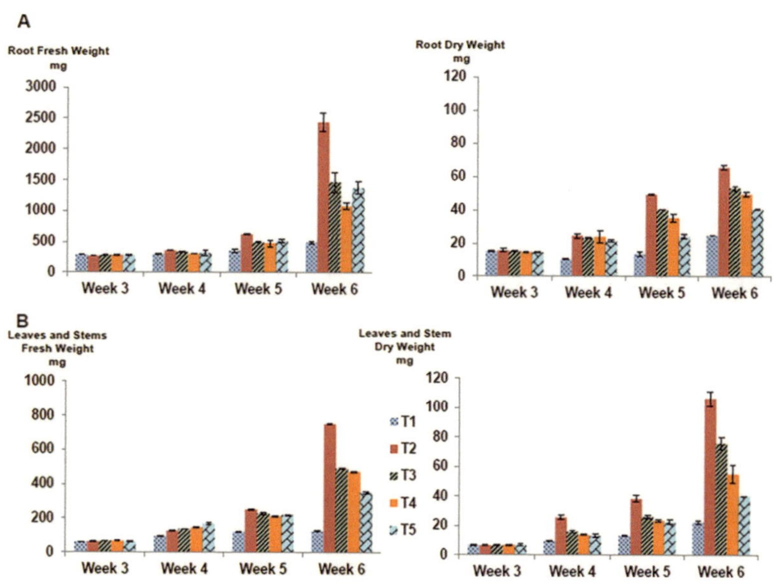


**Figure 2.**
*Effect of bacterial strains on sorghum's roots (A), stems-leaves (B). The values correspond to fresh and dry weight. Treatments: T1-No bacterial treatment, T2-Luteibacter ssp. XiU1292-Nmut, T3- Pseudomonas ssp. XiU1297-Nmut, T4- Acinetobacter ssp. XiU12138-Nut, T5- Daily irrigated plant without bacterial treatment. Water irrigation was discontinued at the end of week 3 and restarted at the end of week 4. Best results were obtained by Luteibacter ssp. XiU1292-Nmut (T2) [22]. Nmut: Strain chemically mutagenized and selected.*

| N | Bacterial strain treatment | Total antioxidant capacity (mmol of TE/g fresh weight) | Lipid peroxidation (nM/g fresh weight) | Hydrogen peroxide assay (µM/g fresh weigh) | Ascorbic acid assay (mg/g) | Glutathione assay (µg/g fresh weight) | MDA (mM/g fresh weight) | OC |
|---|---|---|---|---|---|---|---|---|
| *Maize week 6 drought stress* | | | | | | | | |
| 1 | *Pseudomonas ssp.* XiU1297-Nmut | 8.56 ± 0.14 | 3.52 ± 0.14 | 3.61 ± 0.16 | 3.86 ± 0.14 | 3.56 ± 0.14 | 3.56 ± 0.14 | C |
| 2 | *Luteibacter ssp.* XiU1292-Nmut | 7.75 ± 0.12 | 5.45 ± 0.12 | 3.85 ± 0.12 | 3.55 ± 0.12 | 3.75 ± 0.12 | 3.75 ± 0.12 | C |
| 3 | *Acinetobacter ssp.* XiU12138-Nmut | 7.23 ± 0.15 | 4.23 ± 0.15 | 3.63 ± 0.15 | 3.83 ± 0.15 | 3.23 ± 0.15 | 3.23 ± 0.15 | C |
| 4 | No bacterial treatment drought | 3.75 ± 0.14 | 8.75 ± 0.14 | 6.75 ± 0.14 | 6.55 ± 0.19 | 6.75 ± 0.14 | 6.75 ± 0.14 | OS |
| 5 | No bacterial treatment/water | 6.67 ± 0.14 | 4.08 ± 0.12 | 3.87 ± 0.15 | 3.45 ± 0.14 | 3.21 ± 0.14 | 3.21 ± 0.17 | C |
| *Sorghum week 6 drought stress* | | | | | | | | |
| 1 | *Pseudomonas ssp.* XiU1297-Nmut | 6.56 ± 0.12 | 10.56 ± 0.14 | 3.56 ± 0.14 | 38.56 ± 0.14 | 13.56 ± 0.14 | 3.56 ± 0.13 | C |
| 2 | *Luteibacter ssp.* XiU1292- Nmut | 7.75 ± 0.16 | 13.75 ± 0.12 | 3.75 ± 0.12 | 39.75 ± 0.12 | 14.75 ± 0.12 | 3.75 ± 0.12 | C |
| 3 | *Acinetobacter ssp.* XiU12138-Mmut | 8.23 ± 0.13 | 12.23 ± 0.15 | 3.23 ± 0.15 | 45.23 ± 0.15 | 13.23 ± 0.15 | 3.23 ± 0.10 | C |
| 4 | No bacterial treatment drougt | 4.75 ± 0.19 | 26.75 ± 0.14 | 6.75 ± 0.14 | 16.75 ± 0.14 | 6.03 ± 0.14 | 6.75 ± 0.11 | OS |
| 5 | Not bacterial treatment/water | 6.55 ± 0.11 | 4.42 ± 0.14 | 3.96 ± 0.19 | 3.25 ± 0.24 | 3.29 ± 0.12 | 3.33 ± 0.19 | C |

*Nmut: strain chemically mutagenized and selected; OC: oxidative condition; C: controlled oxidative status; OS: oxidative stress.*

**Table 2.**
*Effect of plant beneficial bacteria on plant antioxidant capacity in maize and sorghum leaves under drought conditions at week 6 of hydric stress. Treatments with Pseudomonas ssp. XiU1297-Nmut, and Luteibacter ssp. XiU1292-Nmut, Acinetobacter ssp. XiU12138-Mnut were evaluated and resulted in better antioxidant defense in maize and sorghum plants (De la Riva G.A. Unpublished data, [22]).*

between ROS activity, antioxidant defense status, and the ability to mitigate drought stress is evidenced. The complementation in activities displayed by selected bacterial strains and the established intercompatibility between them open. They open up the possibility of reconstitute a microbial consortium suitable for application according to the environmental conditions and crop's specificities [158].

## 5. Conclusions

Oxidative stress (OS) is closely related to plant adaptive response to biotic and abiotic stressors. Adverse environmental abiotic stress is a very limiting factor for crop productivity and wild ecosystem wealth and includes drought, salinity, heavy metals, nutrient deficiency, extreme temperatures, high radiation, and even stress caused by agrochemicals. Biotic stress is caused by plant pathogens, fungi, virus, bacteria, nematodes, and weeds. In environmental conditions, both abiotic and biotic stressors can act in sequential or synergically combination enhancing its adverse effects in affected plants. The ROS production within certain limiting doses is essential as a subcellular and cellular signaling system and inductor plant antioxidant defense, plant immunity and adaptive response. ROS overproduction can escape the antioxidant defense causing OS, damaging cellular structures and affecting important physiological functions. Plant microbiota interacts with plants in a complex relation and interdependence. Although many aspects of plant-microbiota functioning are still under research, it is evident its capacity to mitigate the biotic and abiotic stresses. The beneficial microbes such as plant-growth-promoting rhizobacteria (PGPRs) and arbuscular mycorrhizal fungi (AMFs) among other microorganisms can help improve stress tolerance by enhancing plant growth, stimulating the production of phytohormones, siderophores, and solubilizing phosphates, lowering ethylene levels, secreting enzymes, and metabolites with antipathogenic activities and upregulating the expression of antioxidant genes [159].

The plant-microbe interaction leads to the modulation of complex mechanisms in the plant cellular system. Moreover, the residing microbial flora also inhibits the phytopathogens, therefore, it becomes part of plants' innate defense system and this represents an important opportunity to develop suitable strategies to improve ecosystem wealth and crop productivity facing the challenges of climatic changes and growing needs to use more eco-friendly agricultural technologies and safer products for pest control and plant nutrition.

## Acknowledgment

This work was supported by CORFO Chile (Corporación de Fomento de la Producción) through the research project entitled "Research and development of novel endemic microbial consortiums and their validations as high impact biostimulators against challenging conditions provoked by climate change affecting strategic crops (cherry, blueberries, hazelnut)" Grant 23CV12-251461.

## Conflict of interest

The authors declare no conflict of interest.

## Abbreviations

| | |
|---|---|
| OS | oxidative stress |
| ROS | reactive oxygen species |
| RONS | reactive oxygen and nitrogen species |
| SOD | superoxide dismutase |
| GPO | glutathione peroxidase |
| CAT | catalases |
| ETC | electronic transport chain |
| PSI | photosynthetic system I |
| PSII | photosynthetic system II |
| APO | ascorbate peroxidase |
| GOX | glycolate oxidase |
| AO | aldehyde oxidase |
| XOD | xanthine oxidase |
| MDA | malondialdehyde |
| LPO | lipid peroxidation |
| UV | ultraviolet radiation |
| PCD | programmed cell death |
| RBOH | respiratory burst oxidases |
| PRRs | pattern recognition receptors |
| NPR1 | non-expresser of pathogenesis related genes 1 |
| PR | pathogenesis related genes/proteins |
| IAA | indole acetic acid |

## Author details

Gustavo Alberto De la Riva[1,2]*, Juan Carlos Hernández González[3], Rolando Morán Valdivia[4] and Rolando García González[4]

1 SynergiaBio México S.A. de C.V., Copala, Jalisco, Mexico

2 Department of Biochemistry, National Technological Institute of Mexico TecNM- ITLP, La Piedad, Michoacán, Mexico

3 Department of Biochemistry, National Technological Institute of Mexico TecNM- ITM, Morelia, Michoacán, Mexico

4 SynergiaBio/BioTECNOS Ltda, San Javier de Loncomilla, Maule, Chile

*Address all correspondence to: gustavo.riva@synergiabio.com and gustavo.rr@piedad.tecnm.mx

## References

[1] Lyons T, Reinhard CT, Planavsky NJ. The rise of oxygen in Earth's early ocean and atmosphere. Nature. 2014;**506**(7488):307-315. DOI: 10.1038/nature13068

[2] Halliwell B. Free radicals and antioxidants: Updating a personal view. Nutrition Reviews. 2012;**5**(2577):265. DOI: 10.1111/j.1753-4887.1994.tb01453.x

[3] Sies H. Oxidative stress (OS): Concept and some practical aspects. Antioxidants. 2020;**9**:852. DOI: 10.3390/antiox9090852

[4] Massaad CA, Klann E. Reactive oxygen species in the regulation of synaptic plasticity and memory. Antioxidants and Redox Signalling. 2011;**14**(10):2013-2054. DOI: 10.1089/ars.2010.3208

[5] Bild W, Ciobica M, Padurariu M, Bild V. The interdependence of the reactive species of oxygen, nitrogen, and carbon. Journal of Physiology and Biochemistry. 2013;**69**(1):147-154. DOI: 10.1007/s13105-012-0162-2

[6] De la Riva G, Torres-Osornio A, Agüero-Chapín G, Saldaña-Trujillo A, Arellano-Nuñez A, López-Mendoza F. Promising probiotic food supplement based on combination of yeast beta-glucan, bioselenium and lactoferrin for animal health. Biomedical Journal of Scientific & Technical Research. 2020;**32**(4):25105-25113. DOI: 10.26717/BJSTR.2020.32.005271

[7] Swanson S, Gilroy S. ROS in plant development. Physiologia Plantarum. 2010;**138**:384-392. DOI: 10.1111/j.1399-3054.2009.01313.x

[8] Foyer C. Reactive oxygen species, oxidative signaling and the regulation of photosynthesis. Environmental and Experimental Botany. 2018;**154**:134-142. DOI: 10.1016/j.envexpbot.2018.05.003

[9] Mittler R. ROS are good. Trends in Plant Science. 2017;**22**(2017):11-19. DOI: 10.1016/j.tplants.2016.08.002

[10] Berrios L, Rentsch J. Linking reactive oxygen species (ROS) to abiotic and biotic feedbacks in plant microbiomes: The dose makes the poison. International Journal of Molecular Sciences. 2022;**23**:4402. DOI: 10.3390/ijms23084402.11

[11] Czarnocka W, Karpinski S. Friend or foe? Reactive oxygen species production, scavenging and signalling in plant response to environmental stresses. Free Radical Biology & Medicine. 2018;**122**:4-2013. DOI: 10.1016/j.freeradbiomed.2018.01.011

[12] Sies H, Jones D. Reactive oxygen species (ROS) as pleiotropic physiological signalling agents. Nature Reviews. Molecular Cell Biology. 2020;**21**:363-383. DOI: 10.1038/s41580-020-0230-3

[13] Mittler R, Zandalinas S, Fichman Y, Van Breusegem F. ROS signalling in plant stress responses. Nature Reviews. Molecular Cell Biology. 2022;**23**:663-679. DOI: 10.1038/s41580-022-00499-2

[14] Sarvajeet S, Tuteja N. Reactive oxygen species and antioxidant machinery in abiotic stress tolerance in crop plants. Plant Physiology and Biochemistry. 2010;**48**(12):909-930. DOI: 10.1016/j.plaphy.2010.08.016

[15] Huang H, Ullah F, Zhou D, Yi M, Zhao Y. Mechanisms of ROS regulation of plant development and stress responses. Frontiers in Plant Science. 2019;**10**:800. DOI: 10.3389/fpls.2019.00800

[16] Liu H, Li J, Carvalhais L, Percy C, Verma J, Schenk P, et al. Evidence for the plant recruitment of beneficial microbes to suppress soil-borne pathogens. The New Phytologist. 2020;**229**:2873-2885. DOI: 10.1111/nph.17057

[17] Kansman JT, Basu S, Casteel CL, Crowder DW, Lee BW, Nihranz CT, et al. Plant water stress reduce a frontier in ecology and evolution. Frontiers in Ecology and Evolution. 2022;**10**:1-11. DOI: 10.3389/fevo.2022.846908

[18] Cohen I, Zandalinas S, Fritschi F, Sengupta S, Fichman Y, Azad R, et al. The impact of water deficit and heat stress combination on the molecular response, physiology, and seed production of soybean. Physiologia Plantarum. 2021;**172**(1):41-52. DOI: 10.1111/ppl.13269

[19] Zandalinas S, Fritschi F, Mittler R. Global warming, climate change, and environmental pollution: Recipe for a multifactorial stress combination disaster. Trends in Plant Science. 2021;**26**(6):588-599. DOI: 10.1016/j.tplants.2021.02.011

[20] Balfagon D, Teran F, dos Reis De Oliveira T, Santa-Catarina C, Gomez-Cadenas A. Citrus rootstocks modify scion antioxidant system under drought and heat stress combination. Plant Cell Reports. 2022;**41**(3):593-602. DOI: 10.1007/s00299-021-02744-y

[21] Jia Y, Ordon J, Copeland C, Emonet A, Geldner N, Guan R, et al. Coordination of microbe–host homeostasis by crosstalk with plant innate immunity. Nature Plants. 2021;**7**:814-825. DOI: 10.1038/s41477-021-00920-2

[22] Colli J, de la Riva G, Vargas-Sámano C, Pérez-Machado G, Agüero-Chapin G. Plant growth promoting bacteria isolated from a Mexican natural ecosystem induce water stress resistance in maize and sorghum plants. Journal of Microbial & Biochemical Technology. 2017;**9**:209-219. DOI: 10.4172/1948-5948.1000367

[23] Rillig MR, Lehmann A, Aguilar-Trigueros C, Buchert S, Wulf A, Iwasaki A, et al. The role of multiple global change factors in driving soil functions and microbial biodiversity. Science. 2019;**366**(6467):886-890

[24] Nawaz M, Li L, Azeem F, Shabbir S, Zohaib A, Ashraf U, et al. Insight of transcriptional regulators reveals the tolerance mechanism of carpet-grass (*Axonopus compressus*) against drought. BMC Plant Biology. 2021;**21**(1):1-14. DOI: 10.1126/science.aay2832

[25] Castro B, Citterico M, Kimura S, Stevens D, Wrzaczek M, Coaker G. Stress-induced reactive oxygen species compartmentalization, perception and signalling. Nature Plants. 2021;**7**:403-412. DOI: 10.1038/s41477-021-00887-0

[26] Pelaez-Vico M, Fichman Y, Zandalinas S, Van Breusegem F, Karpinski S, Mittler R. ROS and redox regulation of cell-to-cell and systemic signalling in plants during stress. Free Radical Biology and Medicine. 2022;**193**:354-362. DOI: 10.1016/j.freeradbiomed.2022.10.305

[27] Huihui Z, Xin L, Zisong X, Yue W, Zhiyuan T, Meijun A, et al. Toxic effects of heavy metals Pb and Cd on mulberry (*Morus alba* L.) seedling leaves: Photosynthetic function and reactive oxygen species (ROS) metabolism responses. Ecotoxicology and Environmental Safety. 2020;**195**:110469. DOI: 10.1016/j.ecoenv.2020.110469

[28] Hasanuzzaman M, Bhuyan M, Zulfiqar F, Raza A, Mohsin S,

Mahmud J, et al. Reactive oxygen species and antioxidant defense in plants under abiotic stress: Revisiting the crucial role of a universal defense regulator. Antioxidants. 2020;**9**:681. DOI: 10.3390/antiox9080681

[29] Bienert G, Møller A, Kristiansen K, Schulz A, Møller I, Schjoerring J, et al. Specific aquaporins facilitate the diffusion of hydrogen peroxide across membranes. Journal of Biological Chemistry. 2007;**282**:1183-1192. DOI: 10.1074/jbc.M603761200

[30] Quan L, Zhang B, Shi W, Li H. Hydrogen peroxide in plants: A versatile molecule of the reactive oxygen species network. Journal of Integrative Plant Biology. 2008;**50**:2-18. DOI: 10.1111/j.1744-7909.2007.00599.x

[31] Gill S, Tuteja N. Reactive oxygen species and antioxidant machinery in abiotic stress tolerance in crop plants. Plant Physiology and Biochemistry. 2010;**48**:909-930. DOI: 10.1016/j.plaphy.2010.08.016

[32] Miller G, Schlauch K, Tam R, Cortes D, Torres M, Shulaev V, et al. The plant NADPH oxidase RBOHD mediates rapid systemic signalling in response to diverse stimuli. Science Signaling. 2009;**2**:ra45. DOI: 10.1126/scisignal.2000448

[33] Mittler R, Vanderauwera S, Suzuki N, Miller G, Tognetti V, Vandepoele K, et al. ROS signalling: The new wave? Trends in Plant Science. 2011;**16**:300-309. DOI: 10.1016/j.tplants.2011.03.007

[34] Dietz K, Turkan I, Krieger-Liszkay A. Redox- and reactive oxygen species-dependent signalling in and from the photosynthesizing chloroplast. Plant Physiology. 2016;**171**:1541-1550. DOI: 10.1104/pp.16.00375

[35] Hussain S, Ulhassan Z, Brestic M, Zivcak M, Zhou W, et al. Photosynthesis research under climate change. Photosynthesis Research. 2021;**150**:5-19. DOI: 10.1007/s11120-021-00861-z

[36] Strizh I. The Mehler reaction as an essential link between environmental stress and chloroplast redox Signalling. En J. In: Allen, EG. Gantt, & B. Osmond, Photosynthesis. Energy from the Sun. Dordrecht: Springer. 2008. DOI: 10.1007/978-1-4020-6709-9_289

[37] Gelhaye E, Rouhier N, Navrot N, Jacquot J. The plant thioredoxin system. Cellular and Molecular Life Sciences. 2005;**62**(1):24-35. DOI: 10.1007/s00018-004-4296-4

[38] Singh A, Kumar A, Yadav S, Singh I. Reactive oxygen species-mediated signalling during abiotic stress. Plant Gene. 2019;**18**:100-173. DOI: 10.1016/B978-0-323-85665-2.00022-4

[39] Kerchev P, Waszczak C, Lewandowska A, Willems P, Shapiguzov A, Li Z. Lack of GLYCOLATEOXIDASE1, but not GLYCOLATE OXIDASE2, attenuates the photorespiratory phenotype of CATALASE2-deficient Arabidopsis. Plant Physiology. 2016;**171**:1704-1719. DOI: 10.1104/pp.16.00359

[40] Del Río L, López-Huertas E. ROS generation in peroxisomes and its role in cell Signalling. Plant and Cell Physiology. 2016;**57**(7):364-1376. DOI: 10.1093/pcp/pcw076

[41] Navrot N, Rouhier N, Gelhaye E, Jaquot J. Reactive oxygen species generation and antioxidant systems in plant mitochondria. Physiologia Plantarum. 2007;**129**:185-195. DOI: 10.1111/j.1399-3054.2006.00777.x

[42] Hernansanz-Agustín PE. Generation of reactive oxygen species by mitochondria. Antioxidants. 2021;**10**(3):415. DOI: 10.3390/antiox10030415

[43] Janku M, Luhová L, Petrivalský M. On the origin and fate of reactive oxygen species in plant cell compartments. Antioxidants. 2019;**2019**:105. DOI: 10.3390/antiox8040105

[44] Kärkönen A, Kuchitsu K. Reactive oxygen species in cell wall metabolism and development in plants. Phytochemistry. 2015;**112**:22-32. DOI: 10.1016/j.phytochem.2014.09.016

[45] Sharma P, Jha A, Dubey R, Pessarakli M. Reactive oxygen species, oxidative damage, and antioxidative defense mechanism in plants under stressful conditions. Journal of Botany. 2012;**2012**:1-26, 217037. DOI: 10.1155/2012/217037

[46] Hasanuzzaman M, Bhuyan M, Anee T, Parvin K, Nahar K, Mahmud J, et al. Regulation of ascorbate-glutathione pathway in mitigating oxidative damage in plants under abiotic stress. Antioxidants. 2019;**8**:384. DOI: 10.3390/antiox8090384

[47] Corpas F, Gupta D, Palma J. Production sites of reactive oxygen species ROS in organelles from plant cells. In: Gupta DP, editor. Reactive Oxygen Species and Oxidative Damage in Plants under Stress. Cham: Springer; 2015. pp. 1-22. DOI: 10.1007/978-3-319-20421-5_1

[48] Demidchik V. Mechanisms of oxidative stress (OS) in plants: From classical chemistry to cell biology. Environmental and Experimental Botany. 2015;**109**:212-228. DOI: 10.1016/j.envexpbot.2014.06.021

[49] Anjum N, Sofo A, Scopa A. Lipids and proteins—Major targets of oxidative modifications in abiotic stressed plants. Environmental Science and Pollution Research. 2015;**22**:4099-4121. DOI: 10.1007/s11356-014-3917-1

[50] Møller I, Jensen P, Hansson A. Oxidative modifications to cellular components in plants. Annual Review of Plant Biology. 2007;**58**:459-481. DOI: 10.1146/annurev.arplant.58.032806.103946

[51] Kale R, Hebert A, Frankel L, Sallans L, Bricker T, Pospíšil P. Amino acid oxidation of the D1and D2 proteins by oxygen radicals during photoinhibition of photosystem II. Proceedings of the National Academy of Sciences of the United States of America. 2017;**114**:2988-2993. DOI: 10.1073/pnas.1618922114

[52] Saha P, Mukherjee A, Biswas A. Modulation of NaCl induced DNA damage and oxidative stress (OS) in mung bean by pretreatment with sublethal dose. Biologia Plantarum. 2015;**59**:139-146. DOI: 10.1007/s10535-014-0460-3

[53] Choudhury S, Panda P, Sahoo L, Pandspecies signalling in plants under a S. Reactive oxygen abiotic stress. Plant Signaling & Behavior. 2013;**8**:e23681

[54] Banerjee A, Roychoudhury A. Abiotic stress, generation of reactive oxygen species, and their consequences: An overview. In: Singh V, Singh S, Tripathi D, Prasad S, Chauhan D, editors. Reactive Oxygen Species in Plants: Boon or Bane- Revisiting the Role of ROS. Chichester, UK: John Wiley & Sons Ltd.; 2018. pp. 23-50. DOI: 10.1002/9781119324928.ch2

[55] Lehmann M, Laxa M, Sweetlove L, Fernie A, Obata T. Metabolic recovery

of *Arabidopsis thaliana* roots following cessation of oxidative stress (OS). Metabolomics. 2012;**8**:143-153. DOI: 10.1007/s11306-011-0296-1

[56] Ma Y, Días M, Freitas E. Drought and salinity stress response and microbe-induced tolerance in plants. Frontiers in Plant Science. 2020;**11**:591911. DOI: 10.3389/fpls.2020.591911

[57] Fahad S, Bajwa A, Nazir U, Anjum S, Farooq A, Zohaib A, et al. Crop production under drought and heat stress: Plant responses and management options. Frontiers in Plant Science. 2017;**8**:1147. DOI: 10.3389/fpls.2017.01147

[58] Hasanuzzaman M, Nahar K, Anee T, Khan M, Fujita M. Silicon-mediated regulation of antioxidant defense and glyoxalase systems confers drought stress tolerance in *Brassica napus* L. South African Journal of Botany. 2018;**115**:50-57. DOI: 10.1016/j.sajb.2017.12.006

[59] Hasanuzzaman M, Hossain M, Teixeira da Silva J, Fujita M. Plant responses and tolerance to abiotic oxidative stress (OS): Antioxidant defense is a key factor. In: Bandi VS, Shanker C, Mandapaka M, editors. Crop Stress and its Management: Perspectives and Strategies. Berlin: Springer; 2012. pp. 261-316. DOI: 10.1007/978-94-007-2220-0_8

[60] Nahar K, Hasanuzzaman M, Alam M, Rahman A, Mahmud J, Suzuki T, et al. Insights into spermine-induced combined high temperature and drought tolerance in mung bean: Osmoregulation and roles of antioxidant and glyoxalase system. Protoplasma. 2017;**254**:445-460. DOI: 10.1007/s00709-016-0965-z

[61] Hasanuzzaman M, Nahar K, Hossain M, Anee T, Parvin K, Fujita M. Nitric oxide pretreatment enhances antioxidant defense and glyoxalase systems to confer PEG-induced oxidative stress (OS) in rapeseed. Journal of Plant Interactions. 2017;**12**:323-331. DOI: 10.1080/17429145.2017.1362052

[62] Singh D, Singh C, Taunk J, Tomar R, Chaturvedi A, Gaikwad K, et al. Transcriptome analysis of lentil (*Lens culinaris* M) in response to seedling drought stress. BMC Genomics. 2017;**18**:206. DOI: 10.1186/s12864-017-3596-7

[63] Liu J, Hasanuzzaman M, Wen H, Zhang J, Peng T, Sun H, et al. High temperature and drought stress cause abscisic acid and reactive oxygen species accumulation and suppress seed germination growth in rice. Protoplasma. 2019;**256**:1217-1227. DOI: 10.1007/s00709-019-01354-6

[64] Devireddy AR, Tschaplinski TJ, Tuskan GA, Muchero W, Chen JG. Role of reactive oxygen species and hormones in plant responses to temperature changes. International Journal of Molecular Sciences. 2021;**22**(16):8843. DOI: 10.3390/ijms22168843

[65] Siddiqui M, Mostofa M, Akter M, Srivastava A, Sayed M, Hasan M, et al. Impact of salt-induced toxicity on growth and yield-potential of local wheat cultivars: Oxidative stress (OS) and ion toxicity are among the major determinants of salt-tolerant capacity. Chemosphere. 2017;**187**:385-394. DOI: 10.1007/s00709-019-01354-6

[66] Ahmad P, Ahanger M, Alam P, Alyemeni M, Wijaya L, Ali S, et al. Silicon (Si) supplementation alleviates NaCl toxicity in mung bean (*Vigna radiata* L. Wilczek) through the modifications of physio-biochemical attributes and key antioxidant enzymes. Journal of Plant Growth Regulation. 2019;**38**:70-82. DOI: 10.1007/s00344-018-9810-2

[67] Nahar K, Rahman M, Hasanuzzaman M, Alam M, Rahman A, Suzuki T, et al. Physiological and biochemical mechanisms of spermine-induced cadmium stress tolerance in mung bean (*Vigna radiata* L.) seedlings. Environmental Science and Pollution Research. 2016;**23**:21206-21218. DOI: 10.1007/s11356-016-7295-8

[68] El-Amier Y, Elhindi K, El-Hendawy S, Al-Rashed S, Abd-El Gawad A. Antioxidant system and biomolecules alteration in *Pisum sativum* under heavy metal stress and possible alleviation by 5-aminolevulinic acid. Molecules. 2019;**24**:4194. DOI: 10.3390/molecules24224194

[69] Gupta D, Pena L, Romero-Puertas M, Hernández A, Inouhe M, Sandalio L. NADPH oxidases differentially regulate ROS metabolism and nutrient uptake under cadmium toxicity. Plant, Cell & Environment. 2017;**40**:509-526. DOI: 10.1111/pce.12711

[70] Nahar K, Hasanuzzaman M, Rahman A, Alam M, Mahmud J, Suzuki T, et al. Polyamines confer salt tolerance in mung bean (*Vigna radiata* L.) by reducing sodium uptake, improving nutrient homeostasis antioxidant defense, and methylglyoxal detoxification systems. Frontiers in Plant Science. 2016;**7**:1104. DOI: 10.3389/fpls.2016.01104

[71] Hasanuzzaman M, Nahar K, Anee T, Fujita M. Exogenous silicon attenuates cadmium-induced oxidative stress (OS) in *Brassica napus* L. by modulating AsA-GSH pathway and glyoxalase system. Frontiers in Plant Science. 2017;**8**:1061. DOI: 10.3389/fpls.2017.01061

[72] Mahmud J, Hasanuzzaman M, Nahar K, Bhuyan M, Fujita M. Insights into citric acid-induced cadmium tolerance and phytoremediation in *Brassica juncea* L.: Coordinated functions of metal chelation antioxidant defense and glyoxalase systems. Ecotoxicology and Environmental Safety. 2018;**147**:990-1001. DOI: 10.1016/j.ecoenv.2017.09.045

[73] Kabala K, Zboinska M, Głowiak D, Reda M, Jakubowska D, Janicka M. Interaction between the signalling molecules hydrogen sulfide and hydrogen peroxide and their role in vacuolar H+-ATPase regulation. Physiologia Plantarum. 2019;**166**:688-704. DOI: 10.1111/ppl.12819

[74] Mustafa AZ, Naveed M, Ali H, Kintl A, Saeed Q, Kucerik J, Brtnicky M. Nickel (Ni) phytotoxicity and detoxification mechanisms: A review. Chemosphere. 2023;**328**:138574. DOI:10.1016/j. chemosphere.2023.138574

[75] Rai N, Morales L, Aphalo PJ. Perception of solar UV radiation by plants: Photoreceptors and mechanisms. Plant Physiology. 2021;**186**(3):1382-1396

[76] Zha L, Liu W, Zhang Y, Zhou C, Shao M. Morphological and physiological stress responses of lettuce to different intensities of continuous light. Frontiers in Plant Science. 2019;**6**:1440. DOI: 10.3389/fpls.2019.01440

[77] González-Villagra J, Marjorie R-D, Alberdi M, Acevedo P, Loyola R, Tighe-Neira R, et al. Solar UV irradiation effects on photosynthetic performance, biochemical markers, and gene expression in highbush blueberry (*Vaccinium corymbosum* L.) cultivars. Scientia Horticulturae. 2020;**259**:108816. DOI: 10.1016/j. scienta.2019.108816

[78] Li H, Li D, Yang Z, Zeng Q, Luo Y, He N. Flavones produced by mulberry flavone synthase type I constitute a defense line against the ultraviolet-B stress. Plants. 2020;**9**:215. DOI: 10.3390/plants9020215

[79] Robson TK, Urban O, Jansen M. Re-interpreting plant morphological responses to UV-B radiation. Plant, Cell & Environment. 2015;**38**(5):856-866. DOI: 10.1111/pce.12374

[80] Tiwari S, Agrawal M. Effect of ozone on physiological and biochemical processes of plants. In: Tiwari ES, Agrawal M, editors. Tropospheric Ozone and its Impacts on Crop Plants. New York, NY, USA: Springer; 2018. pp. 65-113. DOI: 10.1007/978-3-319-71873-6_3

[81] Apel K, Hirt H. Reactive oxygen species: Metabolism, oxidative stress (OS), and signal transduction. Annual Review of Plant Biology. 2004;**55**:373-399. DOI: 10.1146/annurev.arplant.55.031903.141701

[82] Lee J, Woo S, Kwak M, Park S, Kim H, Lim Y, et al. Effects of elevated temperature and ozone in *Brassica juncea* L.: Growth, Physiology, and ROS accumulation. Forests. 2020;**11**:68. DOI: 10.3390/f11010068

[83] Bhuyan M, Hasanuzzaman M, Mahmud J, Hossain M, Alam M, Fujita M. Explicating physiological and biochemical responses of wheat cultivars under acidity stress: Insight into the antioxidant defense and glyoxalase systems. Physiology and Molecular Biology of Plants. 2019;**25**:865-879. DOI: 10.1007/s12298-019-00678-0

[84] Liu D, Liu M, Liu X, Cheng X, Liang Z. Silicon priming created an enhanced tolerance in alfalfa (*Medicago sativa* L.) seedlings in response to high alkaline stress. Frontiers in Plant Science. 2018;**9**:716. DOI: 10.3389/fpls.2018.00716

[85] Bhuyan M, Hasanuzzaman M, Mahmud J, Hossain M, Bhuiyan T, Fujita M. Unraveling morpho physiological and biochemical responses of *Triticum aestivum* L. to extreme pH: Coordinated actions of antioxidant defense and glyoxalase systems. Plants. 2019;**8**:24. DOI: 10.3390/plants8010024

[86] Bhuyan M, Hasanuzzaman M, Nahar K, Mahmud J, Parvin K, Bhuiyan T, et al. Plants behavior under soil acidity stress: Insight into morphophysiological, biochemical, and molecular responses. In: Hasanuzzaman M, Hakeem K, Nahar K, Alharby H, editors. Plant Abiotic Stress Tolerance. Cham, Switzerland: Springer; 2019. pp. 35-82. DOI: 10.1007/978-3-030-06118-0

[87] Clark R, Baligar V. Acidic and alkaline soil constraints on plant mineral nutrition. In: Wilkinson R, editor. Plant–Environment Interactions. New York, NY, USA: Marcel Dekker Inc.; 2000. pp. 133-177. DOI: 10.1201/9780824746568

[88] Shavrukov Y, Hirai Y. Good and bad protons: Genetic aspects of acidity stress responses in plants. Journal of Experimental Botany. 2015;**67**:15-30. DOI: 10.1093/jxb/erv437

[89] Yang C, Wang P, Li C, Shi D, Wang D. Comparison of effects of salt and alkali stresses on the growth and photosynthesis of wheat. Photosynthetica. 2008;**46**:107-114. DOI: 10.1007/s11099-008-0047-3

[90] Spormann S, Soares C, Fidalgo F. Salicylic acid alleviates glyphosate-induced oxidative stress (OS) in *Hordeum vulgare* L. Journal of Environmental Management. 2019;**241**:226-234. DOI: 10.1016/j.jenvman.2019.04.035

[91] Parween T, Jan S, Mahmooduzzafar S, Fatma T, Siddiqui Z. Selective effect of pesticides on plant-A review. Critical Reviews in Food Science

and Nutrition. 2016;**56**(1):160-179. DOI: 10.1080/10408398.2013.787969

[92] Peillex C, Pelletier M. The impact and toxicity of glyphosate and glyphosate-based herbicides on health and immunity. Journal of Immunotoxicology. 2020;**17**(1):163-174. DOI: 10.1080/1547691X.2020.1804492

[93] Lodha T, Basak. Plant-pathogen interactions: What microarray tells about it? Molecular Biotechnology. 2012;**50**:87-97. DOI: 10.1007/s12033-011-9418-2

[94] Schenk P, Kazan K, Manners JA, Jonathan P, Simpson R, Wilson I, et al. Systemic gene expression in Arabidopsis during an incompatible interaction with *Alternaria brassicicola*. Plant Physiology. 2003;**132**:999-1010. DOI: 10.1104/pp.103.021683

[95] Castejón O, Guillamon A, Huguet G. Phase-amplitude response functions for transient-state stimuli. The Journal of Mathematical Neuroscience. 2013;**3**:13. DOI: 10.1186/2190-8567-3-13

[96] Wang Y, Pruitt RN, Nürnberger T, Wang Y. Evasion of plant immunity by microbial pathogens. Nature Reviews. Microbiology. 2022;**20**:449-464. DOI: 10.1038/s41579-022-00710-3

[97] Mansoor S, Naqvi F. Effect of heat stress on lipid peroxidation (LPO) and antioxidant enzymes in mung bean (*Vigna radiata* L.) seedlings. African Journal of Biotechnology. 2013;**12**:3196-3203. DOI: 10.5897/AJB12.2808

[98] Daudi A, Cheng Z, O'Brien J, Mammarella N, Khan S, Ausubel F, et al. The apoplastic oxidative burst peroxidase in Arabidopsis is a major component of pattern-triggered immunity. Plant Cell. 2012;**24**:275-287. DOI: 10.1105/tpc.111.093039

[99] Kámán-Tóth E, Dankó T, Gullner G, Bozsó Z, Palkovics L, Pogány M. Contribution of cell wall peroxidase- and NADPH oxidase-derived reactive oxygen species to *Alternaria brassicicola*-induced oxidative burst in Arabidopsis. Molecular Plant Pathology. 2019;**20**(4):485-499. DOI: 10.1111/mpp.12769

[100] Kumar S, Zavaliev R, Wu Q, Kumar S, Zavaliev R, Wu Q, et al. Structural basis of NPR1 in activating plant immunity. Nature. 2022;**605**:561-566. DOI: 10.1038/s41586-022-04699-w

[101] Mou Z, Fan W, Dong X. Inducers of plant systemic acquired resistance regulate NPR1 function through redox changes. Cell. 2003;**113**:935-944. DOI: 10.1016/s0092-8674(03)00429-x

[102] Rushton P, Somssich I, Ringler P, Shen Q. WRKY transcription factors. Trends in Plant Science. 2010;**15**(5):247-258. DOI: 10.1016/j.tplants.2010.02.006

[103] Tada Y, Spoel S, Pajerowska-Mukhtar K, Mou Z, Song J, Wang C, et al. Plant immunity requires conformational changes [corrected] of NPR1 via S-nitrosylation and thioredoxins. Science. 2008;**321**(5891):952-956. DOI: 10.1126/science.1156970

[104] Aqa M, Yan C, Mohammad A, Hassan J, Hnin WK, Xiaoxue H, et al. ROS and Oxidative response Systems in Plants under Biotic and Abiotic Stresses: Revisiting the crucial role of Phosphite triggered plants defense response. Frontiers in Microbiology. 2021;**12**:1631318. DOI: 10.3389/fmicb.2021.631318

[105] Zhu S, Dai G, Ma T, Chen C, Chen D, Lü X, et al. Distribution of lignin phenols in comparison with plant-derived lipids in the alpine versus

temperate grassland soils. Plant and Soil. 2019;**439**:325-338 https://www.jstor.org/stable/48703817

[106] Sadeghpour N, Asadi-Gharneh H, Nasr-Esfahani M, Hassanzadeh Khankahdani H, Golabadi M. Antioxidant enzymes associated with resistance to *fusarium oxysporum* F. Sp. *melonis* race 1.2 in melon. Physiological and Molecular Plant Pathology. 2022;**121**:101880. DOI: 10.1071/FP22131

[107] Setty N, Mehrabi R, Lütken H, Haldrup A, Kema G, Collinge D, et al. Role of hydrogen peroxideduring the interaction between the hemibiotrophic fungal pathogen *Septoria tritici* and wheat. The New Phytologist. 2007;**174**:637-647. DOI: 10.1111/j.1469-8137.2007.02026.x

[108] Able A. Role of reactive oxygen species in the response of barley to necrotrophic pathogens. Protoplasma. 2003;**221**:137-143

[109] Wu G, Shortt B, Lawrence E, Levine E, Fitzsimmons K, Shah D. Disease resistance conferred by expression of a gene encoding $H_2O_2$ generating glucose oxidase in transgenic potato plants. The Plant Cell. 1995;7(9):1357-1368. DOI: 10.1105/tpc.7.9.1357

[110] Degousee N, Triantaphylides C, Montillet J. Involvement of oxidative processes in the signalling mechanisms leading to the activation of glyceollin synthesis in soybean (*Glycine max*). Plant Physiology. 1994;**104**:945-952. DOI: 10.1104/pp.104.3.945

[111] Levine A, Tenhaken R, Dixon R, Lamb C. H2O2 from the oxidative burst orchestrates the plant hypersensitive disease resistance response. Cell. 1994;**79**:583-593. DOI: 10.1016/0092-8674(94)90544-4

[112] Derevnina L, Dagdas Y, De la Concepcion J, Bialas A, Kellner R, Petre B, et al. Nine things to know about elicitins. New Phytologist. 2023;**4**:888-895. DOI: 10.1111/nph.14137

[113] Chandra S, Prakash Sharma I. Elicitins as microbe-associated molecular patterns and their role in plant defense. In: Chandra D, Bhatt P, editors. Developments in Applied Microbiology and Biotechnology. Oxford: Academic Press; 2023. pp. 77-86. DOI: 10.1016/B978-0-323-99896-3.00001-1

[114] Kadota Y, Shirasu K, Zipfel C. Regulation of the NADPH oxidase RBOHD during plant immunity. Plant and Cell Physiology. 2015;**56**(8):1472-1480

[115] Qing D, Yang Z, Li M, Wong WS, Guo G, Liu S, et al. Quantitative and functional Phosphoproteomic analysis reveals that ethylene regulates water transport via the C-terminal phosphorylation of aquaporin PIP2;1 in Arabidopsis. Molecular Plant. 2016;**9**(1):158-174. DOI: 10.1016/j.molp.2015.10.001

[116] Rin S, Imano S, Camagna M. Expression profiles of genes for enzymes involved in capsidiol production in *Nicotiana benthamiana*. Journal of General Plant Pathology. 2020;**86**:340-349. DOI: 10.1007/s10327-020-00931-5

[117] Zandalinas S, Mittler R. Plant responses to multifactorial stress combination. New Phytologist. 2022;**234**(4):1161-1167. DOI: 10.1111/nph.18087

[118] Ghosh N, Das A, Chaffee S, Roy R, Sen K. Reactive oxygen species, oxidative damage and cell death. In: Chatterjee S, Jungraithmayr W, Bagchi D, editors. Immunity and Inflammation in Health and Disease. Oxford: Academic

Press; 2018. pp. 45-55. DOI: 10.1016/B978-0-12-805417-8.00004-4

[119] Sharova E, Medvedev S, Demidchik V. Ascorbate in the Apoplast: Metabolism and functions. Russian Journal of Plant Physiology. 2020;**67**:207-220. DOI: 10.1134/S1021443720020156

[120] Averill-Bates D. The antioxidant glutathione. In: Litwack G, editor. Vitamins and Hormones. Vol. 121. Oxford: Academic Press; 2023. pp. 109-141. DOI: 10.1016/bs.vh.2022.09.002

[121] Ali E, Hussain S, Hussain KK, Shah JH, Zaidi R, Jan M, et al. Tocopherol as plant protector: An overview of tocopherol biosynthesis enzymes and their role as antioxidant and signalling molecules. Acta Physiologiae Plantarum. 2022;**44**:20. DOI: 10.1007/s11738-021-03350-x

[122] Štolfa Čamagajevac I, Žuna Pfeiffer T, Špoljarić Maronić D. Abiotic stress response in plants: The relevance of tocopherols. In: Gupta D, Palma J, Corpas F, editors. Antioxidants and Antioxidant Enzymes in Higher Plants. Cha: Springer; 2018. pp. 233-251. DOI: 10.1007/978-3-319-75088-0_11

[123] Uarrota V, Stefen D, Leolato L, Gindri D, Nerling D. Revisiting carotenoids and their role in plant stress responses: From biosynthesis to plant Signalling mechanisms during stress. In: Gupta D, Palma J, Corpas F, editors. Antioxidants and Antioxidant Enzymes in Higher Plants. Cham: Springer; 2018. pp. 207-232. DOI: 10.1007/978-3-319-75088-0_10

[124] Zhang RR, Li T, Tan G, Tao J, Su X, Xu Z, et al. Effects of simulated drought stress on carotenoid contents and expression of related genes in carrot tap roots. Protoplasma. 2021;**258**(2):379-390. DOI: 10.1007/s00709-020-01570-5

[125] Tuladhar P, Sasidharan S, Saudagar P. Role of phenols and polyphenols in plant defense response to biotic and abiotic stresses. In: Jogaiah S, editor. Biocontrol Agents and Secondary Metabolites. Cambridge: Woodhead Publishing; 2021. pp. 419-441. DOI: 10.1016/B978-0-12-822919-4.00017-X

[126] Stephenie S, Chang Y, Gnanasekaran A, Esa N, Gnanaraj C. An insight on superoxide dismutase (SOD) from plants for mammalian health enhancement. Journal of Functional Foods. 2020;**68**:103917. DOI: 10.1016/j.jff.2020.103917

[127] Vanz Borges C, Oliveira Orsi R, Maraschin M, Pereira Lima G. Oxidative stress (OS) in plants and the biochemical response mechanisms. In: Ghorbanpour M, Shahid MA, editors. Plant Stress Mitigator. Oxford: Academic Press; 2023. pp. 455-468. DOI: 10.1016/B978-0-323-89871-3.00022-7

[128] Aslam M, Fakher B, Ashraf M, Cheng Y, Wang B, Qin Y. Plant low-temperature stress: Signalling and response. Agronomy. 2022;**12**(3):702. DOI: 10.3390/agronomy12030702

[129] Mir M, John R, Alyemeni M, Alam P, Ahmad P. Jasmonic acid ameliorates alkaline stress by improving growth performance, ascorbate glutathione cycle and glyoxalase system in maize seedlings. Scientific Reports. 2018;**8**:28-31. DOI: 10.1038/s41598-018-21097-3

[130] Nianiou-Obeidat I, Madesis P, Kissoudis C, Voulgari G, Chronopoulou E, Tsaftaris A, et al. Plant glutathione transferase-mediated stress tolerance: Functions and biotechnological applications. Plant Cell Reports. 2017;**36**:791-805. DOI: 10.1007/s00299-017-2139-7

[131] Trivedi P, Leach J, Tringe S, Sa TS. Plant-microbiome interactions: From community assembly to plant health. Nature Reviews. Microbiology. 2020;**18**(11):607-621. DOI: 10.1038/s41579-020-0412-1

[132] Bacon C, White J. Functions, mechanisms and regulation of endophytic and epiphytic microbial communities of plants. Symbiosis. 2015;**68**:87-98. DOI: 10.1007/s13199-015-0350-2

[133] Khan A, Shin J, Jung H, Lee I. Regulations of capsaicin synthesis in *Capsicum annuum* L. by *Penicillium resedanum* LK6 during drought conditions. Scientia Horticulturae. 2014;**175**:167-173. DOI: 10.1016/j.scienta.2014.06.008

[134] Khan A, Waqas M, Asaf S, Kamran M, Shahzad R, Bilal S. Plant growth-promoting endophyte *Sphingomonas sp*. LK11 alleviates salinity stress in *Solanum pimpinellifolium*. Environmental and Experimental Botany. 2017;**133**:58-69. DOI: 10.1016/j.envexpbot.2016.09.009

[135] Mansour F, Ali E. Evaluation of proline functions in saline conditions. Phytochemistry. 2017;**140**:52-68. DOI: 10.1016/j.phytochem.2017.04.016

[136] Krell V, Unger S, Schoenwandt D, Patel A. Endophytic *Metarhizium brunneum* mitigates nutrient deficits in potato and improves plant productivity and vitality. Fungal Ecology. 2018;**34**:43-49. DOI: 10.1016/j.funeco.2018.04.002

[137] Xu L, Wang A, Wang J, Wei Q, Zhang W. *Piriformospora indica* confers drought tolerance on *Zea mays* L. through enhancement of antioxidant activity and expression of drought-related genes. Crop Journal. 2017;**5**:251-258. DOI: 10.1016/j.cj.2016.10.002

[138] Hubbard M, Germida J, Vujanovic V. Fungal endophytes enhance wheat heat and drought tolerance in terms of grain yield and second-generation seed viability. Journal of Applied Microbiology. 2014;**116**:109-122. DOI: 10.1111/jam12311

[139] Waqas M, Khan AL, Shahzad R, Ullah I, Khan AR, Lee IJ. Mutualistic Fungal endophytes produce phytohormones and organic acids that promote japonica rice groeth under heat stress. Journal of Zhejiang University. Science. B. 2015;**16**:1011-1018. DOI: 10.1631/jzus.B1500081

[140] Bilal S, Shahzad R. Imran MJ, Kim K, Lee I. Synergistic association of endophytic fungi enhances Glycine max L. resilience to combined abiotic stresses: Heavy metals, high temperature and drought stress. Industrial Crops and Products. 2019;**143**:111931. DOI: 10.1016/j.indcrop.2019.111931

[141] Maya M, Matsubara Y. Influence of arbuscular mycorrhiza on the growth and antioxidative activity in cyclamen under heat stress. Mycorrhiza. 2013;**23**:381-390. DOI: 10.1007/s00572-013-0477-z

[142] Mathur S, Sharma M, Jajoo A. Improved photosynthetic efficacy of maize *Zea mays* plants with arbuscular mycorrhizal fungi (AMF) under high temperature stress. Journal of Photochemistry and Photobiology. B. 2016;**180**:149-154. DOI: 10.1016/j.jphotobiol.2018.02.002

[143] Mathur S, Tomar R, Jajoo A. Arbuscular mycorrhizal fungi (AMF) protects photosynthetic apparatus of wheat under drought stress. Photosynthesis Research. 2018;**39**:227-238. DOI: 10.1007/s11120-018-0538-4

[144] Elhindi K, El-Din S, Elgorban A. The impact of arbuscular mycorrhizal

fungi in mitigating salt-induced adverse effects in sweet basil (*Ocimum basilicum* L.). Saudi Journal of Biological Sciences. 2017;24:170-179. DOI: 10.1016/j.sjbs.2016.02.010

[145] Ait-El-Mokhtar M, Laouane R, Anli M, Boutasknit A, Wahbi S, Meddich A. Use of mycorrhizal fungi in improving tolerance of the date palm (Phoenix dactylifera L.) seedlings to salt stress. Scientia Horticulturae. 2019;**253**:429-438. DOI: 10.1016/j.scienta.2019.04.066

[146] Navarro JM, Pérez-Tornero O, Morte A. Alleviation of salt stress in citrus seedlings inoculated with arbuscular mycorrhizal fungi depends on the rootstock salt tolerance. Journal of Plant Physiology. 2014;**171**:76-85. DOI: 10.1016/j.jplph.2013.06.006

[147] Duc N, Csintalan Z, Posta K. Arbuscular mycorrhizal fungi mitigate negative effects of combined drought and heat stress on tomato plants. Plant Physiology and Biochemistry. 2018;**132**:297-307. DOI: 10.1016/j.plaphy.2018.09.011

[148] Glick B. Plant growth-promoting bacteria: Mechanisms and applications. Scientifica. 2012;**2012**:963401. DOI: 10.6064/2012/963401

[149] Wissal C, Barakat A, Zeroual Y, Bargaz A. Phosphate bacterial solubilization: A key rhizosphere driving force enabling higher P use efficiency and crop productivity. Journal of Advanced Research. 2022;**38**:13-28. DOI: 10.1016/j.jare.2021.08.014

[150] Song C, Sarpong C, Zhang X, Wang W, Wang L, Gan Y, et al. Mycorrhizosphere bacteria and plant-plant interactions facilitate maize P acquisition in an intercropping system. Journal of Cleaner Production. 2021;**314**:127993. DOI: 10.1016/j.jclepro.2021.127993

[151] Shalev O, Karasov TL, Lundberg DL, Ashkenazy H, Pramoj Na Ayutthaya P, Weigel D. Commensal pseudomonas strains facilitate protective response against pathogens in the host plant. Nature Ecology & Evolution. 2022;**6**:383-396. DOI: 10.1038/s41559-022-01673-7

[152] Colaianni N, Parys K, Lee H-S, Conway J, Kim N, Edelbacher N, et al. A complex immune response to flagellin epitope variation in commensal communities. Cell Host & Microbe. 2021;**29**:635-649.e9. DOI: 10.1016/j.chom.2021.02.006

[153] Santoyo G. How plants recruit their microbiome? New insights into beneficial interactions. Journal of Advanced Research. 2022;**40**:45-58. DOI: 10.1016/j.jare.2021.11.020

[154] Marcec M, Gilroy S, Poovaiah B, Tanaka K. Mutual interplay of $Ca^{2+}$ and ROS signalling in plant immune response. Plant Science. 2019;**283**:343-354. DOI: 10.1016/j.plantsci.2019.03.004

[155] Lemanceau P, Blouin M, Muller D, Moënne-Loccoz Y. Let the core microbiota be functional. Trends in Plant Science. 2017;**22**:583-595. DOI: 10.1016/j.tplants.2017.04.008

[156] Han G. Origin and evolution of the plant immune system. The New Phytologist. 2019;**222**:70-83. DOI: 10.1111/nph.15596

[157] Tiepo A, Constantino L, Madeira T, Gonçalves L, Pimenta J, Bianchini E, et al. Plant growth-promoting bacteria improve leaf antioxidant metabolism of drought-stressed Neotropical trees. Planta. 2020;**251**:83. DOI: 10.1007/s00425-020-03373-7

[158] Maurya A. Oxidative Stress (OS) in crop plants. In: Hasanuzzaman M, editor. Agronomic Crops. Singapore: Springer; 2020. pp. 349-380. DOI: 10.1007/978-981-15-0025-1_18

[159] Koza NA, Adedayo AA, Babalola OO, Kappo AP. Microorganisms in plant growth and development: Roles in abiotic stress tolerance and secondary metabolites secretion. Microorganisms. 2022;**10**(8):1528. DOI: 10.3390/microorganisms10081528

Section 4

# Clinical and Therapeutic Applications of Antioxidants

Chapter 13

# Biomarker of Oxidative Stress in Erythrocytes: Clinical Significance

*Poonam Tripathi*

## Abstract

Red blood cells (RBCs) are highly specialized and the most abundant cells. RBCs, as oxygen carriers, are constantly exposed to high oxygen tension, both with endogenous and exogenous oxidants. Oxidative stress most commonly is described as the increased production of reactive oxygen species (ROS) and reactive nitrogen species (RNS). It is a condition in which there is a significant decrease in antioxidant capacity, leading to irreversible damage to erythrocytes, resulting in their eventual damage by hemolysis and ultimately their elimination from circulation. These oxidants comprise a large group of molecules with diverse properties. The widely studied clinical biomarkers of oxidative stress and their mechanisms in human erythrocytes are the most interesting and focused issues when describing RBCs' mechanism of action. As mature RBCs are cells that contain no nucleus and other cell organelles, they have no capacity to repair the damage caused by oxidative stress. The effects of oxidative stress in erythrocytes, the biomarkers of oxidative stress, and the effective antioxidative enzyme systems that neutralize reactive oxidants into less reactive species will be discussed in this chapter.

**Keywords:** red blood cell, oxidative stress, reactive oxygen species, antioxidant, hemolysis

## 1. Introduction

Red blood cells, also known as erythrocytes, are the most important cell types in the blood. Erythrocytes play a critical role in mammalian metabolism, they are vital suppliers of oxygen to tissues and transport waste carbon dioxide [1]. Healthy RBCs have a biconcave shape, develop in our bone marrow, and are released into the bloodstream after they fully mature, a process that takes about seven days. These mature RBCs are then released into the bloodstream and survive for about 100 to 120 days.

The mature RBC lacks a nucleus and organelles, but it is pleasingly structured to perform the essential function of delivering oxygen and removing carbon dioxide from all other cells while withstanding the shear stress imposed by navigating small vessels.

Healthy RBCs in circulation are continuously exposed to both endogenous and exogenous sources of reactive oxygen species (ROS) that can damage the RBC and impair its function [2].

Oxidative stress is a serious condition, and it occurs when the generation of reactive oxygen species (ROS), active intermediates of free radicals, exceeds the body's ability to neutralize and eliminate them comparatively decreased [3–5]. Enhanced production of ROS increases levels and leads to an imbalance of oxidant/antioxidant activity [6]. Thus, to reduce the consequences of ROS produced as a resultant of increased oxidative stress condition, for this oxidative stress condition RBCs have a wide-ranging antioxidant system including non-enzymatic low molecular-weight antioxidants like glutathione and ascorbic acid and enzymatic antioxidants including superoxide dismutase (SOD), catalase (CAT), glutathione S Transferase (GST), glutathione peroxidase (GPx).

In normal circumstances, there is a perfect equilibrium between the production of oxidants and antioxidants. But in pathological conditions, there is a modification as the production of oxidants increases, causing oxidative stress [7]. These excess ROS as a result of oxidative stress conditions reacts rapidly with other molecules; inside the cell environment, this causes the defective function of many organelles, principally the cell membrane, leading to cytotoxicity and organ failure [8, 9].

Oxidative stress is caused by prolonged imbalance and antioxidant depletion as well as hyper-production of reactive oxygen species causing damage to mitochondria, lysosomes, lipid membranes, proteins, and DNA. Hence, the antioxidant enzymes, which play an important role in the protection of the cells from oxidative damage and scavenging ROS produced under oxidative stress, are currently the most focused topic for researchers to study [10, 11].

ROS production and accumulation are basic mechanisms of oxidative stress in a number of disease states and can cause severe cellular damage, leading to physiological dysfunction and also cell death. Resetting critical homeostatic parameters is critical in situations when oxidative stress condition occurs as cells function in extreme situations to counteract the oxidant effects and to restore redox balance. Altered cellular activities, may lead to the activation or silencing of genes encoding anti-oxidative enzymes, and transcription factors. Once oxidative stress increases inside the cell, accumulation of oxidative products of lipids, nucleic acids, proteins, sugars, and sterols occurs finally causing cellular dysfunction and making our body prone to external deleterious agents (**Figure 1**).

Defects in antioxidant pathways are connected to several different types of diseases including diabetes, age-related diseases, and cancer [12–14]. A number of previous studies have examined the association between ROS and various polymorphisms of antioxidants and revealed that excessive oxidative stress with decreased antioxidant enzyme activity can worsen the pathologic states of many patients. Moreover, the distribution of these genetic variants varies with ethnic variations.

Since oxidative stress plays a major role in the pathophysiology of disease, it is important to know all the important genetic factors that influence the performance of the antioxidant defense system.

To cope with oxidative stress, humans have evolved with antioxidant systems to protect against free radicals cells and have developed a ubiquitous antioxidant defense system, which consists of enzymatic defense such as superoxide dismutase (SOD), catalase (CAT), glutathione peroxidase (GPx), and glutathione reductase (**Figure 2**) together with a number of non-enzymatic low molecular-weight antioxidants such as the iron-binding proteins transferrin and ferritin, and vitamins A, C, and E and so on. However, the antioxidant defense system might be influenced by several pathological or environmental factors so that a fraction of reactive oxygen species (ROS) may remain affected and form additional reactive hydroxyl radicals. An upsurge in

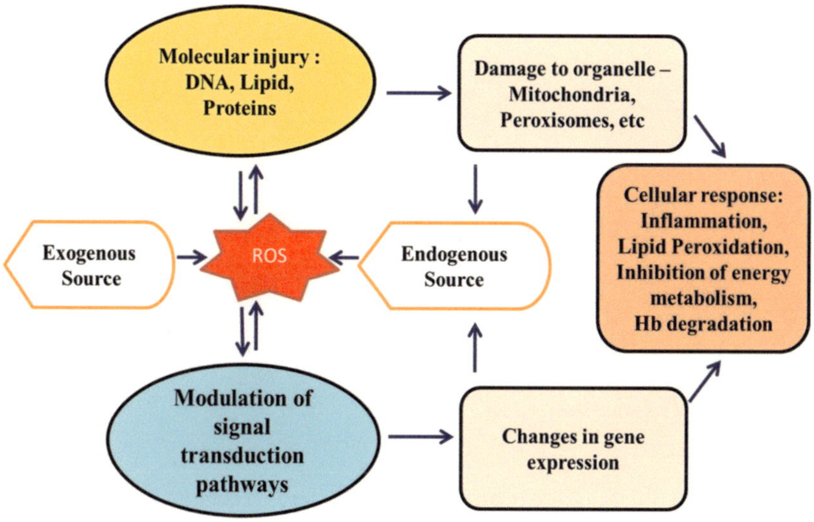


**Figure 1.**
*Mechanism of reactive oxygen species (ROS) generation and stimulation of the molecular pathways ultimately results in an adverse cellular response.*

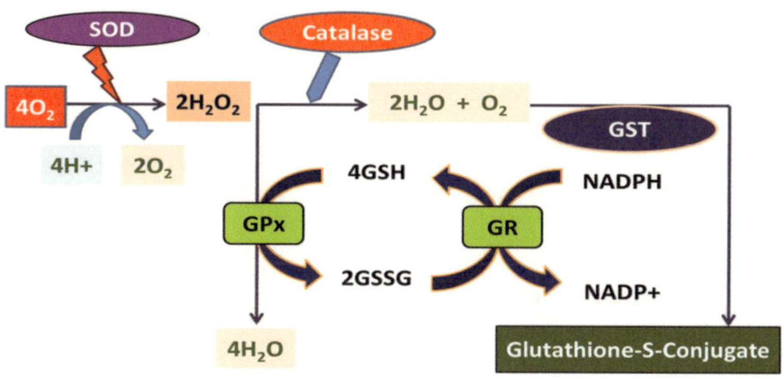


**Figure 2.**
*Antioxidative defense system.*

ROS generated due to oxidative injury to DNA and other biomolecules may ruin the normal functions of tissue cells.

### 1.1 Superoxide dismutase 1 (SOD 1)

Superoxide dismutase (SOD) is an important endogenous antioxidant enzyme that acts as a component of the first line of defense against reactive oxygen species (ROS) (**Figure 3**) [7, 15]. There are three forms of superoxide dismutase present in humans. SOD1 is localized in the cytoplasm, SOD2 in the mitochondria, and SOD3 is extracellular. SOD1 and SOD3 contain copper and zinc, whereas SOD2, the mitochondrial enzyme, contains manganese in its reactive center [16, 17].

The genes of SOD are located on chromosomes 21 (SOD1) (**Figure 4**), 6 (SOD2), and 4 (SOD3), respectively (21q22.1, 6q25.3, and 4p15.3-p15.1). The SOD1 gene consists of five exons interrupted by four introns.

$$2O_2^{\bullet -} + 2H^+ + SOD \rightarrow H_2O_2 + O_2$$

**Figure 3.**
*Reaction catalyze by superoxide dismutase.*

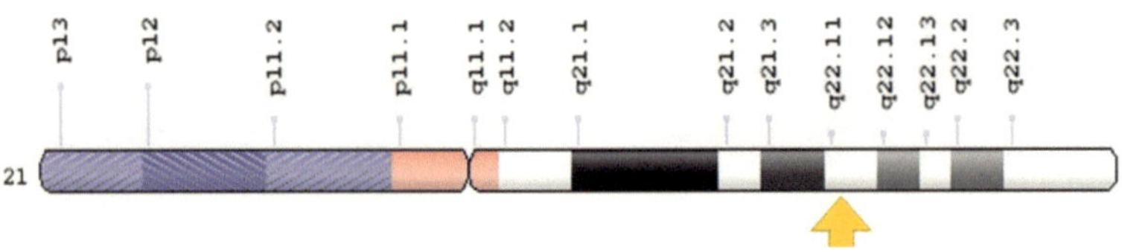


**Figure 4.**
*Chromosomal location of superoxide dismutase.*

Several polymorphisms have been identified in the SOD1 gene that mainly affect the regulatory regions of that gene, including the promoter, UTRs, and introns. In the SOD1 gene, a 50 bp deletion polymorphism (rs 36,232,792) is located 1684 bp upstream of the ATG start codon in the promoter region. It has been reported that this 50 bp deletion decreases promoter activity and also reduces mRNA levels in cells [18–20].

It is well-known that genetic variations in the SOD1 gene coding for the SOD enzymes may be a primary reason for decreased or impaired regulation of SOD enzymatic activity, modifying the ROS detoxification mechanism [21, 22]. Previously published research studies described the genetic and biochemical characterization of SOD1, validating that SOD1 plays a significant role in the occurrence of diseases like thalassemia, heart failure, cancer, diabetes, Down's syndrome, amyotrophic lateral sclerosis, type 2 diabetes, cardiovascular risk factors, temperament, and heroin dependency [23–27].

### 1.2 Glutathione- S-transferases (GST)

Glutathione S-transferases (GST) (EC 2.5.1.18) phase II detoxification enzymes catalyze the conjugation of reduced glutathione to a wide range of substrates, usually resulting in detoxification. They are also responsible for the metabolism of numerous xenobiotics and play a major cellular antioxidant role [28, 29].

The GST gene is located on the long (q) arm of chromosome 22 at position 11.23 (**Figure 5**). These enzymes are encoded by 16 polymorphic genes divided into six classes: alpha (GSTA), pi (GSTP), mu (GSTM), theta (GSTT), zeta (GSTZ), and omega (GSTO) (**Figure 6**), which encompass several genes and isoenzymes in humans [30, 31].

The presence of genetic variants in these genes may result in the lack of production of functional enzymes or result in either increased or decreased metabolic activity [32]. Many hotspots for genetic polymorphisms have been previously reported in GST genes. Still, GSTT1 and GSTM1 are the most broadly studied genes in the GST gene family [33, 34]. GST genes in fact are upregulated in response to oxidative stress. Specifically, the genetic loci of GSTM1 and GSTT1 are found to be highly polymorphic, and the deletion of both (genotype null) affects enzyme activity by increasing exposure to oxidative stress [35, 36]. GSTM1 and GSTT1 are two isoforms of GST. The presence any polymorphisms in these isoforms has been studied by many

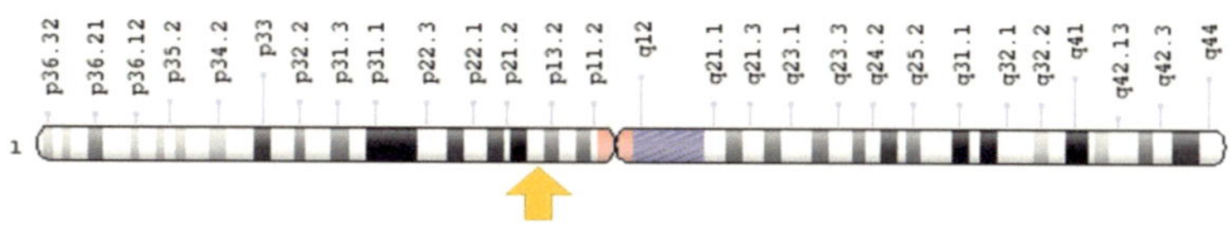


**Figure 5.**
*Chromosomal location of glutathione- S-transferases.*

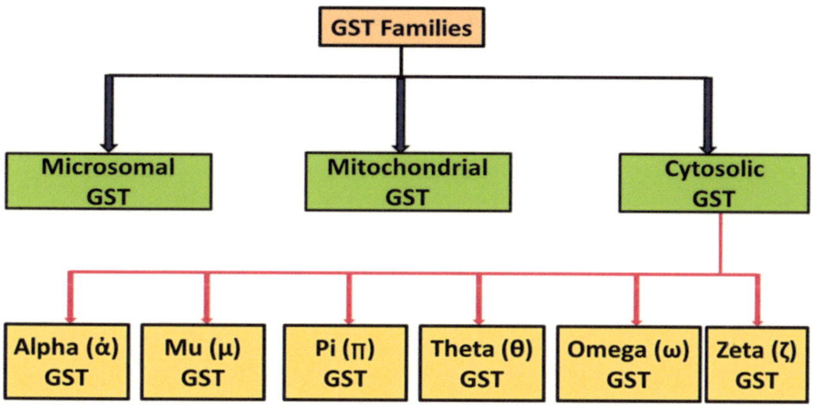


**Figure 6.**
*Different families of GST enzymes.*

researchers for chronic diseases like cardiovascular and neurodegenerative diseases. These allelic variations have been found to be linked to the inflammatory nature of the diseases [35–37].

GSTM1 and GSTT1 genes are among the possible factors that may be involved in the clinical modification process in most patients. The association of GSTM1 and GSTT1 polymorphisms with the clinical severity of diseases has been widely investigated, but few studies have examined their relationship with enzymatic activity. Previously reported findings varied significantly, with some studies supporting and others refuting the associations of the presence and absence of GSTM1 and GSTT1 null polymorphisms with the severity of oxidative injury in diseases like thalassemia major (TM) [38–40]. However, some reports suggested no influence of GSTT1 and GSTM1 polymorphisms in TM patients.

### 1.3 Catalase (CAT)

Catalase (CAT) (EC 1.11.1.6) is an endogenous antioxidant enzyme that plays a major role in controlling hydrogen peroxide concentration in human cells (**Figure 7**) [41, 42]. It decomposes $H_2O_2$ into $H_2O$ and $O_2$, thereby protecting cells from oxidative stress injury [43].

The CAT gene resides on the short (p) arm of chromosome 11 at position 11p13 (**Figure 8**) and consists of 13 exons and 12 introns. Functional polymorphisms in the regulatory regions of the CAT gene encoding the catalase enzyme are responsible for variations in enzyme activity, hence serving as a risk factor for oxidative stress and its associated diseases [21, 44].

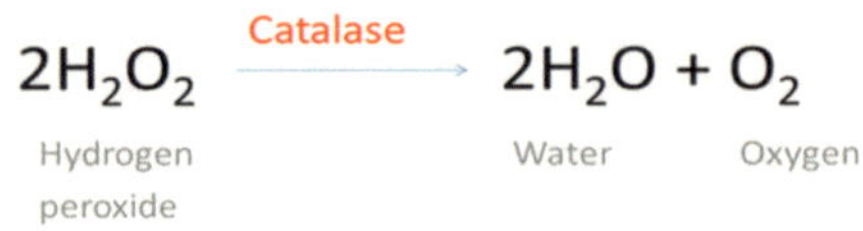


**Figure 7.**
*Reaction catalyze by catalase.*

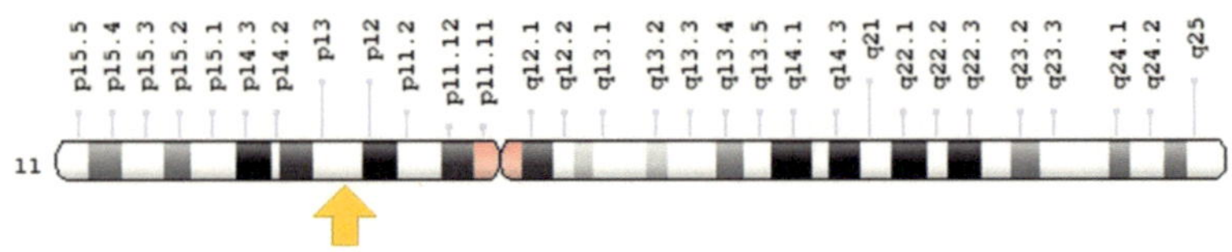


**Figure 8.**
*Chromosomal location of catalase gene.*

In 2012, a total of 245 single-nucleotide polymorphisms (SNPs) were described in the NCBI database in different regions of the CAT gene [45]. Numerous polymorphisms have been identified in the promoter, 5'UTR, 3'UTR, exons, and introns.

### *1.3.1 CAT -262 (C > T) polymorphism*

The CAT gene −262 (C > T) polymorphism (rs1001179) in the promoter region influences transcription factor binding sites, thereby altering the basal transcription mechanism and the expression of the respective enzyme [46, 47]. Thus, a decrease in catalase enzymatic activity is a major reason for the increased formation of hydroxyl radicals. Compared to the wild-type C allele, the variant T allele of the CAT −262 (C > T) gene polymorphism is associated with lower enzyme activity [47, 48].

Decreased activity has been observed to serve as a risk factor in various diseases, including oxidative stress in thalassemic patients, breast cancer development, asbestosis, arsenic-induced hyperkeratosis, acute paraquat intoxication, prostate cancer, and ovarian cancer [40, 45, 47, 49, 50].

### *1.3.2 CAT -89 (a > T) polymorphism*

The CAT gene polymorphism −89 (A > T) (rs7943316), also known as −21 (A > T), is located in the promoter region near the transcription start site (TSS) [51, 52]. This polymorphism is associated with various diseases, including hypertension, age-related cataracts, colorectal cancer, and preeclampsia [53, 54]. This polymorphism reduces the antioxidant capacity and may serve as a risk factor for oxidative stress and its associated diseases.

Few studies have reported a decrease in CAT activity in thalassemia major (TM) patients with this polymorphism [40, 53]. However, most studies from India are done with small sample sizes and have variable findings. Therefore, it is important to understand the effect of this CAT gene polymorphism in patients with reduced enzymatic activity, which makes cells more vulnerable to free radicals and exacerbates disease symptoms.

### 1.4 Glutathione peroxidase (GPx) polymorphism

Glutathione peroxidase 1 (GPx1) (EC 1.11.1.9) is an intracellular soluble seleno-protein located in the cytosol, which converts free hydrogen peroxide ($H_2O_2$) into water and oxygen (**Figure 9**) [55]. "Mills et al. were the first to identify the function of GPx enzymes in protecting hemoglobin from oxidative degradation in red blood cells (RBCs) [56]. There are eight isozymes of GPx found in humans. The gene coding for GPx1 is located on chromosome 3p21.3 (**Figure 10**) and is composed of two exons spanning about 1.42 kb [57].

Most of the genetic variations are present within the introns, exons, and UTR regions of the GPx1 gene and mainly affect the normal functioning of the GPx1 enzyme [57, 58]. GPx gene polymorphism, 198 (C > T; rs1050450), is present at codon 198 in exon 2 of the GPx1 gene, which alters the protein coding sequence due to the substitution of proline in place of leucine within the GPx1 C-terminal region [58, 59]. This leucine amino acid substitution causes a change in the structural conformation of the active site region and modifies the enzyme activity.

Some studies reported a reduced level of GPx enzyme due to the presence of genetic polymorphism, 198 (C > T) in the GPx1 gene. GPx1 gene polymorphism serves as a risk factor for breast cancer, coronary artery disease vitiligo, type 2 diabetes, and relapsing multiple sclerosis [58, 60].

### 1.5 Myeloperoxidase (MPO)

Myeloperoxidase (MPO) (EC 1.11.1.7) is a well-known lysosomal hemeprotein synthesized during myeloid differentiation, making it an important component of neutrophil azurophilic granules [61]. These are endogenous oxidant enzymes that partially generate ROS and are released outside the cell, raising the potential for damage to extracellular tissues [40, 62, 63]. The MPO enzyme mainly catalyzes the reaction between chloride ($Cl^-$) and hydrogen peroxide ($H_2O_2$) to generate hypochlorous acid (HOCl) and other ROS (**Figure 11**).

The human MPO gene is located on chromosome 17q23.1 (**Figure 12**), spans about 13 kb, and consists of 12 exons and 11 introns [64]. There are at least 319 different polymorphisms present in the MPO gene (Available from: http://www.ncbi.nlm.

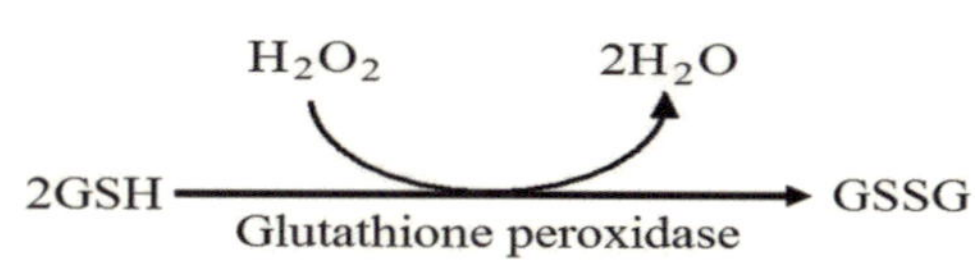


**Figure 9.**
*Reaction catalyze by glutathione peroxidase.*

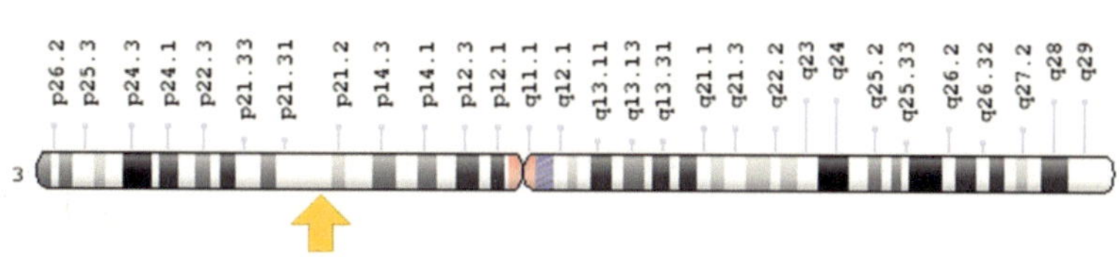


**Figure 10.**
*Chromosomal location of glutathione peroxidase.*

$$H_2O_2 + Cl^- \xrightarrow{\text{Myeloperoxidase}} \boxed{HOCl} + OH^-$$

**Figure 11.**
*Reaction catalyze by myeloperoxidase.*

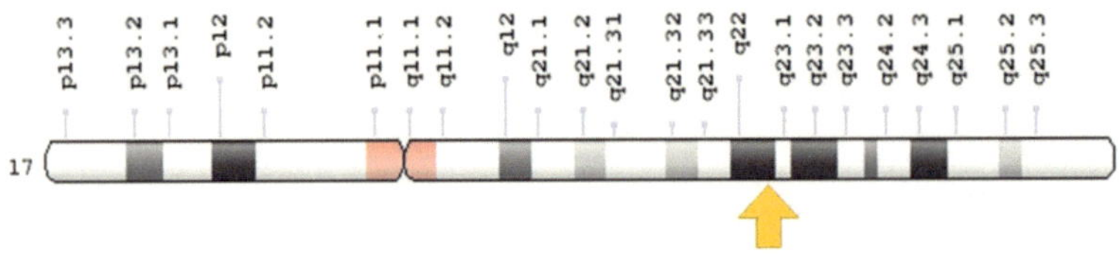


**Figure 12.**
*Chromosomal location of myeloperoxidase.*

nih.gov/projects/SNP). The MPO gene −463 (G > A) polymorphism (rs2333227) is located in the Alu hormone-responsive element of the promoter region, which is 463 bp upstream of the MPO gene [65]. The wild-type G allele is recognized as a transcription factor binding site for stimulatory protein 1 (SP1) and is responsible for increased MPO gene expression [64]. Numerous studies have suggested that patients with mutant (A) allele carriers were found to be associated with lower mRNA expression and transcription activity when compared to wild-type (G) allele in the −463 (G > A) polymorphism [61, 64, 65].

Later on, many studies have investigated the role of this MPO gene polymorphism in the development of cancer of various organs including the lung, hepatic, breast, cervix, ovary cancer, gastric, and bladder, and also contribute as a risk factor in ischemic heart disease [66–68]. Costa et al. [69] in the year 2005 reported an association of the MPO gene, −463 (G > A) polymorphism with infection in sickle cell anemia. Recently, in the year 2018, Yahouédéhou et al. [70] found that this MPO polymorphism was associated with alterations in lipid, inflammatory, renal, hemolytic, and hepatic profiles.

## 2. Conclusion and future prospective

The oxidative stress plays a central role in the pathogenesis of many diseases. Moreover, numerous SNPs were described that too result in variations in the antioxidative enzyme activity, which highly affect the protection mechanism in response to oxidative stress condition. Hence, documentation of gene polymorphism of antioxidative enzymes as risk factors for developing severe manifestations may be helpful in the categorization of the patients, who are hereditarily at risk of developing oxidative injury due to imperfect antioxidative defense mechanism. Therefore, it is essential to identify the whole mechanisms of free radical-induced oxidative stress-related complications in patients before considering it as a prospective diagnostic tool and implying it in the management of the disease. Further, research is needed to explore and better understand the mechanistic basis of the associations.

## Author details

Poonam Tripathi
SGPGIMS, Lucknow, India

*Address all correspondence to: poonam.tripathi90@gmail.com

## References


[1] Ding K, Shameer K, Jouni H, et al. Genetic loci implicated in erythroid differentiation and cell cycle regulation are associated with red blood cell traits. Mayo Clinic Proceedings. 2012;**87**(5):461-474. DOI: 10.1016/j.mayocp.2012.01.016

[2] Grinberg LN, Rachmilewitz EA, Kitrossky N, Chevion M. Hydroxyl radical generation in β-thalassemic red blood cells. Free Radical Biology and Medicine. 1995;**18**:611-615

[3] Ghone RA, Kumbar KM, Suryakar AN, Katkam RV, Joshi NG. Oxidative stress and disturbance in antioxidant balance in beta thalassemia major. Indian Journal of Clinical Biochemistry. 2008;**23**(4):337-340. DOI: 10.1007/s12291-008-0074-7

[4] Ondei Lde S, Estevão Ida F, Rocha MI, et al. Oxidative stress and antioxidant status in beta-thalassemia heterozygotes. Revista Brasileira de Hematologia e Hemoterapia. 2013;**35**(6):409-413. DOI: 10.5581/1516-8484.20130122

[5] Voskou S, Aslan M, Fanis P, Phylactides M, Kleanthous M. Oxidative stress in β-thalassaemia and sickle cell disease. Redox Biology. 2015;**6**:226-239. DOI: 10.1016/j.redox.2015.07.018

[6] Chakraborty D, Bhattacharyya M. Antioxidant defence status of red blood cells of patients with β-thalassemia and Eβ-thalassemia. Clinica Chimica Acta. 2001;**305**:123-129

[7] Ighodaro OM, Akinloye OA. First line defence antioxidants-superoxide dismutase (SOD), catalase (CAT) and glutathione peroxidase (GPX): Their fundamental role in the entire antioxidant defence grid. Alexandria Journal of Medicine. 2018;**54**(4):287-293. DOI: 10.1016/j.ajme.2017.09.001

[8] Livrea MA, Tesoriere L, Pintaudi AM, et al. Oxidative stress and antioxidant status in beta-thalassemia major: Iron overload and depletion of lipid soluble antioxidants. Blood. 1996;**88**:3608-3614

[9] Kattamis C, Kattamis AC. Oxidative stress disturbances in erythrocytes of β-thalassemia. Pediatric Hematology and Oncology. 2001;**18**:85-88

[10] Brittenham GM, Griffith PM, Nienhuis AW, et al. Efficacy of deferoxamine in preventing complications of iron overload in patients with thalassemia major. The New England Journal of Medicine. 1994;**331**(9):567-573

[11] Anderson LJ, Westwood MA, Holden S, et al. Myocardial iron clearance during reversal of siderotic cardiomyopathy with intravenous desferrioxamine: A prospective study using T2* cardiovascular magnetic resonance. British Journal of Haematology. 2004;**127**(3):348-355

[12] Juronen E, Tasa G, Veromann S, Parts L, Tiidla A, Pulges R, et al. Polymorphic glutathione S-Transferases as genetic risk factors for senile cortical cataract in Estonians. Investigative Ophthalmology & Visual Science. 2000;**41**:2262-2267

[13] Forsberg L, Lyrenäs L, de Faire U, Morgenstern R. A common functional C-T substitution polymorphism in the promoter region of the human catalase gene influences transcription factor binding, reporter gene transcription and is correlated to blood catalase levels. Free Radical Biology & Medicine. 2001;**30**:500-505

[14] Tang H, Dong X, Day RS, Hassan MM, Li D. Antioxidant genes, diabetes and dietary antioxidants in association with risk of pancreatic cancer. Carcinogenesis. 2010;**31**:607-613

[15] Giustarini D, Dalle-Donne I, Tsikas D, Rossi R. Oxidative stress and human diseases: Origin, link, measurement, mechanisms, and biomarkers. Critical Reviews in Clinical Laboratory Sciences. 2009;**46**:241-281

[16] Montano MA, Barrio Lera JP, Gottlieb MG, et al. Association between manganese superoxide dismutase (MnSOD) gene polymorphism and elderly obesity. Molecular and Cellular Biochemistry. 2009;**328**(1-2):33-40

[17] Verreth W, De Keyzer D, Davey PC, et al. Rosuvastatin restores superoxide dismutase expression and inhibits accumulation of oxidized LDL in the aortic arch of obese dyslipidemic mice. British Journal of Pharmacology. 2007;**151**(3):347-355. DOI: 10.1038/sj.bjp.0707231

[18] Verreth W, De Keyzer D, Davey PC, et al. Rosuvastatin restores superoxide dismutase expression and inhibits accumulation of oxidized LDL in the aortic arch of obese dyslipidemic mice. British Journal of Pharmacology. 2007;**151**(3):347-355

[19] Milani P, Gagliardi S, Bongioanni P, et al. Effect of the 50 bp deletion polymorphism in the SOD1 promoter on SOD1 mRNA levels in Italian ALS patients. Journal of the Neurological Sciences. 2012;**313**(1-2):75-78

[20] Zendehboodi Z. Association between 50 bp insertion/deletion polymorphism in promoter of the superoxide dismutase-1 and temperament. International Journal of Basic Science in Medicine. 2018;**3**(2):59-62. DOI: 10.15171/ijbms.2018.11

[21] Zhang Y, Zhang L, Sun D, Li Z, Wang L, Liu P. Genetic polymorphisms of superoxide dismutases, catalase, and glutathione peroxidase in age-related cataract. Molecular Vision. 2011;**17**:2325-2332

[22] Rajkumar S, Praveen MR, Gajjar D, Vasavada AR, Alapure B, Patel D, et al. Activity of superoxide dismutase isoenzymes in lens epithelial cells derived from different types of age-related cataract. Journal of Cataract and Refractive Surgery. 2008;**34**:470-474

[23] Flekac M, Skrha J, Hilgertova J, et al. Gene polymorphisms of superoxide dismutases and catalase in diabetes mellitus. BMC Medical Genetics. 2008;**9**(1):30

[24] Chen H, Yu M, Li M, et al. Polymorphic variations in manganese superoxide dismutase (MnSOD), glutathione peroxidase-1 (GPX1), and catalase (CAT) contribute to elevated plasma triglyceride levels in Chinese patients with type 2 diabetes or diabetic cardiovascular disease. Molecular and Cellular Biochemistry. 2012;**363**(1-2):85-91

[25] Wang XL, Duarte N, Cai H, et al. Relationship between total plasma homocysteine, polymorphisms of homocysteine metabolism related enzymes, risk factors and coronary artery disease in the Australian hospitalbased population. Atherosclerosis. 1999;**146**(1):133-140

[26] Joseph N, Pai S, Sengupta S, Bharadwaj S, Dhawan S, Khare K. A clinico-epidemiological study of thalassemia cases in India. Journal of Natural Science, Biology and Medicine. 2018;**9**:236-241

[27] Tripathi P, Agarwal S, Sarangi AN, Tewari S, Mandal K. Genetic variation in SOD1 gene promoter ins/Del and its influence on oxidative stress in Beta thalassemia major patients. International Journal of Hematology-Oncology and Stem Cell Research. 2020;**14**(2):110-117

[28] Pearson WR, Vorachek WR, Xu SJ, Berger R, Hart I, Vannai SD. Identification of class-mu glutathione transferase genes GSTM1-GSTM5 on human chromosome 1p13. American Journal of Human Genetics. 1993;**53**:220-233

[29] Hayes JD, Strange RC. Glutathione S-transferase polymorphisms and their biological consequences. Pharmacology. 2000;**61**:154-166

[30] Strange RC, Spiteri MA, Ramachandran S, Fryer AA. Glutathione-S-transferase family of enzymes. Mutation Research. 2001;**482**:21-26

[31] Prabhu KS, Reddy PV, Jones EC, et al. Characterization of a class alpha glutathione- S-transferase with glutathione peroxidase activity in human liver microsomes. Archives of Biochemistry and Biophysics. 2004;**424**:72-80

[32] Dufour CJ, Svahan A, Bacigalupo DL, Varotto S. Genetic polymorphisms of CYP3A4, GSTT1, GSTM1, GSTP1 and NQO1 and the risk of acquired idiopathic aplastic anemia in Caucasian patients. Haematologica. 2005;**90**:1027-1031

[33] Mannervik B, Awasthi YC, Board PG, et al. Nomenclature for human glutathione transferases. The Biochemical Journal. 1992;**282**:305-306

[34] Hung RJ, Boffetta P, Brockmoller J, Butkiewicz D, Cascorbi I. CYP1A1 and GSTM1 genetic polymorphisms and lung cancer risk in Caucasian non-smokers: A pooled analysis. Carcinogenesis. 2003;**24**:875-882

[35] Nerbert DW, Vasiliou V. Analysis of the glutathione S-transferase (GSDT) gene family. Human Genomics. 2004;**1**:460-464

[36] Sharma V, Kumar B, Saxena R. Glutathione S-transferase gene deletions and their effect on iron status in HbE/beta thalassemia patients. Annals of Hematology. 2010;**89**:411-414

[37] Dasari S et al. Genetic polymorphism of glutathione S-transferases: Relevance to neurological disorders. Pathophysiology. 2018. DOI: 10.1016/j.pathophys.2018.06.001

[38] Ellithy HN, Yousri S, Shahin GH. Relation between glutathione S-transferase genes (GSTM1, GSTT1, and GSTP1) polymorphisms and clinical manifestations of sickle cell disease in Egyptian patients. Hematology. 2015;**20**(10):598-606. DOI: 10.1179/1607845415Y.0000000013

[39] Abo-Shanab AM, El-Desouky MA, Kholoussi NM, El-Kamah GY, Helwa IA, Fahmi AA. The relation between glutathione S-Transferase M1 null-genotype and cardiac problems in beta-thalassemia. Acta Biochimica Polonica. 2016;**63**(2):267-271. DOI: 10.18388/abp.2015_1076

[40] Tripathi P, Agarwal S, Tewari S, Mandal S. Status of catalase, glutathione peroxidase, glutathione S-Transferase and myeloperoxidase gene polymorphisms in beta-thalassemia major patients to assess oxidative injury and its association with enzyme activities. Journal of Pediatric Genetics. 2020. DOI: 10.1055/s-0041-1723961

[41] Ogata M. Acatalasemia. Human Genetics. 1991;**86**:331-340

[42] Ogata M, Wang D-H, Ogino K. Mammalian acatalasemia: The perspectives of bioinformatics and genetic toxicology. Acta Medica Okayama. 2008;**62**:345-361

[43] Von Ossowski G, Hausner, Loewen PC. Molecular evolutionary analysis based on the amino acid sequence of catalase. Journal of Molecular Evolution. 1993;**37**(1):71-76

[44] Saify K, Saadat M. Influence of a 50bp ins/Del polymorphism at promoter of the superoxide dismutase-1 on gene expression and risk of heroin dependency. Environmental Health and Preventive Medicine. 2017;**22**:4. DOI: 10.1186/s12199-017- 0617-8

[45] Crawford A, Fassett RG, Coombes JS, Kunde DA, Ahuja KD, Robertson IK, et al. Relationship between antioxidant enzyme genotype and activity and kidney function: A case-control study. Clinical Nephrology. 2012b;**78**:135-144

[46] Goth L, Vitai M. Polymorphism of 50 of the catalase gene in Hungarian acatalasemia and hypocatalasemia. Electrophoresis. 1997;**18**:1105-1108

[47] Nadif R, Mintz M, Jedlicka A, Bertrand JP, Kleeberger SR, Kauffmann F. Association of CAT polymorphisms with catalase activity and exposure to environmental oxidative stimuli. Free Radical Research. 2005;**39**(12):1345-1350

[48] Mak JCW, Ho SP, Yu WC, Choo KL, Chu CM, Yew WW, et al. Polymorphisms and functional activity in superoxide dismutase and catalase genes in smokers with COPD. The European Respiratory Journal. 2007;**30**:684-690

[49] Ahn J, Gammon MD, Santella RM, Gaudet MM, Britton JA, Teitelbaum SL, et al. Associations between breast cancer risk and the catalase genotype, fruit and vegetable consumption, and supplement use. American Journal of Epidemiology. 2005;**162**:943-952

[50] Franko A, Dolzan V, Arneric N, Dodic-Fikfak M. Asbestosis and catalase genetic polymorphism. Arhiv za Higijenu Rada i Toksikologiju. 2008;**59**:233-240

[51] Miranda-Vilela AL, Alves PC, Akimoto AK, et al. Gene polymorphisms against DNA damage induced by hydrogen peroxide in leukocytes of healthy humans through comet assay: A quasi-experimental study. Environmental Health. 2010;**9**:21. DOI: 10.1186/1476-069X-9-21

[52] Lourdhu Mary A, Nithya K, Isabel W, Angeline T. Prevalence of catalase (−21 a/T) gene variant in south Indian (Tamil) population. BioMed Research International. 2014;**2014**:894237. DOI: 10.1155/2014/894237

[53] Jiang Z, Akey JM, Shi J, Xiong M, Wang Y, Shen Y, et al. A polymorphism in the promoter region of catalase is associated with blood pressure levels. Human Genetics. 2001;**109**:95-98

[54] Mansego ML, Solar Gde M, Alonso MP, Martínez F, Sáez GT, Escudero JC, et al. Polymorphisms of antioxidant enzymes, blood pressure and risk of hypertension. Journal of Hypertension. 2011;**29**:492-500

[55] Arthur JR. The glutathione peroxidases. Cellular and Molecular Life Sciences. 2000;**57**(13-14):1825-1835

[56] Mills GC. Glutathione peroxidase, an erythrocyte enzyme that protects hemoglobin from oxidative damage. The Journal of Biological Chemistry. 1957;**229**:189-197

[57] Ratnasinghe D, Tangrea JA, Andersen MR, Barrett MJ, Virtamo J, Taylor PR, et al. Glutathione peroxidase

codon198 polymorphism variant increases lung cancer risk. Cancer Research. 2000;**60**:6381-6383

[58] Hamanishi T, Furuta H, Kato H, Doi A, Tamai M, Shimomura H, et al. Functional variants in the glutathione peroxidase-1 (GPx-1) gene are associated with increased intima-media thickness of carotid arteries and risk of macrovascular diseases in japanese type 2 diabetic patients. Diabetes. 2004;**53**:2455-2460

[59] Moscow JA, Schmidt L, Ingram DT, Gnarra J, Johnson B, Cowan KH. Loss of heterozygosity of the human cytosolic glutathione peroxidase I gene in lung cancer. Carcinogenesis. 1994;**15**:2769-2773

[60] Tang NP, Wang LS, Yang L, Gu HJ, Sun QM, Cong RH, et al. Genetic variant in glutathione peroxidase 1 gene is associated with an increased risk of coronary artery disease in a Chinese population. Clinica Chimica Acta. 2008;**395**:89-93

[61] Winterbourn CC, Vissers MC, Kettle AJ. Myeloperoxidase. Current Opinion in Hematology. 2000;7:53-58

[62] Austin GE, Lam L, Zaki SR, Chan WC, Hodge T, et al. Sequence comparison of putative regulatory DNA of the 59 flanking region of the myeloperoxidase gene in normal and leukemic bone marrow cells. Leukemia. 1993;7:1445-1450

[63] Klebanoff SJ. Myeloperoxidase. Proceedings of the Association of American Physicians. 1999;**111**:383-389

[64] Piedrafita FJ, Molander RB, Vansant G, Orlova EA, Pfahl M, Reynolds WF. An Alu element in the myeloperoxidase promoter contains a composite SP1-thyroid hormone-retinoic acid response element. The Journal of Biological Chemistry. 1996;**271**:14412-14420

[65] Pecoits-Filho R, Stenvinkel P, Marchlewska A, Heimburger O, Bárány P, Hoff CM, et al. A functional variant of the myeloperoxidase gene is associated with cardiovascular disease in end-stage renal disease patients. Kidney International. Supplement. 2003;**84**:S172-S176

[66] Chu H, Wang M, Wang M, Gu D, Wu D, Zhang Z, et al. The MPO–463G > a polymorphism and cancer risk: A meta-analysis based on 43 case-control studies. Mutagenesis. 2010;**25**(4):389-395

[67] Maddhuri S, Pallapolu P, Bandaru S, Suresh G, Malempati AR, Jyothy A, et al. Analysis of plasma myeloperoxidase levels and functional gene –463G>a and –129G>a polymorphisms with early onset of coronary artery disease in south Indian population. Folia Cardiologica. 2016;**11**(4):272-227

[68] Loria V, Dato I, Graziani F, Biasucci LM. Myeloperoxidase: A new biomarker of inflammation in ischemic heart disease and acute coronary syndromes. Mediators of Inflammation. 2008;**2008**:135625. DOI: 10.1155/2008/135625

[69] Costa RN, Conran N, Albuquerque DM, Soares PH, Saad ST, Costa FF. Association of the G-463A myeloperoxidase polymorphism with infection in sickle cell anemia. Haematologica. 2005;**90**(7):977-979

[70] Yahouédéhou SCMA, Carvalho MOS, Oliveira RM, et al. Sickle cell Anemia patients in use of Hydroxyurea: Association between polymorphisms in genes encoding metabolizing drug enzymes and laboratory parameters. Disease Markers. 2018;**2018**:6105691. DOI: 10.1155/2018/6105691

Chapter 14

# Harnessing Antioxidants to Counter Oxidative Stress-Induced Apoptosis

*Deng Yiran, Wang Jiayi and Ma Lifang*

## Abstract

Oxidative stress is a pathological biochemical process caused by an excessive increase in reactive oxygen species (ROS) compared to the native antioxidant capacity generated during metabolic processes. High ROS levels are cytotoxic to cells, resulting in oxidative damage to DNA, proteins, and lipids and ultimately triggering apoptosis. Antioxidants capture and neutralize oxygen free radicals and play a crucial role in effectively intervening or reducing ROS levels. Excessive ROS has been reported to be tightly linked with events that induce inflammation, cell transformation, angiogenesis, and tumor cell survival and invasion, resulting in tumor development, neurodegenerative diseases, cardiovascular disorders, and other pathologies. This review focuses on the source and regulatory mechanisms of oxidative stress and discusses potential therapeutic strategies for intervening in related diseases using antioxidants.

**Keywords:** oxidative stress, ROS, apoptosis, antioxidants, tumor, neurodegenerative diseases, cardiovascular disorders

## 1. Introduction

The concept of oxidative stress was first proposed in the introductory chapter of the book entitled "*Oxidative Stress,*" published in 1985, describing oxidative stress as "a disturbance in the prooxidant-antioxidant balance in favor of the former, causing excessive oxidative damage [1]." Oxidative stress exists when internal and external reactive oxygen species (ROS) exceed the cellular total clearance capacity. ROS can be oxygen-derived free radicals induced by cellular metabolism and external stimuli and include the superoxide anion ($O_2^{\bullet -}$), hydrogen peroxide ($H_2O_2$), and hydroxyl radical ($OH^{\bullet}$). Although $O_2$ generally does not react with most cellular components, ROS can cause oxidative modifications of biomolecules, such as lipids, proteins, RNA, DNA, and other biomolecules. ROS are highly reactive and are generated in various cellular compartments. Normal cellular functions require ROS to be sustained at physiological concentrations. Abnormally high concentrations of ROS are often deleterious and trigger apoptosis through the intrinsic or extrinsic pathway. For example, ROS binding to DNA causes oxidative DNA damage and subsequent activation of DNA damage response pathways that ultimately lead to apoptosis [2]. ROS also directly impairs mitochondrial function, resulting in a decrease in the mitochondrial membrane


IntechOpen

potential and a change in mitochondrial permeability. This change in membrane properties causes the release of proteins such as cytochrome c, activation of caspase cascades, and, ultimately, apoptosis. Additionally, ROS indirectly regulates apoptosis through various signaling pathways, including the Fas, P53/MPTP, NF-kB, and JNK pathways.

Organisms have evolved an antioxidant system to detoxify oxidants and other electrophiles and thus defend against oxidative stress. These antioxidants can be internally produced or externally provided. They detect ROS and reduce their oxidative impact on cellular molecules, protecting cell tissues from oxidative damage and playing a role in safeguarding the cardiovascular system, anticancer activities, and antiaging effects. Once the antioxidant defense capability against ROS is exceeded, the redox balance is disturbed, which leads to cell death.

## 2. Oxidative stress and apoptosis

### 2.1 Generation process of oxidative stress

Intrinsic and extrinsic factors trigger oxidative stress. As stated above, ROS includes oxygen-generated species such as $O_2^{\bullet-}$, $H_2O_2$, $^{\bullet}OH$, ozone, and singlet oxygen. The use of ROS as a term has led to imprecise statements because the chemical compositions of these species vary significantly.

The primary generation of $O_2^{\bullet-}$ occurs through single-electron reduction of molecular oxygen, which mainly arises *via* electron leakage from the mitochondrial respiratory chain, particularly from electron leakage of ubisemiquinone ($QH^{\bullet-}$) (step 1; **Figure 1**) [3]. NADPH oxidases (NOX4, DuOX1, and DuOX2) and other flavoprotein enzymes reduce $O_2$ to $H_2O_2$ by providing a second electron before $O_2^{\bullet-}$ leaves the active sites [4]. NADPH oxidases catalyze step 2. The predominant source of $H_2O_2$ comes from the dismutation of $O_2^{\bullet-}$, which is a rapid reaction

$$1:\ QH^{\bullet-} + O_2 \longleftrightarrow Q + O_2^{\bullet-}$$

$$2:\ NADPH + 2O_2 \longrightarrow NADP^{+} + H^{+}\ 2O_2^{\bullet-}$$

$$3:\ 2O_2^{\bullet-} + 2H^{+} \longrightarrow H_2O_2 + O_2$$

$$4:\ 2H_2O_2 \longrightarrow 2H_2O + O_2$$

$$5:\ H_2O_2 + 2Trx(SH)_2 \longrightarrow 2H_2O + TrxS_2$$

$$6:\ 2H_2O_2 + 2GSH \longrightarrow GSSG + 2H_2O$$

$$7:\ H_2O_2 + Fe^{2+} \longrightarrow Fe^{3+} + {}^{\bullet}OH$$

$$8:\ LH + {}^{\bullet}OH \longrightarrow L^{\bullet} + {}^{\bullet}OH$$

$$9:\ L^{\bullet} + O_2 \longrightarrow LOO^{\bullet}$$

$$10:\ LOO^{\bullet} + LH \longrightarrow LOOH + L^{\bullet}$$

$$11:\ O_2^{\bullet-} + {}^{\bullet}NO \longrightarrow ONOO^{-}$$

$$12:\ HNOO^{-} \longleftrightarrow {}^{\bullet}NO_2 + {}^{\bullet}OH \longrightarrow NO3^{-} + H^{+}$$

$$13:\ H^{+} + X^{-} + H_2O_2 \longrightarrow HOX + H_2O$$

**Figure 1.**
*The mechanism of oxidative stress.*

(i.e., $2 \times 10^9$ $M^{-1}$ $s^{-1}$) catalyzed by superoxide dismutase (step 3). $H_2O_2$ is primarily reduced by 15 enzymes, including catalase (step 4). Five peroxiredoxins are involved in the use of either thioredoxin (a small protein containing two crucial cysteines, Trx(SH)2) or eight glutathione peroxidases, with peroxiredoxin 6 using glutathione (γ-glutamylcysteinylglycine, GSH) as a tripeptide (steps 5 and 6). $H_2O_2$ reacts rapidly with transition metals such as iron to produce hydroxyl radicals (step 7). Hydroxyl radicals are potent oxidants that oxidize any adjacent molecule rapidly. One reaction triggering oxidative stress is the lipid peroxidation chain reaction initiated by $^{\bullet}OH$ (steps 8–10). Superoxide can release iron from iron-sulfur proteins to catalyze step 7 [5]. However, the main manner that the relatively weak oxidant $O_2^{\bullet-}$ leads to oxidative stress is as a precursor to $H_2O_2$ and peroxynitrite ($ONOO^-$), the latter formed in step 11. Free radicals, $^{\bullet}NO$ and $O_2^{\bullet-}$, do not directly cause oxidative stress. Instead, protonated peroxynitrite or peroxynitrous acid (ONOOH), a non-radical molecule, is the source of oxidative stress. Peroxynitrous acid is a potent oxidant, and its decomposition forms reactive intermediates (step 12). In phagocytic cells, myeloperoxidases catalyze the conversion of $H_2O_2$ to the final oxidant, hypohalous acids (HOX) (step 13). In summary, the intricate processes leading to oxidative stress involve multiple reactive species and their interactions, which contribute to cellular damage and apoptosis. Understanding these mechanisms is crucial for exploring potential therapeutic strategies to mitigate the adverse effects of oxidative stress on cells.

## 2.2 Oxidative stress-induced apoptotic pathways

### *2.2.1 Intrinsic apoptotic pathway*

The intrinsic pathway primarily involves mitochondrial signaling and endoplasmic reticulum (ER) signaling pathways, whereas the extrinsic pathway involves the death receptor pathway [6]. We will focus on describing the process of ROS-triggered intrinsic apoptosis from both the intrinsic and extrinsic pathways.

#### *2.2.1.1 Mitochondrial apoptotic pathway*

Elevated ROS levels are closely associated with the activation of mitochondrial-mediated apoptosis. Leakage from the electron transport chain results in a significant increase in ROS. These ROS can damage mitochondrial DNA and impair the transcription of proteins associated with the electron transport chain. This disruption leads to further ROS generation, manifested by mitochondrial membrane potential loss, impaired ATP synthesis, and increased mitochondrial permeability. Consequently, cytochrome c and apoptosis-inducing factors are released, activating caspase-dependent or caspase-independent apoptosis [7]. Loss of mitochondrial membrane permeation, cytochrome c release to the cytoplasm, and stimulation of caspase-9-mediated caspase-3 activation are all implicated in caspase-dependent apoptosis [8]. Apoptosis-inducing factor, derived from mitochondria, is transferred to the cell nucleus during caspase-independent apoptosis. Moreover, ROS triggers the activation of c-Jun N-terminal kinase (JNK), which phosphorylates and inactivates the anti-apoptotic Bcl-2 and Bcl-XL proteins while concurrently phosphorylating and activating pro-apoptotic proteins.

ROS could oxidize other molecules, such as cardiolipin, in the inner mitochondrial membrane (IMM), where cytochrome c is bound. Under normal circumstances, cytochrome c shuttles electrons between complexes III and IV of the mitochondrial electron transport chain. ROS oxidation of cardiolipin reduces the binding of cytochrome c to the IMM, increasing the level of freely released cytochrome c into the cytosol through the outer mitochondrial membrane (OMM) [9]. This initiation of the apoptotic cascade involves oxidative stress and mitochondrial $Ca^{2+}$ accumulation, triggering the opening of the IMM pore, which is termed the mitochondrial permeability transition (MPT).

Although $Ca^{2+}$ is essential for pore opening, this opening can also be facilitated by ATP reduction, oxidative stress, low pH, and excessive inorganic phosphate. The effects of pore opening appear to depend on the $Ca^{2+}$ concentration, where low concentrations encourage apoptosis, whereas high concentrations favor necrosis [10]. Mitochondria are the main regulatory sites of apoptosis and participate in most apoptotic regulatory processes. Pro-apoptotic factors in the mitochondrial membrane are released into the cytoplasm, activating the apoptosis pathway of mitochondria and causing cell death.

#### *2.2.1.2 Endoplasmic reticulum (ER) signaling pathway*

In the lumen of the endoplasmic reticulum (ER), the redox state is crucial for protein folding and disulfide bond formation. The oxidative environment in the ER lumen, with a higher ratio of oxidized to reduced glutathione (GSSG/GSH) compared to the reducing environment of the cytoplasm, plays a significant role in protein folding. Chaperone proteins, such as protein disulfide isomerase (PDI), the most prevalent chaperone protein in the ER, regulate the formation of disulfide bonds. PDI catalyzes the oxidation of two cysteine residues between peptide substrates while creating disulfide bonds in developing proteins. At its active site, it goes through two rounds of electron reduction. Endoplasmic reticulum oxidoreductin-1 (ERO1) can restore PDI to its oxidized state, but ROS is generated. ERO1 transfers electrons from thiol groups of PDI to $O_2$ through a flavin adenine dinucleotide (FAD)-dependent reaction. As an electron acceptor, FAD reacts with $O_2$ to be reoxidized, generating $H_2O_2$. Moreover, increased glutathione levels in the ER, facilitated by ERO1, can cause excessive $H_2O_2$ production. GSH serves as a substrate for PDI-catalyzed oxidation, and when ERO1 is active, an increase in GSSG levels occurs. The elevated levels of $H_2O_2$ and GSSG disrupt the redox balance in the ER lumen, causing ER stress. Misfolded proteins in the ER, influenced by aberrant disulfide bond formation and reduction, are caught in a futile cycle. This cycle increases the production of $H_2O_2$ derived from ERO1, leading to ER stress and activation of unfolded protein response (UPR) signaling. Furthermore, a reduction in GSH levels because of improper pairing of disulfide bonds in misfolded proteins may increase ROS levels. Approximately 25% of cellular ROS is estimated to be produced during the formation of disulfide bonds in the ER *via* oxidized protein folding [11, 12]. Proteins with numerous disulfide bonds in the ER play a greater role in ROS generation than those with fewer disulfide bonds. As a byproduct of ER protein oxidation, ROS generation may lead to ER stress-mediated cell death. Glutathione peroxidases 7 and 8, which degrade $H_2O_2$, and peroxiredoxin 4 are additional antioxidants that are present in the ER. Because alterations in the redox state can result in oxidative modifications and/or misfolding of newly generated proteins, which can cause ER stress, the ER antioxidant system is essential for preserving ER homeostasis [13, 14]. During protein misfolding, the generation of ROS may increase because of changes in oxidative phosphorylation caused by ER release of $Ca^{2+}$ or energy consumption. In contrast, butylated hydroxyanisole (BHA), an antioxidant, can stop excessive ROS from causing protein misfolding, preventing UPR activation and CHOP generation [15]. Furthermore, by deactivating PDI/ERO1 thiol-disulfide exchange or causing incorrect disulfide bond formation, high ROS generation may obstruct protein folding. ROS and $Ca^{2+}$ are two critical factors influencing cell apoptosis, where high concentrations of ROS promote ER stress. Furthermore, when ER stress activates JNK, ROS inside mitochondria increase, ultimately inducing apoptosis [16]. As ER stress occurs, the ER releases $Ca^{2+}$ to elevate cytoplasmic $Ca^{2+}$ concentrations, leading to an increase in mitochondrial $Ca^{2+}$ levels. As $Ca^{2+}$ concentrations rise, mitochondrial dysfunction occurs because of $Ca^{2+}$ overload in the regulatory network, ultimately resulting in cell apoptosis. As a key signal transduction platform, the ER plays an important role in apoptosis.

#### *2.2.2 Extrinsic apoptotic pathway*

The death receptor pathway is a classical extrinsic apoptotic pathway, and there is a tight connection between ROS and extrinsically induced apoptosis. In the death receptor pathway, the tumor necrosis factor-related apoptosis-inducing ligand (TRAIL) binds to cell surface death receptors, specifically death receptor-4 and death receptor-5. Upon TRAIL binding, the death-inducing signaling complex forms, leading to the recruitment of caspase-8 and subsequent activation of caspase-7, -6, and -3, ultimately resulting in apoptosis [17, 18]. ROS plays a crucial role in regulating the TNF-R1 signaling pathway at low doses. At high levels, ROS can cause direct DNA damage and cell death. Furthermore, overproduction of ROS can disrupt the folding of proteins by blocking thiol–disulfide exchange between PDI and ERO1 or by causing incorrect disulfide bond formation in proteins. The precise processes involving ROS

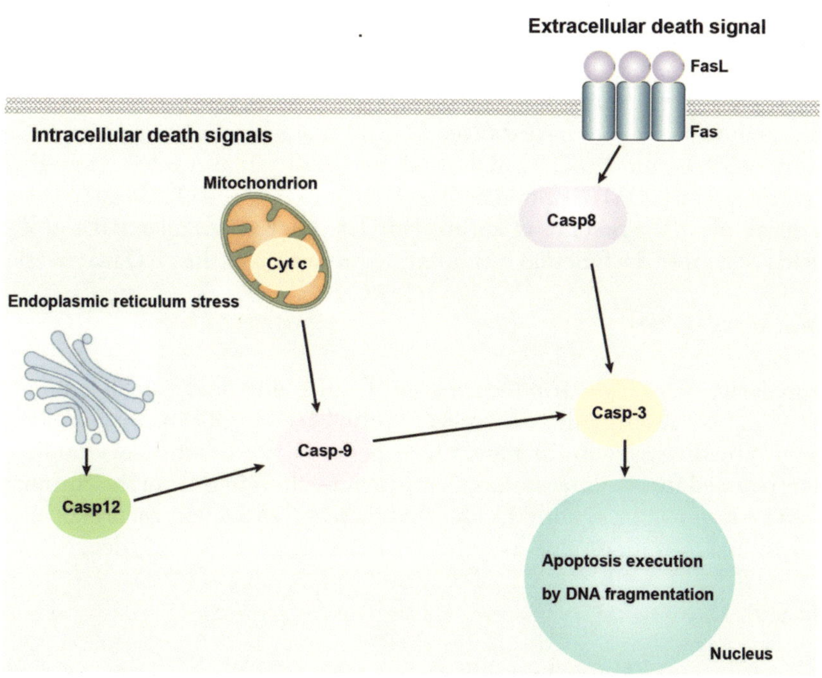


**Figure 2.**
*The extracellular and intracellular apoptosis pathways.*

regulating TRAIL-mediated apoptosis remain unclear. In particular, ROS increases the expression of genes associated with death receptors, such as TRAIL, and this impact can be countered by antioxidants such as N-acetylcysteine. Furthermore, the overexpression of TRAIL by ROS appears to be mediated by p53 and CHOP, and compounds that boost the production of ROS appear to make cells more susceptible to TRAIL-mediated apoptosis (**Figure 2**).

## 3. Antioxidants and their roles in the human body

Antioxidants play a crucial role in the human body by mitigating the harmful effects of ROS during oxidation processes and easing the progression of chronic diseases triggered by lipid peroxidation events. Organisms have developed specific antioxidant systems to prevent oxidative damage from free radicals [19]. Because of the diverse types, mechanisms, and targets of ROS generation in an organism, antioxidants also vary in types, modes of action, and effectiveness. Antioxidants typically possess the following characteristics: (1) Timely arrival at the lesion site: capable of reaching the affected area at an appropriate concentration, (2) rapid reaction with free radicals: antioxidants should quickly react with free radicals, (3) generation of lower toxicity free radicals: after reacting with toxic free radicals, antioxidants should produce new free radicals with lower toxicity, and (4) regeneration capability: antioxidants should revert to their original form for further reactions, thereby continuing their function. Based on their sources and mechanisms of action, antioxidants can be categorized into endogenous (internal) and exogenous (external) antioxidants (**Table 1**).

### 3.1 Superoxide dismutase (SOD)

SOD was first reported by McCord and Fridovich half a century ago and subsequently established as the first line of defense against oxidative stress. Organisms that need oxygen for living express at least one SOD. Three isoforms of SOD possessing different catalytic metal ions have evolved in various organisms, categorized as Cu/Zn SODs, Mn/Fe SODs, and Ni SODs. SOD catalyzes the dismutation of $O_2^{\bullet-}$ into $H_2O_2$ and $O_2$, and subsequently, CAT/GPx metabolizes $H_2O_2$ into $H_2O$ and $O_2$. CuZnSOD enzymes play a crucial role in the first line of antioxidant defense. Extracellular SODs protect endothelial cells from dysfunction by imparting protection against NO inactivation.

### 3.2 Glutathione (GSH)

The majority of total glutathione consists of GSH, with only a small amount existing as GSSG. Glutathione reductase converts GSSG to GSH using NADPH and the flavinic prosthetic group. To maintain exogenous antioxidants like vitamins E and C in their reduced form, glutathione can regenerate vitamin E from the chromanoyl radical and vitamin C from both the ascorbyl radical and dehydroascorbate. Studies have shown that the glutathione-ascorbate cycle contributes to $H_2O_2$ detoxification.

### 3.3 Uric acid

Uric acid, an active free radical scavenger, prevents ROS generation in reactions induced by xanthine oxidase. Urate protects the erythrocyte membrane from lipid oxidation by scavenging free radicals. Some studies have shown that the loss of

| Resources | Physicochemical properties | Antioxidants |
|---|---|---|
| Endogenous | Enzyme | Catalase (CAT) |
| | | Superoxide Dismutase (SOD) |
| | | Glutathione Peroxidase (GSH-Px) |
| | | Peroxidase (POX) |
| | Çofactors | Nicotinamide Adenine Dinucleotide (NADH) |
| | | Nicotinamide Adenine Dinucleotide Phosphate (NADPH) |
| | | Coenzyme Q10 |
| | | Glutathione (GSH) |
| | Other endogenous antioxidants | Trace elements such as selenium |
| | | Uric acid |
| | | Brain peptides |
| | | Melatonin |
| Exogenous | Natural plant compounds | Vitamins A, C, E |
| | | Phytochemicals like phytic acid, flavonoids, lycopene, tea polyphenols. |
| | | β-Carotene and carotenoids |
| | Metal chelators | Deferoxamine |
| | | Ethylenediaminetetraacetic acid (EDTA) |
| | | Phosphate derivatives |
| | Synthetic antioxidants | Butyl Hydroxy Anisole (BHA) |
| | | Butylated Hydroxytoluene (BHT) |
| | | Tert-Butylhydroquinone (TBHQ) |

**Table 1.**
*List of antioxidants by category.*

uricase activity compensates for the disability to synthesize ascorbic acid in humans during prime evolution.

## 3.4 Retinol (vitamin A)

Vitamin A is a type of carotenoid produced in the liver, derived from the breakdown of β-carotene. Over a dozen vitamin A derivatives have been reported. This vitamin is beneficial for the skin, eyes, and internal organs. The antioxidant activity of vitamin A is manifested in its ability to bind to ROO·(peroxyl radicals), preventing their propagation of oxidative reactions in lipids. This capability enables vitamin A to protect the organism from photooxidation. Additionally, vitamin A is capable of scavenging singlet oxygen and peroxyl radicals. The antioxidant property of this vitamin is crucial for maintaining the health of cells and tissues, especially in preventing oxidative stress-induced damage to vital organs such as the skin and eyes.

## 3.5 Tocopherol (vitamin E)

Fat-soluble vitamin E has antioxidant qualities that protect the body against ROS and free radical damage [20]. The name "vitamin E" refers to the group of chemical compounds known as tocopherols, which are members of the phenol class. These compounds are some of the most popular and extensively used antioxidants. Vitamin E is an organic substance in various foods, including leafy greens and high-fat foods like plant oils, seeds, nuts, and egg yolk. Vitamin E exists as eight isomers: four tocopherols and four tocotrienols (i.e., α-, β-, γ-, and δ-forms). α-Tocopherol is the most effective and abundant isomer in biological systems. Tocopherols include a chromanol ring with a hydroxyl group and a polar phytyl side chain, with the prefix α-, β-, γ-, or δ- defining the number and position of methyl groups attached to the chromanol ring. The chromanol head group imparts antioxidant activity to tocopherols, whereas the tail portion has no effect. Tocopherols all have a chromanol ring with an –OH group, providing a hydrogen atom to reduce free radicals and a hydrophobic side chain penetrating biological membranes [21]. α-Tocopherol, for example, is a hydrophobic antioxidant that prevents lipid peroxidation by donating its phenolic hydrogen to ROO, forming tocopheroxyl radicals. Although tocopheroxyl radicals are also radicals, they are unreactive and cannot continue the oxidative chain reaction. α-Tocopherol is the only major lipid-soluble chain-breaking antioxidant found in plasma, red blood cells, and tissues, protecting the integrity of lipid structures, especially membranes [22]. α-Tocopherol is soluble only in organic solvents and membranes, making it challenging to work with in aqueous reaction media. This challenge has been addressed by developing α-tocopherol analogs, such as trolox, where the polyisoprenoid tail of α-tocopherol is replaced by a carboxylic acid moiety, rendering it water-soluble while retaining significant antioxidant activity. These compounds are present in almost all food sources in at least trace amounts. Plant oils typically contain the highest concentration of vitamin E, followed by nuts and seeds, including unprocessed grains. Although α-tocopherol is the most substantial antioxidant in this group of antioxidants, in comparison to other tocopherols, α-tocopherol exhibits weaker antioxidant properties in cooking oils. Potentially, intracellular α-tocopherol is an antioxidant that reduces lipid peroxidation. α-Tocopherol shields the membrane from more oxidation by breaking the chain. The most prevalent subtypes of vitamin E in feed and tissues are α- and γ-tocopherols [7]. However, the most physiologically active and practically useful is α-tocopherol. Like its isomeric counterpart, γ-tocopherol interacts with ROS and RNS, possibly alleviating inflammation. α-Tocopherol is a common form of vitamin E in supplements and European diets, whereas γ-tocopherol is the most common form in American diets. α-Tocopherol is added to foods [23]. The formation of tocopheroxyl radicals is stabilized by the delocalization of electrons in the aromatic ring structure. These substances interact with lipoproteins or membranes with a high lipophilicity. Their ability to prevent lipid peroxidation, scavenge lipid peroxyl radicals, and produce lipid hydroperoxides and tocopherol radicals appears to be their most important antioxidant function. Non-radical byproducts such as stable peroxides, which can be converted to tocopherolquinones and tocopherol dimers, are created during this radical generation process. Hydroperoxide breakdown is also linked to α-tocopherol, reducing oxidative effects and reactions. Research has documented the preventive benefits of α-tocopherol against Parkinson's disease [24, 25].

### 3.6 Ascorbic acid (vitamin C)

L-ascorbic acid and L-dehydroascorbic acid are two vitamin C types with antioxidant properties. Both compounds enter the body through the digestive system and undergo enzymatic conversion. Ascorbic acid has been reported to effectively scavenge $O_2^{\bullet-}$, $H_2O_2$, $OH^{\bullet}$, $I_2$, and reactive $NO_2^{\bullet}$. Considered one of the most potent and least toxic natural antioxidants, ascorbic acid is a water-soluble vitamin found in high concentrations in various foods and plants. Ascorbic acid typically reacts with oxidants, and its unique feature enables the transfer of an electron [26]. Ascorbic acid is an effective reducing agent that terminates chain reactions of free radicals through electron transfer [27, 28]. Moreover, it exhibits characteristics of a conjugated system, especially when electrons are formed in double bonds, hydroxyl pairs, and carbonyl double bonds, presenting itself as ascorbate. The –OH group of ascorbic acid is more acidic than that of standard amino acid hydroxyl groups because this major resonance structure stabilizes the protonated conjugate base of ascorbic acid; that is, ascorbic acid acts as an enol whose deprotonated state is a stable enolate. Studies suggest that ascorbic acid can regenerate α-tocopherol from its tocopherol radicals, restoring its antioxidant activity both *in vivo* and *in vitro*. Additionally, human plasma contains approximately 60 μmol of ascorbic acid. In interactions with reactive oxygen species, ascorbic acid is oxidized to dehydroascorbic acid through intermediate ascorbate radicals [29, 30]. Dehydroascorbic acid is recycled to ascorbic acid through dehydroascorbate reductase. Consequently, the content of dehydroascorbic acid is much lower compared to ascorbic acid. As a scavenger of reactive oxygen species, ascorbic acid is effective against $O_2$, $H_2O_2$, OH,·and $O_2$ in aqueous solutions. Ascorbic acid also efficiently clears reactive nitrogen species. Fruits, particularly citrus fruits, cherries, and melons, and vegetables such as tomatoes, leafy greens, broccoli, cauliflower, Brussels sprouts, and cabbage are important dietary sources of ascorbic acid. More than 1 g/kg of fresh weight may be present in these food sources; for instance, ~1000 mg of vitamin C per kilogram of freeze-dried kiwi extract was reported [31]. At low doses (<100 mg), the bioavailability values of synthetic and food-derived vitamin C are very similar, with absorption efficiency decreasing as the dosage increases [32]. Evidence from *in vitro* studies indicates that vitamin C can regenerate tocopherol from vitamin E to inhibit lipid peroxidation. Although vitamin C is an effective antioxidant that intervenes in many physiological and biochemical reactions, it can also act as an oxidant. This oxidant activity occurs when it binds Fe and Cu, reducing $Fe^{3+}$ to $Fe^{2+}$ or $Cu^{3+}$ to $Cu^{2+}$, thereby reducing $H_2O_2$ to OH. Moreover, some phenolic antioxidants have been reported to lose activity at high concentrations and act as prooxidants by participating in initiating reactions. The procedure permits the radical load to be moved from the lipophilic compartment to the aqueous compartment, where enzyme defense systems efficiently neutralize radicals. Notably, vitamin C may also function in the body as a prooxidant. Lipid peroxidation can be triggered by generating OH·when ascorbic acid and free transition metal ions are present. However, because these metals attach to proteins efficiently, their levels in the body are low. Vitamin C has other recognized biological functions, including as a cofactor for several crucial enzymes, such as hydroxylases.

## 4. Potential therapeutic implications

### 4.1 Neurodegenerative diseases

Neurodegenerative diseases are a heterogeneous group of diseases, including Alzheimer's disease, Parkinson's disease, Huntington's disorder, and amyotrophic lateral sclerosis. These diseases are characterized by the gradual loss of function and death of specific groups of neurons triggered by mitochondrial damage, leading to the clinical manifestations. Pathophysiologically, changes in specific proteins can lead to dysfunction of different cellular pathways, including mitochondrial dysfunction, excitotoxicity, synaptic dysfunction, protein degradation system damage, endoplasmic reticulum stress, DNA damage, inflammation, and other phenotypes, which promote increases in ROS levels. Because of the complexity of the pathways described above, it is challenging to explore the mechanisms of neurotoxicity and cell death; therefore, it is difficult to find effective therapies. Typically, the human brain comprises 2% of the body weight, consumes 15% of the cardiac output, and uses 20% of the entire amount of basal oxygen. The basic tenet of mitochondria in neurodegeneration states that abnormalities in the oxidative phosphorylation system are particularly common in oxidized tissues with high energy requirements. Because the cortex and thalamus absorb three times the amount of blood flow compared to white matter, these brain tissues are the most metabolically active areas of the brain and are therefore most vulnerable to hypoxic-ischemic encephalopathy. This vulnerability is why antioxidant therapy strategies, which involve eliminating excess ROS to restore the redox balance, are considered promising. A wide range of antioxidant treatments primarily target neurodegenerative processes. Over the past years, resveratrol, carotenoids, omega-3 fatty acids, vitamin E, and melatonin supplements have been used to treat Alzheimer's patients in the most recent placebo-randomized clinical trials employing an antioxidant method [33, 34]. Numerous studies have demonstrated the antioxidant, anti-inflammatory, and neuroprotective qualities of resveratrol, a well-known non-flavonoid polyphenol, and its ability to lessen the toxicity and aggregation of Aβ peptides in the hippocampus of Alzheimer's patients. Furthermore, resveratrol inhibits Aβ1–42 from accumulating in the hippocampus and passing through the blood–brain barrier, preventing hippocampal and blood–brain barrier damage and fostering neurogenesis [35, 36]. According to recent research, the combined use of resveratrol and donepezil hydrochloride significantly improves clinical efficacy in the treatment of Alzheimer's disease patients. TNF-α and other inflammatory markers improve in patients throughout the course of a two-month resveratrol treatment [37]. Additionally, cognitive function improves, as assessed by minimal state examination and the Alzheimer's Disease Assessment Scale Cognitive Subscale and Prognosis. Carotenoids, in addition to polyphenols, are also frequently mentioned as neuroprotective substances for treating neurological disorders. These compounds contribute to ameliorating clinical symptoms caused by neurodegeneration in Alzheimer's disease patients. Notably, in Alzheimer's patients, elevated plasma alpha-carotene levels were linked to improved general cognitive function, including higher semantic memory scores [33]. Furthermore, it has recently been reported that carotenoids, omega-3 fatty acids, and vitamin E supplementation enhance working memory in older people. In summary, antioxidant therapy has been shown to play an essential protective role in neuroprotection.

### 4.2 Cancer

ROS are known to be involved in the development of many diseases, including malignancies. Under normal circumstances, the cellular redox state is maintained by balancing antioxidant compound secretion and sequestration of ROS. Although ROS is an important mechanism of the host immune response to prevent bacterial and viral infections, increased production of ROS can lead to an imbalance in the cell redox potential, which leads to alterations in cellular and molecular signaling pathways and oncogenesis. Antioxidant compounds are commonly consumed as dietary supplements, and their effectiveness in preventing different illnesses, including malignancies, has been studied. In addition to antioxidants being used to prevent malignancies, dietary antioxidant compounds are often recommended to patients with malignancies, for example, by clinicians. Lycopene is widely recognized for its high antioxidant properties. Holzapfel et al. demonstrated that lycopene therapy synergistically improves the anticancer properties of paclitaxel and carboplatin [38]. In *in vivo* studies by Tang et al., paclitaxel with vitamin E intravenously administered to mice showed potential efficaciousness against multidrug resistance sensitizers. Although P-gp expression cannot be altered, vitamin E and its derivatives can dramatically upregulate intracellular rhodamine 123 and adriamycin (a P-gp substrate). Moreover, this therapy did not decrease intracellular ATP levels either; however, this treatment did reduce verapamil-mediated PGP-related ATPase activity [39]. The improved safety profile of this medication makes the novel use of vitamin E as a multidrug-resistant sensitizer desirable. Wu et al. demonstrated that vitamin E dramatically reduced lung metastasis *in vitro* and improved anticancer immunity *via* downregulating PD-L1.

### 4.3 Cardiovascular disorders

One of the most energy-dependent organs in the body, the heart, requires over 30 kilograms of ATP daily to pump blood throughout the body. The heart is composed of over 30% mitochondria. For a considerable time, it was believed that the primary locations for ROS generation were mitochondrial ETCs I and II [40]. Cells in humans possess an efficient primary antioxidant defense system that guards against irreversible oxidative damage. A recent study has shown that complexes II and IV also produce ROS in disease states. Cardiomyocytes are readily exposed to mitochondrial oxidative stress, leading to irreversible oxidative damage and cell death [41, 42]. Widespread apoptosis or necrotic death of cardiomyocytes and reduced mitochondrial function of surviving cardiomyocytes lead to the loss of normal cardiac function, which eventually causes heart failure and death. Human cells have an effective primary antioxidant system composed of endogenous antioxidant enzymes to prevent irreversible oxidative damage, including manganese or copper-zinc superoxide dismutase, catalase, glutathione reductase, and glutathione peroxidase. As a second line of defense, the body has an endogenous nonenzymatic antioxidant system, including NADPH and NADH, mercaptan, melatonin, L-arginine, ubiquitin, uric acid, bilirubin, and metal-binding proteins. Moreover, external antioxidants derived from diet or supplementation can either directly eliminate ROS or enhance the curative properties of intrinsic antioxidants. External antioxidants include carotenoids, flavonoids, polyphenols, omega-3/6 fatty acids, and certain vitamins and minerals. For example, some clinical investigations use vitamin E or C because these are relatively simple to obtain commercially. However, the clinical benefits of using these antioxidants remains under evaluation. First, there are insufficient biochemical markers to assess the antioxidant capacity of a drug directly [43]. Second, treating tissue damage caused by oxidative stress in adult patients for longer than 40 years is not possible with the current relatively low dose and short treatment periods. Third, the duration of the follow-up period may not be sufficient to reap the advantages of the treatment.

## 5. Concluding remarks and future perspectives

Organisms possess a set of complex and sophisticated antioxidant systems, including glutathione peroxidase, thioredoxin, catalase, superoxide dismutase, and other antioxidant enzymes, which function to remove $H_2O_2$ and superoxide anions from organisms. In addition, there are nonenzymatic antioxidants, including vitamin C, vitamin E, GSH, carotenoids, and flavonoid analogs. These antioxidant enzymes work together with nonenzymatic antioxidant agents to balance the homeostasis of intracellular ROS. In this review, the sites and mechanisms of ROS were described to provide a theoretical basis for developing targeted antioxidant therapy. Ongoing technological advances and deeper research should lead to the discovery of additional subcellular structures producing ROS. These efforts, in turn, will enhance our understanding of the mechanisms governing intracellular ROS generation, offering guidance in developing new targeted antioxidant drugs and disease treatments.

## Acknowledgements

This study was supported by the National Natural Science Foundation of China (82273139), Shanghai Science and Technology Committee Rising-Star Program (22QA1408300), and Excellent Talents Nurture Project of Shanghai Chest Hospital (2021YNZYY02).

## Author details

Deng Yiran[1], Wang Jiayi[1,2*] and Ma Lifang[1,2*]

1 Department of Clinical Laboratory, Shanghai Chest Hospital, Shanghai Jiao Tong University School of Medicine, Shanghai, China

2 Institute of Thoracic Oncology, Shanghai Chest Hospital, Shanghai Jiao Tong University School of Medicine, Shanghai, China

*Address all correspondence to: jiayi.wang@sjtu.edu.cn; mlf2281@sjtu.edu.cn

## References

[1] Sies H. Biochemistry of oxidative stress. Angewandte Chemie, International Edition. 1986;**25**(12):1058-1071

[2] Orrenius S, Gogvadze V, Zhivotovsky B. Calcium and mitochondria in the regulation of cell death. Biochemical and Biophysical Research Communications. 2015;**460**(1):72-81

[3] Boveris A, Cadenas E, Stoppani AOM. Role of ubiquinone in mitochondrial generation of hydrogen-peroxide. Biochemical Journal. 1976;**156**(2):435-444

[4] Lambeth JD. Nox enzymes and the biology of reactive oxygen. Nature Reviews Immunology. 2004;**4**(3):181-189

[5] Koppenol WH. The centennial of the fenton reaction. Free Radical Biology and Medicine. 1993;**15**(6):645-651

[6] Fan D, Fan T-J. Clonidine induces apoptosis of human corneal epithelial cells through death receptors-mediated, mitochondria-dependent signaling pathway. Toxicological Sciences. 2017;**156**(1):252-260

[7] Yang Y, Yu Y, Wang J, Li Y, Li Y, Wei J, et al. Silica nanoparticles induced intrinsic apoptosis in neuroblastoma SH-SY5Y cells via CytC/Apaf-1 pathway. Environmental Toxicology and Pharmacology. 2017;**52**:161-169

[8] Sinha K, Das J, Pal PB, Sil PC. Oxidative stress: The mitochondria-dependent and mitochondria-independent pathways of apoptosis. Archives of Toxicology. 2013;**87**(7):1157-1180

[9] Zhao R-Z, Jiang S, Zhang L, Yu Z-B. Mitochondrial electron transport chain, ROS generation and uncoupling (review). International Journal of Molecular Medicine. 2019;**44**(1):3-15

[10] Alavian KN, Beutner G, Lazrove E, Sacchetti S, Park H-A, Licznerski P, et al. An uncoupling channel within the c-subunit ring of the F1FO ATP synthase is the mitochondrial permeability transition pore. Proceedings of the National Academy of Sciences of the United States of America. 2014;**111**(29):10580-10585

[11] Ugalde JM, Aller I, Kudrjasova L, Schmidt RR, Schloesser M, Homagk M, et al. Endoplasmic reticulum oxidoreductin provides resilience against reductive stress and hypoxic conditions by mediating luminal redox dynamics. The Plant Cell. 2022;**34**(10):4007-4027

[12] Ponsero AJ, Igbaria A, Darch MA, Miled S, Outten CE, Winther JR, et al. Endoplasmic reticulum transport of glutathione by Sec61 is regulated by Ero1 and Bip. Molecular Cell. 2017;**67**(6):962

[13] ElSayed AI, Rafudeen MS, Gomaa AM, Hasanuzzaman M. Exogenous melatonin enhances the reactive oxygen species metabolism, antioxidant defense-related gene expression, and photosynthetic capacity of *Phaseolus vulgaris* L. to confer salt stress tolerance. Physiologia Plantarum. 2021;**173**(4):1369-1381

[14] Wang H, Cheng Y, Mao C, Liu S, Xiao D, Huang J, et al. Emerging mechanisms and targeted therapy of ferroptosis in cancer. Molecular Therapy. 2021;**29**(7):2185-2208

[15] Farrukh MR, Nissar UA, Afnan Q, Rafiq RA, Sharma L, Amin S, et al. Oxidative stress mediated Ca2+ release manifests endoplasmic reticulum stress

leading to unfolded protein response in UV-B irradiated human skin cells. Journal of Dermatological Science. 2014;**75**(1):24-35

[16] Ajoolabady A, Liu S, Klionsky DJ, Lip GYH, Tuomilehto J, Kavalakatt S, et al. ER stress in obesity pathogenesis and management. Trends in Pharmacological Sciences. 2022;**43**(2):97-109

[17] Germano G, Frapolli R, Belgiovine C, Anselmo A, Pesce S, Liguori M, et al. Role of macrophage targeting in the antitumor activity of trabectedin. Cancer Cell. 2013;**23**(2):249-262

[18] Iurlaro R, Munoz-Pinedo C. Cell death induced by endoplasmic reticulum stress. FEBS Journal. 2016;**283**(14):2640-2652

[19] Sies H, Jones DP. Reactive oxygen species (ROS) as pleiotropic physiological signalling agents. Nature Reviews Molecular Cell Biology. 2020;**21**(7):363-383

[20] Ahsan H, Ahad A, Iqbal J, Siddiqui WA. Pharmacological potential of tocotrienols: A review. Nutrition & Metabolism (London). 2014;**11**(1):52

[21] Lee P, Ulatowski LM. Vitamin E: Mechanism of transport and regulation in the CNS. IUBMB Life. 2019;**71**(4):424-429

[22] Niki E. Role of vitamin E as a lipid-soluble peroxyl radical scavenger: In vitro and in vivo evidence. Free Radical Biology and Medicine. 2014;**66**:3-12

[23] Harris PL, Quaife ML, Swanson WJ. Vitamin-E content of foods. The Journal of Nutrition. 1950;**40**(3):367-381

[24] Hosomi A, Goto K, Kondo H, Iwatsubo T, Yokota T, Ogawa M, et al. Localization of α-tocopherol transfer protein in rat brain. Neuroscience Letters. 1998;**256**(3):159-162

[25] Donoso A, Gonzalez-Duran J, Agurto Munoz A, Gonzalez PA, Agurto-Munoz C. Therapeutic uses of natural astaxanthin: An evidence-based review focused on human clinical trials. Pharmacological Research. 2021;**166**:105479

[26] Li Y, Schellhorn HE. New developments and novel therapeutic perspectives for vitamin C. The Journal of Nutrition. 2007;**137**(10):2171-2184

[27] Frei B, England L, Ames BN. Ascorbate is an outstanding antioxidant in human-blood plasma. Proceedings of the National Academy of Sciences of the United States of America. 1989;**86**(16):6377-6381

[28] Vioque J, Navarrete-Munoz E-M, Gimenez-Monzo D, Garcia-de-la-Hera M, Granado F, Young IS, et al. Reproducibility and validity of a food frequency questionnaire among pregnant women in a Mediterranean area. Nutrition Journal. 2013;**12**:26

[29] Fowler AA III, Syed AA, Knowlson S, Sculthorpe R, Farthing D, DeWilde C, et al. Phase I safety trial of intravenous ascorbic acid in patients with severe sepsis. Journal of Translational Medicine. 2014;**12**:32

[30] Carr AC, Rosengrave PC, Bayer S, Chambers S, Mehrtens J, Shaw GM. Hypovitaminosis C and vitamin C deficiency in critically ill patients despite recommended enteral and parenteral intakes. Critical Care. 2017;**21**(1):300

[31] Redza-Dutordoir M, Averill-Bates DA. Activation of apoptosis signalling pathways by reactive oxygen species. Biochimica et Biophysica

Acta-Molecular Cell Research. 2016;**1863**(12):2977-2992

[32] Luan Q, inventor; Shenyang Chuangda Technology Trading, assignee. Vitamin C extract comprises cherry, kiwi fruit, hawthorn and orange, ensures high extraction rate and good extraction effect patent CN105311173-A

[33] Power R, Nolan JM, Prado-Cabrero A, Roche W, Coen R, Power T, et al. Omega-3 fatty acid, carotenoid and vitamin E supplementation improves working memory in older adults: A randomised clinical trial. Clinical Nutrition. 2022;**41**(2):405-414

[34] Nolan JM, Power R, Howard AN, Bergin P, Roche W, Prado-Cabrero A, et al. Supplementation with carotenoids, Omega-3 fatty acids, and vitamin E has a positive effect on the symptoms and progression of Alzheimer's disease. Journal of Alzheimers Disease. 2022;**90**(1):233-249

[35] Kandimalla R, Manczak M, Yin X, Wang R, Reddy PH. Hippocampal phosphorylated tau induced cognitive decline, dendritic spine loss and mitochondrial abnormalities in a mouse model of Alzheimer's disease. Human Molecular Genetics. 2018;**27**(1):30-40

[36] Shomali N, Baradaran B, Ghodrati MD, Akbari M, Hemmatzadeh M, Mohammadi H, et al. A new insight into thymosin β4, a promising therapeutic approach for neurodegenerative disorders. Journal of Cellular Physiology. 2020;**235**(4):3270-3279

[37] Quadros Gomes BA, Bastos Silva JP, Rodrigues Romeiro CF, dos Santos SM, Rodrigues CA, Goncalves PR, et al. Neuroprotective mechanisms of resveratrol in Alzheimer's disease: Role of SIRT1. Oxidative Medicine and Cellular Longevity. 2018;**2018**:8152373

[38] Holzapfel NP, Shokoohmand A, Wagner F, Landgraf M, Champ S, Holzapfel BM, et al. Lycopene reduces ovarian tumor growth and intraperitoneal metastatic load. American Journal of Cancer Research. 2017;**7**(6):1322-1336

[39] Dastmalchi N, Baradaran B, Latifi-Navid S, Safaralizadeh R, Khojasteh SMB, Amini M, et al. Antioxidants with two faces toward cancer. Life Sciences. 2020;**258**:118-186

[40] Santulli G, Xie W, Reiken SR, Marks AR. Mitochondrial calcium overload is a key determinant in heart failure. Proceedings of the National Academy of Sciences of the United States of America. 2015;**112**(36):11389-11394

[41] Prokudina ES, Naryzhnaya NV, Mukhomedzyanov AV, Gorbunov AS, Zhang Y, Yaggi AS, et al. Effect of chronic continuous Normobaric hypoxia on functional state of cardiac mitochondria and tolerance of isolated rat heart to ischemia and reperfusion: Role of μ and δ2 opioid receptors. Physiological Research. 2019;**68**(6):909-920

[42] Heather LC, Cole MA, Tan J-J, Ambrose LJA, Pope S, Abd-Jamil AH, et al. Metabolic adaptation to chronic hypoxia in cardiac mitochondria. Basic Research in Cardiology. 2012;**107**(3):268

[43] Peh HY, Tan WSD, Liao W, Wong WSF. Vitamin E therapy beyond cancer: Tocopherol versus tocotrienol. Pharmacology & Therapeutics. 2016;**162**:152-169

Chapter 15

# Using Coordination Compounds as Antioxidants

*Ehab M. Abdalla and Samar A. Aly*

## Abstract

To improve antioxidant performance, organic ligands (L) have been complexed with various metal ions of "s-, p- and d-block" in recent decades. Coordination compounds are frequently employed, particularly in the pharmaceutical and medical fields, because of their many positive effects. Since chelation generally improves activity, it is known that the properties of both ligands and metals can result in the creation of molecules with significantly higher activity. Because metal chelates have advantages over conventional synthetic antioxidants in terms of geometry, oxidation state, and coordination number, they can be used in place of standard synthetic antioxidants to support and promote the redox processes linked to antioxidant action. Coordination compounds have been shown to have the most researched antioxidative properties. In addition, they have been shown to scavenge free radicals and shield human bodies from their harmful effects. Numerous interconnected systems can be used to evaluate the antioxidant capacity. The most information about the antioxidant properties of metal chelates is provided by the methodological modification. The methodological modification offers the most knowledge on the antioxidant properties of metal chelates. Colorimetric techniques are the most used, though electron paramagnetic resonance (EPR) is an alternative for metallic compounds, since color does not affect the results.

**Keywords:** antioxidant, ligands, coordination, d-block, pharmaceutical

## 1. Introduction

Human life depends on the metabolic processes, homeostasis, and well-functioning repair systems of our bodies. Reactive molecules with one or more unpaired electrons are known as free radicals [1]. Free oxygen radicals are essential for the regular operation of our bodies [2]. These are relatively stable substances created as a byproduct of metabolism [3]. They can target proteins, fatty acids, and DNA, which can lead to oxidative stress and, consequently, various physical anomalies in humans. However, oxygen ($O_2$) has a double role [4]. It can trigger the production of reactive $O_2/N_2$ species (reactive oxygen species/reactive nitrogen species ROS/RNS), which, when present in excess, can lead to serious health issues and central nervous system degeneration, including aging and Alzheimer's disease [5–7]. It is widely accepted that breathing requires oxygen and that being oxygen-depleted is a death sentence. Appropriate oxygen oxidation is the foundation for all biological functions [8–12].

Chemically speaking (ROS) and (RNS) are characterized as particles with at least one unpaired electron [13–15]. Chemicals known as antioxidants can delay or stop oxidation even in situations where their concentration is much lower than that of an oxygen electrode. The body uses antioxidants as its first line of defense against free radicals [16–18]. However, in order to prevent damage to biomolecules, antioxidants must react swiftly to free radicals [19, 20]. The more exposed one is to free radicals, the more important antioxidants become due to increased oxidative stress. Exposure to free radicals can be heightened by several factors such as pollution, tobacco smoke, drugs, illness, stress, and physical activity, and they are produced by body cells in response to free radicals [21, 22]. Antioxidants are widely used as catalysts in the pharmaceutical sector for anticorrosion and in medicines for anti-inflammatory, antifungal, antibacterial, and antiviral effects [23]. A subgroup of ligands (L) called organic compounds has a range of donor orbitals and favorable coordination arrangements with metals [24, 25]. Because organic molecules are highly stable and soluble in common solvents such as ethanol (EtOH), methanol (MeOH), chloroform ($CHCl_3$), and dimethylformamide (DMF), they are often used as ligands [26–28]. The biological activity of aromatic chemical rings is dependent on the kind of functional group that is attached to them [29]. Due to their extraordinary biological activities—among which are antioxidant, herbicidal, antitumor, anticancer, antibacterial, antitumor, antifertility, and antifungal qualities—organic molecules and their metallic complexes have gained significant attention in recent years. Moreover, organic ligands exhibit fluorescence, photoluminescence, and potentiometric action, anthelmintic, and aggregation properties. Easy to make, aromatic molecules have different oxidation states and symmetries that allow them to bind to a variety of metal ions [30–32]. Because of their availability as precursors and physical diversity, their complexes are acknowledged as containing some of the most important stereo-chemical modeling tools in the main group and transition metal-ligand chemistry [33, 34]. In addition to a little decrease in the cytotoxicity from both ions and Aryl ring, the interaction of metal ions with the organic ligand appears to have a wide variety of applications in analytical chemistry, food industry, dye industry, catalysis, fungicidal, agrochemical, and biological activities [35, 36]. Research on antioxidants produced from metals has attracted a lot of attention and resources in an effort to produce compounds that have a great deal of potential for scavenging free radicals, which are associated with a number of illnesses and disorders that are triggered by ROS [37, 38]. Since synthetic antioxidants are more cost-efficient and effective than natural antioxidants, they are now often used [39–42]. Currently under investigation are several Schiff-base metal chelates that function as potent ROS scavengers and antioxidants, where Schiff bases, which contain nitrogen, sulfur, and oxygen donor atoms, tend to show high selectivity toward the complexation of heavy metal ions under optimal conditions [43, 44]. As a result, no thorough published study that fully outlines the antioxidant benefits of Schiff bases and their metallic chelates is presently available [45]. The findings suggest that further investigation is required to find novel antioxidants that can function inside an organism's defensive mechanisms to both decrease the creation of ROS and destroy them due to their distinct replacement patterns and chemical makeup [46].

### 1.1 Metal complex with antioxidant properties

The synthetic features for metal structures might use different metals and ligands, which would allow for the advancement of the technology for specific applications. Over time, chelate synthesis and performance estimates as physiologically active ligands have gained prominence [47, 48]. It is possible to create antioxidant compounds with several mechanisms of action by combining the redox properties of different ligands and metal ions [49, 50]. Antioxidants that support the body's fight against free radicals are often acquired from food consumption. Rich in fruits, vegetables, wine, tea, seeds, flavonoids, phenolic acids, and cinnamic acids—organic compounds with strong antioxidant capabilities—among other natural products [51, 52]. Because the majority of these compounds are connected to the metal as chelates through oxygen atom complexion, these antioxidants can coordinate toward various metal ions and create stable chelates [53]. They also have an oxo group for flavonoids and flavones and a COOH or OH group in their structural system [54, 55]. Because these phenyl rings allow for bond-strengthening back donation, natural products with phenyl rings in (1) or (2) chains are more likely to form complexes with elements that have lower levels of oxidation and spin. By complexing for the substituent of control solubility and producing stability through the phenoxyl intermediate, which is produced in the substituent electrochemical process with assistance from the metal ion, molecular coordination of the antioxidant ligand to a central metal ion allows [56, 57]. Control of all aspects is made possible by two processes: (1) complexation for the substituent of control solubility and (2) stability via rate, which is produced in the substituent electrochemical process with the help of the metal ion [29, 58]. UV light with a wavelength of 380 nm is absorbed by the heterocyclic structures present in flavonoids, hydroxycinnamic acids, polyphenols, flavones, and carotenoids. Combining organic ligands causes the spectrum to shift to the red region, indicating the creation of a complex. These antioxidants exhibit brightness as well, and the differences in their emissive properties, organic Schiff bases are often utilized to coordinate with antioxidant function in addition to coordinating bioactive compounds [59]. The RN=CH—R0 group is created by the condensation reaction of aldehydes with primary amines. R and R0 are substituent function groups that are linked to the cores. Due to the $N_2$ atom's sp2 hybridization in the azomethine group, it possesses two electrons. Additionally, this azomethine group atom has a site with high electrical density donation due to the double bond's ability to donate electrons and nitrogen's low electronegativity; as a result, the Schiff bases act as suitable ligands for the chelation with metal centers. These are the benefits of forming complexes with Schiff bases and natural antioxidants: The benefits of using chelates as living mechanisms for antioxidant performance include: (1) easy and quick synthesis; (2) numerous methods; (3) the capacity to employ chelates as living mechanisms; and (4) a notable improvement in antioxidant activity in comparison to the complex's free ligand [60]. Because of their antioxidant properties, metal lophthalocyanines have attracted more attention. Phthalocyanines are aromatic due to their four aromatic sp2 $N_2$ atoms, and macrocyclic ligands can be used as a chelate to efficiently match for different metal centers during the complex's formation. Moreover, the conjugated 18 p-electron mechanism of these ligands confers stability and chemical resistance, making them antioxidants [61]. Phthalocyanines are insoluble and have a planar structure. Phthalocyanines need to be effectively synthesized to increase their solubility in order to function as antioxidants; these modifications had no effect on their crystallinity [62]. In this regard, a large number of metal lophthalocyanine sequences were reported. The structural configuration of this family of metal complexes is determined by the location of the metal center of reference with respect to the mean plane of the ligand. This article's contents are mostly chelates that contain Schiff bases, phthalocyanines, and natural ingredients. Additionally, a number of nanoparticle samples are shown to illustrate the various methods in which inorganic materials in general might be employed as antioxidant agents. Additionally, throughout, samples of chelation that either improved or did not improve antioxidant activity are displayed [63]. Additionally, investigations on antioxidant performance degradation or chelation enhancements are included throughout this page. It is clear that the presence of metal, oxidation state, shape, and the ligands' distance from the central metal atom all affect how much the antioxidant activity of free to coordinating ligands varies.

## 2. Main *in vitro* assays to assess antioxidant activity

### 2.1 DPPH (2,2-diphenyl-1-picrylhydrazyl) radical assay

The 2,2-diphenyl-1-picrylhydrazyl (DPPH) radical is a stable radical with an absorption maximum of 517 nm, and its delocalized electron gives it a purple color. It can be detected via electron paramagnetic resonance (EPR) [64, 65]. A downside of this method is the steric barrier that separates DPPH from the antioxidant molecules [66]. Because the radical site is positioned in the middle of the molecule, smaller antioxidant molecules typically show higher activity than larger molecules. As a result, to ensure that the interaction between the antioxidant and DPPH occurs, the reaction solution should be kept in the dark for 45 to 60 minutes [67]. The steric impediment that exists between DPPH and the antioxidant molecules is a shortcoming of this approach. Smaller antioxidant molecules have easier access to the radical site because they are placed in the middle of the molecule, which leads to higher activity than bigger molecules.

#### *2.1.1 UV-Vis detection*

The colorimetric approach is based on an antioxidant's capacity to convert the DPPH radical to hydrazine, which turns a solution's purple hue into yellow [67]. By observing the DPPH absorption band at 517 nm, UV-vis spectroscopy can identify the scavenging of the DPPH radical by an antioxidant [64, 66]. For this test, an alcoholic reaction media (ethanol or methanol) is preferred in order to prevent the stable radical from aggregating. Water should be used sparingly as it can encourage aggregation; aggregation issues do not occur when the ethanol-to-water ratio is 1:1 [68]. The primary constraint of the technique is the DPPH radical's absorption range, which happens in a region where a number of antioxidants also absorb, potentially making it more difficult to notice when the reaction between DPPH and the antioxidant ends [69]. However, The DPPH assay is an easy, precise, and repeatable procedure. The efficiency of metal complexes containing flavonoids with antioxidant qualities is frequently assessed using DPPH.

### 2.2 Lipid peroxidation

Lipid peroxidation is the process by which unsaturated fatty acid side chains are damaged by free radicals. When a hydrogen atom is taken by the radical from close to the double bond of a fatty acid, a conjugated diene is produced. When the unstable lipid radical-conjugated diene combines with oxygen, it forms the lipid peroxyl radical (ROO•) [70]. Since this is a process of fat oxidation, saturated fatty acids are more resistant to radicals than polyunsaturated fatty acids.

### *2.2.1 Lipid peroxidation—EPR detection*

The electron paramagnetic resonance (EPR) assay can be used to performed by tracking the signal of the adduct formed between the radical species and the spin traps, in a manner similar to the Lipid Peroxidation Inhibition Capacity, Superoxide Radical Anion ($O^{2-}$), Hydroxyl Radical ($OH^{-}$), and Singlet Oxygen ($1O^{2}$) assays. EPR can also be used to track the suppression of the DPPH• and ABTS• + signals in the radical scavenging technique. The antioxidant's color has no effect on measurement or result detection when employing EPR in place of spectrophotometry. On the other hand, because color can obstruct detection in the UV–vis region, it might be confusing to measure antioxidant activity using spectrophotometry.

## 2.3 Ferric reducing antioxidant power (FRAP)

This procedure is predicated on an antioxidant's capacity to use the SET mechanism to convert a Fe(III) complex into a Fe(II) complex that contains the 2,4,6-tri (2-pyri dyl)-1,5,5-triazine ligand (TPTZ) [71]. The transition from light to dark blue is accompanied by the reduction of $[Fe(TPTZ)2]^{3+}$ to $[Fe(TPTZ)2]^{2+}$. By observing the absorption band at 593 nm, UV–vis spectroscopy can identify the formation of the reduced complex [72]. In this process, the Fe(III) complex's solubility is preserved by using an acidic reaction media (pH = 3.6). The development of the insoluble solid Fe $(OH)_3$ may be promoted in a reaction media with a pH greater than 7.0. FRAP is an easy, quick, and inexpensive technique [73].

## 2.4 Cupric reducing antioxidant capacity (CUPRAC)

The CUPRAC approach relies on an antioxidant's capacity to use the SET process to convert a Cu(II) complex into a Cu(I) complex with 2,9-dimethylphenanthroline (men) acting as a ligand. This approach is quick and reliable. When $[Cu(dmphen)2]^{2+}$ is reduced to $[Cu(dmphen)2]^{+}$, The color of the reduced complex changes from blue to yellow. By keeping an eye on the absorption band at 450 nm, UV–vis spectroscopy can identify the development of the reduced complex [74]. The CUPRAC reaction is more selective than the FRAP reaction because the Cu(II) ion has a lower redox potential (0.16 V) than the Fe(III) ion [75]. This procedure is carried out in an aqueous solution with neutral pH (7.0–7.4) or ethanol, in contrast to FRAP.

## 2.5 $H_2O_2$ (hydrogen peroxide) scavenging activity

A 2 mg/mL sample of dimethyl sulfoxide (DMSO) was obtained and introduced to 40 mM of hydrogen peroxide in phosphate buffer (pH 7.4) in order to test the hydrogen peroxide-based free radical scavenging activity. The blank was a phosphate solution devoid of hydrogen peroxide, and the absorbance was gauged at 230 nm. Using the general equation in the DPPH technique, the percentage of radical scavenging effects was determined [76].

## 2.6 2,2′-azino-bis(3-ethylbenzothiazoline-6-sulfonic acid (ABTS) assay

The ABTS test was run in accordance with the earlier instructions [77]. To make the reagent, a volumetric ratio of 1:1 was used to mix aqueous solutions of ABTS (c = 7 mmol·dm – 3) and $K_2S_2O_8$ (c = 2.45 $mmol{\cdot}dm^{-3}$). The mixture was then left for

12–16 hours at T = 296.15 K to produce ABTS• + cation radicals. Next, 60 $cm^3$ of methanol was combined with 1 $cm^3$ of the resulting ABTS• + solution.

One centimeter of ABTS• + solution and one centimeter of methanolic solution containing the tested chemicals (made at various concentrations, as previously mentioned) were combined and incubated for 7 minutes at 296.15 degrees Celsius. Instead of the sample, methanol was present in the control. Using methanol as the reference, the absorbance of the solutions was measured at λ = 734 nm using the UV–VIS spectrophotometer [78].

## 3. Overview of metal chelates as an antioxidant

Laila et al. created novel complexes and the DPPH free radical scavenging method was used to assess the in vitro antioxidant activity of the newly synthesized ligand and its complexes. The examined compounds' antioxidant activities were assessed at varying concentrations (10, 25, 50, 100, and 150 μg/mL) and contrasted with L-ascorbic acid. This process relies on the antioxidants' capacity to donate electrons or hydrogen. **Figure 1** displays the percentage of the ligand and its complexes that inhibit the DPPH free radicals. The decrease in DPPH absorbance at 517 nm indicated the rise in stability of the DPPH free radical following the admission of hydrogen or an electron. The obtained data indicate that, when compared to the free ligand and other complexes, the Vanadyl Complex (VOL) complex exhibited the best antioxidant activity and its half-maximal inhibitory concentration $IC_{50}$ (27.0 μg/mL) against the selected free radicals. The tetravalent oxidation state of vanadium (IV), which may make it easier for the free radical to donate an electron in vitro, is the source of this high VOL activity. In addition, compared to the other complexes, it is thought to be the strongest reducing agent (**Figure 2**) [79].

Sumathi created an organic ligand using the amino acid L-histidine in combination with salicylaldehyde and 1,10-phenanthroline. Additionally, the metal chelates of $[ML_1L_2]$ 7–10 (where M = Cd(II), Cu(II), Co(II), and Zn(II)) were prepared as shown in **Figure 3** and tested for in vitro antioxidant activity using DPPH and $H_2O_2$ tests. **Table 1** presents the acquired data. Of the metal chelates that were examined, $Cd^{2+}$ chelate showed the best antioxidant scavenging performance, scoring 86.06 and 84.64% using the DPPH and $H_2O_2$ techniques, respectively [76].

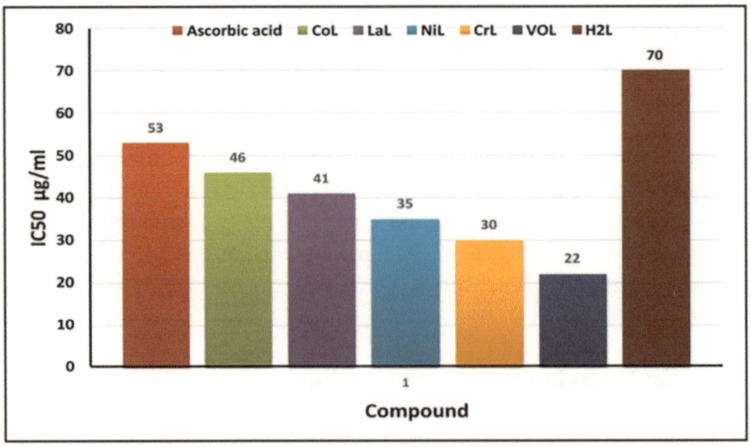


**Figure 1.**
*DPPH scavenging potential of ligand and its Co(II), Ni(II), VO(II), Cr(III), and La(III) metal complexes [79].*

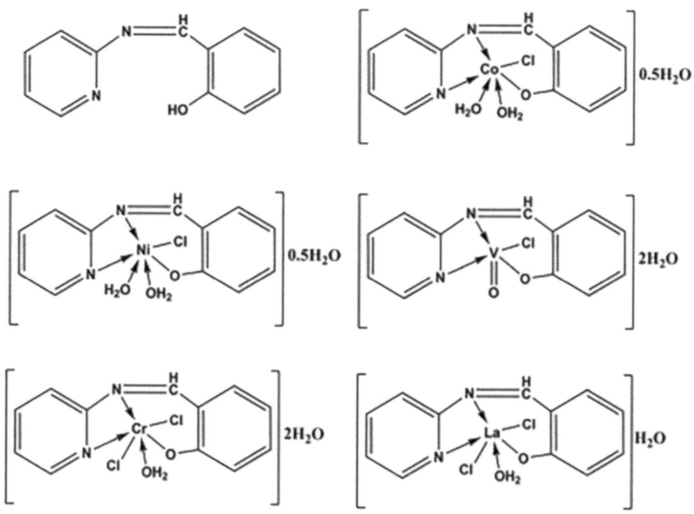

**Figure 2.**
*Suggested structure of the new Schiff base ligand and its complexes [79].*

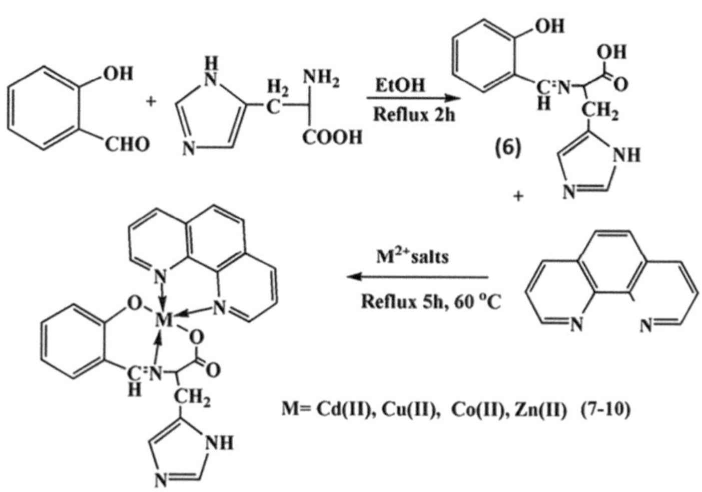

**Figure 3.**
*Synthetic protocol of Schiff base ligand (6) and their metal complexes [76].*

| Complex | % of Antioxidant Activity | |
|---|---|---|
| | DPPH | $H_2O_2$ |
| Cd complex (7) | 86.1 | 80.6 |
| Cu complex (8) | 78.1 | 71.4 |
| Co complex (9) | 74.2 | 72.2 |
| Zn complex (10) | 81.6 | 74.1 |
| α-tocopherol | 89.5 | 83.7 |

**Table 1.**
*Antioxidant performance of metal chelates [76].*

Nongpiur et al. investigated the reaction between $[(arene)MCl_2]_2$ and bidentate 4,5-diazafluorene-9-one (dafo), yielding cationic chelates with the general formula [MLCl(arene)] and the resulting organic ligands ($L_1$–$L_3$) (11–13) in the presence of (NH4)[PF6]. PF6 {M = Ru, arene = benzene (14, 16, 18); M = Ru, arene = p-cymene (15, 17); M = Rh, arene = Cp* (19, 20, 21); M = Ir, arene = Cp* (22, 23, 24); [4,5-diazafluorene-9-one (L1) (11), N-(4,5-diazafluoren-9-ylidene)aniline (L2) (12), N-(4,5-diazafluoren-9-ylidene)phenyl hydrazine (L3) (13)].Antioxidant tests were reviewed in order to assess the investigated substances' biological effectiveness. The results showed that the investigated compounds also exhibited strong antioxidant activity against DPPH radicals. In the same way that replacements operate as sensitive type scavengers, antioxidants may interfere with the oxidation process in order to interact with additional free radicals. The method of DPPH scavenging is frequently used to control the effectiveness of antioxidants. DPPH using ascorbic acid as a benchmark; **Table 2** displays the results. The information shows that the compounds under test performed noticeably in terms of radical scavenging [80].

Ali et al. (2022) produced Cr(III) and Fe(III) heterocyclic chelates by reducing succinic dihydrazide with 5-chloroindoline-2,3-dione in an aqueous medium using the original method in a 1:1:1 ratio with an MW irradiation system. The chelates under the test underwent straining for bioactivity. The human non-small cell carcinoma (HNSC) cell line was used to evaluate anticancer performance, while the DPPH test measured antioxidant performance. Every studied chemical (56–59) demonstrated the ability to scavenge free radicals. $Cl(C_{12}H_{10}N_5O_2Cl)$ $(NO_3)_2$ Out of all the chelates examined, $NO_3$ chelate showed the highest free radical scavenging activity ($IC_{50} \geq 50$ µg) in conjunction with BHA ($IC_{50} \leq 50$ µg). Through MOD, an *in silico* test utilizing EGFR tyrosine kinase was finished. The outcomes showed which of the investigated materials had strong antioxidant and anticancer properties [81].

| Tested Compound | % DRSA $\pm$ Std. Error |
|---|---|
| Ll(11) | $1.60 \pm 0.12$ |
| L2(12) | $78.45 \pm 0.11$ |
| L3(13) | $95.20 \pm 0.14$ |
| [(benzene)Ru(L1)Cl]$PF_6$ (14) | $6.30 \pm 0.25$ |
| [($p$ < cymene)Ru(L2)Cl]$PF_6$ (15) | $7.60 \pm 0.23$ |
| [(benzene)Ru(L2)Cl]$PF_6$ (16) | $8.0 \pm 0.06$ |
| [($p$ < cymene)Ru(L3)Cl]$PF_6$ (17) | $77.0 \pm 0.02$ |
| [(benzene)Ru(L3)Cl]$PF_6$ (18) | $77.0 \pm 0.08$ |
| [Cp*Rh(L1)Cl]$PF_6$ (19) | $3.70 \pm 0.44$ |
| [Cp*Rh(L2)Cl]$PF_6$ (20) | $62.50 \pm 0.12$ |
| [Cp*Rh(L3)Cl]$PF_6$ (21) | $81.20 \pm 0.27$ |
| [Cp*Ir(L2)Cl]$PF_6$ (22) | $40.60 \pm 0.35$ |
| [Cp*Ir(L1)Cl]$PF_6$ (23) | $11.0 \pm 0.08$ |
| [Cp*Ir(L3)Cl]$PF_6$ (24) | $80.30 \pm 0.11$ |
| AA | 100 |

**Table 2.**
*DPPH radical scavenging performance of the studied compounds [80].*

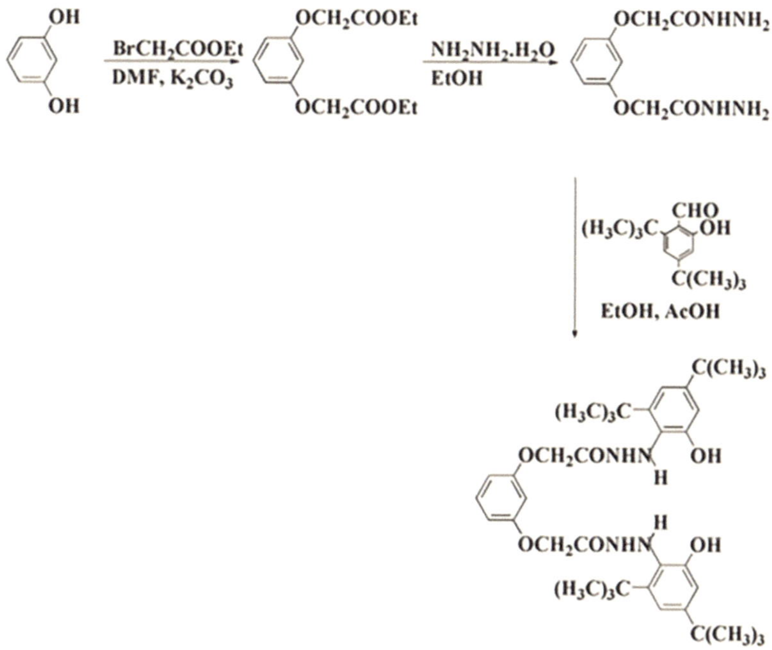


**Figure 4.**
*Source Ref. [82].*

Qasem et al. (2022) Using $Co^{2+}$, $Cu^{2+}$, $Zn^{2+}$, and $Ni^{2+}$ ions (72–75), a new series of bis-hydrazone chelates was produced from (N0E,N000E)-2,2-(1,3-phenylenebis (oxy))bis(N0-(4,5-di-tert-butyl-2-hydroxybenzylidene) acetohydrazide) ligand (SB) (71). **Figure 4**. Additionally, the newly evaluated substances were shown to several cell lines using the MTT assay. Using the DPPH method, the antioxidant performance of the compounds under study in DMSO was calculated. The results indicate that the SB ligand and its metal complexes exhibit superior antioxidant activity.

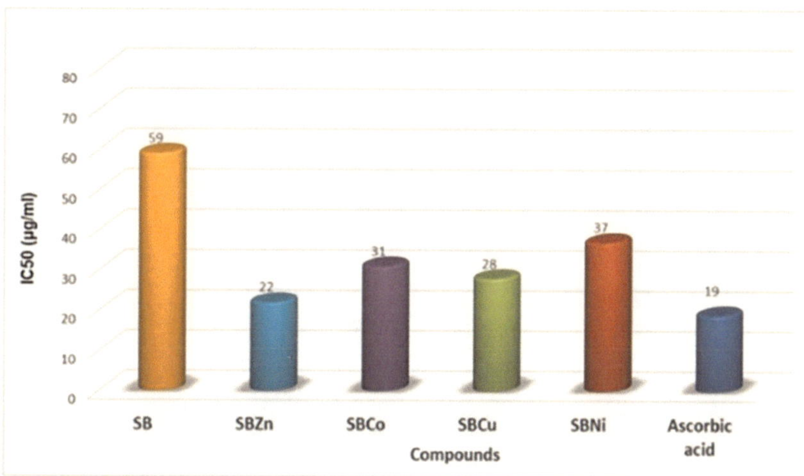


**Figure 5.**
*The suppression of the DPPH radical for tested compounds. Reproduced from the permission of ref. [82].*

**Figure 5** shows the efficiency percentage of inhibitions for the compounds that were evaluated. The studied chelates' comparable antioxidant activity to the DPPH free radical with reference to vitamin C further supported this conclusion. The *in silico* data display the low performance of the free ligand that enhanced the chelation with the Cu(II), in contrast with the practical data [82].

Zn(II), Ni(II), Cu(II), and Co(II) metal complexes prepared by Ikechukwu et al. were moderately and effectively scavengers of the ABTS radical (**Figure 6**), and their activity was comparable to that of common medicines like BHT and rutin (**Table 3**). The test materials with the lowest amounts quenched the ABTS+ radicals in the system more successfully. It was discovered that the Schiff base and its metal complexes had moderate to high activity when it came to scavenging the ABTS+ radical in comparison to the standards (rutin and BHT). Among the synthesized metal complexes, Co(LL) showed the highest activity, with an $IC_{50}$ of around $1.83 \pm 1.08$ μM. The examined chemicals can be classified according to their ability to scavenge ABTS radicals: BHT > Co(LL) > H2LL > Cu(LL) > rutin > Zn(LL) > Ni(LL). The ABTS radical was found to be moderately scavenged by the Schiff base (H2LL), Cu(II), Zn(II), Ni(II), and Co(II) metal complexes in comparison to the DPPH radical, suggesting their potential as chemotherapeutic agents for radical chains termination [83].

| | Compounds | | | | | | |
|---|---|---|---|---|---|---|---|
| | $H_2LL$ | Cu(LL) | Ni(LL) | Co(LL) | Zn(LL) | Rutin* | BHT* |
| $IC_{50} \pm SD$ | $1.86 \pm 2.28$ | $2.15 \pm 1.85$ | $3.66 \pm 1.16$ | $1.83 \pm 1.08$ | $3.08 \pm 0.76$ | $2.83 \pm 1.84$ | $1.64 \pm 1.54$ |
| $R^2$ | 0.845 | 0.992 | 0.975 | 0.986 | 0.953 | 0.983 | 0.919 |

(n = 3, X ± SEM); $IC_{50}$: *inhibitory concentration;* $R^2$: *correlation coefficient.* *Standards.

**Table 3.**
*ABTS radical scavenging activities ($IC_{50}$, μM) of standard drugs and H2LL Schiff bases and their metal complexes [83].*

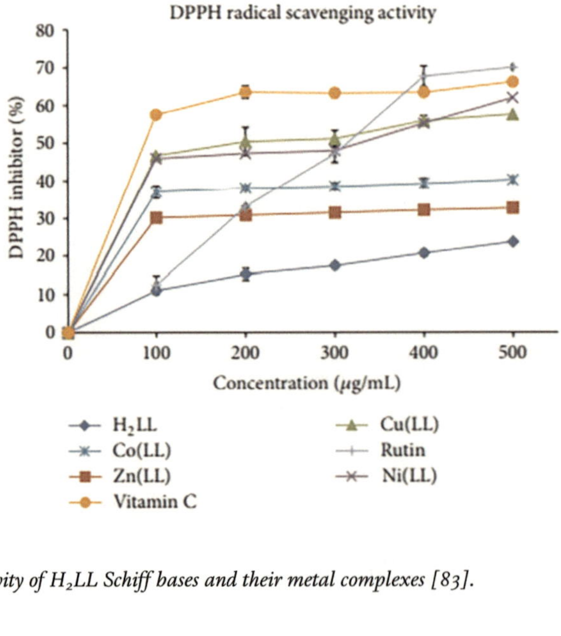


**Figure 6.**
*DPPH scavenging activity of $H_2LL$ Schiff bases and their metal complexes [83].*

| Compounds | $IC_{50}$ (μM) | | | |
|---|---|---|---|---|
| | SASA | XOA | LPI | FRSA |
| Cmplx 1 | 176.48 ± 8.82[a] | 82.06 ± 4.10[a] | 5.90 ± 0.29[c] | 86.58 ± 13.17[a] |
| Cmplx 2 | 64.88 ± 3.24[c] | 53.80 ± 2.69[b] | 8.26 ± 0.17[a] | 75.64 ± 4.43[b] |
| Cmplx 3 | 82.53 ± 4.12[b] | 66.52 ± 3.32[c] | 6.95 + 0.44[b] | 73.68 ± 4.12[b] |
| Trolox | 55.37 ± 2.76[c] | NT | NT | NT |
| Allopurinol | NT | 21.72 ± 1.08[d] | NT | NT |
| BHA | NT | NT | 3.53 ± 0.31[d] | 56.13 ± 2.24[c] |

**Table 4.**
*$IC_{50}$ Values of Test Complexes and Positive Control (Trolox, Allopurinol, and BHA) for Various Antioxidant Activity Parameters.*

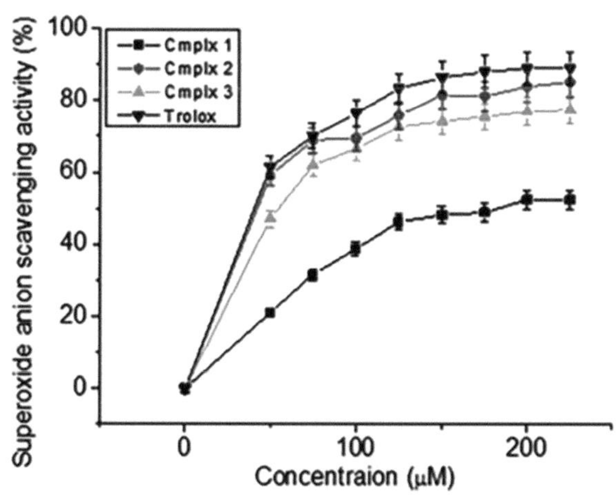


**Figure 7.**
*The superoxide anion radical scavenging activities of organoruthenium complexes and trolox (positive control) in various concentrations. Results are expressed as mean ± standard deviation of triplicate determinations.*

Ibrahim et al. make complexes of organoruthenium. According to **Table 4** and **Figure 7**, Cmplx 2 had the greatest solvent-accessible surface area (SASA) and Cmplx 1 had the lowest. At different dosages, the SASA of Cmplx 2, Cmplx 3, and trolox shown significant activity, exhibiting percentage decreases of 75.73 ± 2.85, 69.19 ± 3.17, and 80.54 ± 1.47%. In contrast, Cmplx 1 displayed a comparatively low SASA of 42.52% ($p < 0.05$). Cmplx 1, Cmplx 2, Cmplx 3, and trolox had $IC_{50}$ values of 176.48 8.82, 64.88 ± 3.24, 82.53 ± 4.12, and 55.37 ± 2.76 μM, in that order. Despite having a lower SASA than trolox and other complexes, the study demonstrated that Cmplx 1 possessed the ability to donate protons and might function as a scavenger or inhibitor of free radicals, potentially functioning as a major antioxidant.

## 4. Conclusions

In recent decades, organic ligands have been complexed with different metal ions of s-p and d-block to enhance antioxidant efficacy. Because of their numerous benefits, coordination compounds are widely used in various fields, especially the pharmaceutical and medical fields. The characteristics of both ligands and metals can lead to the production of molecules with significantly increased activity, as chelation generally enhances activity. Metal chelates can be used in place of standard synthetic antioxidants to support and promote redox processes associated with antioxidant action. They offer advantages over traditional synthetic antioxidants in terms of geometry, oxidation state, and coordination number.

## Conflict of interest

The authors declare no conflict of interest.


## Author details

Ehab M. Abdalla[1*] and Samar A. Aly[2]

1 Faculty of Science, Chemistry Department, New Valley University, Alkharga, Egypt

2 Department of Environmental Biotechnology, Genetic Engineering and Biotechnology Research Institute, University of Sadat City, Sadat City, Egypt

*Address all correspondence to: ehababdalla99@sci.nvu.edu.eg

## References

[1] Sies H, Jones DP. Reactive oxygen species (ROS) as pleiotropic physiological signaling agents. Nature Reviews Molecular Cell Biology. 2020; **21**(7):363-383

[2] Snezhkina AV, Kudryavtseva AV, Kardymon OL, Savvateeva MV, Melnikova NV, Krasnov GS, et al. ROS generation and antioxidant defense systems in normal and malignant cells. Oxidative Medicine and Cellular Longevity. 2019;**2019**:6175804

[3] Neha K, Haider MR, Pathak A, Yar MS. Medicinal prospects of antioxidants: A review. European Journal of Medicinal Chemistry. 2019;**178**: 687-704

[4] Phaniendra A, Jestadi DB, Periyasamy L. Free radicals: Properties, sources, targets, and their implication in various diseases. Indian Journal of Clinical Biochemistry. 2015;**30**:11-26

[5] Dreher D, Junod AF. Role of oxygen free radicals in cancer development. European Journal of Cancer. 1996;**32**(1): 30-38

[6] Vendemiale G, Grattagliano I, Altomare E. An update on the role of free radicals and antioxidant defense in human disease. International Journal of Clinical and Laboratory Research. 1999; **29**:49-55

[7] Ferreira CA, Ni D, Rosenkrans ZT, Cai W. Scavenging of reactive oxygen and nitrogen species with nanomaterials. Nano Research. 2018;**11**:4955-4984

[8] Spitz DR, Hauer-Jensen M. Ionizing Radiation-Induced Responses: Where Free Radical Chemistry Meets Redox Biology and Medicine. NY, USA: Mary Ann Liebert Inc.; 2014. pp. 1407-1409

[9] Peña-Bautista C, Baquero M, Vento M, Cháfer-Pericás C. Free radicals in Alzheimer's disease: Lipid peroxidation biomarkers. Clinica Chimica Acta. 2019;**491**:85-90

[10] Labat-Robert J, Robert L. Longevity and aging. Role of free radicals and xanthine oxidase. A review. Pathologie Biologie. 2014;**62**(2):61-66

[11] Bajpai V, Baek K-H, Kang S. Antioxidant and free radical scavenging activities of taxoquinone, a diterpenoid isolated from Metasequoia glyptostroboides. South African Journal of Botany. 2017;**111**:93-98

[12] Aly SA, Eldourghamy A, El-Fiky BA, Megahed AA, El-Sayed WA, Abdalla EM, et al. Synthesis, spectroscopic characterization, thermal studies, and molecular docking of novel Cr (III), Fe (III), and Co (II) complexes based on Schiff base: *In vitro* antibacterial and antitumor activities. Journal of Applied Pharmaceutical Science. 2023;**13**(2):196-210

[13] Guruvayoorappan C, Afira A, Kuttan G. Antioxidant potential of *Biophytum sensitivum* extract *in vitro* and *in vivo*. Journal of Basic and Clinical Physiology and Pharmacology. 2006; **17**(4):255-268

[14] Demidchik V. Reactive oxygen species and oxidative stress in plants. In: Plant Stress Physiology. UK: CABI Wallingford; 2012. pp. 24-58

[15] Pham-Huy LA, He H, Pham-Huy C. Free radicals, antioxidants in disease and health. International Journal of Biomedical Science: IJBS. 2008;**4**(2):89

[16] Moldogazieva N, Mokhosoev I, Feldman N, Lutsenko S. ROS and RNS

signaling: Adaptive redox switches through oxidative/nitrosative protein modifications. Free Radical Research. 2018;**52**(5):507-543

[17] Ramsay RR. Electron carriers and energy conservation in mitochondrial respiration. ChemTexts. 2019;**5**(2):9

[18] Molyneux P. The use of the stable free radical diphenylpicrylhydrazyl (DPPH) for estimating antioxidant activity. Songklanakarin Journal of Science and Technology. 2004;**26**(2): 211-219

[19] Lobo V, Patil A, Phatak A, Chandra N. Free radicals, antioxidants and functional foods: Impact on human health. Pharmacognosy Reviews. 2010; **4**(8):118

[20] Akbarirad H, Ardabili AG, Kazemeini S, Khaneghah AM. An overview on some of important sources of natural antioxidants. International Food Research Journal. 2016;**23**(3): 928-933

[21] Suleman M, Khan A, Baqi A, Kakar MS, Ayub M. 2. Antioxidants, its role in preventing free radicals and infectious diseases in human body. Pure and Applied Biology (PAB). 2019;**8**(1): 380-388

[22] Gulcin İ, Alwasel SH. Metal ions, metal chelators and metal chelating assay as antioxidant method. Processes. 2022; **10**(1):132

[23] Ibrahim M, Khan A, Ikram M, Rehman S, Shah M, Nabi H, et al. *In vitro* antioxidant properties of novel Schiff base complexes, Asian. Journal of Chemical Sciences. 2017;**2**:1-12

[24] Mohamed GG. Synthesis, characterization and biological activity of bis (phenylimine) Schiff base ligands and their metal complexes. Spectrochimica Acta Part A: Molecular and Biomolecular Spectroscopy. 2006; **64**(1):188-195

[25] Abu-Dief AM, Abdel-Rahman LH, Shehata MR, Abdel-Mawgoud AAH. Novel azomethine Pd (II)-and Vanadium(II) oxide (VO (II))-based metallo-pharmaceuticals as anticancer, antimicrobial, and antioxidant agents: Design, structural inspection, DFT investigation, and DNA interaction. Journal of Physical Organic Chemistry. 2019;**32**(12):e4009

[26] Abu-Dief AM, Abdel-Rahman LH, Abdel-Mawgoud AAH. A robust *in vitro* anticancer, antioxidant and antimicrobial agents based on new metal-azomethine chelates incorporating Ag (I), Pd (II) and VO (II) cations: Probing the aspects of DNA interaction. Applied Organometallic Chemistry. 2020;**34**(2):e5373

[27] Aljohani FS, Omran OA, Ahmed EA, Al-Farraj ES, Elkady EF, Alharbi A, et al. Design, structural inspection of new bis (1H-benzo [d] imidazol-2-yl) methanone complexes: Biomedical applications and theoretical implementations via DFT and docking approaches. Inorganic Chemistry Communications. 2023;**148**:110331

[28] Hassan SS, Aly SA, Al-Sulami AI, Albohy SA, Salem MF, Nasr GM, et al. Synthesis, characterization, PXRD studies, and theoretical calculation of the effect of gamma irradiation and antimicrobial studies on novel Pd (II), Cu (II), and Cu (I) complexes. Frontiers in Chemistry. 2024;**12**:1357330

[29] Abu-Dief AM, El-Khatib RM, Aljohani FS, Alzahrani SO, Mahran A, Khalifa ME, et al. Synthesis and intensive characterization for novel Zn (II), Pd (II), Cr (III) and VO (II)-Schiff base

complexes; DNA-interaction, DFT, drug-likeness and molecular docking studies. Journal of Molecular Structure. 2021;**1242**:130693

[30] El-Lateef HMA, Khalaf MM, Shehata MR, Abu-Dief AM. Fabrication, DFT calculation, and molecular docking of two Fe (III) imine chelates as anti-COVID-19 and pharmaceutical drug candidate. International Journal of Molecular Sciences. 2022;**23**(7):3994

[31] Omar M, Mohamed G, Hindy A. Transition metal complexes of heterocyclic Schiff base: Biological activity, spectroscopic and thermal characterization. Journal of Thermal Analysis and Calorimetry. 2006;**86**(2): 315-325

[32] Alshater H, Al-Sulami AI, Aly SA, Abdalla EM, Sakr MA, Hassan SS. Antitumor and antibacterial activity of Ni (II), Cu (II), Ag (I), and Hg (II) complexes with ligand derived from thiosemicarbazones: Characterization and theoretical studies. Molecules. 2023; **28**(6):2590

[33] Henrici-Olivé G, Olivé S. The Chemistry of the Catalyzed Hydrogenation of Carbon Monoxide. Berlin Heidelberg: Springer Science & Business Media; 2012

[34] Li Y, Yang Z-Y, Wu J-C. Synthesis, crystal structures, biological activities and fluorescence studies of transition metal complexes with 3-carbaldehyde chromone thiosemicarbazone. European Journal of Medicinal Chemistry. 2010; **45**(12):5692-5701

[35] Hranjec M, Starčević K, Pavelić SK, Lučin P, Pavelić K, Zamola GK. Synthesis, spectroscopic characterization and antiproliferative evaluation in vitro of novel Schiff bases related to benzimidazoles. European Journal of Medicinal Chemistry. 2011;**46**(6): 2274-2279

[36] Abu-Dief AM, Díaz-Torres R, Sañudo EC, Abdel-Rahman LH, Aliaga-Alcalde N. Novel sandwich triple-decker dinuclear NdIII-(bis-N, N′-p-bromo-salicylideneamine-1, 2-diaminobenzene) complex. Polyhedron. 2013;**64**:203-208

[37] Ashraf MA, Wajid A, Mahmood K, Maah MJ, Yusoff I. Spectral investigation of the activities of amino substituted bases. International Journal of Chemical Engineering and Applications. 2011; **2**(4):252

[38] Ashassi-Sorkhabi H, Shabani B, Aligholipour B, Seifzadeh D. The effect of some Schiff bases on the corrosion of aluminum in hydrochloric acid solution. Applied Surface Science. 2006;**252**(12): 4039-4047

[39] Consiglio G, Failla S, Finocchiaro P, Oliveri IP. Aggregation properties of bis (salicylaldiminato) zinc (II) Schiff-base complexes and their Lewis acidic character. Dalton Transactions. 2012; **41**(2):387-395

[40] Aljohani ET, Shehata MR, Abu-Dief AM. Design, synthesis, structural inspection of Pd2+, VO2+, Mn2+, and Zn2+ chelates incorporating ferrocenyl thiophenol ligand: DNA interaction and pharmaceutical studies. Applied Organometallic Chemistry. 2021;**35**(4): e6169

[41] Alorabi AQ, Abdelbaset M, Zabin SA. Colorimetric detection of multiple metal ions using Schiff base 1-(2-thiophenylimino)-4-(N-dimethyl) benzene. Chem. 2019;**8**(1):1

[42] Abu-Dief AM, El-Sagher HM, Shehata MR. Fabrication, spectroscopic characterization, calf thymus DNA binding investigation, antioxidant and

anticancer activities of some antibiotic azomethine Cu (II), Pd (II), Zn (II) and Cr (III) complexes. Applied Organometallic Chemistry. 2019;**33**(8): e4943

[43] Al-Abdulkarim HA, El-khatib RM, Aljohani FS, Mahran A, Alharbi A, Mersal GA, et al. Optimization for synthesized quinoline-based Cr3+, VO2+, Zn2+ and Pd2+ complexes: DNA interaction, biological assay and in-silico treatments for verification. Journal of Molecular Liquids. 2021;**339**:116797

[44] Al-Saeedi SI, Abdel-Rahman LH, Abu-Dief AM, Abdel-Fatah SM, Alotaibi TM, Alsalme AM, et al. Catalytic oxidation of benzyl alcohol using nanosized Cu/Ni Schiff-base complexes and their metal oxide nanoparticles. Catalysts. 2018;**8**(10):452

[45] Abu-Dief AM, El-Metwaly NM, Alzahrani SO, Bawazeer AM, Shaaban S, Adam MSS. Targeting ctDNA binding and elaborated in-vitro assessments concerning novel Schiff base complexes: Synthesis, characterization, DFT and detailed in-silico confirmation. Journal of Molecular Liquids. 2021;**322**:114977

[46] Adam MSS, Abu-Dief AM, Makhlouf M, Shaaban S, Alzahrani SO, Alkhatib F, et al. Tailoring, structural inspection of novel oxy and non-oxy metal-imine chelates for DNA interaction, pharmaceutical and molecular docking studies. Polyhedron. 2021;**201**:115167

[47] Štarha P, Trávníček Z. Non-platinum complexes containing releasable biologically active ligands. Coordination Chemistry Reviews. 2019; **395**:130-145

[48] de Fátima Â, de Paula Pereira C, Olímpio CRSDG, de Freitas Oliveira BG, Franco LL, da Silva PHC. Schiff bases and their metal complexes as urease inhibitors–a brief review. Journal of Advanced Research. 2018;**13**:113-126

[49] Galleano M, Verstraeten SV, Oteiza PI, Fraga CG. Antioxidant actions of flavonoids: Thermodynamic and kinetic analysis. Archives of Biochemistry and Biophysics. 2010; **501**(1):23-30

[50] El-Lateef HMA, El-Dabea T, Khalaf MM, Abu-Dief AM. Development of metal complexes for treatment of coronaviruses. International Journal of Molecular Sciences. 2022;**23**(12):6418

[51] Milaeva ER. Metal-based antioxidants-potential therapeutic candidates for prevention the oxidative stress-related carcinogenesis: Mini-review. Current Topics in Medicinal Chemistry. 2011;**11**(21):2703-2713

[52] Gómez FS, Sánchez SP, Gallego Iradi MG, Mohd Azman NA, Almajano MP. Avocado seeds: Extraction optimization and possible use as antioxidant in food. Antioxidants. 2014; **3**(2):439-454

[53] Cherrak SA, Mokhtari-Soulimane N, Berroukeche F, Bensenane B, Cherbonnel A, Merzouk H, et al. In vitro antioxidant versus metal ion chelating properties of flavonoids: A structure-activity investigation. PLoS One. 2016; **11**(10):e0165575

[54] Anwar H, Hussain G, Mustafa I. Antioxidants from natural sources. In: Shalaby E, Azzam GM, editors. Antioxidants in Foods and its Applications. Vol. 3. 2018. DOI: 10.5772/intechopen.75961 Submitted: 08 January 2018 Reviewed: 26 February 2018 Published: 08 April 2018

[55] Kozsup M, Zhou X, Farkas E, Bényei AC, Bonnet S, Patonay T, et al. Synthesis, characterization and

cytotoxicity studies of Co (III)-flavonolato complexes. Journal of Inorganic Biochemistry. 2021;**217**:111382

[56] Khater M, Ravishankar D, Greco F, Osborn HM. Metal complexes of flavonoids: Their synthesis, characterization and enhanced antioxidant and anticancer activities. Future Medicinal Chemistry. 2019; **11**(21):2845-2867

[57] Ejidike IP, Ajibade PA. Transition metal complexes of symmetrical and asymmetrical Schiff bases as antibacterial, antifungal, antioxidant, and anticancer agents: Progress and prospects. Reviews in Inorganic Chemistry. 2015;**35**(4):191-224

[58] Aljohani ET, Shehata MR, Alkhatib F, Alzahrani SO, Abu-Dief AM. Development and structure elucidation of new $VO^{2+}$, $Mn^{2+}$, $Zn^{2+}$, and $Pd^{2+}$ complexes based on azomethine ferrocenyl ligand: DNA interaction, antimicrobial, antioxidant, anticancer activities, and molecular docking. Applied Organometallic Chemistry. 2021;**35**(5):e6154

[59] Yıldırım N, Bilgiçli AT, Alici EH, Arabacı G, Yarasir MN. Formation, characterization, aggregation, fluorescence and antioxidant properties of novel tetrasubstituted metal-free and metallophthalocyanines bearing (4-(methylthio) phenoxy) moieties. Journal of Molecular Structure. 2017;**1144**:66-79

[60] El-Lateef HMA, El-Dabea T, Khalaf MM, Abu-Dief AM. Innovation of imine metal chelates as corrosion inhibitors at different media: A collective study. International Journal of Molecular Sciences. 2022;**23**(16):9360

[61] Amaral GP, Puntel GO, Dalla Corte CL, Dobrachinski F, Barcelos RP, Bastos LL, et al. The antioxidant properties of different phthalocyanines. Toxicology In Vitro. 2012;**26**(1):125-132

[62] Akhtar MJ, Ahamed M, Alhadlaq HA, Alshamsan A. Mechanism of ROS scavenging and antioxidant signalling by redox metallic and fullerene nanomaterials: Potential implications in ROS associated degenerative disorders. Biochimica et Biophysica Acta (BBA)-General Subjects. 2017;**1861**(4):802-813

[63] Sho T, Xu J. Role and mechanism of ROS scavengers in alleviating NLRP3-mediated inflammation. Biotechnology and Applied Biochemistry. 2019;**66**(1): 4-13

[64] Kedare SB, Singh R. Genesis and development of DPPH method of antioxidant assay. Journal of Food Science and Technology. 2011;**48**: 412-422

[65] MacDonald-Wicks LK, Wood LG, Garg ML. Methodology for the determination of biological antioxidant capacity *in vitro*: A review. Journal of the Science of Food and Agriculture. 2006; **86**(13):2046-2056

[66] Ionita P. The chemistry of DPPH·free radical and congeners. International Journal of Molecular Sciences. 2021;**22**(4):1545

[67] Sharma OP, Bhat TK. DPPH antioxidant assay revisited. Food Chemistry. 2009;**113**(4):1202-1205

[68] Sanna D, Delogu G, Mulas M, Schirra M, Fadda A. Determination of free radical scavenging activity of plant extracts through DPPH assay: An EPR and UV–vis study. Food Analytical Methods. 2012;**5**:759-766

[69] Magalhães LM, Segundo MA, Reis S, Lima JL. Methodological aspects about

*in vitro* evaluation of antioxidant properties. Analytica Chimica Acta. 2008;**613**(1):1-19

[70] Nimse SB, Pal D. Free radicals, natural antioxidants, and their reaction mechanisms. RSC Advances. 2015;**5**(35): 27986-28006

[71] Gulcin İ. Antioxidants and antioxidant methods: An updated overview. Archives of Toxicology. 2020; **94**(3):651-715

[72] Jiang GCGC, JiJun YJY, Yu WJWJ, Feng LYLY, Jing XJX, YuGang JYJ. Antioxidant activities of peel, pulp and seed fractions of common fruits as determined by FRAP assay. Nutrition Research. 2003;**23**:1719-1726

[73] Marchi RC, Campos IA, Santana VT, Carlos RM. Chemical implications and considerations on techniques used to assess the *in vitro* antioxidant activity of coordination compounds. Coordination Chemistry Reviews. 2022;**451**:214275

[74] Apak R, Güçlü K, Demirata B, Özyürek M, Çelik SE, Bektaşoğlu B, et al. Comparative evaluation of various total antioxidant capacity assays applied to phenolic compounds with the CUPRAC assay. Molecules. 2007;**12**(7):1496-1547

[75] Apak R, Özyürek M, Güçlü K, Bekdeşer B, Bener M. The CUPRAC methods of antioxidant measurement for beverages. In: Processing and Impact on Antioxidants in Beverages. Elsevier; 2014. pp. 235-244. DOI: 10.1016/b978-0-12-404738-9.00024-6

[76] Abu-Dief AM, Díaz-Torres R, Sañudo EV, Abdel-Rahman LH, Aliaga-Alcalde N. Novel Sandwich Triple-Decker Dinuclear NdIII-(bis-N, N'-p-bromo-salicylideneamine-1,2-diaminobenzene) Complex. Polyhedron. 2013;**64**:203-208. DOI: 10.21203/rs.3.rs-1348908/v1

[77] Re R, Pellegrini N, Proteggente A, Pannala A, Yang M, Rice-Evans C. Antioxidant activity applying an improved ABTS radical cation decolorization assay. Free Radical Biology and Medicine. 1999;**26**(9–10): 1231-1237

[78] Arciszewska Ż, Gama S, Kalinowska M, Świderski G, Świsłocka R, Gołębiewska E, et al. Caffeic acid/Eu (III) complexes: Solution equilibrium studies, structure characterization and biological activity. International Journal of Molecular Sciences. 2022;**23**(2):888

[79] Abdel-Rahman LH, Basha MT, Al-Farhan BS, Shehata MR, Abdalla EM. Synthesis, characterization, potential antimicrobial, antioxidant, anticancer, DNA binding, and molecular docking activities and DFT on novel Co (II), Ni (II), VO (II), Cr (III), and La (III) Schiff base complexes. Applied Organometallic Chemistry. 2022;**36**(1):e6484

[80] Nongpiur CGL, Tripathi DK, Poluri KM, Rawat H, Kollipara MR. Ruthenium, rhodium and iridium complexes containing diazafluorene derivative ligands: Synthesis and biological studies. Journal of Chemical Sciences. 2022;**134**(1):23

[81] Ali S, Singh V, Tripathi V. Microwave assisted synthesis of heterocyclic metal complexes and evaluation of their *in vitro* anticancer activity against oral cancer cells, antioxidant and molecular docking study. Indian Journal of Chemistry (IJC). 2022;**61**(4):385-391

[82] Qasem HA, Aouad MR, Al-Abdulkarim HA, Al-Farraj ES, Attar RM, El-Metwaly NM, et al. Tailoring of some novel bis-hydrazone metal chelates, spectral based characterization and DFT calculations for pharmaceutical

applications and in-silico treatments for verification. Journal of Molecular Structure. 2022;**1264**:133263

[83] Ejidike IP, Ajibade PA. Synthesis, characterization, antioxidant, and antibacterial studies of some metal (II) complexes of tetradentate schiff base ligand: (4E)-4-[(2-(E)-[1-(2, 4-dihydroxyphenyl) ethylidene] amino ethyl) imino] pentan-2-one. Bioinorganic Chemistry and Applications. 2015;**2015**:890734

Chapter 16

# The Power of Antioxidant: Tea Catechin and Body Oxidative Stress

*Baiq Amarwati Tartillah*


## Abstract

Green tea is rich in catechins, particularly (−)-epigallocatechin-3-gallate (EGCG), which act as potent antioxidants and can help to prevent oxidative stress-related diseases. This article revealed the importance of green tea catechins in mitigating the risk of complex diseases such as cardiovascular disease, cancer, and neurological diseases. It also highlighted the potential side effects of excessive green tea consumption, emphasising the need for moderation. The review covered a wide range of potential health benefits of green tea, including its effects on weight loss, diabetes, metabolic syndrome, and cognitive decline. Additionally, the collection of research articles elaborated on the antioxidant and neuroprotective properties of green tea, as well as its potential role in preventing skin cancer and improving cognitive function. Overall, the evidence presented underscores the potential of green tea as a valuable dietary component in inhibiting diseases such as diabetes, cardiovascular, cancer, and infectious illness, while also emphasising the importance of green tea consumption in a balanced manner.


**Keywords:** antioxidant, oxidative stress, green tea, catechin, polyphenols

## 1. Introduction

Oxidative stress occurs when there is an imbalance between antioxidants and free radicals in the human body. Imagine antioxidants and free radicals as players in a delicate game. Free radicals, molecules with an uneven number of electrons, act quickly, creating large chemical chains through a process called oxidation. While oxidation is essential for generating energy and aiding in chemical reactions, it is not always beneficial. Picture antioxidants as the team that donates electrons to stabilise the free radicals, making them less reactive without losing their own stability [1].

Free radicals can be naturally produced from various environmental sources, like pollution, radiation, ozone, smoking, or exposure to chemical substances such as cleaners and pesticides. They can also originate from a diet high in sugar, fat, and alcohol. Imagine the imbalance between free radicals and antioxidants like a seesaw. When free radicals outweigh antioxidants, they can damage three major components of the human body: fats, DNA, and proteins. This damage can lead to various health issues, including diabetes, high blood pressure, atherosclerosis, heart diseases, cancer, and neurodegenerative diseases [1].

To reduce the risk of oxidative stress, we can boost our antioxidant levels, which act as defenders against the formation of harmful free radicals. Including antioxidant-rich foods in our diet, such as tea, can be an effective strategy to prevent various complex illnesses [1]. Notably, green tea, abundant in catechins—a potent antioxidant—holds the potential to combat degenerative diseases like cardiovascular diseases (CVD) [2].

Tea polyphenols, serving as antioxidants, perform various protective roles by scavenging reactive oxygen and nitrogen species, binding to transition metal ions, and stimulating antioxidant enzymes [3]. These actions contribute to safeguarding tissues, cells, and plasma from potential oxidative damage [4, 5].

While there are numerous health benefits associated with consuming antioxidants from tea in general, this article will specifically delve into the antioxidant mechanisms of catechins in preventing diseases related to oxidative stress by incorporating green tea (GT), such as diabetes, cardiovascular disease, cancer, and infectious diseases.

## 2. Green tea catechin and diseases correlation

There are two types of tea plants namely *Camellia sinensis* and Camellia assamica which exhibit distinct characteristics in terms of hardiness and leaf size [6]. *Camellia sinensis* has a robust, small-leaf plant, and resilient to cold weather, while Camellia assamica is more delicate, large-leafed, and thrives in tropical regions. These plants are then processed into the most common four primary tea types—black, green, white, and oolong—that undergo diverse fermentation processes, leading to variations in polyphenol composition [7, 8]. Notably, black tea (BT) is the most produced tea with 78%, while GT makes up about 20% [9]. The fermentation levels also distinguish these teas, with GT being non-fermented, oolong tea semi-fermented, and BT fully fermented. White tea, with a 10–20% fermentation level, stands between green and black teas [10].

This fermentation disparity results in crucial differences in polyphenol balance and sensory characteristics [11] which are indicated by the production of oxidised polyphenolic compounds like theaflavins and thearubigins in BT [12, 13]. In white tea, the least processed one, is often considered the oldest form of tea and also contains unique components like gamma-amino butyric acid (GABA) and L-theanine [14]. The polyphenols found in black and green tea are depicted in **Figure 1** [15, 16].

The physiological effects of tea can be attributed to specific compounds found in different types of tea. BT contains thearubigins and theaflavins, while green tea is rich in catechins. These compounds have various health benefits, including the inhibition of cancer cell growth [17], reduction of inflammation, prevention of platelet aggregation, regulation of glucose and lipid metabolism, DNA repair, stimulation of immune function, and modulation of detoxification enzymes [18, 19]. The antioxidant properties of catechins depend on factors such as the hydroxyl group at the C-3 position or the degree of hydroxylation of the B ring [19]. Among green tea catechins, (−)-epigallocatechin-3-gallate (EGCG) stands out as the most abundant and potent antioxidant, with a notable reputation for its role in cancer prevention [20–22]. In addition to EGCG, that acts as a powerful scavenger of reactive oxygen species [23], caffeine and theanine in tea have stimulating and relaxing properties, working together for improved mental alertness [11]. In various plants, flavonoids with hydroxyl groups

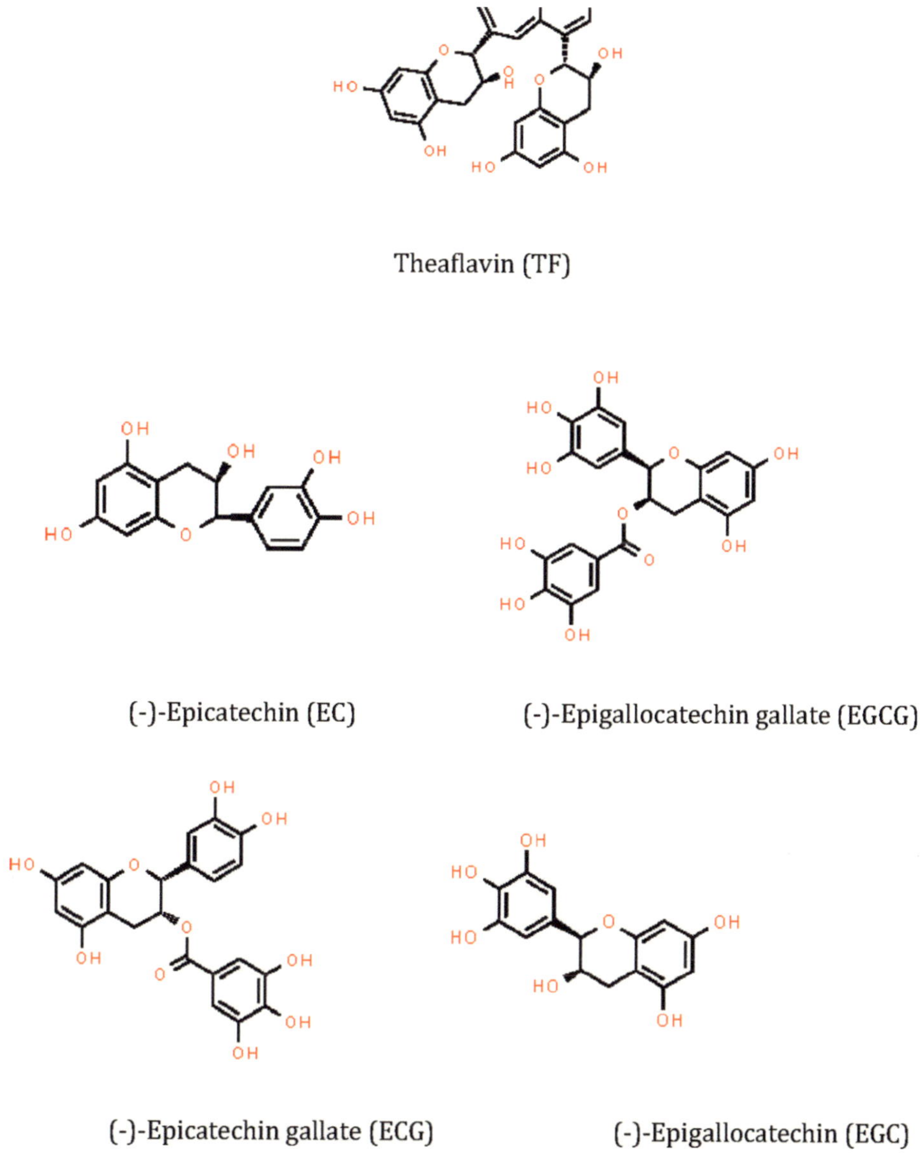


**Figure 1.**
*Polyphenols in black and green tea.*

attached to the ring structure contribute to antioxidant activity, as seen in compounds like vitamin C, β-carotene, and carotenoids [24, 25].

**Table 1** outlines the physical properties of key catechins, typically appearing as colourless crystalline substances with a bitter, astringent taste. Their solubility in polar solvents like water and methanol is influenced by factors such as extraction temperature, time, and the solvent type used [28–30]. In green tea, EGCG is the most concentrated, followed by (−)-Epigallocatechin (EGC), (−)-Epicatechin gallate (ECG), and (−)-Epicatechin (EC) [31]. Esterified catechins, contributing to astringency and a bitter taste, differ from free catechins, which are less astringent and

| | EC | EGC | ECG | EGCG | Reference |
|---|---|---|---|---|---|
| Molecular structure | C15H14O6 | C15H14O7 | C22H18O10 | C22H18O11 | [26] |
| Molecular mass | 290 | 306 | 442 | 458 | [26] |
| Absorbance peak (in nm) | 280 | 269 | 280 | 273 | [26] |
| Melting point (C) | 242 | 218 | 257 | 224 | [26] |
| Dissolution | Time/Temperature/ Solvent Dependent | Time/Temperature/ Solvent Dependent | Time/Temperature/ Solvent Dependent | Time/Temperature/ Solvent Dependent | [27] |
| Solvents Water/ methanol/ethanol | Soluble | Soluble | Soluble | Soluble | http://Chemicalland21.com |
| Chloroform | Not Soluble | Not Soluble | Not Soluble | Not Soluble | http://Chemicalland21.com |
| Taste | Bitter with sweet after taste | Bitter with sweet after taste | Bitter with sweet after taste | Bitter with sweet after taste | [27] |
| Appearance | Yellow to greenish powder | Light yellow powder | Light yellow powder | White to pink powder | http://Chemicalland21.com |

**Table 1.**
*Physical attributes of four catechins.*

slightly sweet [32]. Those with ester binding, like EGCG and ECG, exhibit a greater tendency to form precipitates with enzymes, leading to cream formation [29].

Extraction time, solvent type, and temperature play crucial factors in obtaining higher catechin content such as EGC and EGCG in green tea [27, 33], the same goes with BT in obtaining theaflavin and any compound within [34]. Although theaflavin is the most abundant component in BT, catechin is also available in this type of tea but only a small amount of component. When the catechin is consumed, it has a different function in the body system [35]. In laboratory-based observation showed that catechin in black and green tea could prevent hyperglycaemia by inhibiting the damage of β-cells and increasing insulin activity respectively [35, 36]. Hyperglycaemia refers to elevated blood sugar levels, and these findings suggest that catechins in tea may play a role in managing blood glucose levels.

To measure and identify catechins, UV light absorption is used with the most absorption occurring for EGCG, EGC, ECG, and EC between wavelengths of 210 and 280 nm [37]. Techniques such as liquid chromatography (LC) and capillary electrophoresis (CE) are commonly used to separate, identify, and measure catechins. LC systems often use detectors like UV–vis, diode array (DAD), mass spectrometry (MS), fluorescence (FD), and photodiode array detectors [37].

Not only in green tea, high concentrations of catechins are also present in various sources, including red wine, broad beans, black grapes, apricots, and strawberries. Epicatechin (EC), found in apples, blackberries, broad beans, cherries, black grapes, pears, raspberries, and chocolate, exhibits notable concentrations. The skin of fruits like apples holds flavanols, and peeling reduces catechin levels [38]. Apples contain diverse polyphenols, including quercetin, catechin, phloridzin, and chlorogenic acid, with variations among fruit varieties and changes during maturation. Gallic acid esters of catechin are rare in fruits, found mainly in berries, black currants, and grapes [39]. Strawberries, for instance, possess a complex catechin mixture, including catechin, ECG, EGC, and GC [30]. **Table 2** presents the total catechins found in diverse food sources, while **Table 3** provides a detailed breakdown of the catechin types present in selected food items.

Vegetables like onions, broccoli, kale, lettuce, and tomatoes are rich in flavonols like quercetin and kaempferol. Some vegetables, such as celery, sweet peppers, and lettuce, contain flavones [43]. Catechins and type B procyanidins are absent in leafy greens or root vegetables but present in legumes like broad and green beans [44]. Tomatoes uniquely contain flavanones, naringenin, and hesperetin [45].

It is noticeable that green tea has higher (compared to other teas) antioxidant content because the process does not include fermentation or enzyme activation. It is purposed to prevent autolysis and oxidation of constituents and aid stabilising the components during storage. Compared to BT, the process undergoes fermentation through the enzymatic process, which stimulates oxidation, so most of the catechin is oxidised in this tea [46].

EGCG as an essential constituent in green tea has multiple functions as a strong antioxidant [47] such as reducing agents, hydro-donating antioxidants, singlet oxygen quenchers, and metal chelation. It is classified as an antioxidant because, in low concentration, it prevents autooxidation or free radical-mediated oxidation [25, 48]. Also, during further oxidation, it can stabilise the radical formation after scavenging the electron through intramolecular hydrogen bonding [25, 49].

According to [25], the recommended doses of this component in our dietary is about one gram. However, green tea catechin (GTC) is not a single component that can give health benefits to a human's body. The study conducted by [50] showed that

| Food sources | Catechins[a] | Catechins[b] |
|---|---|---|
| Apple | 20–86 | 10-43 |
| Apricot | 20–50 | 10 |
| Beans | 70–110 | 25 |
| Black tea | 12–100 | 35–55 |
| Blackberry | 9–11 | 6–50 |
| Cherry | 10–44 | 9–11 |
| Chocolate | 23–30 | 5–22 |
| Cider | 8 | 46–61 |
| Grape | 6–35 | 3–17 |
| Green tea | 20–160 | 10–80 |
| Peach | 10–28 | 5–14 |
| Red raspberry | 2–48 | 2–48 |
| Red wine | 8–30 | 8–30 |
| Strawberry | 2–50 | 2–50 |

*Source: [40, 41].*
[a]*Value in mg/ serving.*
[b]*Value in mg/ 100 g sample.*

**Table 2.**
*Distribution of catechins in food sources.*

| | C | EGC | EC | EGCG | ECG | Total catechins |
|---|---|---|---|---|---|---|
| Black current | ND[a] | 1.3 ± 0.01[b] | 3.0 ± 0.01[c] | 6.1 ± 0.01[d] | 3.6 ± 0.03[e] | 13.9 ± 0.02[f] |
| Black tea | 0.7 ± 0.01[a] | 3.0 ± 0.05[b] | 2.3 ± 0.01[c] | 24.9 ± 0.20[d] | 5.9 ± 0.01[e] | 36.9 ± 1.8[f] |
| Cherry | ND[a] | 1.1 ± 0.03[b] | 2.3 ± 0.02[c] | 4.0 ± 0.01[d] | 2.7 ± 0.01[e] | 10.1 ± 0.02[f] |
| Forest fruit | ND[a] | 0.9 ± 0.01[b] | 2.0 ± 0.02[c] | 3.3 ± 0.01[d] | 2.3 ± 0.02[e] | 8.5 ± 0.01[f] |
| Green tea | 1.3 ± 0.02[a] | 19.0 ± 0.09[b] | 4.4 ± 0.03[c] | 23.4 ± 0.05[d] | 5.6 ± 0.03[e] | 53.5 ± 1.5[f] |
| Lemon | ND[a] | 1.1 ± 0.01[b] | 2.0 ± 0.03[c] | 3.4 ± 0.01[d] | 2.7 ± 0.01[e] | 8.6 ± 0.01[f] |
| Orange | ND[a] | 1.0 ± 0.01[b] | 2.5 ± 0.01[c] | 4.2 ± 0.01[d] | 2.9 ± 0.01[e] | 10.6 ± 0.01[f] |
| Strawberry | ND[a] | 1.0 ± 0.02[b] | 2.0 ± 0.01[c] | 3.9 ± 0.01[d] | 2.2 ± 0.02[e] | 9.6 ± 0.01[f] |

*Source: [42].*
*ND: not detected.*
[a,b,c,d,e,f] *values are expressed in mg of dry matter.*

**Table 3.**
*Catechins in selected food sources.*

the combination of a small amount of caffeine from green tea itself gives a mild effect to stimulate exercise performance. This combination aids fat burning by lowering fat absorption and reducing adipocyte lipogenesis, which affects the increase of thermogenesis and fat oxidation [50, 51].

EGCG in green tea is the most potential catechin derivates in suppressing norepinephrine enzyme called COMT (catechol-o-methyl-transferase). Norepinephrine is a neurotransmitter that helps the body's sympathetic nervous system control processes like generating heat (thermogenesis) and burning fat for energy. When we consume catechins in green tea that block the COMT enzyme, it can enable norepinephrine to have a longer-lasting impact on these processes, promoting continued fat burning and heat production [52, 53].

Besides, as a donor for hydrogen bonding, the phenolic groups in catechins bind firmly to proteins, lipids, and nucleic acids. For instance, the binding between EGCG and protein, such as 67-kDa laminin receptor and prolyl cis/trans isomerase, is a mechanism of anti-cancer [46]. Therefore, drinking green tea can reduce the risk of several oxidative stress conditions. However, excessive tea consumption (more than 200–400 mg or 4–8 cups a day) can cause side effects such as headache or nausea due to its function in lowering blood sugar [1].

## 3. Green tea catechin with diabetes

Emerging diseases such as diabetes and obesity become a global health concern, and they are closely associated with metabolic syndrome (MetS) [54]. MetS is a symptom that includes an increase in serum triglyceride, elevated blood pressure, and reduced high-density lipoprotein-associated cholesterol (Ford, 2005). Therefore, the consumption of green tea catechin will be beneficial in countering health effects by reducing body weight and alleviating MetS.

Some studies explained that the consumption of EGCG in green tea as anti-diabetic could significantly reduce adipose tissue, lowering blood glucose or insulin levels, and stimulate insulin sensitivity or glucose tolerance. Insulin resistance is the condition when the body cannot use the hormone insulin properly to metabolites sugar and lead to further weight gain. Hence, by reducing body weight, people who are severe in diabetes can improve glucose tolerance [46, 50, 54].

This evidence was validated in an earlier *in-vitro* study by using genetically obese/ diabetic samples [55], showing that EGCG intake can reduce not only body weight but also visceral fat weight. A similar result was also obtained in the experiment using western-style diet, where EGCG attenuated insulin resistance, plasma cholesterol, and monocyte chemoattractant protein concentrations [52, 55].

Also, tea catechin can deal with obesity because it can alter bile acids (BAs) as the regulator of body cholesterol. BAs are steroid molecules produced in the liver, essential for digesting and absorbing dietary fats. They aid in emulsifying fats in the small intestine, facilitating enzymatic digestion. After digestion, BAs are reabsorbed in the small intestine and recycled to the liver, contributing to efficient digestion. Beyond their digestive role, BAs also serve as signalling molecules, influencing physiological processes such as lipid metabolism and glucose homeostasis. Abnormalities in BA metabolism can lead to disorders like gallstones and liver diseases.

So, when there is an oral exposure to catechin, the major proportion of physiological cholesterol will be cascaded in the system through the metabolic process in hepatocytes [56, 57]. Complex diseases can be significantly induced by gut microflora through metabolic processes that involve BAs and the intervention of bioactive component like catechin can either directly affect BAs synthesis in the liver or disturb microflora by increasing the level of BAs [58]. Due to its role in metabolism, high levels of BAs can remove fat and then suppress the diet-induced obesity [59] and are responsible for inclining total body expenditure and dissipation of energy in the form of heat [57].

On the other hand, the study about diabetes is also associated with the reduction of BMD (bone mineral density), which seems contradictive with the way catechin aids in reducing body weight. The potential effects of EGCG on BMD may be linked to its interaction with various biological pathways. On one hand, EGCG has antioxidant and anti-inflammatory properties, which could theoretically be beneficial for bone health by reducing oxidative stress and inflammation. These factors are known to influence bone metabolism. On the other hand, certain studies have reported that high doses of EGCG may have adverse effects on bone health. One possible explanation is that, in excess, EGCG might interfere with the activity of osteoblasts, cells responsible for bone formation. Additionally, EGCG may affect the absorption of minerals like calcium, which is essential for bone mineralisation [13].

The contradictory results in different studies may be attributed to variations in study design, dosage, duration, and the specific populations being studied. Factors such as the source of EGCG, individual variations in response, and the presence of other compounds in green tea may also contribute to the inconsistent findings. Therefore, while some studies suggest that EGCG may have a positive impact on BMD due to its antioxidant and anti-inflammatory properties, conflicting results and potential adverse effects at high doses indicate that more research is needed to understand the complex relationship between EGCG and bone health.

In general, the prevention of diabetes can be achieved by controlling body weight, particularly addressing obesity. One effective approach is supplementing with EGCG, known to attenuate insulin resistance, lower blood sugar, and increase the level of BAs. This is crucial for enhancing insulin sensitivity, allowing cells to respond more efficiently to insulin and improving glucose uptake. EGCG has additional benefits, such as inhibiting the absorption of glucose in the intestine, contributing to more stable blood sugar levels post-meals. Its anti-inflammatory properties also play a role in alleviating chronic inflammation, a factor linked to insulin resistance. Furthermore, EGCG shows a protective effect on pancreatic beta cells, essential for insulin production, and may activate AMP-activated protein kinase (AMPK), an enzyme associated with increased glucose uptake and improved insulin sensitivity. While these mechanisms suggest a potential role for EGCG in managing insulin resistance, further research, including clinical trials, is necessary for concrete recommendations. Individuals with diabetes or insulin resistance should seek guidance from healthcare professionals before making significant dietary changes or incorporating supplements [36, 52].

## 4. Green tea and cardiovascular disease

Cardiovascular disease is a significant global health concern, with lifestyle factors, including dietary choices, playing a crucial role in its development [60, 61]. The latest research highlights the potential protective effects of tea against cardiovascular issues, specifically through tea polyphenols. These compounds exhibit various cardioprotective properties, such as reducing cholesterol absorption, lowering blood cholesterol levels, and mitigating risks of thrombosis, inflammation, and oxidative damage [62, 63].

Tea polyphenols go beyond by regulating vascular reactivity, influencing plasma lipid profiles, and demonstrating anti-thrombotic, antioxidant, anti-inflammatory, and anti-proliferative activities, all contributing to improved vascular function [64–66]. Green tea polyphenols, in particular, have shown promise in preventing various cardiovascular diseases. Meanwhile, BT polyphenols have potential benefits, such as reducing postprandial hypertriacylglyceremia and enhancing endothelial function [67, 68].

First and foremost, these compounds act as potent antioxidants, neutralising free radicals that can otherwise damage blood vessels and contribute to cardiovascular ailments. Furthermore, EGCG demonstrates anti-inflammatory properties, countering chronic inflammation associated with heart-related conditions. By positively influencing lipid metabolism, EGCG aids in improving the lipid profile, reducing levels of low-density lipoprotein (LDL) cholesterol and triglycerides. Notably, it enhances endothelial function, promoting optimal blood vessel dilation and contraction. Additionally, studies suggest a potential mild hypotensive effect, contributing to overall blood pressure regulation [67].

Polyphenols in BT offer their own set of advantages. Notably, they play a role in reducing postprandial hypertriacylglyceremia, addressing elevated triglyceride levels after meals. This effect contributes to a more stable post-meal lipid profile. Additionally, BT polyphenols have been associated with enhancing endothelial function, crucial for maintaining healthy blood vessels and optimal blood flow. This suggests that incorporating tea into one's lifestyle may play a positive role in cardiovascular health. Thus, tea polyphenols present a compelling strategy for lowering the risk of cardiovascular disease, making them a valuable inclusion in discussions about maintaining a heart-healthy lifestyle [68].

## 5. Green tea catechin with the cancers

Tea polyphenols, known for their ability to prevent cancer, protect against different stages of cancer development by reducing inflammation and slowing down tumour growth. They act as antioxidants, neutralising harmful free radicals, which help to induce cell death and stop the growth of cancer cells. Both black and green tea polyphenols are important in blocking cancer-causing enzymes, activating protective enzymes, and preventing damage to DNA, cell death, blood vessel formation, and the spread of cancer, providing defence against various types of cancers [69–72].

Ovarian cancer, ranking seventh in global cancer-related deaths, is influenced by familial, age, genetic, dietary, and lifestyle factors. Some conflicted results exist regarding tea consumption, especially green tea, with ovarian cancer risk. In general, green tea polyphenols demonstrate promise by reducing protein expression linked to ovarian cancer and inducing apoptosis, but BT polyphenols show the contrary effect [73, 74].

While some studies suggest a reduction in risk, epidemiological studies indicate an inverse relationship between tea consumption and ovarian cancer [75–77]. This lack of clarity may arise from variations in tea-drinking habits, types of tea consumed, and the duration of consumption across different study populations, as well as a failure to control for potential confounding factors. Existing literature exploring the relationship between the type and duration of tea consumption and ovarian cancer risk has not provided consistent findings. However, a particular study [78] suggests that an increase in the frequency and duration of tea drinking, especially green tea, may reduce the risk of ovarian cancer. It is important to note that further investigation is required to determine the protective effects of BT and oolong tea.

It is proven in another study [77] that the correlation between caffeinated beverages including green tea and ovarian cancer remains inconclusive. Interestingly, there was no elevated risk observed for coffee or caffeinated soft drinks. While overall tea consumption did not show an association with ovarian cancer risk, a nuanced analysis considering the type of tea revealed an increased risk among exclusive BT drinkers. However, no excess risk was found for those exclusively consuming green tea. These patterns persisted when examining post-menopausal women, with the association with BT consumption mainly noted in the endometrioid histotype.

In addition to that, there is a positive effect of oestrogen modulation, which is the risk factor for ovarian cancer, when reacts with green tea polyphenols, impacting oestrogen metabolism. Meanwhile, BT consumption is associated with increased oestrogen levels in postmenopausal women, influencing ovarian cancer risk. Unless the TF3 (theaflavin-3) in BT, synergistically works with CDDP (cisplatin, a chemotherapy drug commonly used for ovarian cancer treatment). TF3 and CDDP together exerted synergistic cytotoxicity against cisplatin-resistant ovarian cancer cells. TF3 and CDDP synergistically induced apoptosis (programmed cell death) and G1/S cell cycle arrest in ovarian cancer cells. Additionally, the combination treatment with TF3 and CDDP synergistically down-regulated Akt phosphorylation (a signalling pathway associated with cell survival) in ovarian cancer cells [76].

Lung cancer, the primary cause of global cancer-related deaths, is potentially influenced by dietary factors, including tea consumption. Tea polyphenols exhibit protective effects against lung cancer by inducing apoptosis, inhibiting tumour progression, and regulating key factors in lung cancer cells. Observational studies suggest a dose-dependent protective effect of green tea against lung cancer, while BT shows no significant impact [79]. White tea is proposed to have chemo-preventive properties against lung tumour formation due to its anti-carcinogenic attributes [80].

Skin cancer, with an alarming increase in incidence globally, is linked to solar ultraviolet radiation. Polyphenols in black and green tea emerge as potential chemopreventive agents, inhibiting UV radiation penetration and reducing DNA damage by countering skin inflammation through antioxidant properties. Green tea, specifically, offers skin photoprotection, preventing photo-carcinogenesis, inducing DNA repair, stimulating immune responses, inhibiting angiogenesis, and influencing various UVB-carcinogenesis biomarkers. Some studies find no association between tea consumption and skin cancer risk, while others suggest an inverse association [81, 82].

Breast cancer, the most frequent cancer in women, exhibits lower incidence in Asian countries, prompting hypotheses linking higher green tea consumption to reduced risk. Green tea polyphenols impact oestrogen biosynthesis enzymes, which are significant in breast cancer development. Despite findings supporting green tea's protective role, some studies report weak or no relationships between green tea and breast cancer risk, occasionally noting positive correlations with BT [83, 84].

Colorectal cancer, a prevalent and deadly form of cancer, has drawn attention regarding potential protective effects from dietary and lifestyle factors. Findings on tea consumption's impact present contradictions, with some studies indicating no association between green tea and colorectal cancer risk and others reporting a reduced risk with regular tea consumption. EGCG, a green tea polyphenol, plays a role in inhibiting key factors integral to colon cancer development [85].

Lastly, endometrial cancer, a prevalent gynaecological cancer, is influenced by dietary habits that can either increase or decrease its long-term risk factors. Studies present conflicting findings on the association between tea consumption and endometrial cancer risk. Some indicate no or minimal association, while others suggest a reduced risk with tea consumption. In contrast to the protective effect of green tea, BT consumption is positively associated with the risk of endometrial cancer [86].

The lack of clear evidence for black tea's cancer-reducing potential might be due to variations in the types and amounts of compounds present in BT compared to green tea. Both teas come from the same plant, *Camellia sinensis*, but they undergo different processing methods. Green tea is minimally processed, preserving more of its natural compounds, including polyphenols like EGCG. BT, on the other hand, undergoes fermentation, altering its composition [18].

Polyphenols, known for their potential health benefits, are more abundant in green tea. These compounds have antioxidant properties and may play a role in reducing the risk of certain cancers. The fermentation process in BT might result in lower levels of specific polyphenols compared to green tea, potentially affecting its observed protective effects against cancer [20].

It's essential to note that research in this area is ongoing, and factors like individual variations, study designs, and lifestyle differences can contribute to the complexity of the findings [85, 86].

## 6. Tea catechin and neurological disease

Tea polyphenols demonstrate significant neuroprotective effects against major neurological diseases, such as Alzheimer's and Parkinson's disease [87]. Lifestyle factors, particularly dietary choices, play a crucial role in maintaining central nervous system health and combating neurological diseases [17]. Studies suggest that regular tea consumption is associated with improved cognitive function [16, 88], better verbal fluency scores [89], and a reduced risk of cognitive decline [90] and impairment [91].

Bioactive components in green tea, such as catechins and theanine, have been linked to improved cognition in rodent models. Their anti-oxidative and anti-inflammatory properties contribute to cognitive benefits. Green tea's neuroprotective mechanisms include inhibiting acetylcholinesterase, regulating stress hormone secretion, and modulating neurotransmission in the brain. Human studies reported enhanced parieto-frontal connectivity, crucial for working memory processing, as a potential mechanism behind green tea's positive impact on cognitive function, proven by the improved cognitive dysfunction scores. However, further clarification requires long-term, large-scale controlled studies [88]. Also, by exhibiting anti-inflammatory and antioxidant activities, tea polyphenols induce iron-chelating effects, modulate cell survival and signalling pathways, regulate stress hormone secretion, enhance neurotransmitter system levels, and reduce oxidative stress linked to age-related brain disorders [92, 93].

A small amount of caffeine in tea also contributes to protective effects against neurological diseases like Alzheimer's and Parkinson's by stimulating the central nervous system [94]. Recent research suggests that caffeine or caffeine-containing products may protect against the degeneration of dopaminergic neurons, impacting the progression of Parkinson's Disease (PD). Furthermore, these findings indicate that caffeine could improve motor function in PD patients. Additionally, drugs like istradefylline, which block adenosine A2A receptors, have shown promise in reducing OFF time and dyskinesia linked to standard dopamine treatments for PD [95]. The caffeine content in green tea can vary, but on average, it typically contains about 20 to 45 mg of caffeine per 8-ounce cup. These values can differ based on factors such as the type of green tea, brewing time, and other variables. It's essential to note that green tea generally has lower caffeine content compared to BT and coffee [94].

Furthermore, in PD, tea polyphenols offer neuroprotective effects with anti-aggregation, anti-chelating, anti-inflammatory, and antioxidant properties, inhibiting a-synuclein aggregation and modulating intracellular signalling pathways [70, 96]. Tea consumption is associated with a lower risk of PD and delays its onset, making it a promising dietary modification to slow down age-related neurodegenerative diseases [27, 97].

In Alzheimer's disease, characterised by progressive neurodegeneration, tea polyphenols, particularly in green tea, reduce the risk by lowering toxic levels of brain Amyloid β (Aβ) peptide and inhibiting its production, a trigger for Alzheimer's disease. Amyloid-β (Aβ) is associated with problems in the energy-producing part of cells, known as mitochondria, which are linked to the onset and progression of Alzheimer's disease (AD). Seeking to enhance mitochondrial function as a therapeutic approach, researchers have explored plant-derived compounds, particularly flavonoids found in green tea like EGCG and luteolin. These compounds not only lower harmful Aβ levels in the brain but also show potential in safeguarding the energy production of nerve cells affected by Alzheimer's [9].

Tea polyphenols also exhibit protective effects against associated conditions like diabetes, depression, and hypercholesterolemia. During depression, there is an activation of inflammatory pathways, dysfunction in mitochondria, an increase in indicators of oxidative stress, and a decline in the body's antioxidant capacity. Around 30% of depressed individuals do not respond well to traditional medications. Preclinical studies suggest that phenolic compounds found in green tea may have the potential to alleviate depressive symptoms by influencing factors linked to oxidative stress, neuroinflammation, and the balance of intestinal microorganisms. These polyphenols can impact the makeup of gut bacteria, which is associated with health advantages, and the gut microbes can transform polyphenols into bioactive substances that offer therapeutic effects [98].

## 7. Green tea and infectious diseases

As effective antimicrobial agents, tea polyphenols demonstrate a broad-spectrum antimicrobial effect against various microorganisms both gram-positive and gram-negative bacteria, viruses, fungi, and parasites. Notably, strains such as Methicillin-resistant *Staphylococcus aureus* (MRSA) and extended-spectrum beta-lactamases (ESBLs), which pose significant health threats, have demonstrated vulnerability to green tea extracts. Another study revealed that the efficacy of (−)-epigallocatechin (EGC) in green tea worked against multi-drug-resistant *Escherichia coli* strains from urinary tract infections. However, it is crucial to consider variations in green tea types, preparations, and compositions when interpreting these results [99, 100].

Green tea catechins exert their direct effects by interacting with the lipid bilayer cell membrane of bacteria. This interaction causes damage to the membrane, leading to antimicrobial effects. When *Escherichia coli* is exposed to green tea polyphenols, it results in changes to the regulation of 17 genes, causing significant damage to the bacterial cell membrane. This effect is more noticeable in gram-negative bacteria due to the negatively charged lipopolysaccharide (LPS) outer membrane in gram-negative bacteria. The compromised bacterial cell membrane disrupts essential functions, such as binding to host cells and forming biofilms, which play a crucial role in the pathogenesis of infections. Additionally, the damage to the bacterial membrane prevents the secretion of toxins, further contributing to the antimicrobial action of green tea catechins [101].

EGCG also has shown antiviral effects against various virus families, including Retroviridae, Orthomyxoviridae, and Flaviviridae. This includes well-known human pathogens like HIV [102], influenza A [103], and hepatitis C [104]. Additionally, EGCG disrupts the replication cycle of DNA viruses, such as hepatitis B [105], herpes simplex, and adenovirus [106]. The compound exhibits different modes of action against these viruses, highlighting its potential as a broad-spectrum antiviral agent. In general, to prevent infective diseases, EGCG exhibits antiviral properties by acting in various ways. It serves as a protective barrier, preventing viruses like HIV from entering our cells. Additionally, it disrupts the virus's ability to replicate, essentially hindering its reproduction process. EGCG also plays a role in enhancing our immune system's response to infected cells, aiding in their identification and elimination. Moreover, it helps to reduce inflammation, preventing an exaggerated response to the infection. Furthermore, EGCG can directly interact with viruses, rendering them ineffective in infecting cells. While promising in laboratory studies, more research is needed to determine its potential as a real treatment, and its effectiveness may vary depending on the specific virus [107].

In influenza virus, catechins work by changing how the influenza virus infects cells, specifically by interacting with viral hemagglutinin (HA) and affecting the synthesis of viral RNA in cells. Additionally, catechins hinder the activity of viral RNA polymerase, a key enzyme for the virus. The galloyl group in catechins plays an important role in this process. This suggests that green tea and its catechins could be useful in protecting against influenza. Even the leftover parts of green tea after extraction, known as by-products, can still be a strong source of anti-influenza properties due to the substantial amount of catechins they contain. Specifically, catechins highly concentrated in the ethyl acetate-soluble fraction show effective inhibitory effects against various stages of the influenza virus lifecycle, including transcription and release [108].

Furthermore, the catechin has the ability to fight against harmful fungi like *Candida albicans*, which can cause infections in humans. While the exact way EGCG works is not completely clear, some evidence suggests that it attaches to the protective outer layer of fungi, disrupting their function. Additionally, EGCG seems to interfere with the way bacteria and fungi manage folic acid, a vital nutrient for their growth, by inhibiting a key enzyme called dihydrofolate reductase. This disruption can weaken the fungi and make them less harmful. However, more research is needed to fully understand these processes and their effectiveness in real-life situations [109].

Additionally, EGCG shows promise in addressing anthrax infections caused by *Bacillus anthracis*, a spore-forming gram-positive bacterium that is associated with high mortality [110]. This is considered one of the most powerful biological warfare agents because its spores are highly resistant to natural conditions and can persist for many decades in the environment. *B. anthracis* spores can enter the body through a skin lesion (cutaneous anthrax), the lungs (pulmonary anthrax), or the gastrointestinal route (gastrointestinal anthrax). They then germinate, leading to the development of the active vegetative form. Anthrax is a public health concern in many countries, especially those relying on agriculture as their primary source of income [111].

EGCG from green tea works as a potent inhibitor of anthrax lethal factor by preventing LF-induced death of cells. It does this by interfering with the activity of LF, which is a major contributor to the development of anthrax. EGCG has been found to protect cells and animals from LF-induced toxicity, making it a valuable tool for studying the mechanisms of LF action and potentially preventing and treating anthrax. The exact mechanism of how EGCG achieves this inhibition is not fully understood, but it is thought to involve the direct interaction between EGCG and LF, leading to the prevention of LF-induced cytotoxicity. The multifaceted antimicrobial properties of tea polyphenols underscore their potential as therapeutic agents against a wide range of infectious diseases [110].

## 8. Conclusion

One of the key takeaways is the potential of green tea catechins, particularly EGCG, in combating oxidative stress. Oxidative stress, characterised by an imbalance between antioxidants and free radicals in the body, is implicated in the development of numerous health issues. By acting as powerful antioxidants, green tea catechins can help to restore this balance and thereby prevent oxidative stress-related diseases. This highlights the importance of incorporating antioxidant-rich diets, such as green tea, into daily consumption patterns to promote overall health and well-being.

Furthermore, the multifaceted benefits of green tea catechins extend beyond oxidative stress prevention. These benefits include the potential to reduce body weight, alleviate MetS, improve glucose tolerance, and regulate BAs. Additionally, the protective effects of green tea catechins against cardiovascular disease, cancer, and neurological diseases are underscored, indicating the wide-ranging impact of green tea consumption on human health.

It is crucial to mention that ongoing and detailed research is being conducted to delve deeper into the specific mechanisms and outcomes of green tea consumption in relation to the diseases discussed. Researchers are exploring nuanced aspects, such as optimal dosage, variations in individual response, and the long-term effects of sustained green tea intake.

It is important to note, however, the potential side effects of excessive green tea consumption, such as headaches or nausea. This highlights the need for moderation in consumption (around 200–400 mg a day) and the importance of seeking a balanced approach to reaping the benefits of green tea without experiencing adverse effects.

In conclusion, the scientific evidence presented in the passage strongly supports the notion that green tea catechins, particularly EGCG, can play a significant role in reducing the risk of various complex diseases by acting as potent antioxidants. The findings underscore the potential of green tea as a valuable dietary component in promoting overall health and well-being. However, it is essential to approach consumption in moderation and in conjunction with a balanced diet to maximise the benefits while minimising potential adverse effects. Ongoing research endeavours aim to provide more nuanced insights into the specific effects of green tea on various diseases, refining our understanding and optimising its integration into preventive healthcare strategies.

## Acknowledgements

I extend my sincere gratitude to the IntechOpen for providing me with the opportunity to undertake and publish my previous writing. Also, the support and resources offered by the university have been instrumental in the successful completion of this work.

## Conflict of interest

The authors declare no conflict of interest.

## Author details

Baiq Amarwati Tartillah
University of Cenderawasih, Jayapura, Indonesia

*Address all correspondence to: tartilbaiq@gmail.com

## References

[1] Legg JT. Everything You Should Know About Oxidative Stress. 2017. Available from: https://www.healthline.com/health/oxidative-stress [Accessed: May 20, 2019]

[2] Beecher GR, Warden BA, Merken H. Analysis of tea polyphenols. Proceedings of the Society for Experimental Biology and Medicine. 1999;**220**:267-270

[3] Frei B, Higdon JV. Antioxidant activity of tea polyphenols in vivo: Evidence from animal studies. The Journal of Nutrition. 2003;**133**:3275S-3284S

[4] Hollman PC, Feskens EJ, Katan MB. Tea Flavonols in cardiovascular disease and Cancer epidemiology. Proceedings of the Society for Experimental Biology and Medicine. 1999;**220**:198-202

[5] Huang Q, Wu L-J, Tashiro S-I, Gao H-Y, Onodera S, Ikejima T. (+)-catechin, an ingredient of green tea, protects murine microglia from oxidative stress-induced DNA damage and cell cycle arrest. Journal of Pharmacological Sciences. 2005;**98**:16-24

[6] Üstün Ç, Demirci N. The Plant Of Tea (Camellia Sinensis L.) Historical Development And Medical Evaluation-Çay Bitkisinin (Camellia Sınensıs L.) Tarihsel Gelişimi Ve Tıbbi Açıdan Değerlendirilmesi. Mersin Üniversitesi Tıp Fakültesi Lokman Hekim Tıp Tarihi Ve Folklorik Tıp Dergisi. 2013;**3**:5-12

[7] Butt M, Imran A, Sharif M, Ahmad RS, Xiao H, Imran M, et al. Black tea polyphenols: A mechanistic treatise. Critical Reviews in Food Science and Nutrition. 2014;**54**:1002-1011

[8] Vernarelli JA, Lambert JD. Tea consumption is inversely associated with weight status and other markers for metabolic syndrome in US adults. European Journal of Nutrition. 2013;**52**:1039-1048

[9] Dragicevic N, Smith A, Lin X, Yuan F, Copes N, Delic V, et al. Green tea Epigallocatechin-3-gallate (EGCG) and other flavonoids reduce Alzheimer's amyloid-induced mitochondrial dysfunction. Journal of Alzheimer's Disease. 2011;**26**:507-521

[10] Kim YC, Choi SY, Park EY. Anti-melanogenic effects of black, green, and White tea extracts on immortalized melanocytes. Journal of Veterinary Science. 2015;**16**:135-143

[11] Ahmed S, Stepp JR. Green tea: The plants, processing, manufacturing and production. In: Tea in Health and Disease Prevention. 2013. pp. 19-31. DOI: 10.1016/B978-0-12-384937-3.00002-1

[12] Samanta T, Cheeni V, Das S, Roy AB, Ghosh BC, Mitra A. Assessing biochemical changes during standardization of fermentation time and temperature for manufacturing quality black tea. Journal of Food Science and Technology. 2015;**52**:2387-2393

[13] Shen C-L, Chyu M-C, Yeh JK, Felton CK, Xu KT, Pence BC, et al. Green tea polyphenols and Tai chi for bone health: Designing a placebo-controlled randomized trial. BMC Musculoskeletal Disorders. 2009;**10**(1):110

[14] Malongane F, Mcgaw LJ, Mudau FN. The synergistic potential of various teas, herbs and therapeutic drugs in health improvement: A review. Journal of the Science of Food and Agriculture. 2017;**97**:4679-4689

[15] Lorenz M. Cellular targets for the beneficial actions of tea polyphenols. The American Journal of Clinical Nutrition. 2013;**98**:1642S-1650S

[16] Yong Feng W. Metabolism of green tea catechins: An overview. Current Drug Metabolism. 2006;**7**:755-809

[17] Gomez-Pinilla F, Gomez AG. The influence of dietary factors in central nervous system plasticity and injury recovery. PM&R. 2011;**3**:S111-S116

[18] Katiyar S, Elmets CA, Katiyar SK. Green tea and skin Cancer: Photoimmunology, angiogenesis and DNA repair. The Journal of Nutritional Biochemistry. 2007;**18**:287-296

[19] Tenore GC, Campiglia P, Giannetti D, Novellino E. Simulated gastrointestinal digestion, intestinal permeation and plasma protein interaction of White, green, and black tea polyphenols. Food Chemistry. 2015;**169**:320-326

[20] Chan EW, Soh EY, Tie PP, Law YP. Antioxidant and antibacterial properties of green, black, and herbal teas of Camellia sinensis. Pharmacognosy Research. 2011;**3**:266

[21] Du G-J, Zhang Z, Wen X-D, Yu C, Calway T, Yuan C-S, et al. Epigallocatechin gallate (EGCG) is the Most effective Cancer Chemopreventive polyphenol in green tea. Nutrients. 2012;**4**:1679-1691

[22] Johnson JJ, Bailey H, Mukhtar H. Green tea polyphenols for prostate Cancer chemoprevention: A translational perspective. Phytomedicine. 2010;**17**:3-13

[23] Higdon JV, Frei B. Tea Catechins And Polyphenols: Health Effects, Metabolism, And Antioxidant Functions. 2003

[24] Harborne JB. The Flavonoids: Advances in Research since 1980. NY: Springer; 2013

[25] Rice-Evans CA, Miller NJ, Paganga G. Structure-antioxidant activity relationships of flavonoids and phenolic acids. Free Radical Biology and Medicine. 1996;**20**(7):933-956

[26] Ninomiya M, Unten L, Kim M. In: Yamamoto T, Juneja LR, Chu DC, Kim M, editors. Chemistry and Application of Green Tea. Boca Raton: CRC Press; 1997. pp. 23-36

[27] Hu G, Bidel S, Jousilahti P, Antikainen R, Tuomilehto J. Coffee and tea consumption and the risk of Parkinson's disease. Movement Disorders: Official Journal of the Movement Disorder Society. 2007;**22**:2242-2248

[28] Ho C-T, Lin J-K, Shahidi F. Tea and Tea Products: Chemistry and Health-Promoting Properties. Boca Raton: CRC Press; 2008

[29] Sekiya J, Kajiwara T, Monma T, Hatanaka A. Interaction of tea catechins with proteins: Formation of protein precipitate. Agricultural and Biological Chemistry. 1984;**48**:1963-1967

[30] Yamamoto T, Juneja LR, Kim M. Chemistry and Applications of Green Tea. Boca Raton: CRC Press; 1997

[31] Yang J, Liu RH. The phenolic profiles and antioxidant activity in different types of tea. International Journal of Food Science & Technology. 2013;**48**:163-171

[32] Hara Y. Green Tea: Health Benefits and Applications. Boca Raton: CRC Press; 2001

[33] Qu W, Pan Z, Ma H. Extraction Modeling and activities of antioxidants

from pomegranate Marc. Journal of Food Engineering. 2010;**99**:16-23

[34] Turkmen N, Velioglu YS, Sari F, Polat G. Effect of extraction conditions on measured Total polyphenol contents and antioxidant and antibacterial activities of black tea. Molecules. 2007;**12**:484-496

[35] Sanlier N, Gokcen BB, Altuğ M. Tea consumption and disease correlations. Trends in Food Science & Technology. 2018;**78**:95-106

[36] Anderson RA, Polansky MM. Tea enhances insulin activity. Journal of Agricultural and Food Chemistry. 2002;**50**(24):7182-7186

[37] Ananingsih VK, Sharma A, Zhou W. Green tea catechins during food processing and storage: A review on stability and detection. Food Research International. 2013;**50**:469-479

[38] Kondo S, Tsuda K, Muto N, Ueda J-E. Antioxidative activity of apple skin or flesh extracts associated with fruit development on selected apple cultivars. Scientia Horticulturae. 2002;**96**:177-185

[39] Cantos E, Espin JC, Tomás-Barberán FA. Varietal differences among the polyphenol profiles of seven table grape cultivars studied by LC- DAD- MS- MS. Journal of Agricultural and Food Chemistry. 2002;**50**(20):5691-5696

[40] Manach C, Scalbert A, Morand C, Rémésy C, Jiménez L. Polyphenols: Food sources and bioavailability. The American Journal of Clinical Nutrition. 2004;**79**:727-747

[41] Santos-Buelga C, Scalbert A. Proanthocyanidins and tannin-like compounds–nature, occurrence, dietary intake and effects on nutrition and health. Journal of the Science of Food and Agriculture. 2000;**80**:1094-1117

[42] Khokhar S, Magnusdottir S. Total phenol, catechin, and caffeine contents of teas commonly consumed in the United Kingdom. Journal of Agricultural and Food Chemistry. 2002;**50**:565-570

[43] Holden JM, Bhagwat SA, Patterson KY. Development of a multi-nutrient data quality evaluation system. Journal of Food Composition and Analysis. 2002;**15**:339-348

[44] Andersen OM, Markham KR. Flavonoids: Chemistry, Biochemistry and Applications. Boca Raton: CRC Press; 2005

[45] Buta J, Spaulding D. Endogenous levels of phenolics in tomato fruit during growth and maturation. Journal of Plant Growth Regulation. 1997;**16**:43-46

[46] Yang CS, Zhang J, Zhang L, Huang J, Wang Y. Mechanisms of body weight reduction and metabolic syndrome alleviation by tea. Molecular Nutrition & Food Research. 2016;**60**(1):160-174

[47] Coimbra S, Castro E, Rocha-Pereira P, Rebelo I, Rocha S, Santos-Silva A. The effect of green tea in oxidative stress. Clinical Nutrition. 2006;**25**(5):790-796

[48] Halliwell B, Gutteridge JM, Cross CE. Free radicals, antioxidants, and human disease: Where are we now? The Journal of Laboratory and Clinical Medicine. 1992;**119**(6):598-620

[49] Ferriola PC, Cody V, Middleton E Jr. Protein kinase C inhibition by plant flavonoids: Kinetic mechanisms and structure-activity relationships. Biochemical Pharmacology. 1989;**38**(10):1617-1624

[50] Jówko E. Green tea catechins and sport performance. In: Antioxidants in Sport Nutrition. Boca Raton, US: CRC Press/Taylor & Francis; 2015

[51] Wolfram S, Wang Y, Thielecke F. Anti-obesity effects of green tea: From bedside to bench. Molecular Nutrition & Food Research. 2006;**50**(2):176-187

[52] Chen N, Bezzina R, Hinch E, Lewandowski PA, Cameron-Smith D, Mathai ML, et al. Green tea, black tea, and epigallocatechin modify body composition, improve glucose tolerance, and differentially Alter metabolic gene expression in rats fed a high-fat diet. Nutrition Research. 2009;**29**(11):784-793

[53] Lambert JD, Elias RJ. The antioxidant and pro-oxidant activities of green tea polyphenols: A role in Cancer prevention. Archives of Biochemistry and Biophysics. 2010;**501**(1):65-72

[54] Zaveri NT. Green tea and its polyphenolic catechins: Medicinal uses in cancer and noncancer applications. Life Sciences. 2006;**78**:2073-2080

[55] Bose M, Lambert JD, Ju J, Reuhl KR, Shapses SA, Yang CS. The Green Tea Polyphenol,(−)-Epigallocatechin-3-Gallate, Inhibits Obesity and Metabolic Syndrome in High-Fat Fed Mice. San Francisco: Federation Of American Societies For Experimental Biology; 2008

[56] Ma H, Patti ME. Bile acids, obesity, and the metabolic syndrome. Best Practice & Research Clinical Gastroenterology. 2014;**28**(4):573-583

[57] Sun L, Xu H, Ye J, Gaikwad NW. Comparative effect of black, green, oolong and White tea intake on body weight gain and bile acid metabolism. Nutrition. 2019;**10**:1-10

[58] Li G, Lin W, Araya JJ, Chen T, Timmermann BN, Guo GL. A tea catechin, Epigallocatechin-3-gallate, is a unique modulator of the Farnesoid X receptor. Toxicology and Applied Pharmacology. 2012;**258**(2):268-274

[59] Hartley L, Flowers N, Holmes J, Clarke A, Stranges S, Hooper L, et al. Green and black tea for the primary prevention of cardiovascular disease. Cochrane Database of Systematic Reviews. 2013;(6)

[60] Eilat-Adar S, Sinai T, Yosefy C, Henkin Y. Nutritional recommendations for cardiovascular disease prevention. Nutrients. 2013;**5**:3646-3683

[61] Tarride J-E, Lim M, Desmeules M, Luo W, Burke N, O'Reilly D, et al. A review of the cost of cardiovascular disease. Canadian Journal of Cardiology. 2009;**25**:E195-E202

[62] Hodgson JM. Tea flavonoids and cardiovascular disease. Asia Pacific Journal of Clinical Nutrition. 2008;**17**:288-290

[63] Zhong W, Huan XD, Cao Q, Yang J. Cardioprotective effect of Epigallocatechin-3-gallate against myocardial infarction in hypercholesterolemic rats. Experimental and Therapeutic Medicine. 2015;**9**:405-410

[64] Anandh Babu PV, Liu D. Green tea catechins and cardiovascular health: An update. Current Medicinal Chemistry. 2008;**15**:1840-1850

[65] Deka A, Vita JA. Tea and cardiovascular disease. Pharmacological Research. 2011;**64**:136-145

[66] Moore RJ, Jackson KG, Minihane AM. Green tea (Camellia sinensis) catechins and vascular

function. British Journal of Nutrition. 2009;**102**:1790-1802

[67] Bhardwaj P, Khanna D. Green tea catechins: Defensive role in cardiovascular disorders. Chinese Journal of Natural Medicines. 2013;**11**:345-353

[68] Dias PM, Changarath J, Damodaran A, Joshi MK. Compositional variation among black tea across geographies and their potential influence on endothelial nitric oxide and antioxidant activity. Journal of Agricultural and Food Chemistry. 2014;**62**:6655-6668

[69] Amin AR, Zhang H, Shin DM. Molecular aspects of Cancer prevention by green tea: An overview. Tea in Health and Disease Prevention. 2013:751-766

[70] Dutta D, Mohanakumar KP. Tea and Parkinson's disease: Constituents of tea synergize with antiparkinsonian drugs to provide better therapeutic benefits. Neurochemistry International. 2015;**89**:181-190

[71] Kumar Y, Bhatia A. Polyphenols and skin cancers. In: Polyphenols in Human Health and Disease. Amsterdam: Elsevier; 2014

[72] Nagini S, Murugan RS. Cancer chemoprevention by black tea polyphenols: Emerging evidence and molecular targets. Tea in Health and Disease Prevention. 2013;**33**:737-750

[73] Gao Y, Rankin GO, Tu Y, Chen YC. Inhibitory effects of the four Main theaflavin derivatives found in black tea on ovarian Cancer cells. Anticancer Research. 2016;**36**:643-651

[74] Trudel D, Labbé DP, Bairati I, Fradet V, Bazinet L, Têtu B. Green tea for ovarian Cancer prevention and treatment: A systematic review of the in vitro, in vivo and epidemiological studies. Gynecologic Oncology. 2012;**126**:491-498

[75] Baker J, Boakye K, Mccann S, Beehler G, Rodabaugh K, Villella J, et al. Consumption of black tea or coffee and risk of ovarian Cancer. International Journal of Gynecologic Cancer. 2007;**17**:50-54

[76] Leung AC, Cook LS, Swenerton K, Gilks B, Gallagher RP, Magliocco A, et al. Tea, coffee, and caffeinated beverage consumption and risk of epithelial ovarian cancers. Cancer Epidemiology. 2016;**45**:119-125

[77] Zhan X, Wang J, Pan S, Lu C. Tea consumption and the risk of ovarian Cancer: A Meta-analysis of epidemiological studies. Oncotarget. 2017;**8**:37796

[78] Zhang M, Binns CW, Lee AH. Tea consumption and ovarian Cancer risk: A case-control study in China. Cancer Epidemiology Biomarkers & Prevention. 2002;**11**:713-718

[79] Wang Y, Yu X, Wu Y, Zhang D. Coffee and tea consumption and risk of lung Cancer: A dose–response analysis of observational studies. Lung Cancer. 2012;**78**:169-170

[80] Mao JT, Nie W-X, Tsu I-H, Jin Y-S, Rao JY, Lu Q-Y, et al. White tea extract induces apoptosis in non–small cell lung Cancer cells: The role of peroxisome proliferator-activated receptor-Γ and 15-lipoxygenases. Cancer Prevention Research. 2010;**3**:1132-1140

[81] Asgari MM, White E, Warton EM, Hararah MK, Friedman GD, Chren M-M. Association of tea consumption and cutaneous squamous cell carcinoma. Nutrition and Cancer. 2011;**63**:314-318

[82] Rees JR, Stukel TA, Perry AE, Zens MS, Spencer SK, Karagas MR. Tea consumption and basal cell and squamous cell skin Cancer: Results of a case-control study. Journal of the American Academy of Dermatology. 2007;**56**:781-785

[83] Larsson SC, Bergkvist L, Wolk A. Coffee and black tea consumption and risk of breast Cancer by Estrogen and progesterone receptor status in a Swedish cohort. Cancer Causes & Control. 2009;**20**:2039-2044

[84] Wu Y, Zhang D, Kang S. Black tea, green tea and risk of breast Cancer: An update. Springerplus. 2013;**2**:1-5

[85] Moseley VR, Morris J, Knackstedt RW, Wargovich MJ. Green tea polyphenol epigallocatechin 3-gallate, contributes to the degradation of DNMT3A and HDAC3 in HCT 116 human Colon Cancer cells. Anticancer Research. 2013;**33**:5325-5333

[86] Butler LM, Wu AH. Green and black tea in relation to Gynecologic cancers. Molecular Nutrition & Food Research. 2011;**55**:931-940

[87] Firoz C, Jabir NR, Khan MS, Mahmoud M, Shakil S, Damanhouri GA, et al. An overview on the correlation of neurological disorders with cardiovascular disease. Saudi Journal of Biological Sciences. 2015;**22**:19-23

[88] Ide K, Yamada H, Takuma N, Park M, Wakamiya N, Nakase J, et al. Green tea consumption affects cognitive dysfunction in the elderly: A pilot study. Nutrients. 2014;**6**:4032-4042

[89] Shen W, Xiao Y, Ying X, Li S, Zhai Y, Shang X, et al. Tea consumption and cognitive impairment: A Cross-sectional study among Chinese elderly. PLoS One. 2015;**10**:E0137781

[90] Noguchi-Shinohara M, Yuki S, Dohmoto C, Ikeda Y, Samuraki M, Iwasa K, et al. Consumption of green tea, but not black tea or coffee, is associated with reduced risk of cognitive decline. PLoS One. 2014;**9**:E96013

[91] Ma Q-P, Huang C, Cui Q-Y, Yang D-J, Sun K, Chen X, et al. Meta-analysis of the association between tea intake and the risk of cognitive disorders. PLoS One. 2016;**11**:E0165861

[92] Feng L, Ng T-P, Kua E-H, Lee T-S, Preedy VR. Tea and cognitive health: A focus on community-based studies. In: Diet and Nutrition in Dementia and Cognitive Decline. Amsterdam: Elsevier; 2015

[93] Schmidt A, Hammann F, Wölnerhanssen B, Meyer-Gerspach AC, Drewe J, Beglinger C, et al. Green tea extract enhances Parieto-frontal connectivity during working memory processing. Psychopharmacology. 2014;**231**:3879-3888

[94] Cappelletti S, Daria P, Sani G, Aromatario M. Caffeine: Cognitive and physical performance enhancer or psychoactive drug? Current Neuropharmacology. 2015;**13**:71-88

[95] Prediger RD. Effects of caffeine in Parkinson's disease: From neuroprotection to the management of motor and non-motor symptoms. Journal of Alzheimer's Disease. 2010;**20**:S205-S220

[96] Caruana M, Vassallo N. Tea polyphenols in Parkinson's disease. Natural Compounds as Therapeutic Agents for Amyloidogenic Diseases. 2015;**863**:117-137

[97] Seidl SE, Santiago JA, Bilyk H, Potashkin JA. The emerging role of nutrition in Parkinson's disease. Frontiers in Aging Neuroscience. 2014;**6**:36

[98] Winiarska-Mieczan A, Kwiecień M, Jachimowicz-Rogowska K, Donaldson J, Tomaszewska E, Baranowska-Wójcik E. Anti-inflammatory, antioxidant, and neuroprotective effects of polyphenols—Polyphenols as an element of diet therapy in depressive disorders. International Journal of Molecular Sciences. 2023;**24**(3):2258

[99] Reygaert W, Jusufi I. Green tea As an effective antimicrobial for urinary tract infections caused by Escherichia Coli. Frontiers in Microbiology. 2013;**4**:162

[100] Seebaluck-Sandoram R, Mahomoodally F. Management of infectious diseases in Africa. In: Medicinal Spices and Vegetables from Africa. Amsterdam: Elsevier; 2017

[101] Reygaert WC. The antimicrobial possibilities of green tea. Frontiers in Microbiology. 2014;**5**:434

[102] Carneiro BM, Batista MN, Braga ACS, Nogueira ML, Rahal P. The green tea molecule EGCG inhibits zika virus entry. Virology. 2016;**496**:215-218

[103] De Oliveira A, Adams SD, Lee LH, Murray SR, Hsu SD, Hammond JR, et al. Inhibition of herpes simplex virus type 1 with the modified green tea polyphenol palmitoyl-epigallocatechin gallate. Food and Chemical Toxicology. 2013;**52**:207-215

[104] Ye P, Zhang S, Zhao L, Dong J, Jie S, Pang R, et al. Tea polyphenols exerts anti-hepatitis B virus effects in a stably HBV-transfected cell line. Journal of Huazhong University of Science and Technology [Medical Sciences]. 2009;**29**:169-172

[105] Fukazawa H, Suzuki T, Wakita T, Murakami Y. A cell-based, microplate colorimetric screen identifies 7, 8-benzoflavone and green tea gallate catechins As inhibitors of the hepatitis C virus. Biological and Pharmaceutical Bulletin. 2012;**35**:1320-1327

[106] Li X, Zhu X. Tea: Types, Production, And Trade. 2016

[107] Matsumoto K, Yamada H, Takuma N, Niino H, Sagesaka YM. Effects of green tea catechins and theanine on preventing influenza infection among healthcare workers: A randomized controlled trial. BMC Complementary and Alternative Medicine. 2011;**11**:1-7

[108] Lee H, Lee Y, Youn H-N, Lee D, Kwak J, Seong B, et al. Anti-influenza virus activity of green tea by-products in vitro and efficacy against influenza virus infection in chickens. Poultry Science. 2012;**91**:66-73

[109] Steinmann J, Buer J, Pietschmann T, Steinmann E. Anti-infective properties of Epigallocatechin-3-gallate (EGCG), a component of green tea. British Journal of Pharmacology. 2013;**168**:1059-1073

[110] Goel AK. Anthrax: A disease of biowarfare and public health importance. World Journal of Clinical Cases: WJCC. 2015;**3**:20

[111] Pillai SK, Huang E, Guarnizo JT, Hoyle JD, Katharios-Lanwermeyer S, Turski TK, et al. Antimicrobial treatment for systemic Anthrax: Analysis of cases from 1945 to 2014 identified through a systematic literature review. Health Security. 2015;**13**:355-364